Plant Diseases Management for Sustainable Agriculture

The Editor

Shahid Ahamad, *Ph.D.* born on July 1,1972 in a village Gaharauli,Teh. Maudaha, Distt. Hamirpur, U.P. Dr. Ahamad completed his B.Sc (Ag.) from N.D.U.A.T. Kumarganj, Faizabad in 1994; M.Sc (Ag.) Plant Pathology from C.S.Azad Unversity of Agri. and Technology, Kanpur, U.P. in 1996 and Ph.D. (Botany) from C.S.J. M. Univ. Kanpur, U.P. in 2003. He worked as a Research Associate with Prof. A.K.Sarbhoy for three years (1997-2000) in the Division of Plant Pathology, I.A.R.I., Pusa, New Delhi. He was appointed as Jr. Scientist (Plant Pathology) in 2001 in the SKUAST-Jammu at Regional Agril. Research Station, Rajouri, J.&K. During short tenure of ten years, he has published more than 80 publications in his credit. He has published more than 30 research papers in the journals of national and international repute, 20 popular articles in different magazines, 10 book chapters, 1 review/ conceptual articles, 7 bulletins, 10 research abstracts and one book. Recently he has been conferred with SPPS Fellow-2005 by the Society of Plant Protection Sciences, Division of Nematology, IARI, New Delhi. Presently he is working as I/c Associate Director Research, RARS, & I/c Programme Coordinator, KVK, SKUAST-J, Rajouri.

Plant Diseases Management for Sustainable Agriculture

Editor
Dr. Shahid Ahamad
Scientist (Plant Pathology)
I/c Associate Director Research
Regional Agriculture Research Station, Rajouri – 185 131, J&K
&
I/c Programme Coordinator
Krishi Vigyan Kendra, , Rajouri – 185 131, J&K
Sher-e-Kashmir University of Agricultural Sciences and Technology

2017
Daya Publishing House®
A Division of
Astral International Pvt. Ltd.
New Delhi - 110 002

ISBN 9789351241485 (Int. Edition)

Publisher's Note:

Every possible effort has been made to ensure that the information contained in this book is accurate at the time of going to press, and the publisher and author cannot accept responsibility for any errors or omissions, however caused. No responsibility for loss or damage occasioned to any person acting, or refraining from action, as a result of the material in this publication can be accepted by the editor, the publisher or the author. The Publisher is not associated with any product or vendor mentioned in the book. The contents of this work are intended to further general scientific research, understanding and discussion only. Readers should consult with a specialist where appropriate.

Every effort has been made to trace the owners of copyright material used in this book, if any. The author and the publisher will be grateful for any omission brought to their notice for acknowledgement in the future editions of the book.

Published by : **Daya Publishing House®**
A Division of
Astral International Pvt. Ltd.
– ISO 9001:2015 Certified Company –
4736/23, Ansari Road, Darya Ganj
New Delhi-110 002
Ph. 011-43549197, 23278134
E-mail: info@astralint.com
Website: www.astralint.com

Laser Typesetting : **Classic Computer Services,** Delhi - 110 035

Printed at : **Replika Press Pvt. Ltd.**

Foreword

Plant Protection is an integral part of crop protection. Modern agriculture is highly input intensive and greatly dependent on fertilizers and pesticides. If the input–oriented agriculture has helped to achieve the green revolution, it has created the problem of sustainability of the production system due to abuse *i.e.* excessive and indiscriminate use of the chemical inputs. Hence, in achieving second green revolution it is necessary to look into alternative management strategies which are environment friendly but long term sustainable. Thus, this book is timely thought and themes are well projected.

Diseases caused by fungi, bacteria and viruses are prime causal organisms which restrict the crop quality and quantity produce in our country. Enormous losses have been caused every year due to the pre-and post harvest diseases of crops in India. Due to crop failure by insect pests and diseases, several families of farmers had to take extreme steps such as suicides and burning of their crops.

This book edited by Dr. Ahamad is the culmination of the wishes expressed by many plant pathologists of India. It deals with the diseases of various crops by a number of pathogens occurring in India. All aspects of diseases *i.e.* symptomatology, etiology, transmission, epidemiology, characterization, detection, identification, and diseases management practices have been described comprehensively.

The book will be very useful as a reference book by the plant pathologists, teachers, students, extension workers and progressive farmers alike.

I would like to congratulate Dr. Shahid Ahamad for his efforts and hard work in editing this important book.

Prof. Nagender Sharma

Vice Chancellor

Sher-e-Kashmir University of Agricultural Sciences and Technology of Jammu, RARS, Rajouri

Preface

Crop protection plays a pivotal role in agricultural production. During the past few decades, the science of crop protection has developed rapidly resulting in many diverse techniques to manage a variety of pests, pathogens and nematodes. Currently the prospects of chemical control do not appear good because of accidental and incidental trauma encountered on the field of chemical pesticides. Therefore, vigorous research efforts for exploring newer but programmatic tools for the control of diseases have become imperative.

A recent survey by the Economist revealed that the world population has increased by 90 per cent in the past 40 years while food production has increased only by 25 per cent per head. With an additional 1.5 billion mouth to feed by 2020, farmers worldwide have to produce 39 per cent more. The new millennium promises excitement and hope for the future by new advancement in ecofriendly technologies in integrated disease management of the crops. During past twentieth century, Plant Pathology has witnessed a dramatic advancement in management of plant diseases through in depth investigations of host–pathogen interactions, development of molecular diagnostic tools, integration of new concepts, principles and approaches.

The diseases of economic importance caused by fungi, bacteria, viruses and virus like organisms of each crop are covered, describing their history, distribution, losses incurred, symptoms, latest diagnostic tools, epidemiology and integrated management approaches including cultural, chemical, genetic resources, and bio control agents being adopted world-wide.

There has been a long felt need for a book exclusively on the *"Plant Diseases Management for Sustainable Agriculture"*. An effort has been made to collect information on various aspects of diseases and their management. Chapters in this book have been contributed by noted Plant Pathologists of India. The Scientists, teachers, students, scholars, administrators and policy makers dealing with disease management in particular and plant protection in general expected this book will be very useful and informative. The book *"Plant Diseases Management for Sustainable Agriculture"*, not only provides references but also serves as guide and inspiration for future research into the realm of biological, chemical, physical and quarantine aspects and onslaughts of modern agriculture on it. This book also serve as a very good references book on the IDM of Plant diseases. I am grateful and indebted to all the learned galaxy of contributors who have spent their considerable time in contributing the chapters on various important crops.

I am also thankful to *Shri Anil Mittal, Daya Publishing House*, New Delhi for his technical guidance and help in bringing out this book.

Finally, I wish to extend regards to late Mustaque Ahmad (Father) who was the source inspiration and Drs. Deepak Kher, M.H. Chisti, Anshuman Kohli, and Jamaluddin for their continuous support and encouragement in bringing out this book.

Shahid Ahamad

Contents

List of Contributors

A. Mishra
Advance Centre for Plant Virology, IARI, New Delhi

A. Naseema
Department of Plant Pathology, College of Agriculture, Vellayani, Thiruvananthapuram – 695 522, Kerala

A. Vaid
Division of Plant Pathology, SKUAST, Jammu (J & K)

A.A. Jafri
National Centre for Integrated Pest Management, IARI Campus, New Delhi – 110 012

Abdul Khalid
Department of Plant Pathology, GBPUAT, Pantnagar – 263 145 (Uttaranchal)

Amit Kumar Jain
Department of Microbiology, Janta Vedic College, Baraut (Baghpat), U.P.

Ashok Kumar
National Bureau of Plant Genetic Resources, New Delhi – 110 012

Ashraf Ali Khan
DMR, IARI Campus, Pusa, New Delhi – 110 012

Atul Kumar
Department of Plant Pathology, Birsa Agricultural University, Ranchi – 834 006 (Jharkhand)

B.K. Goswami
Amity Centre for Biocontrol and Plant Disease Management, Amity University, U.P.

B.N. Reddy
Mycology and Plant Pathology Laboratory, Department of Botany, Osmania University, Hyderabad – 500 007

Brajesh Kumar
S.D.J. Post Graduate College, Chandeshwar, Azamgarh, U.P.

C. Ravindran
RARS, SKUAST-J, Rajouri – 185 131

C.R. Raghavender
Mycology and Plant Pathology Laboratory, Department of Botany, Osmania University, Hyderabad – 500 007

C.S. Kalha
Division of Plant Pathology, FOA-Udheywalla, SKUAST-J, Jammu (J&K)

D. Gupta
Department of Plant Pathology, Dr. Y.S. Parmar University of Horticulture and Forestry, Solan (H.P.)

D. Gupta
Temperate Horticultural Research Station, Kotkhai, Shimla – 171 202 (H.P.)

Gunjeet Kumar
National Bureau of Plant Genetic Resources, New Delhi – 110 012

H.C. Lal
Department of Plant Pathology, Birsa Agricultural University, Ranchi – 834 006 (Jharkhand)

H.S. Chaube
Department of Plant Pathology, B.A.U., Ranchi–834 006 (Jharkhand)

H.S. Chaube
Department of Plant Pathology, GBPUAT, Pantnagar – 263 145 (Uttaranchal)

J.C. Rana
NBPGR Regional Station, Shimla

J.D.S. Panwar
Indian Institute of Pulses Research, Kalyanpur, Kanpur, U.P.

J.N. Srivastava
Regional Horticulture Research Station, Bhaderwah, Distt. Doda, SKUAST-Jammu

J.P. Upadhyay
Department of Plant Pathology, Rajendra Agricultural University (Dholi Campus), Pusa, Samastipur – 848 125, Bihar

Jameel Akhtar
Department of Plant Pathology, Birsa Agricultural University, Ranchi – 834 006 (Jharkhand)

Kabindra Singh Rathour
Division of Nematology, Indian Agricultural Research Institute, New Delhi – 110 012

Kumud Jarial
Department of Mycology and Plant Pathology, Dr. YSPUHF, Nauni, Solan (H.P.)

M. Azam Wani
Division of Plant Pathology, Sher-e-Kashmir University of Agricultural Sciences and Technology of Kashmir, Shalimar Campus, Srinagar–191 121 (J&K)

Manjul Pandey
Division of Crop Protection, Indian Institute of Pulses Research, Kanpur – 208 024 (U.P.)

Meena Shekhar
Senior Scientist, Directorate of Maize Research, Pusa Campus, IARI, New Delhi – 110 012

Mushtaq Ahmad
Division of Plant Pathology, Sher-e-Kashmir University of Agricultural Sciences and Technology of Kashmir, Shalimar Campus, Srinagar–191 121 (J&K)

Nasim Ahmad
National Centre for Integrated Pest Management, IARI Campus, New Delhi – 110 012

O.M. Bambawale
National Centre for Integrated Pest Management, IARI Campus, New Delhi – 110 012

P.K. Sharma
ICAR Research Complex for North Eastern Hill Region, Manipur Centre, Imphal

P.P. Thirumalaisamy
DMR, IARI Campus, Pusa, New Delhi – 110 012

Pratibha Sharma
Division of Plant Pathology, IARI, New Delhi – 110 012

R.G. Chaudhary
Division of Crop Protection, Indian Institute of Pulses Research, Kanpur – 208 024 (U.P.)

R.K. Prajapati
Subject Matter Specialist (Plant Protection), Krishi Vigyan Kendra, Mandsaur (M.P.)

Rajesh Kumar Pandey
Amity Centre for Biocontrol and Plant Disease Management, Amity University, U.P.

Rakesh Kumar Singh
Division of Crop Protection, Indian Institute of Pulses Research, Kanpur – 208 024 (U.P.)

Ram Dutta
Directorate of Maize Research, Pusa Campus, New Delhi – 110 012

S. Kumar
Department of Plant Pathology, CSK Himachal Pradesh Krishi Vishvavidyalaya, Palampur – 1760

S.K. Mishra
National Bureau of Plant Genetic Resources, New Delhi – 110 012

S.K. Sharma
National Bureau of Plant Genetic Resources, New Delhi – 110 012

Sachin Gupta
Division of Plant Pathology, FOA-Udheywalla, SKUAST-J, Jammu (J&K)

Sangit Kumar
Directorate of Maize Research, Pusa Campus, New Delhi – 110 012

Shahid Ahamad
Regional Agricultural Research Station, SKUAST-J, Rajouri – 185 131

Sheoraj Singh
Division of Plant Pathology, Indian Agricultural Research Institute, New Delhi – 110 012

Sudhir Kumar
Indian Institute of Pulse Research, Kanpur

Sunita Chandel
Department of Mycology and Plant Pathology, Dr. Y.S. Parmar University of Horticulture and Forestry, Nauni, Solan – 173 230, H.P

T.K. Kotha
Division of Plant Pathology, Sher-e-Kashmir University of Agricultural Sciences and Technology of Kashmir, Shalimar Campus, Srinagar–191 121 (J&K)

U.R. Sangle
ICAR Research Complex for North Eastern Hill Region, Manipur Centre, Imphal

V.K. Girija
Department of Plant Pathology, College of Agriculture, Vellayani, Thiruvananthapuram – 695 522, Kerala

V.K. Razdan
Division of Plant Pathology, FOA-Udheywalla, SKUAST-J, Jammu (J&K)

Vishal Gupta
Division of Plant Pathology, FOA-Udheywalla, SKUAST-J, Jammu (J&K)

Plant Diseases Management for Sustainable Agriculture *Pages* **1–10**
Editor: **Dr. Shahid Ahamad**
Published by: **DAYA PUBLISHING HOUSE**

Chapter 1

Concepts and Practices in Integrated Plant Disease Management

Atul Kumar, H.C. Lal, Jameel Akhtar and H.S. Chaube

Department of Plant Pathology, Birsa Agricultural University,
Ranchi – 834 006 (Jharkhand)

Introduction

The country must continue to produce more food and other agricultural commodities per unit of land, water, energy and time to meet the demands of the increasing population. In this process, it has to be ensured that there is no damage to the ecological foundations of agriculture, such as land, water, forests, bio-diversity and the atmosphere. Thus, the greatest challenge to the Indian agriculture is to produce adequate food to feed the growing population without degrading the natural resource base and the environment. Intensive agriculture spread by the green revolution actually led to newer problems due to excessive and ultimately use of irrigation water, replacement of a rich diversity of traditional varieties with fewer high yielding varieties, and inappropriate use of fertilizers and pesticides (Paroda, 1999).

An impetus to the integrated pest management was perhaps given after pesticides, particularly the insecticides, came under the direct attack by environmentalists in the 1960's as our dependence on the insecticides in controlling crop pests had created various problems.

The publication of classic book *Silent Springs* by Rachel Carson essentially brought two issues into sharp focus: (*a*) chemical pesticides can be dangerous to human being as well as the environment and should be used as the last resort, (*b*) there are biologically based more safe alternatives to synthetic pesticides, and these eco-friendly methods need to be studied and exploited. The message though reasonable, when first introduced whipped up reactions particularly by the people who had vested interests in chemical industry. But the 'indiscriminate' use of pesticides brought home the message of Carson, and recent policy makers are serious concerned. Thus, the policy adopted by the World Bank since 1980s is to finance projects that do not seriously compromise on public health or cause irreversible environmental deterioration. The World Bank objective is focused on the IPM approach for sustainable agriculture.

The Concept

Several definitions of sustainable agriculture are available. No definition is perfect, but the one adopted by the Technical Advisory Committee (TAC) of the CGIAR seems satisfactory. It reads *"sustainable agriculture should involve the successful management of resources for agriculture to satisfy changing human needs while maintaining or enhancing the quality of the environment and conserving natural resources"* (TAC, 1989; Flint and Gouveia, 2001).

The term, pest management was first envisaged at the beginning of the 1905s as a concept of integrating the use of biological and other methods of pest control. This was later broadened to include the coordinated use of available biological, cultural, and chemical methods. Subsequently, various authors advocated, a total system approach incorporating pest management practices together with increased production objectives. This approach needs to be a part of an overall crop production system. The present IPM thesis is a composite of disease management and integrated crop management (ICM) (Figure 1.1).

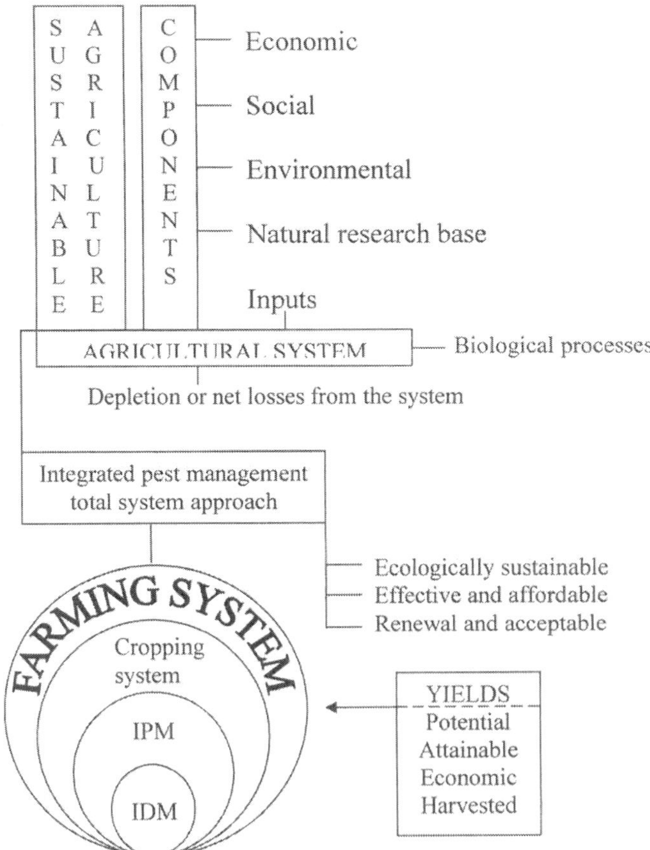

Figure 1.1: Components, Processes and Levels of Various Systems of Sustainable Agriculture and IPM

The Total System Approach

Essentially the IPM paradigm is: (*i*) eradication should not necessarily be the goal, but only application of minimal level of control that will maintain the pest below economic damage threshold, (*ii*) it accepts multiple techniques in preference to following only the pest eradication strategy, (*iii*) the no-option goal in reducing the levels of pesticide use and least or no damage to the non target organisms and ecosystem, (*iv*) the need for indigenous knowledge of agricultural system and pest life-cycles. IPM may thus be an excellent choice for industrial agriculture, but its transfer to resource poor farmers is problematic. Furthermore, IPM programme so far operated only to limit pest populations. There is need for more simple solutions to determine the economic threshold for each crop location set up. The terms pest control or pest management or crop management prefixed with the word 'Integrated' endorse a need for a total system approach wherein there is a meeting of minds, materials, and methods.

Subsystems of IPM

The IPM system is ecologically safe, as it emphasizes judicious use of pesticides along with non-chemical methods like cultural practices and biological control. Various subsystems and approaches for development and recommendation of IPM modules include surveillance, forecasting, habitat analysis, crop husbandry, impact analysis, technology transfer, human resource besides conservation of parasites, predators, parasitoids, etc. (Figure 1.2).

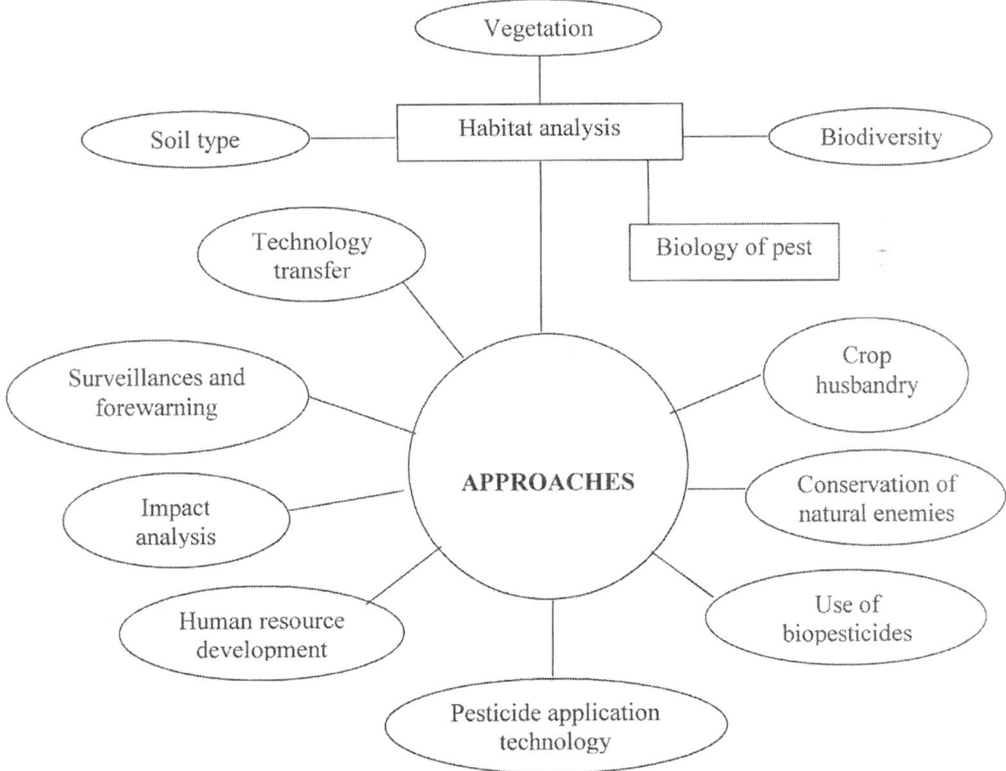

Figure 1.2: Subsystems of IPM

IPM Strategy

To achieve ecofriendly and sustainable management of plant disease, the inputs of both productions as well as protection technologies are to be applied appropriately (Figure 1.3). For healthy crop the management practices that reduce inoculum potential (sanitation, removal of weeds, ploughing, etc.) and avoid the conditions that predispose host are selected. In protection technology, practices that reduce inoculum load, and slow down the disease progress are invariably adopted.

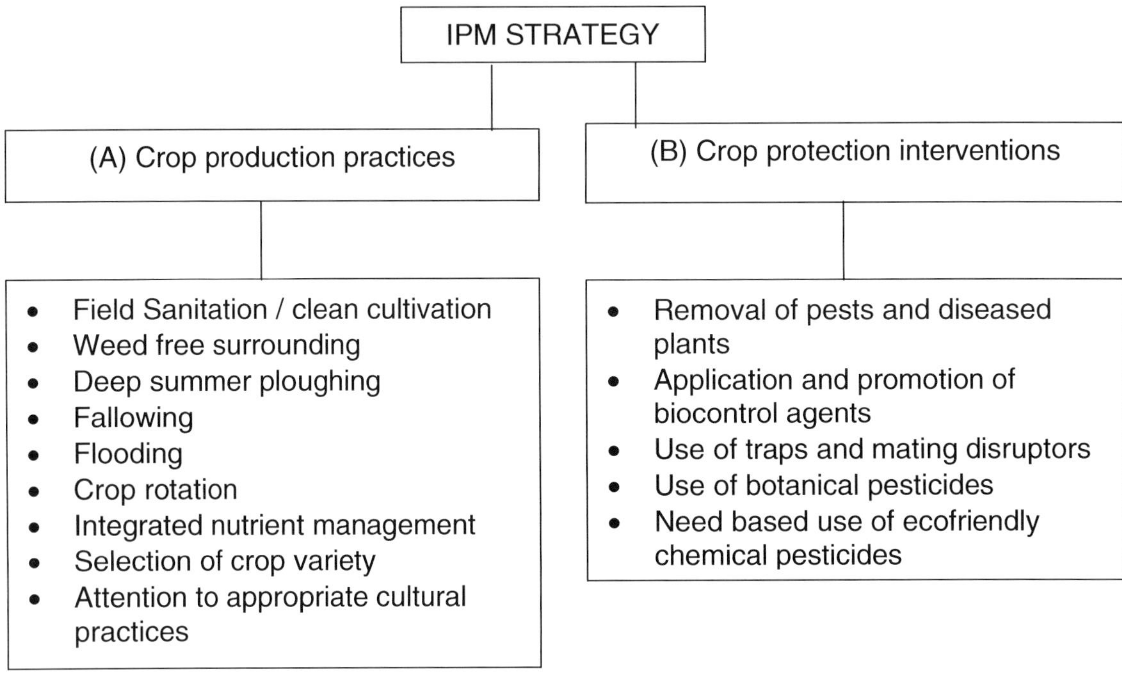

Figure 1.3: IPM Strategy

Integration of Practices

The variety of methods for management of plant diseases listed in the preceding chapters suggests that there exists a big 'arsenal of weapons' to combat the pathogens. However, there is need of a well planned strategy making best use of the 'weapons' in a co-coordinated/integrated manner with proper application technology (Figure 1.4). In order to achieve this, there will have to be integration of methods directed against the pathogen, in favour of the host and for modification of the environment. This is what is known as integrated disease management.

Benefits and Limitations

IPM embodies an ecological approach to the pest problem with the objective of minimizing use of chemical pesticides while sustaining production levels. The gains are reduction in costs of production, more economic access of food and conservation of the resilience and integrity of the ecosystem. The modules developed and validated have to be either crop or location specific (Figure 1.5) and therefore, lack universal application.

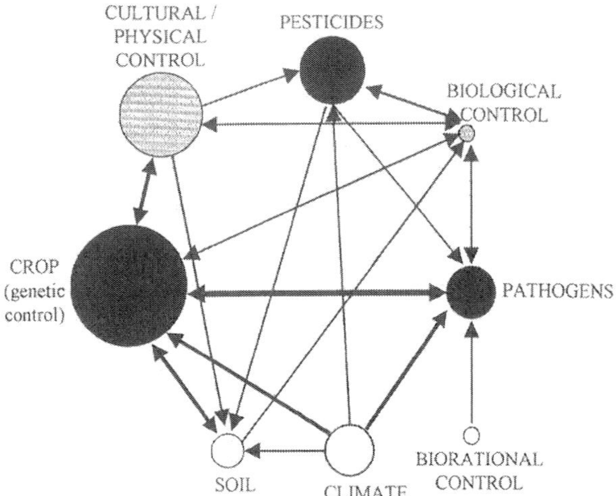

Figure 1.4: Diagrame Depicting the IPM Paradigm for Management of Pathogens

The size of the circles represents relative contribution of the tactics and the weight or thickness of the arrows represents the relative impact of the tactics.

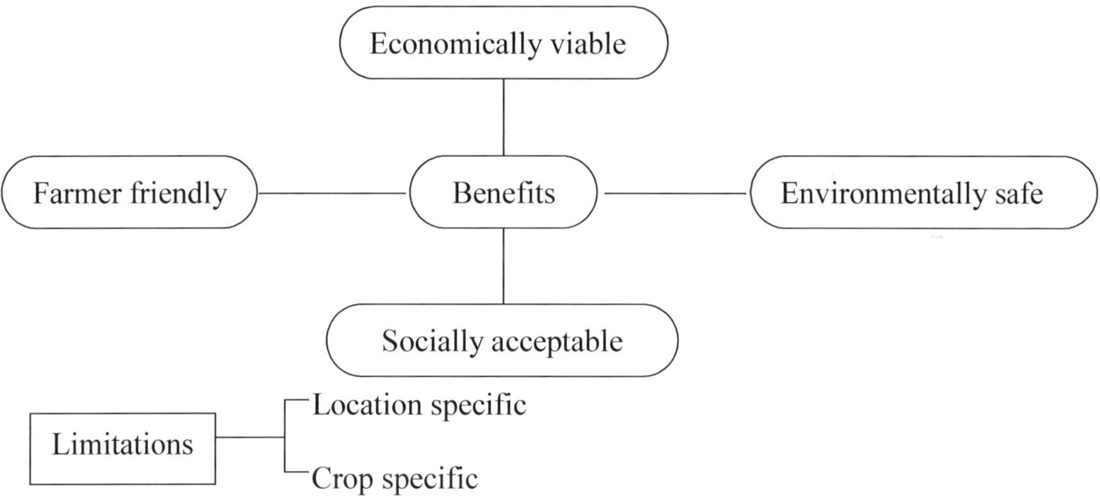

Figure 1.5: IPM Benefits and Limitation

Integrated Management of Wheat Diseases

Wheat is grown throughout the subtropical and temperate regions of the world. Although several diseases caused by fungi, bacteria, nematodes and viruses are known to infect wheat, all are not equally important in every locality. Major diseases of wheat, their mode of carryover, secondary spread and control measures are summarized Table 1.1.

Table 1.1: Important Diseases of Wheat

Sl.No.	*Disease/ Causal Agent*	*Primary Inoculum and Secondary Inoculum*	*Management Practices*
1.	Black/stem rust (*P. graminis*) Brown/leaf rust (*P. recondita*) Yellow/stripe rust (*P.striiformis*)	Inoculum from distant sources, secondary spread through uredinio-spores	Resistant varieties and spray of the fungicides
2.	Loose smut (*U.segatum*)	Internally seed-borne, Infection of ovaries in standing crop	Rouging and seed treatment (physical or chemical)
3.	Karnal bunt (*T. indica*)	Soil/seed borne inoculum, secondary spread through sporidia	Crop rotation, healthy seeds, treatment with fungicides, spray of fungicides at ear emergence
4.	Hill bunt (*T. tritici, T. laevis*)	Seed and soil borne, no secondary spread	Crop rotation, healthy seeds/certified seeds, seed treatment
5.	Flag smut (*U. agropyri*)	Contaminated seeds, soil and straw sources of primary inoculum, no secondary spread	Resistant varieties, rotation, sanitation, rouging, hot weather ploughing and seed treatment
6.	Powdery mildew (*B. graminis*)	Inoculum from distant sources, secondary spread by conidia	Resistant varieties, spray of fungicides
7.	Ear cockles (*A. tritici*)	Soil and seed contamination with cockles, no secondary spread	Hot weather ploughing, crop rotation, healthy seeds, soil treatment with nematicides

In order to manage these diseases, the following practices are integrated to develop IDM module.

Soil borne inoculum can be managed by burning crop debris, hot summer ploughing, crop rotation with legume, maize and mustard, and with fallow or greening manuring or organic amendment (Jha *et al.*, 2007). But, rust, powdery mildew and loose smut diseases can not be managed by this method.

Seed borne inoculum can be reduced by use of clean/healthy/certified seeds of resistant or tolerant varieties and by treating seeds with systemic fungicides in combination with protectant one.

Secondary spread of the pathogen can be reduced by the use of resistant or tolerant varieties, or by selecting the varieties of different maturity durations or by prophylactic sprays of fungicides/biocontrol agents (Chaube and Akhtar, 2002; Lal *et al.*, 2006).

Integrated Management of Nursery Diseases

Most horticultural crops (fruits and vegetables) are first raised in nurseries and thereafter the seedlings are transplanted. Several soil-borne fungal pathogens such as *Pythium*, *Phytophthora*, *Fusarium*, *R. solani*, *S. rolfsii* cause seed rot and damping-off. Soil-borne insects, mites, nematodes also damage the seedlings raised in infested soil (Sharma and Sain, 2003). Very often the weeds also affect the crop health. To ensure optimum production both population and health of the seedlings are of prime importance. In the following flow chart 1.1, the practices that provide desired population and sound health are given.

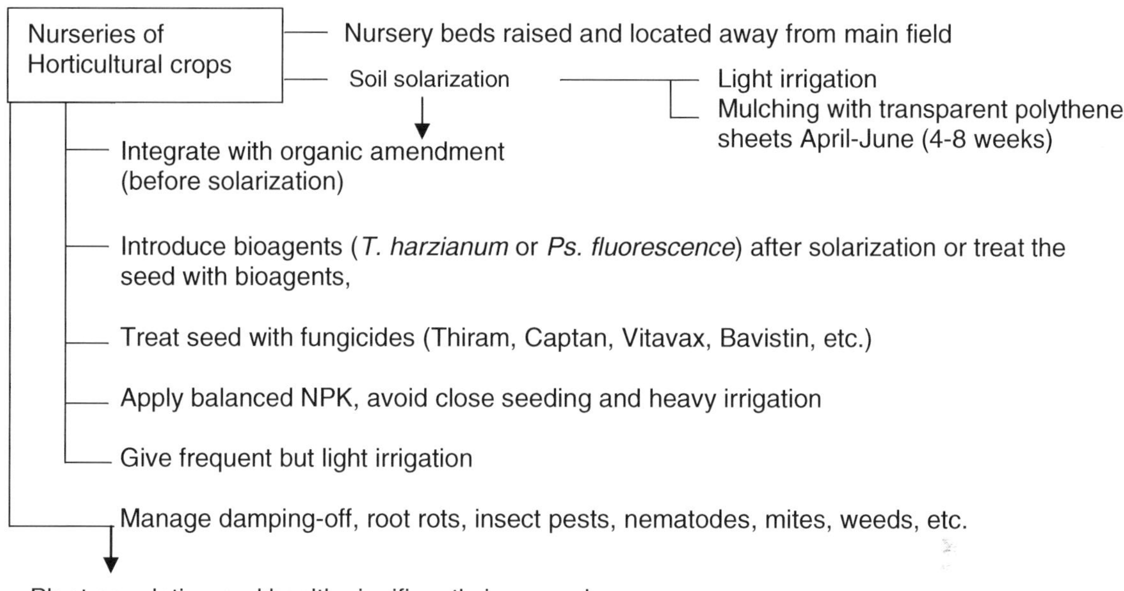

Flow Chart 1.1

Integrated Management of Potato Diseases

As mentioned earlier, all the recorded diseases of a single crop do not occur simultaneously in any locality. However, the possibility of occurrence of any one or more of them in the field does exist. IDM in potato crop is discussed here. Wherever the crop is grown, late and early blight diseases, black scurf/stem canker, bacterial wilt and brown rot, root knot, mosaic and leaf curl are quite common diseases of this crop. Due to vegetative propagation, all the pathogens are mainly seed (tuber) borne, while some of them are soil borne also. Spread by soil, insects, water and wind is common. Therefore, improvement of quality and quantity of the yield can be brought about by adopting following schedule:

1. Selection of field. Field with record of serious incidence of soil-borne diseases in the near past should be avoided. The land must have proper drainage and high fertility level.
2. Selection of seed. Certified seed from reliable agency should be used for planting. Additional precaution should be taken by treating seed tubers with 3 per cent boric acid. This may help in eradication of residual inoculum from the seeds and protects during germination and emergence.
3. The date of planting should be decided according to prevailing soil temperature and moisture.
4. Protective steps to prevent secondary infection are essential. One prophylactic spray of Mancozeb @ 0.2 per cent or copper oxychloride @ 0.3 per cent should be given at 30-40 days after planting.
5. Plants showing viral infection are properly removed and the crop is sprayed with metasystox or other insecticides when there is build-up of population of aphid vectors.

6. In addition, need based spraying with fungicides such as zineb, mancozeb, copper oxychloride, etc. is continued to check early and late blight pathogens.

IDM in Orchards (Perennial Crop)

In an integrated disease management program of an orchard like apple, mango, peach, citrus etc., one must first consider the nursery stock to be used and the location where it is to be planted. The nursery stock (both the root stock and scion) must be free from biotic and mesobiotic agents. Even after obtaining them from a nursery where the saplings are inspected and certified, the stock must be appropriately fumigated, especially to eliminate nematodes. The next step is selection of planting sites. Several pathogens like *Armillaria, Fusarium, Phytophthora*, etc., and several nematodes are known to survive in soil. If experience and history suggest their presence in soil, the soil should be treated with desired pesticide, particularly with fumigants before planting. As far as possible root-stock resistant to these pathogens should be preferred. The drainage of the location should be checked and improved, if necessary. Finally the young trees should not be planted between or next to old trees that are heavily infected with canker fungi, bacteria, insect transmitted viruses and phytoplasmas, pollen transmitted viruses or with other pathogens.

Once the trees have established and until they begin to bear fruits, they should be fertilized, irrigated, pruned, and sprayed for the most common insects and diseases so that they will grow vigorously and free of infections. Any tree that develops symptoms of a disease caused by a systemic pathogen, should be removed at the earliest.

The Following Line of Treatment Ensure Health of Fruit Bearing Trees

1. Removal through pruning and destruction of dead twigs, branches and fruits.
2. Pruning equipment should be disinfested before moving to new trees.
3. A dormant spray containing a fungicide-bactericide (Bordeaux mixture) or a plain fungicide plus an acaricide/insecticide is applied before the buds break.
4. After the buds open, blossoms and leaves must be protected with sprays containing a fungicide and/or a bactericide and an insecticide (that do not harm bees).
5. Young fruits should be protected by appropriate chemical sprays.
6. Wounding of fruits during harvesting and handling must be avoided.

Proper post harvest care (transit and storage) is required for profitable marketing of perishable produce.

Constraints

1. IPM practitioners should be aware of the key pests and the natural enemies of pests in relation to the crop and changing environmental conditions. The monitoring and intervention at different steps requires intensive training and educational programmes to develop the desired human resources.
2. Lack of timely expert advice under changing crop/weather/pest scenario needs to be addressed by the use of information technology for creating awareness as well as for expert advisory system.

3. Microbial and herbal biopesticides preparations are expected to play a major role in IPM, their timely availability and quality must be ensured.
4. IPM technologies have been synthesized for almost every crop. But, these have been validated for a few crops only. The technology needs large scale popularization for adoption at newer locations throughout the country.
5. As IPM research and development are multi-disciplinary, multi organizational, and multi locational, it is necessary to harness synergy among multiple players.

References

Agrios, G. N. (1997). *Plant Pathology*. Academic Press, New York, pp. 635.

Akhtar, J., Abdulmajid Ansari, Kumud Rani Tiu and Chaube, H.S. (2007). Potential of soil solarization in the management of seedling diseases of vegetable nurseries. In: *Ecofriendly Management of Plant Diseases,* (Eds.) Shahid Ahamad and Udit Narain. Daya Publishing House, New Delhi, p. 80–94.

Allen, G.E. and Bath, J.E. (1980). The conceptual and institutional aspects of integrated pest management. *Bioscience*, 30: 658.

Bolkan, H.A. and Reinert, W.R. (1994). Developing and implementing IPM strategies to assist farmers: An industry approach. *Plant Dis.*, 78: 545.

Chaube, H.S. and Singh, U.S. (1990). *Plant Disease Management: Principles and Practices*. CRC Press, Boca Raton, USA.

Chaube, H.S. and Akhtar, J. (2002). Potential of biopesticides in agriculture: Status and outlook. *Plant Disease Research*, 17(1): 160.

Chaube, H.S. and Akhtar, J. (2005). Role of cultural practices in plant disease management. In: *Integrated Pest Management: Principle and Applications Vol. 1,* (Eds.) Amerika Singh, O.P. Sharma and D.K. Garget.CBS Publishers and Distributors, Bangalore, p. 70–89.

Eversmeyer, M.G. and Kramer, C.L. (1987). Components and techniques of integrated pest management threshold determination for pathogens. *Plant Dis.*, 71: 456.

Ferris, H. (1987). Components and techniques of integrated pest management threshold determination for soil borne pathogens. *Plant Dis.*, 71: 452

Flint, M.L. and Gouveia, P. (2001). *IPM in Practice: Principles and Methods of Integrated Pest Management.* University of California Press, pp. 296.

Gliessman, S.R. (1995). Sustainable Agriculture: An agroecological prospectives. *Advances in Plant Pathology*, 11: 45.

Lal, H.C., Upadhyay, J.P. and Jha, A.K. (2006). An integrated approach to the management of lentil rust. *J. Mycol. Pl. Pathol.*, 36(2): 205–208.

Norris, R.F., Caswell-Chen, E.P. and Kogan, M. (2002). *Concepts in Integrated Pest Management.* Prentice Hall of India Pvt. Ltd., New Delhi, pp. 586.

Paroda, R.S. (1999). For a food secure future. *The Hindu: Survey of Indian Agriculture.*

Sharma, P. and Sain, S. (2003). Evaluation of commercial and laboratory formulations of bioagents and plant nutrients against wilt of tomato and damping-off of cauliflower. *Indian J. Plant Pathol.*, 21(1&2): 105–109.

Singh, R.S. (2000). *Plant Disease Management*. Oxford and IBH Publishing Co. Pvt. Ltd., New Delhi.

Singh, A. (2004). Sustainable crop protection through IPM strategies: Current status and prospects in India. Lecture Delivered in the *Training on Biocontrol Strategies for the Management of Plant Pathogens* at IARI, New Delhi, p. 1–23.

Thurston, H.D. (1991). *Sustainable Practices for Plant Disease Management in Traditional Farming Systems*. Westview, Boulder, Colorado.

Van Bruggen, A.H.C. (1995). Plant disease severity in high input compared to reduced input and organic farming systems. *Plant Diseases*, 79: 976.

Plant Diseases Management for Sustainable Agriculture *Pages* **11–25**
Editor: **Dr. Shahid Ahamad**
Published by: **DAYA PUBLISHING HOUSE**

Chapter 2

Emerging and Threatening Diseases of Plants: Detection, Diagnosis and their Management

B.N. Reddy* and C.R. Raghavender

Mycology and Plant Pathology Laboratory,
Department of Botany, Osmania University, Hyderabad – 500 007

Introduction

More than 800 million people do not have adequate food; 1.3 billion live on less than $1 a day and at least 10 per cent of global food production is lost to plant disease (Christou and Twyman, 2004; FAO, 2000; James, 1998). Plant pathologists cannot ignore the coincidence of these figures for food shortage and the damage to food production caused by plant pathogens (Strange and Scott, 2005). Tens of thousands of diseases affect cultivated and wild plants. On an average, each kind of crop plant can be affected by a hundred or more plant diseases. Some pathogens affect only one variety of a plant. Other pathogens affect several dozen or even hundreds of species of plants (Agrios, 2005).

Emerging and Threatening Diseases

New plant diseases caused by various pathogens such as fungi, bacteria, nematodes, insects, viruses and phytoplasmas are appearing at an alarming rate worldwide. Plant pathologists are describing them in scientific journals regularly. In the light of concerns for biosecurity–new, emerging and threatening diseases of plants have received more attention from all interested parties. New and emerging plant diseases are caused by pathogens that: (*i*) have increased incidence, geographical or host range; (*ii*) have changed pathogenesis; (*iii*) have newly evolved; or (*iv*) have been discovered or

---

* *Corresponding author*. E-mail: reddybn1@yahoo.com.

newly recognized (Lederberg *et al.*, 1992; Daszak *et al.*, 2000). Interest in these has focused on those affecting humans (Binder *et al.*, 1999), livestock (Cleaveland *et al.*, 2001) and wildlife (Daszak *et al.*, 2000). Some emerging diseases of plants along with their hosts are tabulated (Table 2.1). Plant diseases impact negatively on human well-being through agricultural and economic loss, and also have consequences for biodiversity conservation. The underlying cause of most emerging plant diseases is the anthropogenic introduction of parasites, although severe weather events are also important drivers of disease emergence. Much is known about crop plant diseases, but there is little information about wild-plant diseases, suggesting that their impact on conservation is underestimated (Anderson *et al.*, 2004).

Table 2.1: Some Emerging Infectious Diseases of Plants Listed According to Host Type with Reference to Agricultural Development (adapted with modification from Anderson *et al.*, 2004)

Disease	*Pathogen*	*Hosts*	*Factors driving emergence*
		Staple crops	
Potato late blight	*Phytophthora infestans*	Potato	Repeated anthropogenic introductions into non-native regions
Rice stripe necrosis	*Rice stripe necrosis virus*	Rice	Hypothesized anthropogenic introduction with rice germplasm
Rice blast	*Magnaporthe grisea*	Rice, barley, wheat, pearl millet and turf grasses	Thought to have spread worldwide by the exchange of seeds
Karnal bunt	*Tilletia indica*	Common wheat, durum wheat and triticale	Dispersal appears to be primarily by contaminated seed
High Plains disease	*Maize high plains virus*	Maize, wheat, barley and several grasses	Possibly has a large distribution, but only recently recognized
		Cash crops, secondary staples and non-food crops	
Cassava mosaic	*EACMV-Ug 3Svr*	Cassava	Interspecific recombination of begomo viruses possibly driven by pathogen pollution
Moko disease	*Ralstonia solana-cearum race 2*	Banana and several weeds	Anthropogenic introduction of crop plant into non-native region
Tomato yellow leaf curl	*Tomato yellow leaf curl virus*	Tomato and common bean	Anthropogenic introduction of propagation material
Sugarcane orange rust	*Puccinia kuehnit*	Sugarcane	Evolution of new strain that broke Q124 resistance
Citrus canker	*Xanthomonas axonopodis pv. citri*	Citrus	Probably arrived with workers or equipment that had been poorly decontami-nated, or via imported infected fruits
		Wild plants	
Dutch elm disease	*Ophiostoma* spp.	Ulnus	Anthropogenic introduction of pathogen and vector in imported timber
Chestnut blight	*Cryphonectria parasitica*	American chestnut	Anthropogenic introduction host and pathogen
Jarrah dieback or root rot	*Phytophthora cinnamomi*	Members of Proteaceae, Papilonaceae, Mimosaceae etc.	Anthropogenic introduction and dissemination within regions
Dogwood anthracnose	*Discula destructive*	Flowering dogwood	Anthropogenic introduction of pathogen

Moreover, plant pathogens are enormous, both in kind and number, and they are in direct competition with us for our food and our cash crops. They are therefore a cause of malnutrition, starvation and poverty. They also deplete our food reserves by spoiling stored produce and ruin our environment by destroying vegetation. 'Know your enemy' is a good adage in most walks of life and is particularly relevant to that of the plant pathologist. The range of organisms which cause plant disease and the techniques for their identification are therefore very much necessary. The correct identification of plant pathogens can hardly be over-emphasized since it is fundamental to control. Those with an historical inclination may know that a disease of potato was the cause of a great famine in the 1840s in which Ireland lost over a quarter of its population through starvation and emigration. Similarly, death by starvation of an estimated two million people living in Bengal (now the Province of West Bengal in India and Bangladesh) is thought to have been caused by a disease of rice in the 1940s. As an example of the devastation of a cash crop, the growing of coffee in Sri Lanka became uneconomic in the last part of the 19th century owing to coffee rust; in consequence, the farmers turned to growing tea and the British, obligingly, to drinking it! Finally, of the many organisms that cause spoilage of produce in storage, fungi of the *Aspergillus flavus* group are perhaps the most notorious; not only are the infected or infested plant products such as maize and peanuts toxic, they are also highly carcinogenic (Strange, 2003).

Significance of Plant Disease Diagnosis

Detection is fundamental to quarantine procedures (Waterworth, 1993) and identification allows the experimenter to draw on the published experiences of other workers with the pathogen or its close relatives (Strange, 2003). Identification of affected plants is one of the first steps in diagnosing a plant disease. Knowing the identity of the plant species affected allows the pathologist to utilize various resources that contain lists of plant diseases associated with specific plants. These lists are very helpful in suggesting possible pathogenic agents. An example of such a list is found on the American Phytopathological Society (APS) web site as a part of it's the online resources (http://www.apsnet.org/online/common/top.asp) (Riley *et al.*, 2002).

Traditional Techniques of Disease Diagnosis

There is a range of diagnostic tests that are traditionally regarded as conventional. These techniques are generally considered to include the plant diseases that involve the progression of symptoms that can vary significantly. The progression of symptoms is one of the most important characteristics associated with problems caused by biotic agents. Diseases can result in primary and secondary symptoms. Variations in symptoms expressed by diseased plants may lead to an improper diagnosis. Sometimes neither these symptoms nor signs provide enough specific or characteristic information to decide the cause of an infectious plant disease. For example, many plant pathogens cause chlorosis, necrosis and wilts so these symptoms, although they may be dramatic, are generally of little diagnostic use. In contrast, the occurrence of unorganized tumorous tissue at the crown of a plant strongly suggests that *Agrobacterium* may be the cause. In such cases, it may be necessary to bring a sample back to the laboratory for further tests to isolate and identify the causal agent. This can be a time-consuming and laborious process that requires specialized skills (Riley *et al.*, 2002). One of the first steps when getting back to the laboratory may be to place a sample of the diseased tissue under conditions that will allow an infectious agent to grow and possibly induce sporulation. This can be accomplished by placing a leaf in a moist chamber (Shutleff and Averre, 1997; Waller *et al.*, 1998). Isolation of fungi usually requires that pieces of infected plant tissue be placed on various nutrient media (Shutleff and

Averre, 1997). The organism that grows out of this tissue is then isolated in pure culture (Agrios, 2005; Waller *et al.*, 1998).

Once an organism is isolated, is that organism the true cause of the problem? Conducting Koch's postulates (Agrios, 2005), which involves the inoculation of healthy plants, may be necessary to conclusively answer this question, especially if the organism has not been previously reported as a plant pathogen on that host. Koch's postulates are seldom conducted for routine diagnoses, but may be extremely important for new diseases and for research. Inoculation of a healthy host and obtaining the symptoms originally observed in the field may be difficult. This may be due to problems in replicating the conditions through which the host was inoculated and also in reproducing the environmental conditions present when the host became infected. It is often impossible to replicate in the laboratory the original conditions present during the disease development in natural conditions.

Conventional plant pathological techniques need substantial expertise and experience of a wide array of specialized technologies. All these features tend to make them less attractive for routine use than the techniques based on the intrinsic microbial biochemistry of the pathogens. Despite most common established preconceptions, although the assessment of symptoms has the potential to be simple, albeit usually at a late stage, in practice it is often inherently unreliable (Fox and Narra, 2006).

Modern Techniques of Disease Diagnosis

Recent advances in molecular biology and biotechnology are being applied to the development of convenient, effective, rapid, specific and sensitive tools for the detection of phytopathogens. Several techniques are available for the identification of the causal agent includes substrate metabolism, fatty acid profiles, protein analysis, serological techniques including ELISA. Nucleic-acid hybridization- and immunological reaction-based detection of phytopathogens are gaining increasing importance in recent times. In this review molecular techniques are described under three broad categories *i.e.* (1) Biochemical markers, (2) Immunological reactions and (3) Nucleic acid techniques (Figure 2.1).

Biochemical Markers

All organisms have distinctive biochemical features and these can be used for diagnosis. Some characteristics are shared by large groups while, at the opposite end of the scale, others are unique to individual populations. The choice of character is therefore paramount in determining the taxonomic level to which an organism is defined. Bacteria have long been identified by their metabolic functions, such as their ability to metabolize certain substrates and more recently, by analysis of their fatty acid profiles. Additionally, soluble protein analysis by gel electrophoresis has been adopted both for bacteria and fungi. All three methods *i.e.* substrate metabolism, fatty acid profiles and protein analysis, rely upon gene expression and, since this may be regulated by environmental factors, care has to be taken to standardize these (Strange, 2003).

In the recent past, research on substrate metabolism is attracting the plant pathologists worldwide. Goor and co-workers (1984) were among the first to investigate the applicability of the 'galleries' of biochemical tests contained in the API (Appareils et Procedes d' Identification) systems to the identification of phytopathogenic bacteria *i.e. Erwinina* and *Pseudomonas*. The test provided an adequate reference system and available as a useful alternative to conventional procedures.

Identification of phytopathogens by fatty acid methyl ester analysis is usually performed on pure cultures of the organism. For the analysis of fatty acid profiles of the organism, about 40 mg of wet cells are saponified and methylated. The fatty acid methyl esters are extracted in an ether-hexane mixture and analysed by gas chromatography. Areas of the resulting peaks on the chromatographs are

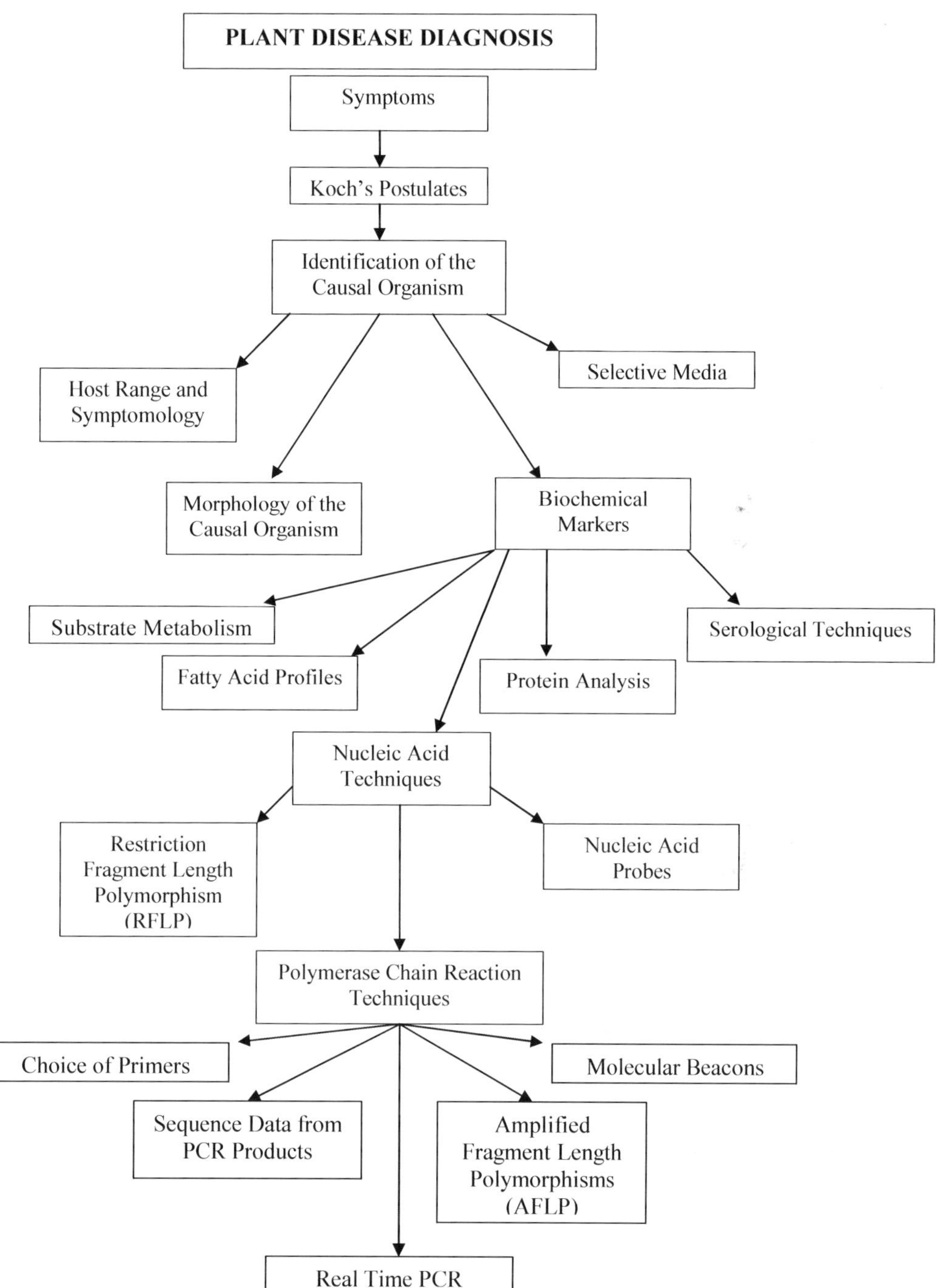

Figure 2.1: Diagram to Show the Range of Techniques Available for the Detection and Diagnosis of Plant Diseases and Plant Pathogens (*Source*: Strange, 2003)

calculated and compared with profiles of known reference strains by computer programs (Roy, 1988). By using this technique Norman *et al.* (1997) differentiated three species of *Xanthomonas* and some pathovars of *Stenotrophomonas maltophilia* with 100 per cent accuracy.

The protein analysis technique is widely used for the diagnosis of bacteria, fungi and also for the identification of viral strains. This technique relies on electrophoresis of soluble proteins from phytopathogens which often gives rise to complex patterns and these can be used for identification purposes. Instead of using a general protein stain, such as Coomassie blue, particular proteins, which for example might have enzymic activity, may be revealed by appropriate staining methods. However, a major disadvantage of the technique is that, when compared with immunological and nucleic acid methods, relatively large quantities of the pathogen must be grown in order to obtain the enzymes either from culture filtrates or from extracts of the organism itself (Bonde *et al.*, 1993). In contrast, peptide profiling may be performed on comparatively small amounts of pathogen and this technique has been used to define virus strains (Strange, 2003).

Immunological Reactions

Immunological methods were amongst the first, and the simplest to use and interpret, of the techniques based on the measurement of molecular rather than the observation of visible criteria. Immunological assays are being used routinely for the detection of plant pathogens in vegetatively propagated material and seeds in conjunction with quarantine seed testing, seed certification and pathogen indexing programmes (Halk and DeBoer, 1985; Hamilton, 1983; Lawson, 1986; Mink and Aichlee, 1984; Raju and Olson, 1985; Sheppard, 1983; Torrance and Dolby, 1984; van Vuurde and Maat, 1983; van Vuurde *et al.*, 1983). While the majority of these programmes are directed towards the detection of viruses and bacteria and also to detect fungi in seeds (Johnson *et al.*, 1983; Mitchell and Sutherland, 1986).

ELISA (Enzyme-Linked Immunosorbent Assay)

ELISA (Enzyme-Linked Immunosorbent Assay) has become widely accepted as an immunodetection method for use with plant pathogens that provides high sensitivity, ease of use, speed and the ability to quantify pathogen biomass in plant tissue and other matrices. Viral identification is often accomplished utilizing ELISA, which is based on the binding of an antibody produced to a specific virus with the virus in the infected plant material (Agrios, 2005). Several ELISA protocols have been devised for different purposes, particularly the monoclonal antibody ELISA is being used in taxonomy (Benhamou *et al.*, 1984; Estrada-Garcia *et al.*, 1989), fine structure of pathogens (Hardham *et al.*, 1985; Day *et al.*, 1986) and detection of their toxins (Benhamou *et al.*, 1985). The monoclonal antibodies that have been raised against fungi (Wong *et al.*, 1988) including seed-borne fungi (Mitchell and Sutherland, 1986) form the basis of many effective diagnostics.

Immunoassay Technology

Immunoassay technology is used for the detection of pathogens, particularly plant viruses for many years (Schaad, 1979; Torrance and Jones, 1981) but more recently they have been applied to the diagnosis of fungi (oomycetes), phytoplasmas and bacteria. Advances in immunoassay technologies have been used to detect *Ralstonia solanacearum* race 3, a pathogen of potato, tomato and egg plant for which many countries have quarantine regulations (Griep *et al.*, 1998). The other serological tests such as Bacterial extracellular polysaccharides (EPS), Immunosorbent Dilution Plating (ISDP), Immunodiffusion and Radio-immuno Assay (RIA) techniques are used more frequently in clinical microbiology laboratories (Tsu and Herzenberg, 1980) but infrequently for the detection in plant pathology (Ghabrial and Shepherd, 1980; Savage and Sall, 1981).

Immunological methods are usually accurate, always quick and invaluable for diseases with inconsistent or undeveloped symptoms. For this reason, many commercial kits are now used on a large scale by plant breeders and propagators when disease-indexing propagating material (Miller and Martin, 1988). There is clearly considerable potential for the use of antibodies in plant disease epidemiology now that many applications to detect phytopathogens are well established and will probably continue to expand considerably (Barnes, 1986; Klausner, 1987).

Nucleic Acid Techniques

The genome of every organism consists of a unique sequence of nucleotides but the degree of conservation of different domains is variable. Highly-conserved regions are found among many organisms which are distantly related whereas some regions such as satellite DNA sequences which are repetitive and usually non-transcribed are highly variable. In view of the variation in the genetic make-up, many diagnostic methods based on the detection of similarities between nucleic acids have been successfully adapted for identification of pathogens. Most of these techniques depend on the use of restriction enzymes to cleave DNA into fragments at or near a defined recognition sequence. This allows the enzymes to be used in the differentiation of non-clonal organisms, as each inherits a unique distribution of restriction sites (Fox and Narra, 2006).

Restriction Fragment Length Polymorphism (RFLPs)

Restriction Fragment Length Polymorphism (RFLPs), a combination of hybridization, Southern blot and restriction mapping forms the basis for the genetic analysis and characterization of pathogens (Botstein *et al.*, 1980). After the DNA has been digested into fragments, these are separated by agarose gel electrophoresis and then stained with ethidium bromide, which can be seen under UV light. The separated bands that result represent a genomic 'fingerprint' which can be analysed in a way that is somewhat analagous to 'reading' a 'bar chart code'. Although their chief advantage over antibodies is that they can detect any part of the genome of the pathogen, however large, nucleic acid techniques are more common for viral (Robinson, 1988) rather than fungal or bacterial plant pathogens as the banding pattern becomes more complex as the genome becomes larger. These techniques have been particularly popular for detecting unidentified viruses or viroids in plants (Fox and Narra, 2006).

Nucleic Acid Probes

Nucleic acid probes are selectively used on the basis of their complementarity to unique nucleotide sequences of the pathogen to be detected and will therefore bind wherever these sequences occur. Recognition that binding has occurred is achieved by labeling the probe. Frequently the label is ^{32}P or ^{33}P which enables binding to be recognized by autoradiography but non-radioactive probes are being developed which may have greater applicability in areas where elaborate laboratory facilities are not available (Miller and Martin, 1988).

Polymerase Chain Reaction (PCR)

Polymerase Chain Reaction (PCR) technique has facilitated advances in many areas of biology and the diagnosis of plant disease is no exception. Essentially the technique provides a method for amplifying in vitro specific nucleotide sequences from the organisms of interest (Strange, 2003). As the choice of primers determines whether the selectivity of the PCR is either specific or broad. This flexibility *i.e.* the detection of variation in pathogen populations in the detection of specific pathogens has proved especially useful in disease detection (Henson and French, 1993). Real-time (RT-PCR) has improved and simplified methods for PCR-based quantification. It has many advantages over conventional PCR being faster, efficient and high throughput more so post reaction processing is not required as the amplified products are detected by a built-in flourimeter thereby eliminating the risk of contamination (Fox and Narra, 2006).

Random Amplified Polymorphic DNA (RAPD)

Random Amplified Polymorphic DNA (RAPD) utilizes short oligonucleotides of about 10 oligomers are used as primers which should be complimentary to sequences in the target genome (Williams *et al.*, 1990). Although no prior knowledge of the actual sequences in the target genome is necessary, the pathogens must be purified by culturing before fingerprinting (Crowhurst *et al.*, 1991; Guthrie *et al.*, 1992; Welsh *et al.*, 1992). Therefore, this method is unsuitable for the direction detection of pathogens in situ in plants or soil. This procedure has also been termed Arbitrarily Primed PCR (AP-PCR) by Welsh and McClelland (1990) and produces relatively simple fingerprints. DNA Amplification Fingerprinting (DAF) uses DNA silver stains and polyacrylamide gel electrophoresis (PAGE) to produce more complex fingerprints (Cactano-Anolles *et al.*, 1991). Inter Simple Sequence Repeats (ISSRs) are an anonymous to RAPDs have been used to address a wide array of issues (Fox and Narra, 2006). Using ISSR analysis considerable variability was observed among isolates from different geographical locations and hosts in different fungal populations (Han *et al.*, 2002; Mishra *et al.*, 2003). This technique will also enable a great understanding of the movement of economically important pathogens around the world.

Amplified Fragment Length Polymorphisms (AFLP)

Amplified Fragment Length Polymorphisms (AFLP) is a DNA fingerprinting technique like RAPD but it does not require sequence knowledge. It is a versatile technique that has been used to perform fingerprinting analyses, mapping and other genetic studies in a broad range of organisms (Garzon *et al.*, 2005). This technique has been successfully used in genetic studies of oomycetes, including species of *Peronospora* (Rehmany *et al.*, 2000) and *Phytophthora* (Lamour and Hausbeck, 2001; van Der Lee *et al.*, 1997; van Der Lee *et al.*, 2001; Werres *et al.*, 2001), and serves as a promising source of information for studies on *Pythium*.

DNA Sequencing

DNA sequencing gained its importance in pathogen diagnosis and paved their way into the disciplines of fungal, bacterial and viral systematics and proved capable of resolving the problems related to classification, detection and discrimination between and within the closely related genera and species (Fox and Narra, 2006). Ribosomal DNA (rDNA) technique is mainly targeted for the development of diagnostic markers because of their presence in all organisms at high copy number that allow very sensitive detection. Some of the rDNA regions are well conserved throughout the evolutionary process and other parts are variable even within a species and based on these variations multiple strategies have been adopted to characterize a wide spectrum of phytopathogens (Louws *et al.*, 1999). rDNA sequences of different organisms are available in databases (Ward *et al.*, 2004). DNA microarrays offer the latest technological advancement for multi-gene detection and diagnostics. These are being used in the detection of multiple pathogens based on the differences in 16S rDNA sequences (Call, 2001). This technique is now gaining increasing importance in the diagnosis of plant pathogens and has been deployed in detecting several plant viruses (DeYong *et al.*, 2005), viroids (Hadidi *et al.*,

2004), bacteria (Call *et al.*, 2003) and also to study host-microbe interaction (Cummings and Relman, 2001).

Molecular Beacons

Molecular beacons are another technique that employs fluorochromes and quenchers which was originally developed by Tyagi and Kramer (1996). A beacon consists of single-stranded DNA with a sequence complementary to that of the target sequence and a stem loop formed by the annealing of the 5′ and 3′ arm sequences. A fluorochrome is attached to the 5′ end and a quencher to the 3′ end of the beacon. The beacon is added to a solution of a target DNA heated 80°C and allowed to cool. When the beacon anneals with the complementary DNA the fluorochrome and its quencher are moved apart resulting in fluorescence. Since different fluorochromes may be used in the same reaction mixture it is possible to detect multiple target sequences simultaneously. Using this technique Eun and Wong (2000) were able to detect both cymbidium mosaic virus and odontoglossum ringspot virus in orchids. Since these are RNA viruses it was necessary to perform RT-PCR on specific sequences of the two viruses first and use the molecular beacons to detect these cDNA molecules.

The advantage of nucleic acid analysis unlike other techniques described so far they are not dependent on gene expression and therefore, are independent of environmental factors. A wide array of techniques based on the nucleic acids have been developed many of which are extremely ingenious and revolutionizing the detection and assay of plant pathogens.

Management of Menacing Diseases

Plant diseases create challenging problems in commercial agriculture and pose real economic threats to both conventional and organic farming systems. Plant pathogens are difficult to manage for several reasons. First of all, plant pathogens are hard to identify because they are so small. The positive identification of a pathogen often requires specialized equipment and training, and in some cases accurate diagnosis in the field is difficult (Koike *et al.*, 2000). Plant pathogens are constantly changing and mutating, resulting in new strains and new challenges to farmers. Also, given the local, regional, and international movement of seed, plant material, and farming equipment, new and introduced pathogens periodically enter the new areas to cause new disease problems.

Regardless of the plant host or particular type of disease that you encounter, the concepts of disease prevention and management are fundamentally the same. Management of plant diseases involves a two-step process that first requires accurate diagnosis and assessment of the severity of the problem. This is followed by implementing strategies to minimize the impact of the disease (Douglas, 2002). A common misconception to disease control is that chemical sprays, dusts, and soil drenches are the only effective means of reducing the effects of plant disease. However, chemical control is only one component of a multifaceted approach that includes culture, sanitation, resistance, biological, and chemical components.

It is important to realize that the goal of disease management is not necessarily to completely eliminate diseases, but to manage them such that they remain at acceptable levels. Additionally, in most cases, prevention is the best strategy for disease control.

Culture

A key opportunity for disease management focuses attention to cultural manipulations that help to minimize conditions favorable for disease development. These include numerous methods that modify the plant's growing conditions in order to optimize growth and vigour.

Sanitation

This option for disease management focuses on minimizing the introduction of disease agents through plant selection and by eradication of diseased plants or plant parts as a means to reduce the potential for spread of biotic agents.

Resistance

This management option utilizes resistant or tolerant cultivars or species of plants to minimize or avoid disease. When available, genetic resistance is probably the most desirable and effective management tool since it circumvents the need for additional controls. It is especially important for diseases caused by viruses, nematodes, and by soil borne and wilt pathogens since these are all extremely difficult to control with other means. Although genetically resistant plants are not available for all plants and all diseases, breeding programs are underway and the availability of these types of plants is expected to increase in the near future.

Biological

This management tool employs living agents (usually antagonists or competitors of the causal agent) to control plant diseases. Effective biological controls take advantage of the natural competition of living organisms for limited resources or ecological niches. Thus, two organisms cannot occupy the same space at the same time, they cannot consume the same resource (*e.g.*, food source) at the same time, and, in some cases, one organism produces compounds that are inhibitory to the growth and development of the other organism. The availability of biological controls is somewhat limited at present, although this option for disease management shows considerable promise for the future.

Chemical: Although it is possible to successfully manage many disease problems without the use of pesticides, there are situations where pesticide usage is important and highly successful. Chemical disease control uses pesticides (fungicides, bactericides, and nematicides) to limit the effects of biotic agents. Fungicides are the most common chemicals used for disease control. In most cases, however, the degree of control depends on the proper selection, timing, and method of application of the compound. In this regard, selection of the appropriate fungicide is contingent upon accurate diagnosis of the problem since fungicides vary in their efficacy; some fungicides are toxic to all or most kinds of fungi, whereas others affect only specific types of fungi. Another way of looking at pesticides is as "plant medicines"; these are compounds used to protect or cure plants from infectious agents.

Conclusions

The past few decades have seen the development of an increasingly secure global system of plant health inspection and quarantine (phytosanitary system) that balances agriculture protection against international trade (Mc Gee, 1997). However, our understanding of the risk of pathogen introductions is still poor and there is an urgent need to improve surveillance strategies. Global surveillance and reporting of plant emerging infectious diseases requires an international coalition that does not yet exist. Currently, the United Nations Food and Agriculture Organization (FAO) mandates countries to report plant pests and diseases to the International Plant Protection Convention (FAO, 1995), but economic consequences often hinder accurate reporting. There is a lack of standardization in testing protocols for seed-borne diseases and significant variation in seed protection treatments among countries. Strengthening the global phytosanitary system requires improved information, standardized seed testing, improved access to seed treatment protocols and revised pathology research priorities (Mc Gee, 1997), but funding for adequate diagnostic facilities and programs is often lacking.

Recent developments in molecular detection technology resulting in more convenient, effective and specific assays have opened new avenues to greater use of these tests for detecting plant pathogens. These techniques will help to free growers, crop consultants and plant health care professionals from having to rely excessively on symptomatology and/or time-consuming diagnostic procedures and permit early detection of phytopathogens (Miller and Martin, 1988). With the advancement of the technology, the sensitivity of detection increases, many pathogens may be found at levels below economic thresholds. Where inoculum-threshold data have been developed, assays that quantify pathogen biomass can be used to determine the point at which control measures should be implemented.

Disease management is complicated by the presence of multiple types of pathogens. For any one crop the grower must deal with a variety of fungi, bacteria, viruses, and nematodes. The world market continues to be extremely competitive and continues to require that farmers supply high-quality, disease-free produce with an acceptable shelf life. Disease management is therefore a critical consideration in safe and healthy plant production.

References

Agrios, G.N. (2005). *Plant Pathology*, Fifth Edition. Elsevier Academic Press, USA, 922 pp.

Anderson, P.K., Cunningham, A.A., Patel, N.G., Morales, F.J., Epstein, P.R. and Daşzak, P. (2004). Emerging infectious diseases of plants: Pathogen pollution, climate change and agrotechnology drivers. *Trends in Ecol. Evolution*, 19(10): 535–544.

Barnes, L.W. (1986). The future of phytopathological diagnostics. *Plant Dis.*, 70: 180.

Benhamou, N., Ouellete, G.B., Lafontaine, J.G. and Joly, J.R. (1984). Monoclonal antibodies to a glycopeptide produced by *Ophiostoma ulmi*. *Can. J. Plant Pathol.*, 6: 260.

Benhamou, N., Ouellette, G.B., Lafontaine, J.G. and Joly, J.R. (1985). Use of monoclonal antibodies to detect a phytotoxic glyopeptide produced by *Ophiostoma ulmi*, the Dutch elm disease pathogen. *Can. J. Bot.*, 63: 1177–1184.

Binder, S., Levitt, A.M., Sacks, J.J. and Hughes, J.M. (1999). Emerging infectious diseases: public health issues for the 21st century. *Science*, 284: 1311–1313.

Bonde, M.R., Micales, J. A. and Peterson, G.L. (1993). The use of isozyme analysis for identification of plant-pathogenic fungi. *Plant Dis.*, 77: 961–968.

Botstein, D., White, R.L., Skolnick, M. and Davis, R.W. (1980). Construction of a genetic linkage in man using restrction fragment length polymorphism. *Amer. J. Human Genet.*, 32: 314–331.

Caetano-Anolles, G., Bassam, B.J. and Gresshoff, P.M. (1991). DNA amplification fingerprinting using very short arbitary oligonucleotide primers. *Biotechnol.*, 9: 553–557.

Call, D. (2001). DNA microarrays: Their mode of action and possible applications in molecular diagnostics. *Vet. Sci. Tomorrow*, 3: 1–9.

Call, D.R., Borucki, M.K. and Loge, F.J. (2003). Detection of bacterial pathogens in environmental samples using DNA microarrays. *J. Microbiological Methods*, 53: 235–243.

Christou, P. and Twyman, R.M. (2004). The potential of genetically enhanced plants to address food insecurity. *Nutr. Res. Rev.*, 17: 23–42.

Cleaveland, S., Lauerenson, M.K. and Taylor, L.H. (2001). Diseases of humans and their domestic mammals: Pathogen characteristics, host range and the risk of emergence. *Philos. Trans. R. Soc. Lond. Series B: Biological Sciences*, 356: 991–1000.

Crowhurst, R.N., Hawthorne, B.T., Rikkerink, E.H.A. and Templeton, M.D. (1991). Differentiation of *Fusarium solani* f.sp. *cucurbitae* races 1 and 2 by random amplification of polymorphic DNA. *Curr. Genet.*, 20: 391–396.

Cummings, C.A. and Relman, D.A. (2000). Using DNA microarrays to study host microbe interactions. *Emerging Infectious Diseases*, 6: 513–525.

Daszak, P., Cunningham, A.A. and Hyatt, A.D. (2000). Emerging infectious diseases of wildlife: Threats to biodiversity and human health. *Science*, 287: 443–449.

Day, A.W., Gardinier, R.B., Smith, R., Svircev, A.M. and McKeen, W.E. (1986). Detection of fungal fimbrae by protein A gold immunocytochemical labelling in host plants infected with *Ustilago heufleri* or *Peronospora hiyoscami f.sp. tabacina*. *Can. J. Microbiol.*, 32: 577–584.

DeYong, Z., Willingmann, P., Heinze, C., Adam, J., Pfunder, M., Frey, B. and Frey, J.E. (2005). Differentiation of Cucumber mosaic virus isolates by hybridization to oligonucleotides in a microarray format. *J. Virological Methods*, 123: 101–108.

Douglas, S.M. (2002). Introduction and historical impact of plant health problems. In: *The Connecticut Agricultural Experiment Station, Department of Plant Pathology and Ecology, Fact Sheet,* 2/02R: PP080.

Estrada-Garcia, M.T., Green, J.R., Booth, J.M., White, J.G. and Callow, J.A. (1989). Monoclonal antibodies to cell surface components of zoospores and cysts of the fungus *Pythium aphanidermatum* reveal species-specific antigens. *Exp. Mycol.*, 13: 348–356.

Eun, A.J.C. and Wong, S.M. (2000). Molecular beacons: A new approach to plant virus detection. *Phytopathology*, 90: 269–275.

FAO (1995). International Standards for Phytosanitary Measures: Principles of Plant Quarantine as related to International Trade, Food and Agriculture Organisation of the United Nations. *World Agricultural Information Centre* (WAICENT), Publication No. 1.

FAO (2000). The state of food insecurity in the world (SOFI). Rome, Italy: FAO, UN. www.fao.org/FOCUS/E/SOFI00/sofi001–e.htm.

Fox, R.T.V and Narra, H.P. (2006). Plant disease diagnosis. In: *The Epidemiology of Plant Diseases,* (Eds.) Cooke, B.M., Jones, D.G. and B. Kave. Springer, The Netherlands, pp.1–42.

Garzón, C.D., Geiser, D.M. and Moorman, G.W. (2005). Diagnosis and population analysis of *Pythium* species using AFLP fingerprinting. *Plant Dis.*, 89: 81–89.

Ghabrial, S.A. and Shepherd, R.J. (1980). A sensitive radioimmunosorbent assay for the detection of plant viruses. *J. Gen. Virol.*, 48: 311–317.

Goor, M.C. *et al.* (1984). The use of API systems in the identification of phytopathogenic bacteria. *Mededelingen Van De Faculteit Landbouwtenschappen Rijksuniversiteit, Gent*, 49: 499–507.

Griep, R.A., van Twisk, C., van Beckhoven, J.R.C.M., van der Wolf, J.M. and Schots, A. (1998). Development of specific recombinant monoclonal antibodies against the lipopolysaccharide of *Ralstonia solanacearum* race 3. *Phytopathology*, 88: 795–803.

Guthrie, P.A.I., Magill, W.W., Federicksen, R.A. and Odvody, G.N. (1992). Random amplified polymorphic DNA markers: A system for identifying and differentiating isolates of *Colletotrichum graminicola*. *Phytopathology*, 82: 832–835.

Hadidi, A., Czosnek, H. and Barba, M. (2004). DNA microarrays and their potential applications for the detection of plant viruses, viroids, and phytoplasmas. *J. Plant Pathol.*, 86: 97–104.

Halk, E.L. and DeBoer, S.H. (1985). Monoclonal antibodies in plant disease research. *Ann. Rev. Phytopathol.*, 23: 321–350.

Hamilton, R.I. (1983). Certification schemes against seed-borne viruses in leguminous hosts, present status and further areas of research and development. *Seed Sci. and Technol.*, 11: 1051–1062.

Han, Q., Inglis, G.D. and Hausner, G. (2002). Phylogenetic relationships among strains of the entomopathogenic fungus, *Nomuraea rileyi*, as revealed by partial β-tubulin sequences and intersimple sequence repeat (ISSR) analysis. *Lett. Appl. Microbiol.*, 34: 376–383.

Hardham, A.R., Suzaki E. and Perkin, J.L. (1985). The detection of monoclonal antibodies specific for surface components on zoospores and cysts of *Phytophthora cinnamomi. Exp. Mycol.*, 9: 254–268.

Henson, J.M. and French, R. (1993). The polymerase chain reaction and plant disease diagnosis. *Ann. Rev. Phytopathol.*, 31: 81–109.

James, C. (1998). Global food security. *Abstr. Int. Congr. Plant Pathol.*, 7th Edinburgh, UK, Aug. No. 4.1GF. http://www.bspp.org.uk/icpp98/4/1GF.html

Johnson, M.R., Anderson, R.L., Kryscio, R.J. and Siegel, M.R. (1983). Sampling procedure for determining endophyte content in tall fescue seed lots by ELISA. *Phytopathology*, 73: 1406–1409.

Klausner, A. (1987). Immunoassays flourish in new markets. *Biotechnol.*, 5: 551–556.

Koike, S.T., Gaskell, M., Fouche, C., Smith, R. and Mitchell, J. (2000). Plant disease management for organic crops. In: *University of California, Division of Agriculture and Natural Resources, Publication* 7252: 1–6.

Lamour, K. H. and Hausbeck, M. K. (2001). The dynamics of mefenoxam insensitivity in a recombining population of *Phytophthora capsici* characterized with amplified fragment length polymorphism markers. *Phytopathology*, 91: 553–557.

Lawson, R.H. (1986). Pathogen detection and elimination. In: *Tissue Culture as a Plant Production System for Horticultural Crops*, (Ed.) R.H. Zimmerman. Nijhoff, Dordrecht, pp.97–117.

Lederberg, J., Shope, R.E. and Oaks, S.E. Jr. (1992). *Emerging Infections: Microbial Threats to Health in the United States.* Institute of Medicine, National Academy Press, Washington D.C., USA, 312 pp.

Louws, F.J., Rademaker, J.L.W. and de Bruijn, F.J. (1999). The three Ds of PCR-based genomic analysis of phytobacteria: diversity, detection and disease diagnosis. *Ann. Rev. Phytopathol.*, 37: 81–125.

McGee, D.C. (1997). *Plant Pathogens and the Worldwide Movement of Seeds*. APS Press, St. Paul, MN, 109 pp.

Miller, S.A. and Martin, R.R. (1988). Molecular diagnosis of plant disease. *Ann. Rev. Phytopathol.*, 26: 409–432.

Mink, G.S. and Aichlee, M.D. (1984). Detection of prunus necrotic ringspot and prune dwarf viruses in *Prunus* seed and seedlings by enzyme-linked immunosorbent assay. *Plant Dis.*, 68: 378–381.

Mishra, P.K., Fox, R.T.V. and Culham, A. (2003). Inter-simple sequence repeat and aggressiveness analyses revealed high genetic diversity, recombination and long-range dispersal in *Fusarium culmorum*. *Annals Appl. Biol.*, 143: 291–301.

Mitchell, L.A. and Sutherland, J.R. (1986). Detection of seed-borne *Sirococcus strobilinus* with monoclonal antibodies in an enzyme-linked immunosorbent assay. *Can. J. For. Res.*, 16: 945–948.

Norman, D.J., Chase, A.R., Hodge, N.C. and Stall, R.E. (1997). Differentiation of three species of *Xanthomonas* and *Stenotrophomonas maltophilia* using cellular fatty acid analyses. *Eur. J. Plant Pathol.*, 103: 687–693.

Raju, B.C. and Olson, C.J. (1985). Indexing systems for producing clean stock for disease control in commercial floriculture. *Plant Dis.*, 69: 189–192.

Rehmany, A. P., Lynn, J. R., Tor, M., Holub, E. B. and Beynon, J. L. (2000). A comparison of *Peronospora parasitica* (Downy mildew) isolates from *Arabidopsis thaliana* and *Brassica oleracea* using amplified fragment length polymorphism and internal transcribed spacer 1 sequence analyses. *Fungal Genet. Biol.*, 30: 95–103.

Riley, M. B., Williamson, M.R. and Maloy, O. (2002). *Plant Disease Diagnosis: The Plant Health Instructor.* DOI: 10.1094/PHI–I–2002–1021–01.

Robinson, D.J. (1988). Prospects for the application of nucleic acid probes in plant virus detection. In: *Proceedings 1988 Brighton Crop Protection Conference on Pests and Diseases*, 2: 805–810.

Roy, M.A. (1988). Use of fatty acid for the identification of phytopathogenic bacteria. *Plant Dis.*, 72: 460.

Savage, S.D. and Sall, M.A. (1981). Radioimmunoassay for *Botrytis cinerea*. *Phytopathology*, 71: 411–415.

Schaad, N.W. (1979). Serological identification of plant pathogenic bacteria. *Ann. Rev. Phytopathol.*, 17: 123–147.

Sheppard, J.W. (1983). Detection of seed-borne bacterial blights of bean. *Seed Sci. and Technol.*, 11: 561–567.

Shutleff, M. C. and Averre, C.W. (1997). *The Plant Disease: Clinic and Field Diagnosis of Abiotic Diseases.* American Phytopathological Society, St. Paul, MN.

Strange, R.N. (2003). *Introduction to Plant Pathology.* John Wiley and Sons Limited, West Sussex, England, 480 pp.

Strange, R.N. and Scott, P.R. (2005). Plant disease: A threat to global food security. *Annu. Rev. Phytopathol.*, 43: 83–116.

Torrance, L. and Dolby, C.A. (1984). Sampling conditions for reliable routine detection by enzyme–linked immunosorbent assay of three ilarviruses in fruit trees. *Ann. Appl. Biol.*, 104: 267–276.

Torrance, L. and Jones, R.A.C. (1981). Recent developments in serological methods suited for use in routine testing for plant viruses. *Plant Pathol.*, 30: 1–24.

Tsu, T.T. and Herzenberg, L.A. (1980). Solid phase radioimmunoassays. In: *Selected Methods in Cellular Immunology*, (Eds.) Mishell, B.B. and S.M. Shiigi. Freeman, San Francisco, pp. 383–393.

Tyagi, S. and Kramer, F.R. (1996). Molecular beacons: Probes that fluoresce upon hybridization. *Nature Biotechnol.*, 14: 303–308.

Van Der Lee, T., Dewitte, I., Drenth, A., Alfonso, C. and Govers, F. (1997). AFLP linkage map of the oomycete *Phytophthora infestans*. *Fungal Genet. Biol.*, 21: 278–291.

Van Der Lee, T., Robold, A., Testa, A., Van't Klooster, J. W. and Govers, F. (2001). Mapping of avirulence genes in *Phytophthora infestans* with amplified fragment length polymorphism markers selected by bulked segregant analysis. *Genetics*, 157: 949–956.

Van Vuurde, J.W.L. and Maat, D.Z. (1983). Routine application of ELISA for the detection of lettuce mosaic virus in lettuce seeds. *Seed Sci. and Technol.*, 11: 505–513.

Van Vuurde, J.W.L., Van Den Bovenkamp, G.W. and Birnbaum, Y. (1983). Immunofluorescence microscopy and enzyme-linked immunosorbent assay as potential routine tests for the detection of *Pseudomonas syringae* pv. *phaseolicola* and *Xanthomonas campestris* pv. *phaseoli* in bean seed. *Seed Sci. and Technol.*, 11: 547–559.

Waller, J. M., Ritchie, B.J. and Holderness, M. (1998). *Plant Clinic Handbook.* CAB International, New York, USA, 104pp.

Ward, E., Foster, S.J., Fraaije, B.A. and McCartney, H.A. (2004). Plant pathogen diagnostics: immunological and nucleic acid–based approaches. *Annals Appl. Biol.*, 145: 1–16.

Waterworth, H. (1993). Processing foreign plant germplasm at the National–Plant– Germplasm Quarantine Center. *Plant Dis.*, 77: 854–860.

Welsh, J. and McClelland, M. (1990). Fingerprinting genomes with arbitary primers. *Nucl. Acids Res.*, 18: 7213–7218.

Welsh, J., Chada, K., Dalal, S.S., Cheng, R., Ralph, D. and McClelland, M. (1992). Arbitary primed PCR fingerprinting of RNA. *Nucl. Acids Res.*, 20: 4965–4970.

Werres, S., Marwitz, R., Veld, W., de Cock, A. W. A. M., Bonants, P. J. M., De Weerdt, M., Themann, K., Ilieva, E. and Baayen, R. P. (2001). *Phytophthora ramorum* sp nov., a new pathogen on *Rhododendron* and *Viburnum*. *Mycol. Res.*, 105: 1155–1165.

Williams, J.G.K., Kubelik, A.R., Livak, K.J., Rafalski, J.A. and Tingey, S.V. (1990). DNA polymorphisms amplified by arbitary primers are useful as genetic markers. *Nucl. Acids Res.*, 18: 6531–6535.

Wong, W.C., White, M. and Wright, I.G. (1988). Production of monoclonal antibodies to *Fusarium oxysporum* f.sp. *cubense* race 4. *Lett. Appl. Microbiol.*, 6: 39–42.

Plant Diseases Management for Sustainable Agriculture *Pages* **26–37**
Editor: **Dr. Shahid Ahamad**
Published by: **DAYA PUBLISHING HOUSE**

Chapter 3

Prospects of Biocontrol Agents in Plant Disease Management

H.C. Lal, J.P. Upadhyay*, Atul Kumar and Jameel Akhtar

Department of Plant Pathology, Birsa Agricultural University, Kanke, Ranchi – 834 006

Introduction

Biological control is the control of one organism by another. This control may be expressed as any condition under which, or practice whereby, survival and activity of a pathogen is reduced through the agency of any other living organism (except man himself) with the result that there is a reduction in the incidence of disease caused by the pathogen. Biological control was defined by Baker and Cook (1974) as the reduction of inoculum density or disease producing activities of a pathogen or parasite in its active and dormant state, by one or more organisms, accomplished naturally or through manipulation of the environment, host, or antagonists or by mass introduction of one or more antagonists. Subsequently, the definition was revised to "biological control is the reduction of the amount of inoculum or disease producing activity of a pathogen accomplished by one or more organisms other than man" by Baker and Cook in 1983. Biological control depends on knowledge of biological interactions at ecosystem, organism, cellular and molecular levels and is often more complicated to manage compared with physical and chemical methods. Biocontrol as a practice for reducing crop losses from disease is not new to agriculture. Crop rotation, burrowing disease infested crop residues and fertilizing with organic manures were practiced by farmers even in earlier days. These and many other traditional practices give effective disease control by allowing the time and opportunity for biological destruction of disease causing pathogens (Upadhyay *et al.*, 2002).

* Department of Plant Pathology, Rajendra Agricultural University (Dholi Campus), Bihar, Pusa, Samastipur – 848 125.

Increasing use of chemical pesticides for the management of soil-borne plant pathogens had caused pollution of soil, surface and ground water besides affecting the crop produce and ecosystem. The pendulum of public opinion once so favourable to chemical control has gone against chemicals and advocated phasing out the chemical pesticides. Therefore, plant pathologists around the globe are now experimenting to control plant diseases by innovative approaches with limited use of chemicals. During 1980's, several biocontrol agents available in nature have been tested against plant pathogens but *Trichoderma* and *Gliocladium* have gained maximum success and popularity. It is now well established that certain bioagents have tremendous potential and can be exploited successfully in modern agriculture for management of plant diseases. Now a days good number of *Trichoderma/ Gliocladium* based bioformulations such as Binab-T, Root Pro, Root Shield, Plant Shield, T22, T 22G, Planter box, Trichodex, Trichopel, Primostop, Soil guard, etc. are available in the international market for the management of plant diseases. Even in Indian market, *Trichoderma* based bioformulations developed by government/semi-government agencies like agricultural Universities, ICAR institutions or private companies. The talc based formulation of *T. viride* such as Bioderma, Bioguard, Biocon, Biocure-F, Defense-SF, Ecoderma, Ecofit, Funginil, Sanjeevani, Tricho-X, Trichoguard and Nisarg are marketed by private companies for the control of primarily soil-borne diseases of wide variety of agricultural, horticultural and agroforestry (Chaube and Akhtar, 2002; Singh, 2004; Lal *et al.*, 2006). Literature review on *Trichoderma* and *Gliocladium* clearly indicated that efficacy of these bioagents in controlling plant diseases improved when used in conjunction with sublethal dose of chemicals. Bioagents in conjunction with small amount of fungicides may compete well with chemical pesticides in controlling plant diseases. So far, there has been an inadequate amount of technological development in integrated control with biological control as the main component. In this chapter, prospects of biocontrol agents in integrated management of pathogens/diseases have been discussed.

Mechanism of Action

The biocontrol agents provide protection against plant diseases either by direct action against the pathogen (*i.e.* antagonism) and/or indirectly by reducing host susceptibility towards the pathogen.

As an antagonist, biocontrol agent may directly kill the pathogen by mycoparasitisms and/or antibiosis. It may adversely affect the growth and development of the pathogen either by antibiosis or by competing for the nutrients, oxygen or space. Antagonist like *Trichoderma* is not only capable of colonizing and killing the hyphae but also hard sclerotia of the fungal plant pathogens.

Parasitism

This complex process involves tropic growth of the biocontrol agent towards the target organism, coiling and finally dissolution of the cell wall/cell membrane by the activity of enzymes (Weindling, 1934, Chet, 1987, Elad, 1995). Recent findings indicated that hyperparasitism and antibiosis are the main mechanism involved in the antagonism of *Trichoderma* as biocontrol agent (Howell, 1987, Mukhopadhyay, 1994, Sharon *et al.*, 2001; Lal and Upadhyay, 2002; Akhtar and Dwivedi, 2000 and 2006).

Antibiosis

The antibiotics and other metabolites released by the antagonistic organisms have harmful effect on the pathogen and their growth are inhibited. *Trichoderma* sp. produce three antibiotics like Trichodermin, Gliotoxin and Viridin in culture (Upadhyay and Mukhopadhyay, 1986, Howell, 1998). *Aspergillus* spp. produces compounds of hydroxamate and catecholate group (Sen, 2000) while pyoverdin and 2, 4-diacetylphloroglucinol are produced by *Pseudomonas fluorescens.*

Competition

The antagonistic micro-organisms compete with the pathogen for food and essential elements and space, making them unavailable for the pathogen thereby suppressing the growth of the latter (Chet, 1987).

Antagonist	Pathogen
Fusarium sp. (Non-pathogenic)	*F. oxysporum* f.sp. *melonis*
Trichoderma harzianum	
Pseudomonas putida	
Pseudomonas spp.	

Bacteria such as *Pseudomonas, Enterobacter, Acaligens* and *Arthnobacter* compete with the plant pathogenic fungi and reduce the disease.

Siderophores

Many bacteria produce siderophores which are low molecular weight ferric ion-transport agent. Siderophores supply iron to the bacterial cell and selectively complex with iron with very high affinity and make it unavailable to other micro-organisms including plant pathogens, *P. fluorescens* and *P. putida* produce siderophore which controls soft rot of potato caused by *Erwinia carotavora.* The siderophores of *Pseudomonas* sp. is fluorescent and is called pseudobactin. It has been found to control take-all disease of wheat and barley and flax wilt disease. The addition of pseudobactin to the soils convert pathogen conducive soil into pathogen suppressive soils, thereby, controlling wilt diseases.

Induced Systemic Resistance

Evidence that plants could be protected against infection by prior infection with another pathogen began to accumulate. This phenomenon is known as induced or acquired resistance to diseases. Induced resistance can be expressed either locally or systematically. In local induced resistance, plant tissues that respond to an initial infection attempt by a virulent pathogen or an avirulent form of pathogen or non-pathogen (biocontrol agents) become resistant to subsequent infection by virulent pathogen inoculated at or very near to the site of pathogen/non-pathogen attack. In systemic induced resistance, the localized infection of one part of a plant by non-pathogen or a necrotic lesion inducing pathogen results in enhanced state of resistance throughout the plants (Singh *et al.,* 2002).

T. harzianum produces cellulose that acts as elicitor by triggering peroxidase and chitinase activity to induce SAR (Elad, 2000). Wilt and powdery mildew infection on greenhouse melon plants were reduced in plants treated with cellulose.

Howell *et al.* (1999) have demonstrated the presence of SAR induced by *T. virens*. Effective native strains of *T. virens* usually produce gliotoxin and gliovirin. Mutants were developed that lacked both mycoparasitic ability and the capacity to produce these or other detectable antibiotics. Surprisingly, many of the mutants were effective or more effective than the parent strains in biocontrol of *R. solani* (Howell, 1987; Howell and Stipanovic, 1995). This high level of protection was associated with significantly enhanced levels of various terpenoid phytoalexins known to be involved in disease resistance in cotton (Howell *et al.,* 1999). Elad *et al.* (1999) considered induced resistance to be primary method whereby T-39 controls powdery mildew. Similar mode of action has been advocated by Harman (2000) against this pathogen by *T. harzianum* T22.

Growth Promotion

Both fungal and bacterial biocontrol agents are reported to induce the growth of various crops owing to suppression of deleterious root microflora including those not causing obvious disease, production of growth stimulating factors like hormones or growth regulators and increased nutrient uptake through solubilization and sequestration of nutrients and/or enhanced root growth. The enhanced rooting by *T. harzianum* T 22 probably also induces tolerance to diseases that it does not directly controls (Harman, 2000). *Trichoderma* preparations have been reported to increase vigour and emergence of radish seedling (Henis *et al.,* 1978). *Aspergillus niger* AN-27 was reported to produce growth promotory compounds that were directly responsible for increased root and shoot length and biomass of the crop plants (Sen, 2000).

Potential of Biocontrol Agents

Soil Application

Augmentation of soil with *T. harzianum* by adding large amount of rye grass seed and soil preparation of *Trichoderma* is the pioneer work of Wells *et al.* (1972) on biological control of *Sclerotium rolfsii* on tomato transplants. Backman and Rodriguez-Kabana (197) used a diatomaceous earth granule impregnated with 10 per cent molasses solution as food base for growth *T. harzianum* and applied the granules by hand over rows of peanuts. At 140 kg/ha, *T. harzianum* significantly reduced *S. rolfsii* damage and increased yield over a 3 year test period in peanuts. Control by this method was equivalent to those obtained with PCNB.

Trichoderma species grown on various solid substrates like Grain bran (Wells *et al.,* 1972), wheat bran (Henis *et al.,* 1978), wheat bran and saw dust (Elad *et al.,* 1980), lignite and stillage (Jones *et al.,* 1984), Molasses and Brewer's yeast (Papavizas *et al.,* 1984) and sorghum grains (Upadhyay and Mukhopadhyay, 1986) have been tested against several soil borne pathogens with varying degree of success in India and abroad. Some notable examples are white rot of onion (Abd-El-Moity and Shatla, 1981), Sclerotium root-rot of sugarbeet (Upadhyay and Mukhopadhyay, 1986), Pythium damping-off of cauliflower (Mukherjee *et al.,* 1989), Rhizoctonia damping-off and blight (*S. rolfsii*) of several crops (Elad *et al.,* 1980) and Rhizoctonia fruit rot of cucumber (Lewis and Papavizas, 1980). *Gliocladium* species especially *G. roseum* and *G. virens* grown on solid substrates have also been tested successfully against Black rot of cucumber (Gindrat *et al.,* 1977), damping-off of cotton caused (Howell *et al.,* 1982) and wilt of chickpea (Upadhyay and Singh, 1995). The efficiency of biocontrol with solid inoculants depended on temperature, type of inoculum used, time of introduction of antagonist to soil and the rate of application (inoculum density) of the antagonist.

Seed Treatments

Application of biocontrol agents *Trichoderma, Gliocladium* and others to seeds was suggested as an alternative approach to introducing them into soil, as large amount of biological material is required for application in soil. The seed treatment requires smaller amount of biological material than in furrow or in broadcast applications and reach in the rhizosphere region alongwith the seed. Spore suspension as well as dry powder of fungal antagonists have been used to coat the seeds. The general procedure of biological seed treatment is to harvest the conidia from a well sporulated culture, making a spore suspension of desired concentration with or without sticker, soaking the seeds in the suspension and drying them in shade before sowing. Combining biological seed treatment with solid matrix priming improves the degree of disease protection by pre-establishing the antagonist on the seed coat (Harman and Taylor, 1988).

Evidences of successful disease management through biological seed treatment are ample and the volume of literature expanding continuously. Damping-off caused by *R. solani* and *Pythium* spp. by treating seeds of peas and radishes with conidia of *T. hamatum* isolated from a *Rhizoctonia* suppressive Columbian soil (Harman *et al.*, 1980), *R. solani* by *Trichoderma* spp. in cotton (Elad *et al.*, 1982a), *Pythium* spp. by *T. harzianum* and *T. koningii* in bean, pea and cucumber (Hadar *et al.*, 1984), *Fusarium oxysporum* by *T. harzianum* in cotton, wheat, muskmelon and tomato (Sivan and Chet, 1986; Sivan *et al.*, 1987), *R. solani, S. rolfsii* and *F. oxysporum* in chickpea (Mukhopadhyaya and Mukherjee, 1991) are some examples of promising results by seed treatment with biological agents.

The trends of results on the management of soil-borne pathogens by the use of biological antagonists across the country clearly indicated that biocontrol agents have tremendous potential in reducing the survival and activity of soil-borne pathogens.

Pesticides Tolerance of Antagonists and Induction of Tolerant Strains

Wild strains of various fungal antagonists have the capacity to tolerate some of the more commonly used old and new pesticides in modern agriculture. Named isolates of *Trichoderma* and *Gliocladium* showed tolerance to several fungicides, *viz.* carboxin, metalaxyl, captan, copper oxychloride, quintozene, oxyadixyl, copper sulphate, foltaf and carbox L (Mukhopadhyay *et al.*, 1982; Dey and Mukhopadhyay, 1994 and Gupta, 1995). Benomyl and bavistin were strongly inhibitory to these antagonists (Abd-El-Moity *et al.*, 1982; Singh *et al.*, 1993). This natural tendency of tolerance to pesticides in biocontrol agents have opened new dimension in the integration study.

The level of tolerance in bioagents may be increased further by exposing them to increasing concentrations of fungicides. The induction of tolerance to pesticides in *Trichoderma* and *Gliocladium* and the selection of stable biotypes for the use in combination with pesticides is one of the best options available for dividing an effective control scheme. *Trichoderma* thus developed by adaptation technique tolerated higher concentration of fungicides metalaxyl, ronilam, captan, chlorothalonil, PCNB, procymidone, iprodione and vinclozolin in culture media (Abad-El-Moity *et al.*, 1982). However, several fungicide tolerant biotypes lost their ability to tolerate the fungicide after repeated transfer to fungicide free culture media (Papavizas, 1985).

Mutation in *T. harzianum* can be induced by UV-irradiation. Mutation improved the tolerance of *T. harzianum* to benomyl as well as its survival in soil. These induced biotypes can tolerate between 50 and 500 mg/ml benomyl of the medium (Papavizas and Lewis, 1983). New benomyl resistant biotypes of *T. viride* and *T. harzianum* have possessed better biocontrol capabilities than that of wild types. Howell (1982) while studying the effect of *G. virens* on *Pythium ultimum* reported higher production of antibiotics "glioviridin" by the mutant strains of *G. virens* than that of the wild type strains. Mukherjee and Mukhopadhyay (1993) generated seven stable mutants of *G. virens* by exposing the slant culture to 125 Kr of gamma radiation. One of the mutants, M-7 was more aggressive than the wild type in antagonizing *Sclerotium rolfsii* and *F. oxysporum* f.sp. *ciceri.*

The development of pesticide resistance by genome modification may be a more promising approach to biological control because the tolerance resulted from non-genetic adaptations or training in *Trichoderma* and *Gliocladium* may be unstable.

Soil Solarization

The effect of solar heating of soils on the soil-borne pathogens has been extensively studied in Israel (Chet *et al.*, 1980; Elad *et al.*, 1980; Katan *et al.*, 1981). This technique involves the mulching of moist soil with transparent polythene sheets during the hot season, thereby trapping the incoming solar radiation, increasing soil temperatures and killing propagules of the pathogens. Under field conditions, soil mulching was extended to about five weeks, during which time, the maximum temperature were 55°C at 5 cm and 45°C at 20 cm depth, respectively. Soil temperature in non-mulched plots were 7-14°C lower. The solarization of soil makes the pathogen more vulnerable to antagonism by the biocontrol agents. Solar heating of soil by polythene mulching or the fumigation with methyl bromide, when tested under field conditions significantly reduced the disease caused by *Rhizoctonia solani* and *Verticillium dahliae*. Combining solar heating or methyl bromide with *T. harzianum* improved their efficiency and also resulted in the enhanced control of *S. rolfsii*. Incidence of disease caused by *R. solani* in potato remained low throughout the season may be attributed to a shift in biological equilibrium in the soil in favour of the antagonists. Such a shift in solar heated soil may explain the reduction in inoculum build up and a more efficient control when this treatment is combined with *T. harzianum* (Elad *et al.*, 1980; Chaube *et al.*, 2003).

In India, the effect of soil solarization has been studied against diseases caused by *Fusarium udum* on pigeonpea and *Fusarium oxysporum* f. sp. *ciceri* on chickpea and promising results were reported (Chauhan *et al.*, 1988). Manmohandas and Sivaprakasam (1993) reported 14 days of solarization after amending the soil with neem cake and irrigation recorded the lowest incidence of pre-emergence damping-off of chilli (39 per cent).

Integrated Use of Biocontrol Agents

Although it is well established that certain biocontrol agents have tremendous potential of controlling plant diseases, it is understood that the integration of several control technologies wherein biological control component would be pre-dominant may be required to increase the yield and minimize environmental hazards. In the past, the integrated use of biocontrol agents with other management practices has been inadequately studied. Literature indicated that *Trichoderma* or *Gliocladium* exhibited synergistic or additive effect when used in conjunction with sublethal doses of pesticides. Such approach can be successful if the biocontrol agents are tolerant or insensitive to pesticide of compatible with other control practices.

Control of *Armillaria mellea* with a combination of methyl bromide and *Trichoderma* is a classical example of integrated control (Or *et al.*, 1973). Other examples of integrated use of pesticide are *T. harzianum* and PCNB against *R. solani* on radish and eggplant (Henis *et al.*, 1978; Elad *et al.*, 1980), *T. harzianum* and chlorothalonil against *R. solani* on cucumber (Levis and Papavizas, 1980), *T. harzianum* and PCNB against *Sclerotium rolfsii* on bean seedlings (Chet *et al.*, 1979), *T.harzianum* and Ridomil against *Phytophthora capsici* and *Pythium aphanidermatum* on pepper and cauliflower (Papavizas and Lewis, 1981; Mukherjee *et al.*, 1989), *Gliocladium virens* and foltaf against *Fusarium oxysporum* f. sp. *ciceri* on chickpea (Singh *et al.*, 1993), *G. virens* and Thiram or *G. virens* and Apron against *P. aphanidermatum* on tomato (Dey and Mukhopadhyay, 1994), *G. virens* and Fytolan against *Fusarium udum* on pigeonpea (Upadhyay and Roy, 1997). During the recent past, *T. harzianum* and *G. virens* have been widely used in integrated control programme with maximum success.

The biological seed treatment technique developed by Mukhopadhyay *et al.* (1992) involving treatment of seeds first with antagonist *G. virens* (10^6 conidia/ml) and then with the fungicide carboxin (0.1 per cent) also indicated that the integration of biological and fungicidal seed treatment improved the disease control potential to a great extent. This method, demonstrated in research trials and farmers' fields, has proven to be highly effective in managing several soil borne pathogens, including *S. rolfsii, R. solani* and *F. oxysporum* in chickpea, lentil, rajmash and groundnut (Jha *et al.*, 2007).

Induction of resistance to pesticides in *Trichoderma* or *Gliocladium* or other antagonists and selection of stable biotypes for use in conjunction with pesticides is another technological advancement for the use of biocontrol agents in integrated control programmes. Some biotypes with induced resistance to pesticides showed better disease control ability. *T. harzianum* biotype Th-1 (Ron 2M) adapted to 200 mg a.i. Ronilan per ml of the medium suppressed the white rot of onion more effectively than their respective wild type (Locke *et al.*, 1984; Papavizas *et al.*, 1982).

Several ultra violet induced biotypes of *Trichoderma* were consistently more effective in suppressing damping-off (*P. ultimum*) of peas, root rot (*R. solani*) in cotton and radish and damping-off and blight (*S. rolfsii*) of bean (Beagle-Ristano and Papavizas, 1984; Locke *et al.*, 1984).

Lewis and Papavizas (1980) used *Trichoderma* sp. (isolate VT 6) in integration with mechanical ploughing to control Rhizoctonia fruit rot of cucumber and reported that use of antagonist along with ploughing controlled the disease more effectively than that of either *Trichoderma* or fungicide.

Soil solarization, steaming and other control methods can also modify the ecosystem to a great extent and then may allow the proliferation of antagonists *Trichoderma* and *Gliocladium* (Katan, 1981). Several workers have observed enhanced control of the diseases caused by *R. solani* and *S. rolfsii* by integrated use of *T. harzianum* and solarization (Elad *et al.*, 1980; Chet *et al.*, 1982; Akhtar *et al.*, 2007). Kraft and Papavizas (1983) used host resistance, *Trichoderma* and fungicide to control *Pythium ultimum* on peas. They reported maximum yield of peas under field conditions when the seeds were treated with matalaxyl and *T. harzianum*.

The examples discussed here lead to the conclusion that biocontrol agents have a better prospect in the integrated disease management systems. Due to this important reason only, in recent years much emphasis has been placed on integrated biological control of plant diseases. Moreover, this approach of disease control may prove eco-friendly and more effective in controlling a plant disease where single component is ineffective in reducing a particular disease.

Constraints

The success of biological control depends on the selection of potential antagonist, method of application and its establishment in the ecological niche. Since the biocontrol technology is not popular among farmers there is need to provide incentive or subsidy on the biopesticide preparations to increase large scale use. Moreover, biocontrol practices exert a long term effect on plant pathogens as compared to short term kill by fungicidal control. Therefore, assessing the economics of biocontrol is also difficult. Another constraint is the variability performance of the antagonist in different locations due to change in soil conditions like structure, texture, temperature, moisture, organic matter content, oxygen and carbondioxide besides micro-flora and fauna of the soil.

Future Prospects

There is a need for selection of potential strains of antagonists which can survive well in variable soil conditions, may tolerate pesticides and be used for commercial production for multiple disease control. Biocontrol agents like *Trichoderma* and *Gliocladium* could now be genetically manipulated to incorporate desired traits through protoplast fusion, recombinant DNA technology, mutation, adaptation and UV-irradiation. There is tremendous scope for boosting the effectiveness of mycofungicides by genetic engineering. Recent advances in molecular biology and genetic engineering have opened new avenues for engineering the organism with lethal arsenals against target organism.

Since the performance of bioagents is not stable in diverse soil conditions, increased attention is needed in the following areas of biocontrol research:

1. Genetic manipulation to enhance the pesticide tolerance potentiality and survivability by mutation and biotechnology for constant and multiple disease control in diverse soil conditions.
2. Integration of biocontrol agents with other control components to develop integrated management systems for multiple disease control.
3. Technologies for large scale production of biofungicides may be worked out.
4. To understand the effect of micro-environment on survival and proliferation of biocontrol agents, computerized models need to be developed.

References

Abd-El Moity, T.H. and Shatla, M.N. (1981). Biological control of white rot disease of onion (*Sclerotinia sclerotiorum*) by *Trichoderma harzianum*. *Phytopathology*, 2(100): 29–35.

Abd-El Moity, T.H., Papavizas, G.C. and Shatla, M.N. (1982). Induction of new isolates of *Trichoderma harzianum* tolerant to fungicides and their experimental use for biological control of white rot of onion. *Phytopathology*, 72: 396–400.

Akhtar, J. and Dwivedi, R.R. (2006). *In vitro* screening of bioagents against *Colletotrichum graminicola* causing anthracnose of sorghum. *Annals of Plant Protection*, 14(2): 505.

Akhtar, J. and Dwivedi, R.R. (2000). Foliar application of oil-cakes and Bioagents: Strategy for management of Anthracnose of Sorghum. *J. Mycol. Pl. Pathol.*, 30(2): 256.

Backman, P.A. and Rodriguez-Kabana, R. (1975). A system for the growth and delivery of biological control agents to the soil. *Phytopathology*, 65: 819–821.

Beagle-Ristaino, J. and Papavizas, G.C. (1984). Reduction of *Rhizoctonia solani* in soil with fermentor preparations of *Trichoderma* and *Gliocladium*. *Phytopathology*, 74: 836 (Abstr.).

Chaube, H.S. and Akhtar, J. (2002). Potential of biopesticides in agriculture: Status and outlook. *Plant Disease Research*, 17(1): 160.

Chaube, H.S., Akhtar, J. and Vishwakarma, S.N. (2003). Effect of soil solarization on microbiological properties, disease suppressiveness and fertility of soil. *Indian Phytopath.*, 56(3):353.

Chauhan, Y.S., Nene, Y.L., Johnsen, C., Haware, M.P., Saxena, N.P., Sardar Singh, Sharma, S.B., Sahrawat, K.L., Burford, J.R., Rupela, O.P., Kumar Rao, J.V.D.K. and Sithanantham, S. (1988). Effect of soil-solarization on pigeonpea and chickpea. *Research Bulletin*, No. 11, ICRISAT Patancheru, A.P., 502, 324 p. 16.

Chet, I. (1987). *Trichoderma*: Application, mode of action, and potential as biocontrol agent of soil borne plant pathogenic fungi. In: *Innovative Approaches to Plant Disease Control*, (Ed.) I. Chet. John Wiley and Sons, New York, p.137–160.

Chet, I. and Henis, Y. (1985). *Trichoderma* as a biocontrol agent against soil borne root pathogens. In: *Ecology and Management of Soil Borne Plant Pathogens*, (Eds.) Parker, C.A. *et al.* American Phytopathological Society, St. Paul. Minn, pp. 110.

Chet, I. and Katan, J. (1980). Effect of solar heating of soils by transparent polythene mulching on their chemical properties. *Soil Sci.*, 130: 271–277.

Chet, I., Hadar, Y., Elad, Y., Katan, J. and Henis, Y. (1979). Biological control of soil-borne pathogen by *Trichoderma harzianum* in soil borne. In: *Plant Pathogens*, (Eds.) Schipper, B. and W. Academic Press London, pp. 585.

Chet, I., Elad,Y., Kalfon, Y., Hadar,Y. and Katan, J. (1982). Integrated control of soil borne and bulb-borne pathogens in iris. *Phytoparasitica*, 10: 229–236.

Dey, R.K. and Mukhopadhyay, A.N. (1994). Biological control of tomato damping off by *Gliocladium virens*. *J. Biol. Control.*, 8(1): 34–40.

Elad, Y., Chet,I. and Katan, J. (1980). *Trichoderma harzianum*: A biocontrol agent effective against *Sclerotium rolfsii* and *Rhizoctonia solani*. *Phytopathology*, 70: 119–121.

Elad, Y., Katan, J. and Chet, I. (1980). Physical, biological and chemical control of soil borne diseases in potatoes. *Phytopathology*, 70: 418–422.

Elad, Y., Kalfon, J. and Chet,I. (1982). Control of *Rhizoctonia solani* in cotton by seed coating with *Trichoderma* spp. spores. *Plant and Soil*, 66: 279–281.

Elad, Y., Hadar, Y., Chet, I. and Henis, Y. (1982). Prevention with *Trichoderma harzianum* Rifai aggr. of reinsfestation by *Sclerotium rolfsii* Sacc. and *Rhizoctonia solani* Kuhn of soil fumigated with methyl bromide and improvement of disease control in tomatoes and plant. *Crop Prot.*, 1: 199–211.

Elad, Y., David, D.R., Levi, T., Kapat, A., Krishner, B., Guvrin, E. and Levine, A. (1999). *Trichoderma harzianum* T–39 mechanism of biocontrol of foliar pathogens, In: *Modern Fungicides and Antifungal Compounds II*, (Ed.) H. Lyr. Intercept Ltd., Andovr, Hampshire, UK, p. 459–467.

Gindrat, D., Van der Hoeven, E. and Moody, A.R. (1977). Control of *Phomopsis sclerotioides* with *Gliocladium* or *Trichoderma*. *Nat. J. Plant Pathol.*, 83(Suppl. 1): 429–438.

Grinstein, A., Orione, D., Greenberger, A. and Katan, J. (1979). Solar heating of soil for the control of *Verticillium dahliae* and *Patylenchus thornei* in potatoes in soil borne In: *Plant Pathogens*, (Eds.) Schipper, B. and W. Academic Press London, pp. 431–437.

Gupta, S.K. (1995). Studies on the management of chickpea wilt complex. *M.Sc.(Ag.) Thesis*, Rajendra Agril. Univ., Bihar, Pusa, pp.45.

Hadar, Y., Harman, G.E. and Taylor, A.G. (1984). Evaluation of *Trichoderma koningii* and *Trichoderma harzianum* from New York soils for biological control of seed rot caused by *Pythium spp. Phytopathology*, 74: 106–110.

Harman, G.E. (2000). Myths and dogmas of biocontrol. *Plant Disease*, p. 377, 393–525.

Harman, G.E. and Taylor, A.G. (1988). Improving seedling performance by integration of biological control agents at favourable pH levels with solid matrix priming. *Phytopathology*, 78: 520–525.

Harman,G.F., Chet,I. and Baker, R. (1980). *Trichoderma hamatum* effects on seed and seedling disease induced in radish and pea by *Pythium* spp. or *Rhizoctonia solani*. *Phytopathology*, 70: 1167–1172.

Harman, G.E., Chet, I. and Baker, R. (1981). Factors affecting *Trichoderma hamatum* applied to seeds as a biocontrol agent. *Phytopathology*, 71: 569–572.

Henis, Y., Ghaffar, A. and Baker, R. (1978). Integrated control of *Rhizoctonia solani* damping off of radish: Effect of successive plantings, PCNB and *Trichoderma harzianum* on pathogen and disease. *Phytopathology*, 68: 900–907.

Howell, C.R. (1982). Effect of *Gliocladium virens* on *Pythium ultimum* and damping off of cotton seedlings. *Phytopathology*, 75: 496–498.

Howell, C.R. (1987). The role of mycoparasitism in the biological control of *Rhizoctonia solani* by *Gliocladium virens. Phytopathology*, 77: 992–994.

Howell, C.R. and Stipanovie, R.D. (1995). Mechanisms in the biocontrol of *R. solani* induced cotton seedling disease by *Gliocladium virens. Antibiosis, Phytopathology*, 85: 469–472.

Howell, C.R. Hanson, L.E., Stipanovic, R.D. and Puckhaber, L.S. (1999). Induction of terpenoid synthesis in cotton roots and control of *R. solani* by seed treatment with *Trichoderma virens. Phytopathology*, 90: 248–252.

Jha, A. K, Upadhyay, J. P. and Lal, H. C..(2007).Evaluation of Organic amendment against White Mold of *Phaseolus vulgaris. J. Mycol. Pl. Pathol.*, 37(1):141–142.

Jones, R.W. Pettit, R.E. and Tabet, R.T. (1984). Lignite and stillage: carrier and substrate for application of fungal biocontrol agents to the soil. *Phytopathology*, 74: 1167–1170.

Katan, J. (1981). Solar heating (solarization) of soil for control of soil borne pests. *Ann. Rev. Phytopathol.*, 19: 211–236.

Klasson, W. (1981). The role of biological control in integrated pest management systems. In: *Biological Control in Crop Production*, (Eds.) G.C. Papavizas. Allenheld, Totowa, N.J., p.433.

Kraft, J.M. and Papavizas, G.C. (1983). Use of host resistance *Trichoderma* and fungicide to control soil borne diseases and increase seed yield of peas. *Plant Dis.*, 67: 1234–1237.

Lal, H.C. and Upadhyay, J.P. (2002). Biological control of Alternaria leaf blight of pigeonpea. *J.Biol.Control*, 16(2):141–144.

Lal, H.C., Upadhyay, J.P. and Jha, A.K.(2006). An integrated approach to the management of lentil rust. *J. Mycol. Pl. Pathol.*, 36 (2): 205–208.

Lewis, J.A. and Papavizas, G.C. (1980). Integrated control of *Rhizctonia* fruit rot of cucumber. *Phytopathology*, 70: 85–89.

Lewis, J.A. and Papavizas, G.C. (1984). A new approach to stimulate population proliferation of *Trichoderma* species and other potential biocontrol fungi introduced into natural soil. *Phytopathology*, 74: 1240–1244.

Lewis, J.A. and Papavizas, G.C. (1984). Effect of fumigant metham on *Trichoderma* spp. *Can. J. Microbiol.*, 30: 739–745.

Locke, J.C., Marois, J.J. and Papavizas, G.C. (1984). Biological control of *Fusarium* wilt of green house grown chrysanthemums. *Plant Dis.*, 69: 167–169.

Manomohandas, T.P. and Sivaprakasam, K. (1993). Biological control of damping off disease in chilli nursery. In: *Crop Diseases Innovative Techniques and Management*, (Eds.) K. Sivaprakasam and K. Seetharaman, p. 199–203.

Mukherjee, P.K. and Mukhopadhyay, A.N. (1993). Induction of stable mutants of *Gliocladium virens* by gamma radiation. *Indian Phytopath.*, 46: 393–397.

Mukherjee, P.K., Upadhyay, J.P. and Mukhopadhyay, A.N. (1989). Biological control of *Pythium* damping off of cauliflower by *Trichoderma harzianum*. *J. Biol. Control*, 3(2): 119–124.

Mukhopadhyay, A.N. (1994). Biocontrol of soil borne fungal plant pathogens: Current status, future prospects and potential limitations. *Indian Phytopath.*, 47: 119–126.

Mukhopadhyay, A.N. and Mukherjee, P.K. (1991). Innovative approaches in biological control of soil borne diseases in chickpea. In: *Proc. Fourth Int. Trichoderma and Gliocladium Workshop*, Belgirate, Italy, 17–20 July, *Petria*, 1: 146.

Mukhopadhyay, A.N., Shreshtha, S.M. and Mukherjee, P.K. (1992). Biological seed treatment for control of soil borne plant pathogens. *FAO Plant Prot. Bull.*, 40/1–2: 21–30.

Or, H.D., Munnecke, D.E. and Bricker, J.L. (1973). The interaction of *Armillaria mellea* and *Trichoderma* spp. as modified by methyl bromide. *Phytopathology*, 63: 965–973.

Papavizas, G.C. (1985). *Trichoderma* and *Gliocladium:* Biology, ecology and potential for biocontrol. *Ann. Rev. Phytopathol.*, 23: 23–54.

Papavizas, G.C. and Lewis, J.A. (1983). Physiological and biocontrol characteristics of stable mutants of *Trichoderma viride* resistant to MBC fungicides. *Phytopathology*, 73: 407–411.

Papavizas, G.C., Lewis, J.A. and Abd-El Moity, T.S. (1982). Evaluation of new biotypes of *Trichoderma harzianum* for tolerance to benomyl and enhanced biocontrol capabilities. *Phytopathology*, 72: 126–132.

Papavizas, G.C., Dunn, M.T., Lewis, J.A. and Beagle Restaino, J. (1984). Liquid fermentation technology for experimental production of biocontrol fungi. *Phytopathology*, 74: 1171–1175.

Sen, B. (2000). Biocontrol: A success story. *Indian Phytopath.*, 53(1): 1–16.

Singh, R.N., Upadhyay, J.P. and Ojha, K.L. (1993). Management of chickpea wilt by fungicides and *Gliocladium. J. Appl. Biol.*, 3(1–2): 46–51.

Singh, U.S., Mishra, D.S., Rohilla, R., Singh, A., Vishwanath, and Pal, B.P. (2002). Induced resistance: Present status and future prospects as disease management strategy. In: *Biopesticides and Pest Management*, (Ed.) O. Koul. 1: 262–302.

Singh, R.S. (2004). Biological management of soil borne diseases. In: Lecture Delivered in the *Training on Biocontrol Strategies for the Management of Plant Pathogens* at IARI, New Delhi, p. 1–8.

Sivan, A., Ucko, O. and Chet, I. (1987). Biological control of *Fusarium* crown rot of tomato by *Trichoderma harzianum* under field conditions. *Plant Dis.*, 71: 587–592.

Upadhyay, J.P. and Mukhopadhyay, A.N. (1986). Biological control of *Sclerotium rolfsii* by *Trichoderma harzianum* in sugarbeet. *Trop. Pest Management*, 32: 215–220.

Upadhyay, J.P. and Singh, R.N. (1995). Integrated approach for management of chickpea wilt. In: *Recent Advance in Phytopathological Research*, (Eds.) A.K. Roy and K.K. Sinha. HD Publication Pvt. Ltd., New Delhi, pp. 85–91.

Upadhyay, J.P. and Roy, S. (1997). Third annual report of ICAR adhoc scheme Biological control of pigeonpea wilt, submitted to ICAR, p. 35.

Upadhyay, J.P., Lal, H.C. and. Ojha, K.L. (2002). Biological control of some soil-borne pathogens in pulse crops. In: *Advances in Management of Biotic and Abiotic Stresses in Pulse Crops*, (Eds.) Ali, M., Asthana, A.N., Rathore, Y.S., Gurha, S.N., Chaturvedi, S.K. and S. Gupta. ISPRD, Kanpur, p. 21–25.

Weindling, R. (1934). Studies on a lethal principle effective in the parasitic action of *Trichoderma lingorum* on *R. solani* and other soil fungi. *Phytopathology*, 24: 1153–179.

Wells, H.D., Bell, D.K. and Jawarski, C.A. (1972). Efficacy of *Trichoderma harzianum* as a biological control for *Sclerotium rolfsii. Phytopathology*, 62: 442–447.

Plant Diseases Management for Sustainable Agriculture *Pages* **38–48**
Editor: **Dr. Shahid Ahamad**
Published by: **DAYA PUBLISHING HOUSE**

Chapter 4

Corm Rot Disease of Saffron (*Crocus sativus* L.) and its Management

M. Azam Wani, T.K. Kotha and Mushtaq Ahmad

Division of Plant Pathology, Sher-e-Kashmir University of Agricultural Sciences and Technology of Kashmir, Shalimar Campus – 191 121 (J&K)

Saffron (*Crocus sativus* L.) is a prize crop, wonder of nature and gift to mankind. The crop though grown in limited area of a few countries of world, is not spared by disease causing organisms world over.

Saffron is cultivated in China, England, Iran, Italy and Spain. Geographically, its cultivation is confined to 0°-90° longitude and 30°-45° N latitude and it is believed to have originated in Greece, Asia minor and Persia from where it is believed to have spread towards Kashmir valley and China (Warburg, 1957).

It is extensively grown in upland and Karewa areas of Kashmir valley and parts of Kishtwar in Jammu Division of India. Initially saffron has been a prime crop of Pampore and adjoining areas but now its cultivation is being extended to many other suitable areas of district Anantnag and Budgam. Experimental cultivation of saffron has also been undertaken in Himachal Pradesh and Uttaranchal but Jammu and Kashmir remains a leading state as far area and production of saffron in India is concerned. The State of Jammu and Kashmir has near monopoly in this crop (Plates 4.1 and 4.2).

Occurrence

Bilgrami *et al.* (1981) compiled and updated the list of fungal flora available in India, were not able to record the occurrence of any fungus from India on saffron corms, foliage or saffron of commerce. According to Zadoks (1981), Duttamel in 1728 has been given the first experimental evidence of corm diseases in his published paper "The treatise on the violet root rot of *Crocus*". However, survey of

Plate 4.1: Healthy Saffron Flower

Plate 4.2(a): View of Saffron FIeld (Healthy)

Plate 4.2(b): Saffron Bed Showing the Corm Rot Disease

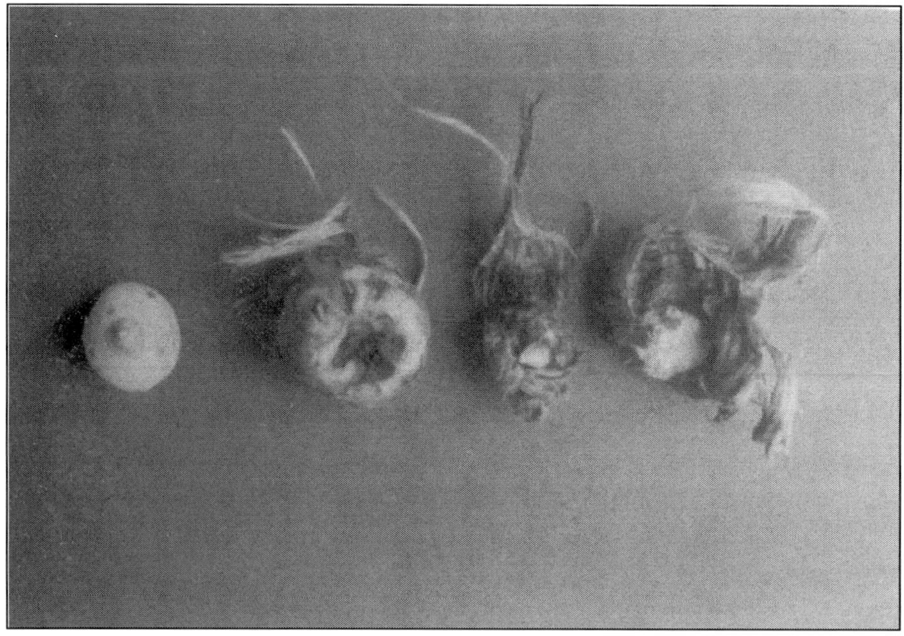

Plate 4.3: Saffron Corm and Symptoms of Corm Rot

literature is revealed that dry rot of corm was first noticed in Kyota district of Japan by Abe (1933) and since then its occurrence has been made from many saffron growing countries of the world. Prevalence of this disease has been reported from USA, Canada, UK, Holland, Germany and New Zealand (Drayton, 1934), Spain and France (Madan *et al.*, 1967), Japan (Yamamoto *et al.*, 1956), Spain (Alarcon and Sanchez, 1968), Italy (Carta *et al.*, 1982, Cappelli *et al.*, 1991, Cappelli, 1994) and China, (Xu and Ge, 1990).

From India this disease was first reported by Shah and Srivastava (1984) from Chabautia, Almora (UP) and later on by Thakur *et al.* (1992) from Kishtwar in Jammu region and Dhar (1992) from Pampore in Kashmir, while Wani (2004) has also been conducted the survey of entire valley of Kashmir division and revealed that it has been of higher order in traditional Pampore area and comparatively of low order in other newer areas. Rhizosphere studies that involve association of fungi and nematodes provided useful information as far as the etiology is concerned. Pathogenicity of fungi belonging to genera *Fusarium* namely *Fusarium solani, Fusarium oxysporum* and *Fusarium moniliforme* has been worked out and amongst them *Fusarium solani* was found to be most predominant and virulent, which is only species available at corm digging and harvesting stages, while other two species were encountered only from the infected corms collected at storage stage. There too *F. solani* was the most predominant.

Status of Disease

According to Abe (1933), the dry rot disease in Japan was mostly noticed during May and June, when the digging of the saffron corms is attempted. Alarcon and Sanchez (1968) have been noticed severe crop losses in Spain due to corm rot. Zadoks (1981) reported that the saffron cultivation is abandoned in France and England partly because of frequent outbreaks of this disease. Severe damage to the crop due to disease in Italy has also been reported by Cappelli (1994). In the most severely affected areas of Italy, Cappelli *et al.* (1991) found 40-50 per cent corms infected with dry rot disease.

Cappelli and Minco (1999) considered the disease as the major limiting factor of stigma yield of saffron in Italy. From India, Thakur *et al.* (1992) recorded disease incidence in the range of 30-40 per cent in Kishtwar of Jammu division. Dhar (1992) reported that none of the saffron growing areas in Kashmir valley were found free from the disease and recorded disease incidence in the range of 6.7 to 15.2 per cent. However, Thakur (1997) has also been reported the corm rot incidence to the magnitude of 70-80 per cent from saffron growing fields of Kashmir. Wani *et al.* (2007) has been observed that corm rot of saffron is one of the serious problems of saffron in Kashmir valley. It has mainly contributed for the decline in productivity per unit area year after year. Corm rot is a soil borne problem and leads to rotting of corms above root region, which not only stops flower bearing but also decreases the potential of corm multiplication during vegetative phase of the crop. The incidence of corm rot ranged from 4.66 to 42.00 per cent and intensity 0.8 to 17.43 per cent.

Causal Organism

The fungal nature of the disease was first reported by Abe (1933) and identified the causal fungus as *Fusarium bulbigenum* var *blasticola* but Francesconi (1974) has found *Pencillium cyclopium* involved in the corm rot disease in Italy. Recently Guzhen *et al.* (1997) from China described the causal fungus as *Penicillium corymbiferum*. However, Cappelli (1994) from Italy, and Shah and Srivastava (1984) from India identified causal fungus as *F. oxysporium* f.sp. *gladioli.* Dhar (1992) found *F. moniliforme* var. *intermedium* and an unsporulating basidiomycetous fungi associated with the disease. Meanwhile, Wani (2004) has been found a complex problem caused by *F. solani, F. oxysporum* and *F. moniliforme.* But Cappelli and Donato (1994) isolated *Fusarium oxysporum and Penicillium corymbiferum* from the severely damaged corms. However, Sud *et al.* (1999) have been reported *F. solani* as the causal agent of saffron corm rot from the Himachal Pradesh. Recording the disease and pest status of saffron in the Spain and France, Madan *et al.* (1967) found *Rhizoctonia crocorum* (*Helicobasidium crocorum*) and *Phoma crocophila* responsible for causing the rot diseases of corms but earlier Zadoks (1981) found *Rhizoctonia crocorum* as the only pathogenic fungus associated with the disease in the saffron belts of France. However, earlier Alarcon and Sanchez (1968) reported *Sclerotinia bulborum* besides *R. crocorum* as the causal agents of the rot disease in saffron fields of Spain.

Carta *et al.* (1982) from Sarvonia (Italy) and Thakur *et al.* (1992) from Jammu (India) isolated *Macrophomina phaseolina* from the rotted corms. Similarly, Mir *et al.* (1989) from Kashmir isolated *Trichoderma polysporum* from the infested corms. However, Thakur (1997) reported that many fungal pathogens like *Pythium* spp., *Rhizoctonia* spp., *Phoma* spp., *M. phaseolina* and *F. solani* are associated with the corm dry rot.

The bacterial nature of saffron corm rot has been reported by Xu and Ge (1990) from China and identified the strain of *Pseudomonas gladioli* as the cause of disease. However, earlier Mizusava (1923) had identified the causal bacterial species as *Bacillus croci* from Japan. All these diseases have been reported to inflict heavy crop losses either singly or in association with the dry rot diseases.

Symptomatology

While studying the infection, Madan *et al.* (1967) noted that the pathogenic fungi responsible for corm rot usually attacks the corms after penetrating through protective sheaths and transforms the flesh of corms from white to yellow and ultimately to black resulting death and destruction of the invaded corms. Besides above symptoms Shah and Srivastava (1984) even observed chlorotic lesions on the leaves of saffron plants affected with corm rot fungus. Cappelli *et al.* (1991) have been characterized by poor growth and wilting of saffron plants besides damping off, basal stem rot, drooping and wilting of shoots. Dhar (1992) observed the corm rot symptoms, in the form of brown to dark sunken irregular patches at any point of infection below the corm scales, lesions usually 1mm deep, have raised margins and mostly located in root and the bud regions, affected corms showed yellow foliage after bloom and got detached easily. Detailed disease symptoms has also been reported by Thakur (1997) and according to him the rotting is dry and there is no symptoms of corm softening and dry depressed, brownish to black coloured spots appears at advanced stage of infection. Detailed symptoms and environmental relationship of fungal corm rot have also been presented by Carta *et al.* (1982) and that of bacterial dry rot by Xu and Ge (1990).

Wani (2004) has observed that the foliage was yellow dull, drooping with tip burn symptoms. These symptoms are not shown any fungal or bacterial infection. The flowering period lasted just for 5 weeks and the flower emergence was erratic. The reproductive phase becomes short by one month and the dormancy phase gets prolonged and rotten (dry). The tunic was loose and the corm diameter in diseased is undersize (Plate 4.2b).

Initially the rot lesions on the corms are appeared beneath the outer tunic in the form of brown to dark brown, sunken, irregular patches at any side but frequently on buds. The lesions were usually 1-2 mm deep having raised margins. As rot advanced, saffron plants manifested damping off, basal stem rot, root rot, drooping and wilting of plants. The affected corm plants generally showed discoloured foliage and foliage got detached easily, the outer tunic was observed free of any rot symptoms (Plate 4.3).

Predisposition

Poor soil aeration due to water stagnation, injuries due to transplanting and hailstorm to be predisposing the host to fungal infection. Excessive soil moisture during hot summer months is favourable for the development and spread of the corm rot disease in Kashmir especially when untimely showers are received by the crop in the field (Anonymous, 1988). Injured corms are most susceptible to the disease and found warm temperatures and damp conditions favourable for disease development. Earlier, Weber (1924) had reported that over 70 per cent humidity helps in disease development. Francesconi (1974) while studying the infection, Abe (1933) noted that incidence and severity of infection in saffron corm was much higher in arid than in humid soils and reached to maximum at soil temperature of 32°C but at below 20°C the symptoms were negligible. He recorded optimum temperature for mycelial growth as 28°C and pH ranging from 3.5 to 5.6.

Morphology of Pathogens

The morphological and cultural characters of the three pathogenic species isolated are described as under:

	Fusarium solani	*Fusarium oxysporum*	*Fusarium moniliforme*
Growth rate	Fast	Fast	Fast
Colony character	Green to bluish brown, little aerial mycelium	White with purple tinge aerial mycelium, pigmented purple to blue	White granulated to purple, reverse of culture colourless to purple.
Microconidia	Abundant, 0-1 septate, 8.5–16 x 2.50-3.60 µm, oval with thicker wall, hyaline.	Generally abundant, round 6-10 x 2.0-3.5 µm, non-septate, ellipsoidal and straight.	Rare, 5-12 x 2.5-3.4 µm, non-septate, formed in chains.
Macroconidia	Moderately curved with short blunt apical cell and pedicellate basal cell, 3 septate, 29-42 x 4-6 µ m.	Fusiform, moderately curved, pointed at both ends, basal cell pedicellate, 22-25 x 3-4 µ m.	Clavate–non-septate with flattened base, 4.5-11 x 2.3 µm. Formed in chains basal cell bearing 2-3 apical phialids.
Conidiophores	Distinct, simple, elongated, narrow slightly towards apex, later sparsely shorter branched vertically forming sporodochium.	Sparsely branched, simple terminal, hyaline smooth, 6-12 µm.	Branched.
Chlamydospores	Produced singly or in pairs in terminal, lateral or inter colony position, hyaline, smooth 6.5-10 µm.	Sparse, hyaline, smooth walled both inter calary and terminal, 7.75 x 7.31 µ m.	Absent.

Management

Organic Soil Amendments

Several workers have suggested organic amendments as one of the effective methods of management of soil borne diseases.

The effect of decomposing crop residues and other organic additions on fungal pathogens are stimulating and inhibitory depending upon the source and age of residues (Meir, 1959 and Synder *et al.*, 1959).

While working on effects of oil cakes on general population of *Fusarium* sp. in soil, Bhalla (1966) observed the pathogenic fungal population decreased with the use of organic soil amendments especially at higher temperatures.

Jordan *et al.* (1972) observed significant decrease in the number of viable propagules of *Verticillium dahlae* in the soil and in disease severity of strawberry when organic soil amendments were used. It has been observed that Hardwood bark compost suppressed *Phytophthora cinnamoni* root rot pathogen (Hoitink *et al.*, 1976).

Wani (2004) has also found that soil amendments FYM, poultry manure, mushroom compost and spent mushroom compost in the ratio of 4:1 and 5:1 were beneficial but FYM and poultry manure in the ratio of 4:1 were most effective in controlling the corm rot disease.

Biological Control

Biological methods of plant disease management mainly involve the usage of micro-organisms to check the harmful micro-organisms that cause plant diseases without disturbing the ecological balance. Use of biological control agents along with solarization have been reported to provide synergistic effect in reducing incidence of many plant diseases (Mohapatra and Dash, 1990; Rao and Krishnaappa, 1995 and, Ristaino *et al.*, 1996). Application of *T. hamatum* reduced the Aubergene wilt to 38 per cent followed by *T. viride* compared to 74 per cent observed in untreated check (Sheela *et al.*, 1995). Thakur (1997) observed that combined treatment of saffron corms before sowing with selected fungicides and antagonistic micro-organisms (*Trichoderma viride; T. harzianum* and *Gliocladium virens* and three *Bacillus* spp.) was found most effective and economical in the corm rot disease management. Anith *et al.* (2000) reported that the integration of soil solarization and biological control agents produced positive effect on the management of bacterial wilt of ginger. Dubey and Patel (2001) evaluated successfully the fungal antagonists like *T. viride; T. harzianum* and *Gliocladium roseum* against *R. solani* causing web blight of urd and mung bean.

Wani (2004) reported bioagents like *Trichoderma viride, T. harzianum, Gliocladium roseum* and *Aspergillus niger* have antagonistic effect against *Fusarium solani.* Amongst all the bioagents, *Trichoderma viride* was the most successful bioagent.

Soil Solarization

Soil solarization has been investigated by a number of workers for controlling soil borne plant pathogens, insect pests and weeds.

Katan (1980) described soil solarization management method a safe and non-chemical, does not produce phytotoxic residues, is relatively inexpensive and simple to use. Katan *et al.* (1983) studied the effect of soil solarization and reported that the treatment resulted in reducing the fungal population in soil, decreased disease incidence, depressed *Fusarium* proliferation wilt and increased crop yields for as long as three years after treatment. Population of soil borne fungal and bacterial pathogens remained significantly depressed in solarized soil after 6-12 months (Stapleton and Devay, 1982). Dwivedi and Dubey (1987) observed that maximum temperature in polyethylene sheeted soil with full sunlight at a depth of 1 cm reduced the inoculum of *Macrophomina phaseolina.*

Satish and Lodha (1989) observed that solar heating reduced the population of *Macrophomina phaseolina* in arid soils. They recorded reduction of 57-81 per cent and 21-64 per cent at 15 cm and 30 cm soil depths respectively due to increase in soil temperature by using soil solarization. Arora *et al.*, 1989) reported that solarization of soil resulted into marked reduction of propagules of *Fusarium solani* and weeds in their field experiments.

According to Lozarovits *et al.* (1991) solarization is an effective soil dis-infestation procedure for the top 10 cm of soil. The population level of *Fusarium* sp. in soil progressively declined after 10, 20 and 30 days of soil solarization treatment (Dwivedi, 1991). Lodha (1995) concluded that solarization together with amendment of soil with neem cake or organic matter reduced diseases caused by soil borne fungi and nematodes.

Wani (2004) has been suggested that soil solarization, a novel concept for managing soil borne pathogens was investigated. Soil solarization in combination with FYM and irrigation have positive effect in checking corm rot of saffron.

Chemical Control

Madan *et al.*, 1967) reported that the corm rot disease could be avoided by choosing good healthy corms for planting and by peeling off the outer most protective sheaths besides giving cormlets a bath in 5 per cent solution of copper sulphate. Alarcon and Senchez (1968) suggested the use of pentachloronitrobenzene (PCNB) @ 300 kg/ha to ward off the damage inflicted to crop by corm rot fungi (*R. crocorum* and *S. bulborum*). Shah and Srivastava (1984) controlled the disease successfully with difolatan, bavistin or benomyl, each at 0.2 per cent, by dipping the corms for 20 minutes before planting in September. Dhar (1992) achieved best results with thiophenate methyle, carbendazim (0.1 per cent), captafol and copper oxychloride (0.3 per cent). Removal of outer sheaths of the corm and dipping in the 5 per cent $CuSO_4$ solution before planting have widely been recommended for disease management (Anonymous, 1988). Cappelli (1994) achieved disease prevention by using healthy reproductive material and recommended use of benzimidazoles in some cases.

References

Abe, T. (1933). Studies on a new dry rot disease of bulb of *Crocus sativus* L. caused by *Fusarium bulbigenerm* cke. Et. Mass var. *blasticola* (Rostr.). *Wr. Transaction Tottori Society of Agriculture Science*, 4(3): 212–228.

Alarcon, J. and Sanchez, Y.A. (1968). Elazatran hega civnlgadora 13-68-H. *Ministeria de Agriculture*, Madrid, pp. 68.

Anith, K.N., Manmohandas, T.P., Jayarajan, M., Vasanthakumar, K. and Aipe, K.C. (2000). Integration of soil solarization and biological control with a fluorescent *Pseudomonas* sp. for controlling bacterial wilt *Ralstonia solanacearum* (E.F. Smith) Yabuuchi *et al.* of ginger. *Journal of Biological Control*, 4 (1): 25–29.

Anonymous (1988). *Saffron the Golden Condiments*. Directorate of Extension Education, SKUAST (K), Shalimar, pp. 8.

Arora, D.K., Pandey, A.K. and Agrihotri, V.P. (1989). Soil solarization for control of soil borne diseases: Theory and application, In: *Perspectives in Phytopathology*, (Eds.) Singh, N., Chaube, H. S., Singh, U.S. and T.S. Dwivedi. Today and Tomorrow's Printers and Publishers, New Delhi, India, pp. 429–438.

Bhalla, H.S. (1966). Population of Fusaria in soils amended with oil cakes. *Master's Thesis*. Agriculture University, Pantnagar, India. [cf. Phytopathologische Zeitschrift ed. 69].

Bilgrami, K.S., Jamaluddin and Rizvi, M.A. (1981). *Fungi of India, Part I and II*. Today and Tomorrow's Printers and Publishers, Karol Bagh, New Delhi, pp. 467.

Cappelli, C. and Donato, G.D. (1994). Attack of *Fusarium oxysporum* Schlecht. in crops of saffron in Abruzzo, Italy. *Informatore Agrario,* 50 (25): 75–76.

Cappelli, C. and Munco, G.D. (1999). Three years of trials on saffron disease in Abruzzo Italy, *Unformatore. Fitopatologico,* 49(9): 27–32.

Cappelli, C., Buonaurio, R. and Polverari, A. (1991). Occurrence of *Penicillium corymbiferum* on saffron in Italy. *Plant Pathology,* 40 (1): 148–149.

Cappellic, C. (1994). Occurrence of *Fusarium oxysporum* f. sp. *gladioli* on saffron in Italy. *Phylopathologia Medetarranea,* 33(1): 93–94.

Carta, C., Flori, M. and Franceschini, A. (1982). Charcoal rot of saffron (*Crocus sativus* L.) bulbs–institute dh Patologia vegetable dell University disassare Italy, 29: 193–197.

Dhar, A.K. (1992). Bioecology and control of corm rot of saffron (*Crocus sativus* L). *Master Thesis.* Division of Plant Pathology, Sher-e-Kashmir University of Agricultural Sciences and Technology Srinagar, p. 109.

Drayton, F.L. (1934). The Gladiolus dry rot caused by *Sclerotinia gladioli* (Massey) n. Comb. *Phytopathology,* 14(4): 397–404.

Dubey, S.C. and Patel, B. (2001). Evaluation of fungal antagonists against *Thanatephorus cucumeris* causing web blight of urd and mung bean. *Indian Phytopathology,* 54(2): 206–209.

Dwivedi, R.S. and Dubey, R.C. (1987). Effect of soil solarization on the population dynamics of *Macrophomina phaseolina* (Tassi) Goid and general soil mycoflora. *International Journal of Tropical Plant Diseases,* 5: 67–73.

Dwivedi, S.K. (1991). Studies on population dynamics of *Fusarium oxysporum* f. sp. *lycopersici* in solar heated soil. *National Academy Science Letters,* 14(6): 235–237.

Francesconi, A. (1974). The rotting of bulbs of *Crocus sativus* L. by *Penicillium cycloprium.* Werstling. *Annali de Botanica University Rome, Italy,* 32: 63–70.

Guzhen Fang, Zhi-yue-e and Guzf Zhi, Y. (1997). A study on biological characteristics of *Penicillium corymbiferum* in crocus. *Acta Agricultural Shanchiai,* 13(4): 71–74.

Hoitink, H.A.J., Heer, L.J. and Schmittenner, A.F. (1976). Survival of some plant pathogens during composting of hardwood tree bard. *Phytopathology,* 66: 1369–1372.

Jordan, V.W. L., Sneh, B. and Eddy, B.P. (1972). Influence of organic soil amendments on *Verticillium dahliae* and on the microbial composition of the strawberry rhizosphere. *Annals of Applied Biology,* 70(2): 139–148.

Katan, J. (1980). Solar pasteurisation of soils for disease control: Status and Prospects. *Plant Disease,* 64(5): 450–454.

Katan, J., Fishler, G. and Grinstein, A. (1983). Short and long term effects of soil solarization and crop sequence on *Fusarium* wilt and yield of cotton in Israel. *Phytopathology,* 73(8): 1251–1219.

Lazarovits, G., Hawke, M.A., Tomlin, A.D., Othof, Th.H.A. and Squire, S. (1991). Soil solarization to control *Verticillium dahliae* and *Pratylenchus penetrans* on potatoes in Central Ontario. *Canadian Journal of Plant Pathology,* 13: 116–123.

Lodha, S. (1995). Soil solarization, summer irrigation and amendments for the control of *Fusarium oxysporum* f. sp. *ciceri* and *Macrophomina phaseolina* in arid soil. *Crop Protection,* 14: 215–219.

Madan, C.L., Kapoor, B.M. and Gupta, U.S. (1967). *Saffron Economic Botany.* 20(4): 377–385.

Meir, C.R. (1959). Effect of crop residues on bean root rot pathogen. *Plant Disease Reporter,* 43: 1027–1030.

Mir, N.M., Dar, G.M. and Dar, G.H. (1989). Record of *Trichoderma polysporum* on saffron corms. *Indian Journal of Mycology and Plant Pathology,* 19(2): 243–244.

Mizusawa, Y. (1923). A bacterial rot disease of saffrons. *Annals of Phytopathological Society of Japan,* 1(5): 1–12.

Mohapatra, P.D. and Dash, S.C. (1990). Efficacy of different methods of controlling *Sclerotium rolfsii* in betelvine. *Orrisa Journal of Agricultural Research,* 3: 126–129.

Rao, V.K. and Krishnappa, K. (1995). Integrated management of *Meloidogyne incognita-Fusarium oxysporum* f. sp. *ciceri* wilt disease complex in chickpea. *International Journal of Pest Management,* 41: 234–237.

Ristaino, J.B., Perry, K.B. and Lumsden, R.D. (1996). Soil solarization and *Gliocladium virens* reduce the incidence of southern blight (*Sclerotium rolfsii*) in bell pepper in the field. *Biocontrol Science and Technology,* 6: 583–593.

Satish, L. and Lodha, S. (1989). Soil solarization for the control of *Microphomina phaseolina* in arid soils. *Indian Phytopathology,* 42(1):170.

Shah, A. and Srivastava, K.K. (1984). Controlling of corm rot of saffron. *Progressive Horticulture,* 16: 141–143.

Sheela, J., Sivaprakasam, K., Seetharaman, K. (1995). Biological control of *Fusarium* wilt of egg plant. *Madras Agricultural Journal,* 82(3): 199–201.

Stapleton, J.J. and Devay, J.E. (1982). Effect of soil solarization on populations of selected soil borne micro-organisms and growth of deciduous fruit tree seedlings. *Phytopathology,* 72(3): 323–326.

Sud, A.K., Paul, Y.S. and Thakur, B.R. (1999). Corm rot of saffron and its management. *Indian Journal of Mycology and Plant Pathology,* 29(3): 380–382.

Synder, W.C., Schroth, M.N. and Christon, T. (1959). Effect of plant residues on root rot of bean. *Phytopathology,* 66: 755–756.

Thakur, R.N. (1997). Corm rot in saffron and its control. In: *Supplement to Cultivation and Utilization of Aromatic Plants,* (Eds.) Handa, S.S. and M.K. Kaul. Regional Research Laboratory, Jammu Tawi, pp. 447–448.

Thakur, R.N., Singh, C. and Koul, B.L. (1992). First report of corm rot in *Crocus Sativus. Indian Phytopathology,* 45(2): 278.

Wani, Azam M. (2004). Studies on corm rot of saffron *Crocus sativus. Ph.D. Thesis,* Division of Plant Pathology, S.K. University of Agriculture Sciences and Technology of Kashmir Shalimar Campus, J&K (India).

Wani, Azam M., Dar, G.M. and Kotha, T.K. (2007). Status of corm rot of saffron (*Crocus sativus* L) in Kashmir. *Agricultural and Biological Research,* 23(1): 24–35.

Warburg, 1957. *Crocuses endeavour*, 16(64): 209–216.

Weber, A. (1924). Diseases of cultivated flowering bulbs. *Gartnertidende,* pp. 2 [cf: RAM (1925) 4: 5].

Xu, C.X. and Ge, Q.X. (1990). A preliminary study on corm rot in *Crocus sativus* L. *Acta Agriculture Universities Zhegangensi*, 16: 241–246.

Yamamoto, W., Maeda, M. and Oyasu, N. (1956). Studies on the *Penicillium* diseases occurring on cultivated plants. *Scientific Reports of the Hyogo University of Agriculture,* 2: 23–28 [cf. RAM (1958) 37: 170].

Zadoks, J.C. (1981). Treature on the violet root of saffron crosus. Mededelingen land bonwhoge school. *Wageningen Naderland*, 81(7): 1–8.

Plant Diseases Management for Sustainable Agriculture *Pages* **49–59**
Editor: **Dr. Shahid Ahamad**
Published by: **DAYA PUBLISHING HOUSE**

Chapter 5

Fungal Diseases of Zeera [*Bunium persicum* (Boriss) Fedtsch] and their Management

T.K. Kotha, M. Azam Wani and Mushtaq Ahmad

Division of Plant Pathology, S.K. University of Agricultural Sciences and Technology of Kashmir, Srinagar – 191 121

Introduction

Zeera, [*Bunium persicum* (Boriss) Fedtsch] or "Kalazeera" is glabrous, branched perennial herb with globular tuberous roots and profuse branches, attaining a height of 50-60 cm and canopy of 40-50 cm. It has dissected leaves with narrow segments, while flowers are small, white in compound umbels. It bears oblong, 3-4 mm long, darkish brown and ribbed fruit with pointed ends having deep aroma. The plant belongs to family umbellifereae. It is cultivated through seed and subsequent regeneration is possible through tubers. The tubers after harvesting of crop in June-July remain dormant till February-March, subsequent cycle of vegetative and reproductive phase is completed in the next 3-4 months. The tubers possess a productive span of 10-12 years.

"Kalazeera" grows wild in Western Himalayas extending over Pakistan, Afghanistan and North India in the temperate zones at altitudes of 1818-3100 m above sea level, the occurrence being restricted to mountainous forest ecosystems and grassy slopes (Sharma, 2003). In India, "Kalazeera" is cultivated in hilly pockets of Jammu and Kashmir, Himachal Pradesh and Uttaranchal. In Jammu and Kashmir, its cultivation is confined to the karewas of Kashmir, Kishtiwar, Padder and Ladakh over an area of 225 ha with an annual production of about 290 q (Masoodi, 2003).

Jammu and Kashmir State which comprises mainly of western Himalayas irregular mountainous ranges with low-lying valley at altitudes 1818.18 m above sea level (lesser Himalayas) is the most ideal habitat for the crop (Anonymous, 1999).

"Kalazeera" seeds are used as condiments and are preferred in north Indian meat dishes especially in Kashmiri *Wazwan*. The seeds are regarded as stimulants, carminative diuretic, anti-asthama, anticonvulsion and anti-helmintic in indigenous system of medicine. It is also useful in curing diarrhoea, dyspepsia, fever, flatulence, stomach-ache, haemorrhoids; digestive disorders and urinary tract and gynaecological disorders (Boskabady and Maghaddar, 2003).

Quality production of "Kalazeera" is hampered due to various constraints like untimely rains, drought, poor management, attack of pests and pathogens, etc. These factors contribute significantly to reduction in yield and quality produce (Sharma *et al.*, 1993; Hakim, 2006 and Masoodi, 2003). Some of the important pathogens causing the severe diseases like umbel blight [*A. alternata* (Fr.) Keissler], septoria blight (*Septoria bunicae*), rust (*Puccinia bulbocastani*) and tuber rot [*Mucor cicinelloids* Van Tieghan and *Fusarium solani*] significantly reduce the production and quality of "Kalazeera". Among these, umbel blight and tuber rot are the most destructive, which have dissuaded the farming community to cultivate "Kalazeera" crop and thus the plants in the wild habitat are endangered due to ruthless extraction (Pandotra and Raina, 1991).

Crop Loss

Patel and Pillai (1983) reported the severity of *Alternaria* blight of cumin in early planting. Sharma *et al.* (1993) determined the incidence and severity index of several diseases of "Kalazeera" and related them to environmental factors. Sprouting and bolting stages of crop growth were prone to rust (*Puccinia bulbocastani*) and blight (*Alternaria alternata*)., whereas powdery mildew (*Erysiphe polygoni*) affected flower bud and grain development stages. The incidence of rust, blight and powdery mildew varied in the ranges of 2.5-15.4, 6.2-30.0 and 2.9-9.5 per cent, respectively, while the incidence of tuber rot (*Fusarium solani*) was in the range of 12.0-38.0 per cent. Kumar *et al.* (2002) found the outbreaks of *Alternaria burnsii* blight only after flowering on cumin in Hisar (Haryana) which caused yield losses upto 10 per cent. Further, the disease is favoured by cloudy and humid conditions.

Status

Pandotra and Raina (1991) reported umbel blight [*Alternaria alternata* (Fr.) Keissler] on "Kalazeera" from Kishtwar, Jammu and Kashmir, India. According to them, symptoms mostly appear on blossom resulting in distortion and reduction in umbel size which turn greyish. They also observed dry rot of tubers of "Kalazeera" caused by (*Mucor circinelloids* Van Tieghan,) for the first time in India from Atholi (Jammu and Kashmir). The disease is observed during the months of November-December and causes wilting and drooping of the plants. Such tubers at the time of dig out are shrivelled with hard flesh inside. Singh (2000) has summarised the diseases of "Kalazeera" and reported that foliage and branches of blight affected "Kalazeera" plants become dark brown to black followed by defoliation and drying of plants. However, in case of Cumin blight (*Alternaira burnsii*) the disease appears as small, isolated, whitish necrotic areas on aerial parts of the plants especially on tips of young leaves. Further, in Funnel, the blight (*Alternaria tenius*) appears as brown spots on the tips of young leaves which increase in size with age and ultimately turn dark. However, Mondal *et al.* (2002) reported Alternaria blight on "Kalazeera" at flowering stage as dark brown lesions on leaves and stems which eventually spread towards the flower head and infected plants were unable to bear seeds.

Hakim (2006) also reported the occurrence of diseases like leaf blight and umbel blight (*Alternaria alternata*) and tuber rot (*Fusarium solani*) from temperate zones of Kashmir which restrict commercial cultivation of Kalazeera. The disease progress with advancement in crop growth. Blight appears first on leaves and stem and later umbels are also affected. The umbels turn pale yellow, then dark brown and finally greyish. Tuber showing rot symptoms are reduced in size and had dry skin only, which did not bear above-ground biomass.

Fusarium solani in association with white grub has also been reported as causal pathogen of tuber rot of "Kalazeera". The infected tubers show rot symptoms, die under severe infections and emit putrifying smell (Sharma *et al.*, 1993).

Bhardwaj and Sharma (1990) observed rust (*Puccinia bulbocastani*) on "Kalazeera" at Research Farm Sangla in Himachal Pradesh which appeared in severe proportion on plant leaflets and petioles exhibited small yellowish pustules which increased in length with time. Further, the affected leaflets and petioles were malformed due to hypertrophy and hyperplasia.

Septoria blight (*Septoria bunicae*) on Kalazeera has been reported by Paul *et al.* (2000) from dry temperate zone of Himachal Pradesh, India. The symptoms on stem include scattered pinhead sized brown spots which enlarge along the stem and turn dark brown to black in colour with age, while on leaf margins small necrotic spots appear which enlarge, coalesce and cover large areas resulted in defoliation under severe conditions.

Morphological Characters of Major Pathogens

Alternaria burnsii isolated from blighted leaves of cumin was observed to have olive-green conidia (7-23 x 25-119 µm) with one to nine transverse septa (mostly three to five) with none to a few longitudinal septa per segment. Conidia in culture are borne singly or in chains of 2-10 (Davis and Raid, 2002).

Tahvonen (1978) has described the morphology of *Alternaria petroselini* isolated from parsley that fungal hyphae are sub-hyline to olive-brown, septate and 4-10 µm wide. Conidiophores are olive-brown, septate and upto 80 µm long, broadly ovoid to broadly ellipsoid and 28-40 x 16-24 µm in length with three to four transverse septa and one to four longitudinal septa per segment. Conidia are generally borne singly. Colonies appear green-black to black with woolly mycelia and even margins.

Domsch *et al.* (1980) described the morphology of *Alternaria alternata* (Fr.) Keissler isolated from different hosts and revealed that the colonies reach 6 cm in diameter in seven days at 22-30°C. Conidiophores and conidia in the medium golden brown. The conidiophores were usually simple, straight or curved, 1-3 septate, upto 50 µm long, 3-6 µm wide with one or several apical conidial pores. The conidia were formed in long but often branched chains, ovoid, obclavate obpyriform or more rarely ellipsoidal with a conspicuous basal pore, with or without a short conical apical beak not exceeding one third of the conidial length, medium brown, smooth walled, slightly constricted at three to eight transverse septa, with one or two longitudinal septa, 18-63 x 7-18 µm.

Hakim (2006) has characterised cultural and morphological characters of *Alternaria alternata* and appeared colony on PDA in fluffy and compact showing sparse radial growth which, at an average, attained 4 cm growth in 7 days. The colony was initially off white in colour which turned dark olive grey and finally to black colour, after six days of growth (Plate 5.4a). Microscopic observations of the fungus revealed that the mycelium was septate. Hyphae was light brown irregularly branched and measured 3-5 µm in width. The conidiophore were dark brown in colour, measured 31.4-64.4 µm in length and 3.5-18.6 µm in width (Av. 47.5 x 11.05 µm), simple septate and branches arise singly or in group bearing conidia at the apex (Plate 5.4b). The conidia were olive green to brown in colour with

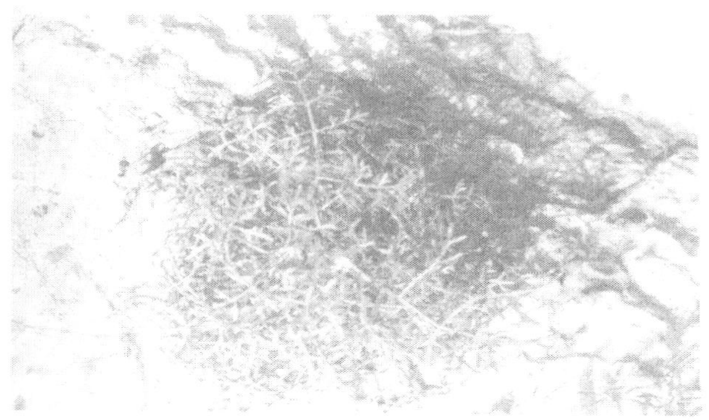

Plate 5.1a: Leaf Blight

Plate 5.1b: Stem Blight

Plate 5.1c: Umbel Blight at Flowering and Maturity Stage

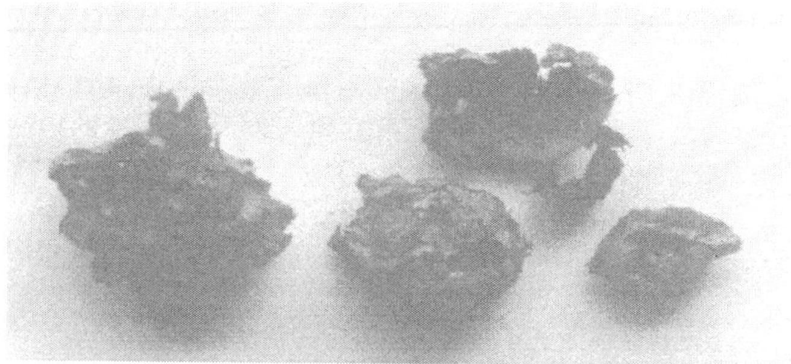

Plate 5.1d: Tuber Rot (Dry Tunica)

Plate 5.2a: *Alternaria alternata* Inoculated on Umbel

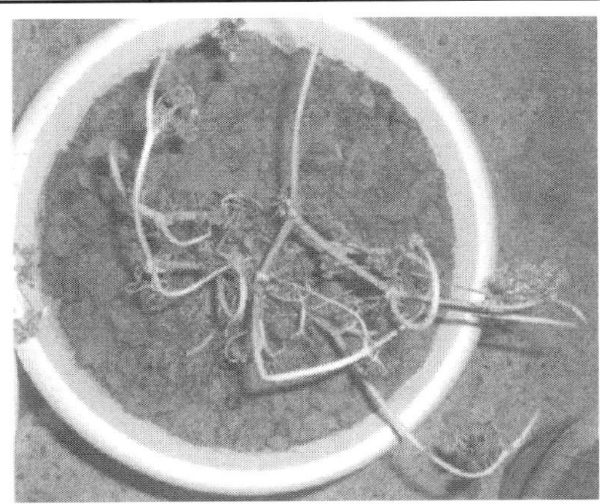

Plate 5.2b: *In vivo* Inoculation of *Alternaria alternata*

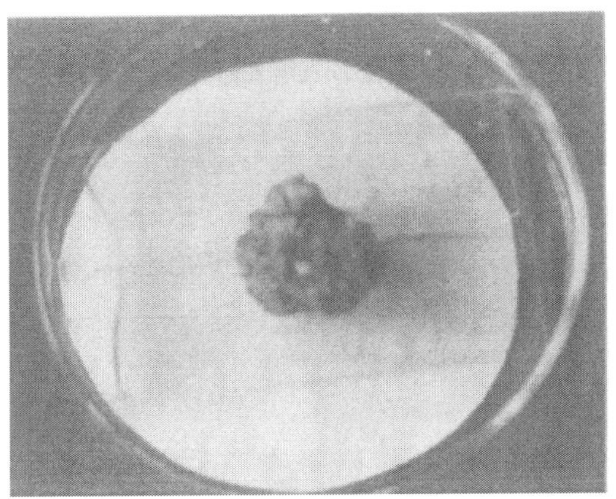

Plate 5.2c: *Alternaria alternata* Inoculated on Injured Tuber

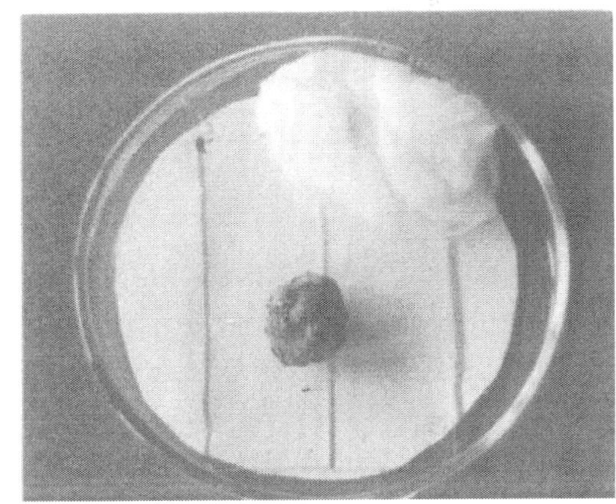

Plate 5.2d: Control

round base and tapered apex with beak (obclavate), measured 10.4-42.4 µm in length and 3.6-18.5 µm in width (Av. 26.4 x 11.05 µm). Conidia were septate having 2-4 transverse and 0-3 longitudinal septa. The conidia were simple, straight or curved, geniculate. Conidia were mostly borne in chains of 3-12. The beak of the conidia measured 3.5-13.5 µm in length (Av. 8.5 µm), equal to 1/3rd of the conidial length, olive green and was aseptate (Plate 5.4c).

Saccardo (1881) observed that the hyphae of *Fusarium solani* isolated from tubers were septate and hyaline. The conidiophores were simple, macroconidia borne on short conidiophore that form sporodochia were stout, thick-walled, usually 3-5 septate and measured 4-6 x upto 65 µm in size. Micro conidia borne on monophialldes were one to three celled, 2-5 x 8-16 µm in size and occurred in clusters at the tip of phiallide.

The fungus isolated from diseased tubers produced scanty aerial mycelium on potato dextrose agar plates. Aerial mycelium was initially colourless and became creamy with age, reaching 3.2 cm in

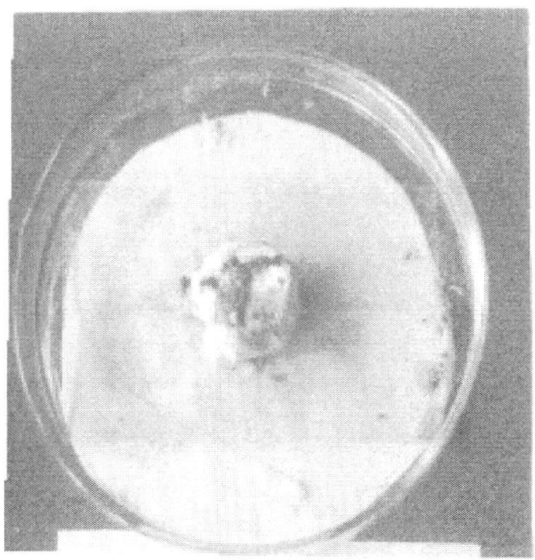

Plate 5.3a: *Fusarium solani* Inoculated on Injured Tuber

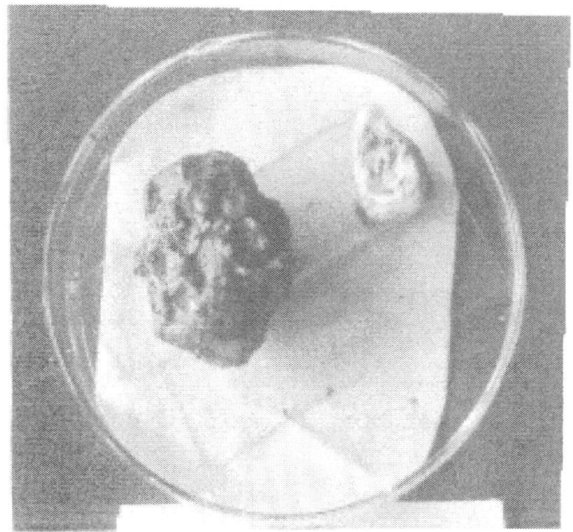

Plate 5.3b: *Fusarium solani* Inoculated on Uninjured Tuber

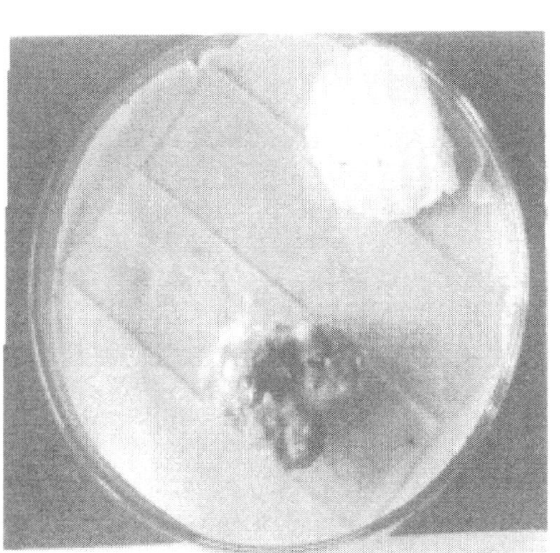

Plate 5.3c: *Fusarium solani* and *Alternaria alternata* (1:1) Inoculated on Injured Tuber

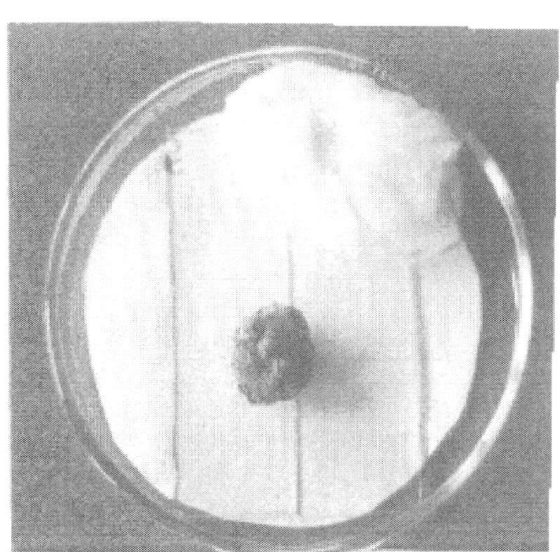

Plate 5.3d: Control

diameter in 4 days and covered almost the entire surface of the plate within 7 days at 25±1°C (Plate 5.5a). Microscopic observations revealed that mycelium was hyaline septate having hyphael width of 1.5-2.1 μm (Av. 1.8 μm). Three types of spores *viz.* microconidia, macroconidia and chlamydospores were observed. The microconidia were hyaline, ovoid in shape, aseptate, produced on aerial mycelium from elongated lateral phialides and were abundant, measured 2.4-4.1 x 6.1-15.2 μm (Av. 3.25 x 10.65 μm) in length and width, respectively. Macroconidia were sickle shaped, hyaline, septate, produced

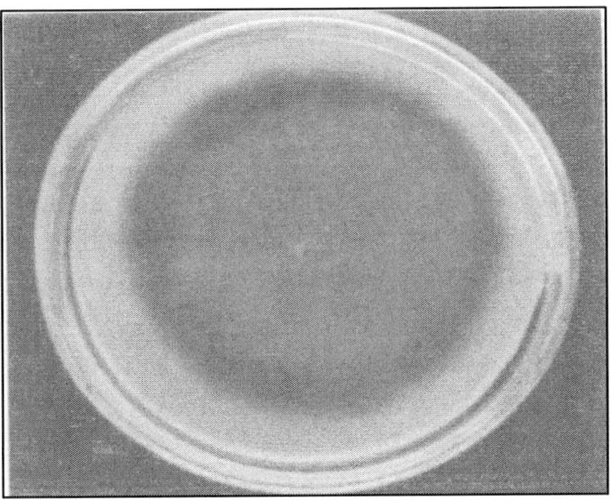

Plate 5.4a: Culture Plate of *Alternaria alternata*

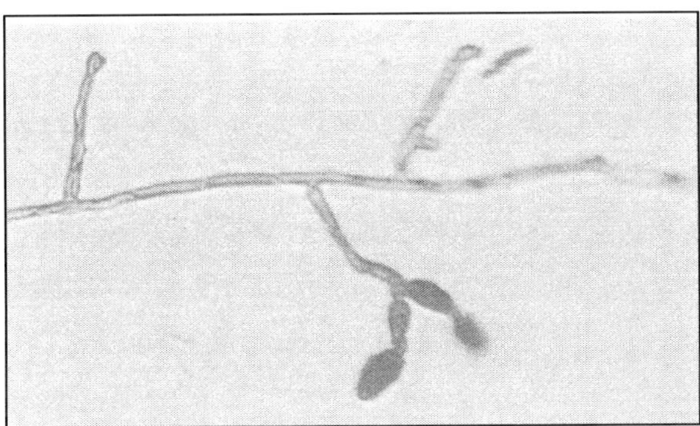

Plate 5.4b: Coniodiophore Shownig Conidia in Acropetal Fashion

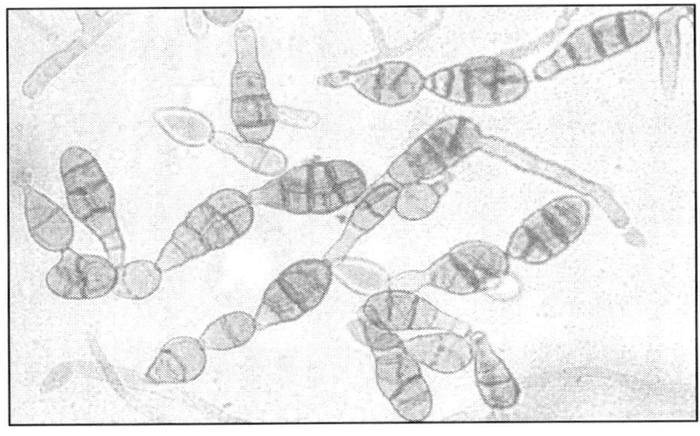

Plate 5.4c: Conidia of *Alternaria alternata*

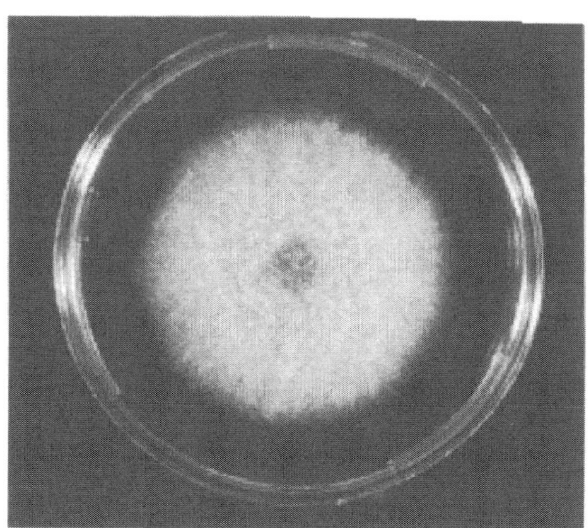

Plate 5.5a: Culture Plate of *Fusarium solani*

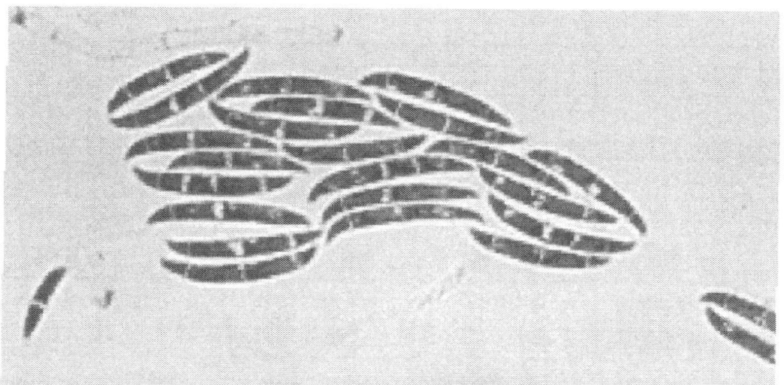

Plate 5.5b: Macro-conidia of *Fusarium solani*

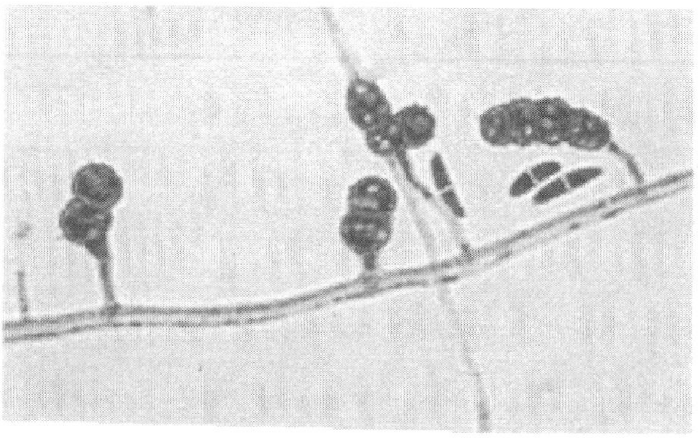

Plate 5.5c: Micro-conidia and Chlamydospores of *Fusarium solani*

on well-developed short multi-branched conidiophores having blunt apical and pedicellate basal cells and measured 3.4-5.4 µm in width and 24-34 µm in length (Av. 4.4 x 25 µm) [Plate 5.5b]. Chlamydospores were hyaline, globose, aseptate, smooth to rough walled, borne singly or in pairs on short lateral hyphal branches or they intercalary, measured 5-9 µm in diameter (Plate 5.5c).

Symptoms

Hakim (2006) reported status of diseases encountered at different phenological stages of crop growth in different agro-ecological locations in Kashmir valley. The disease symptoms were observed on leaf, umbel and tuber (Plates 5.1–5.3). The leaf blight was noticed first in ending March to Ist week of April and the disease incidence showed gradual increase with increase in plant age. Hakim (2006) established that *A. alternata* produced disease symptoms within two days on injured leaves and umbel by spraying method (Plate 5.2a) *in vitro* and *in vivo* (Plate 5.2b).

Small pinhead sized yellowish dots appeared on lower pinnate leaves in ending March in low lying locations (Budgam, Pulwama) and in the first week of April, 2005 in higher locations (Baramulla, Srinagar), which coalesced covering the entire leaf surface, turned brown with age and finally grey (Plate 5.1a). Severely infected plants failed to produce umbel stalk.

The yellowish pinhead sized dots appeared on stem in ending April, which coalesced, turned brownish and finally greyish black, in the form of irregular patches resulting in the weakening of stem. The photosynthetic area got reduced (Plate 5.1b).

Yellowish pinhead sized dots appeared on the blossoms in the first week of May onwards, which coalesced, turned brownish and finally greyish, resulting in the distortion and reduction in umbel size. Severely infected umbels failed to produce seeds (Plate 5.1c), while severely affected umbels had immature, shrivelled and flavourless seeds (Plate 5.1c). The tuber rot was characterised by mummification, change in pulp colour, emitting putrefying smell. Severely infected tubers had only dry tunic (Plate 5.1d).

The rot was observed from upper side, basal side and from lateral sides. The rotting proportionately increased with weight and size of the tuber. The size and number of umbelets and umbles were found less in wild plants, than in cultivated plants.

The co-inoculation of *Alternaria alternata* and *Fusarium solani* produced diseased symptoms on the plant parts tested. With symptom express on leaves and umbel in two days, on injured tubers in 6-7 days and on uninjured tubers in 12-17 days. Data revealed rapid symptom development in both injured and uninjured tubers by co-inoculation of pathogens as compared to individual inoculation (Plates 5.3c–d).

Management

Pandotra and Raina (1991) observed (Blitox 50 WP) to be most effective against *Mucor circinelloids* and *A. alternata,* isolated from "Kalazeera". They also observed ziram and ferbam effective against *A. alternata. In vivo* evaluation of fungicides against *Alternaria burnsii,* isolated from cumin indicated that mancozeb was superior followed by propiconazole, penconazole, hexaconazole, tridemorph, thiophonate-methyl triadimefon and dinocap while sulphur dust proved ineffective at recommended concentrations (Akbari *et al.,* 1996). According to Mishra and Rath (1988) Bavistin and Benlate are most effective in inhibiting the mycelial growth of *Fusarium* spp, isolated from decayed vegetables, in comparison to captan 50 WP, difolatan, mancozeb 75 WP, biltox 50 WP, thiophanate methyl and thiram. Anwar and Askari (1991) evaluated 6 fungicides under *in vitro* conditions against *Fusarium solani,* and found that benomyl provided 100 per cent inhibition even at 20 ppm while captan 50 WP was superior in inhibiting mycelial growth on 8th day of inoculation followed by Diathane M-45, Bayleton and Hardron (Mamatha and Ravishankar, 2004). Topsin M showed 11 per cent inhibition at 50 ppm whereas Emisan and Bavistin did so at 100 ppm concentration Sharma *et al.* (2004). Moreover, Hakim (2006) suggested that fungitoxicants hexaconazole 5 EC @ 300-400 $\mu g\ ml^{-1}$ was most effective in completely inhibiting the mycelial growth and spore germination of both the fungi. The efficacy of various systemic fungitoxicants was in order of hexaconazole 5 EC > difenaconazole 25 EC > myclobutanil 10 WP whereas, amongst non-systemic fungitoxicants was mancozeb 75 WP > captan 50 WP > copper oxychloride 50 WP. Seed tuber, field sanitation and protectant and or therapeutic applications with effective fungitoxicants are believed to provide a protection cover against all the major pathogens expecially *Alternaria blight* and tuber rot of Kalazeera to enable harvest a healthy disease free crop at reasonably low cost with least addition.

References

Akbari, L.F., Dhruj, I.U., Khandar, R.R. and Vaishnav, M.V. (1996). Management of cumin blight through fungicides. *Plant Disease Research,* 11(1): 103–105.

Anonymous (1999). *Gazetteer of India, Jammu and Kashmir State, Kashmir Region, Vol. 1,* (Ed.) Bashir Ahmad. State Gazetteers Unit, Govt. of Jammu and Kashmir, Srinagar, pp. 2–4.

Anwar, N., Ahmad, S.I. and Askari, A. (1991). Laboratory evaluation of six systemic fungicides for the control of root rot in *Duboisia leichhardtii* F. Muell. *Pakistan Journal of Scientific and Industrial Research,* 34(10): 402–403.

Bhardwaj, L.N. and Sharma, P.N. (1990). Occurrence of rust on Black zira in Himachal Pradesh. *Indian Journal of Mycology Plant Pathology,* 20(3): 270.

Boskabady, M.H. and Moghaddar, A. (2003). Anthistaminic effect of *Bunium persicum* on guinea pig tracheal chains, p. 1. Department of Physiology, Ghaem Medical Centre, Mashhad University of Medical Sciences, Mashhad, Iran [cf: http:www.ikisan.com/links/ap_ regulators html].

Davis Michael, R. and Raid Richard, N. (2002). Alternaria leaf blight of cumin. In: *Compendium of Umbelliferous Crop Diseases.* University of Florida, Everglades Research and Education Centre, USA, p. 16.

Domsch, K.H., Gams, W. and Anderson, T.H. (1980). Alternaria. In: *Compendium of Soil Fungi, Vol. 1.* Academic Press, London, New York, Toronto, Sydney, San Francisco, pp 34–35.

Hakim, S.A. (2006). Studies on disease of Kalazeera (*Bunium persicum* (Boiss) Fedtsch. *M.Sc. Thesis,* Division of Plant Pathology, S.K. University of Agricultural Sciences and Technology of Kashmir Srinagar, J&K, India.

Kumar, S., Rakesh, M., Srivastava, M.P. and Panwar, R.C. (2002). Epidemiology of Alternaria blight of cumin. *Indian Journal of Plant Pathology,* 20(1&2): 72–74.

Mamatha, T. and Ravishankar, R.V. (2004). Evaluation of fungicides and plant extracts against *Fusarium solani* leaf blight in *Terminalia cateppa. Journal of Mycology and Plant Pathology,* 34(2): 306–307.

Masoodi, M.A. (2003). Regionwise area and production in Jammu and Kashmir. In: *Agriculture in Jammu and Kashmir: A Perspective.* Mohisarw Book Series, Srinagar, pp. 110–119.

Mishra, D. and Rath, G.C. (1988). *In vitro* evaluation of chemicals against *Fusarium* spp. causing post harvest decay in vegetables. *Pesticides,* 22(2): 44–47.

Mondal, K.K., Rana, S.S., Sood, P. and Singh, Y. (2002). Kalazeera: A new host for *Alternaria burnsii*. *Indian Phytopathology*, 55(4): 532.

Pandotra, V.R., Raina, A.K. (1991). Two new diseases of *Bunium persicum* (Boiss) Fedtsch from India. *Indian Journal of Forestry*, 14(4): 325–326.

Patel, R.K. and Pillai, S.N. (1983). Age of crop and the sowing period in the incidence of cumin blight. *Indian Journal of Mycology and Plant Pathology*, 13: 79.

Paul, Y.S., Singh, A. and Dhancholia, S. (2000). Taxonomy of septoria species collected from temperate zone of Himachal Pradesh. In: *Proceeding of Annual Meeting of the Society of Mycology and Plant Pathology*, 3–4th May, Solan (H.P).

Saccardo, Martius (1881). Doctor fungus *Fusarium solani* species [cf: http://www.v.saskca/agriculture/plants_sci/vegetable/med/n.cum.].

Sharma, P.N., Panwar, K.S., Kapoor, A.S. and Sagwal, J.C. (1993). Diseases of Kalazeera in Himachal Pradesh. *Indian Cocoa, Arecanut and Spices Journal*, 17: 1–2.

Sharma, R. (2003). *Medicinal Plants of India: An Encyclopaedia*. Daya Publishing House, Delhi, p. 38.

Singh, D. (2000). Diseases of spice crops, In: *Spice Crops of India*, (Ed.) P.S. Arya. Department of Plant Pathology, Himachal Pradesh Krishi Vishvavidyalaya Palampur. Kalyani Publishers, New Delhi, pp 155–157.

Tahvonen, R. (1978). Seed borne fungi on parsley. *Journal of Science on Agricultural Society*, Finland, 50: 91–102.

Plant Diseases Management for Sustainable Agriculture *Pages* ***60–71***
Editor: **Dr. Shahid Ahamad**
Published by: **DAYA PUBLISHING HOUSE**

Chapter 6

Black Sigatoka: A Threatening Disease of Banana

A. Naseema and V.K. Girija

Department of Plant Pathology, College of Agriculture, Vellayani, Thiruvananthapuram – 695 522, Kerala

Banana is an important fruit crop that contributes to half the daily calorie intake of half-a-billion people in Africa and Asia. India is the world's biggest banana grower, with an annual production of more than 16.8 million tonnes. Twenty per cent of total world output of 72.6 million tonnes recorded in 2005 was contributed by our country. In spite of the ever growing demand for this fruit, there had been a decline in banana yields in Africa, Asia and Central America owing to mainly the incidences pests and diseases. Black sigatoka or black leaf streak disease (BLSD) caused by the fungus *Mycosphaerella fijiensis* is a devastating leaf spot disease of banana worldwide. It gets its name from appearing as a different form of yellow Sigatoka. It has black colour in the young streaks and spots and lacks the distinct yellow halo that is present in young streaks of yellow sigatoka. Black sigatoka is a regular and recurring problem and significantly reduces the yield by 35-50 per cent and the quality of banana and plantain through widespread destruction of foliage. The disease reduces the fruit number and size. The productive life of banana plants also gets considerably reduced from 30 years to merely two to three years. The Cavendish banana types are the worst affected, though all the other commonly grown cultivars are also not spared of the disease.

Distribution

Black sigatoka was first reported from Sigatoka valley located on the South Eastern coast of Viti Levu of Fiji in 1963 (Rhodes,1964). However, the published work of several scientists as well as the specimens submitted to herbariums indicate that the disease could have been prevalent in several parts of Asia and Pacific islands even before. Subsequently, the disease has been reported throughout the Pacific (Torres Strait and Cape York Peninsula region of Australia, Papua New Guinea, Solomon Islands, Vanuatu, Newaledonia, Norfolk Island, Federal States of Micronesia, Tonga, Western Samoa,

Niue, Cook Islands, Tahiti and Hawaii). BLSD has also been found in Asia (Bhutan, Taiwan, and Southern China including Hainan Island, Vietnam, Philippines, West Malaysia and Sumatra in Indonesia). However, the distribution of the disease in Southeast Asia requires further clarification, especially in the Indonesian archipelago.

The black sigatoka disease made its appearance in the African continent in Zambia in the year 1978 and later in Gabon in 1978. The disease further spread to other African countries as well as Sub-Saharan Africa. BLSD spread along the West Coast to Cameroon, Nigeria, Benin, Togo, Ghana and Côte d'Ivoire. The disease occurs in the Congo and its eastward spread most likely occurred across Democratic Republic of Congo (ex Zaire) to Burundi, Rwanda, Western Tanzania, Uganda, Kenya, and Central African Republic. An introduction occurred on the island of Pemba around 1987 and from there to Zanzibar and coastal areas of Kenya and Tanzania. BLSD is also found in Malawi and the Comoro Islands. The present distribution is undoubtedly under estimated. BLSD and yellow sigatoka are disseminated locally by ascospores and conidia. Distance spread is believed to be by the movement germplasm (infected suckers, diseased leaves) and wind borne ascospores. In the Pacific and in lowland Latin America and Africa, BLSD was reported to have replaced yellow sigatoka in less than 3 years. However, yellow sigatoka is more adapted to wide temperature range and is often dominant at altitudes over 1400 m where BLSD is rarely seen.

BLSD was identified in Latin America for the first time in 1972 in Honduras. It spread northwards to Guatemala, Belize and southern Mexico and southwards to El Salvador, Nicaragua, Costa Rica, Panama, Colombia, Ecuador, Peru and Bolivia. BLSD has recently been reported in Venezuela, Cuba, Jamaica and Dominican Republic from where it threatens the rest of the Caribbean Islands.

The disease has come to be established in the banana growing countries of the continents of Africa, Asia and the Americas (Ploetz and Mourichan, 1999).

Losses and Economic Impact

Black sigatoka attack the leaves to cause black and brown lesions interfering with photosynthesis and depleting the plant of its energy reserves which in turn is reflected in the reduction of fruit production by 35- 50 per cent. Almost all the cultivated varieties are susceptible to this disease. It is very severe on Cavendish bananas, ladyfingers and several other dessert bananas. The fungus reduced yields by 40 per cent in a year in Uganda, the world's second largest producer. Banana production in Brazil, the world's fourth largest producer, was likely to fall by 70 per cent. To combat this devastating disease fungicidal spraying has become a necessity. Annually, a typical plantation requires 38-50 fungicide sprays, and these sprays may account for around 30 per cent of the production costs. In Central America, black Sigatoka may account for 27 per cent of the total production cost, whereas other diseases and pests may account for only 3-5 per cent of the total production cost. In India, many different cultivars that include resistant and susceptible ones are grown together in area in order to meet regional and sub-regional demands. This reduces the risk of severe economic damage due to black sigatoka as compared to that of the total yield loss due to monoculture as in African countries. However the disease worst upset the cultivation of the susceptible dwarf Cavendish varieties meant for export.

Symptom

The first detailed description of the symptoms of the disease was published in 1969 by Meredith and Lawrence. Foure (1985) studied in detail the symptom and development of the disease and categorized them into six stages as detailed below:

1. Stage 1 (initial speck stage): Symptoms occurs initially as faint, minute (up to 0.25 mm in diameter), reddish brown specks on the lower surface of the leaf, which first appear on the third, fourth or older leaves and are often more abundant near the margin, particularly towards the tip.
2. Stage 2 (first streak stage): The specks are elongate, becoming slightly wider, and forming narrow, reddish brown streaks up to 20 x 2 mm, with the long axis parallel to the leaf venation. The streaks are more clearly visible on the lower surface of the leaf than the upper surface and its distribution on the leaf is variable. Numerous such streaks can coalesce to form larger streaks.
3. Stage 3 (second streak stage): The streaks change colour from reddish brown to dark brown or almost black, sometimes with a purplish tinge, becoming clearly visible on the upper surface of the leaf. When they are numerous and more or less evenly distributed, the entire leaf blackens.
4. Stage 4 (first spot stage): The streak broadens and becomes somewhat fusiform or elliptical and develops a light brown, water-soaked border around the spot.
5. Stage 5 (second spot stage): The dark brown or black central area of the spot becomes slightly depressed and the water-soaked border becomes more pronounced due to darkening. At this stage, a slight yellowing of the leaf tissue immediately surrounding the water-soaked border may occur.
6. Stage 6 (third or mature spot stage): The centre of the spot dries out, becoming light grey or buff-coloured, and further depressed. The spot is surrounded by a narrow, well-defined, dark brown or black border. Between this border and the normal green colour of the leaf, there is often a bright yellow transitional zone. After the leaf has collapsed and withered, spots remain clearly visible with the light-coloured centres and dark borders.

Small brown specks less than 1mm in diameter are first visible on the undersurface of the third and fourth fully opened leaves. This appears within 14 to 20 days after infection which varies according to the cultivar and severity of infection (Fullerton, 1994; Jacome and Schuh, 1992; Mayorga, 1990). Infection occurs on the youngest leaves of the plant during and immediately after unfurling. Older leaves are not readily infected. The symptoms on the young newly emerged leaves are often oval and surrounded by yellow margin. The necrotic areas become elliptical or fusiform on enlargement and darken to produce characteristic black streaks on the leaves. The central part of the leaves collapse. The black leaf streak further intensifies as the lesions coalesce and cover almost the entire leaf which succumbs to death within three to four weeks after the initiation of the disease.

Eighteen hours of leaf wetness after inoculation is optimum for symptom expression. With shorter periods of leaf wetness the symptom development was delayed 7-14 days (Jacome and Schuh, 1992). On resistant cultivars symptom development is very slow and may be restricted to the first spot stage (stage 4) until the natural senescence of the leaf (Fullerton, 1994). However on a susceptible cultivar, the chlorotic flecks soon develop into narrow rusty brown streaks and gradually change to brown or black. The sides of the necrotic lesions are delimited by veins and are provided with pointed ends. The colour of the lesion changes to red, brown or black and may have purplish tinges. Meredith and Lawrence (1969) observed that the distinctive feature of black sigatoka was the absence of yellow streaks as in Yellow sigatoka. However, Graham (1968) found yellow streaks during the cooler months in Fiji.

Causal Organism

Black leaf streak is caused by the fungus, *Mycosphaerella fijiensis*, a distinct strain of *Mycosphaerella musicola*. The major morphological differences between *M. fijiensis* and *M. musicola* are in the conidial state, *Pseudocercospora* (Leach, 1964). Fruiting bodies are amphigenous, predominantly hypophyllous, dark brown and scattered on the lesion. Stroma may not or moderately be developed, substomatal, erumpent, or on young spermatia. Conidiophores occur in loose or dense fascicles, continuous or septate, smooth, cylindric, simple, olivaceous brown, up to 25 x 3-4 µm. According to Meredith and Lawrence (1969), conidiophores are 16.5-62.5 x 4-7 µm. They first develop during stage 2 and continue to emerge up to the stage 5. On old spots, they are found in the central dried area. Conidiogenous cells are apical, sometime sessile, straight or geniculate, simple, smooth, olivaceous brown, and sympodial. Conidial ontogeny though holoblastic, the proliferation is enteroblastic and with 1-3 moderately visible scars on the apices or on geniculations; up to 25 µm long, 3-4 µm wide at the base and 2-4 µm wide at the apex. Conidia solitary, obclavate or cylindric-obclavate, subhyaline, rarely guttulate, base truncate with a moderately visible scar, 1-6 septate, smooth, schizolytic secession, 10-120 x 1.5-2 µm in size, (Pons 1987). Although some authors described the conidial scars as conspicuous, the conidiogenous locus of this species is actually confined to a narrow, thickened rim around the scar (Pons, 1987). It was considered as a key characteristic for segregation of the genus *Paracercospora* from *Pseudocercospora* until recently. However, the differentiation between both genera is often difficult, and there are numerous intermediate taxa. Crous *et al.* (2002) have considered these two genera as synonymous based on the results of molecular analyses.

The perithecial state of this pathogen was identified as *M. fijiensis* (Meredith and Lawrence, 1969). Stover (1969) observed that the sexual stage of black sigatoka fungus has identical morphologically with *M. musicola*, the perfect stage of yellow sigatoka pathogen and hence *M. fijiensis* could be recognized as a variety of *M. musicola*. Perithecia are flask shaped or globose, submerged, spherical, papillate ostiole, slightly erumpent, cells polygonal, with thick dark brown wall that emerges piercing pseudothecia formed within the mature lesions. The size of the perithecia measures on an average 65×54 µ.The perithecia contain 10-27 asci which measures on an average 38×11 µ. The asci are oblong to club shaped, bitunicate, saccate and contain eight ascospores arranged in two rows. Pseudoparaphysis or sterile hyphae are absent in the pseudothecia. Ascospores measure 14-20 x 4-6 µm, and are colourless, two celled with one of the cells slightly broader than the other. There is a distinct constriction at the septum. Mulder and Stover (1976) proposed the name *M. fijiensis* var. *difformis*, based on the presence of a sporadic stroma as the origin for conidiophores in contrast with the absence of stroma for *M. fijiensis* var. *fijiensis*. Pons (1987) examined the holotype specimens of both varieties and observed the presence of stroma on both *M. fijiensis* var. *fijiensis* and *M. fijiensis* var. *difformis*. He concluded that they are one and the same fungus and amended the description of the species *M. fijiensis*, considering *M. fijiensis* var. *difformis* as a synonym. Therefore *M. fijiensis* produces stromata. However, this is an inconsistent feature and many times the stroma is absent and the conidiophores arise singly or in fascicles from hyphae present in the substomatal chamber, emerging through the stoma.

Diagnosis

Sometimes, black Sigatoka and yellow Sigatoka (*M. musicola*) are difficult to distinguish based on the symptoms. The perfect state of *M. fijiensis* is virtually identical with *M. musicola* and distinction can only be made by examination of its imperfect state. In the former, conidiophores possess sub-conspicuous conidial scars. The conidium has a thickened basal hilum. These features do not exist in *M. musicola*. Moreover, conidiophores of *M.fijiensis* are formed mainly on the lower surface of the leaf, contrarily those of *M. musicola*, which in addition has bottle-shaped conidiophores (Mulder and Holliday 1974). The two species can also be distinguished by a PCR technique (Johanson *et al.*, 1994, Henderson *et al.*, 2006).

Table 6.1: Comparison of Yellow and Black Sigatoka

Distinguishing Features	*Yellow Sigatoka*	*Black Sigatoka*
Pathogen		
Teleomorph	*Mycosphaerella musicola*	*Mycosphaerella fijiensis*
Anamorph	*Pseudocercospora musae*	*Pseudocercospora fijiensis*
Sporulation	Abundant on both leaf surfaces but more on upper surface	More on lower leaf surface
Conidiophore	Conidiophores formed in dense clusters/fascicles (sporodochia) on dark stromata	Conidiophores formed singly or in small groups (2-8), sporodochia absent
Nature of conidiophore	Conidiophores 5- 25x 2-3.5 μ, straight, usually non-septate and unbranched, no spore scars	Conidiophores 16.5–62.5x 4–7.0 μ straight or bent, 0-5 septate and rarely branched near the base, well defined, slightly thickened spore scars
Conidia	Conidia averaging 59x 3 μ in size, pale olivaceous 0-6 septate, cylindrical, straight or curved, no distinct basal scar.	Conidia averaging 72.5x 4 μ in size, mostly hyaline 1-10 septate, obclavate, distinct basal scar.
Colony characteristics	On culturing, the colony surface of the fungus light grey and smooth. Sporulation abundant	On culturing, the colony surface of the fungus grey and very smooth. Sporulation more abundant
Stomatal penetration of germ tube	Less	More, result in greater spotting
Germination of conidia in V8 juice agar	Germ tube straight, branching of germ tube slow and sparse	Germ tube curved, branching of germ tube is quick and extensive
Hosts	Bananas (AAA) generally susceptible; most cooking bananas and plantains (AAB and ABB) moderately to highly resistant–pathogen virulent	Most dessert bananas, cooking bananas, and plantains susceptible–Pathogen more virulent and affect wide range of genotypes
Symptoms	Early streak pale yellow	Early streak dark brown
	Streaks appear on leaf numbers 4-5 (unsprayed Cavendish)	Streaks appear on leaf numbers 2-4 (unsprayed Cavendish)
Epidemiology	More common in cooler environments	More common in warmer environments
	Inoculum consists of both conidia (water-dispersed) and ascospores (wind-dispersed)	Windborne ascospores are the major inoculum
	Conidia first appear in the mature spot stage	Conidia first appear in early streak stage
	Produce more than 30,000 conidia per spot	Produce about 1200 conidia per spot
	Conidia not dislodged by wind	Conidia both water- and wind-dispersed
	Mature ascospores produced 4 weeks after the appearance of streaks	Mature ascospores produced 2 weeks after the appearance of streaks

Epidemiology

Conidia and ascospores play an important role in the spread of the disease. Rainfall pattern is important in effective dissemination. Three components of weather determine the production and movement of inoculum of the disease: rainfall, dew and temperature. Profuse sporulation of black sigatoka pathogen occurs on the affected tissues during tropical and sub-tropical summers or under conditions of high humidity and rainfall, especially if there is a film of free water on the leaves. Conidial production is largely noticed in the initial stage of the disease, *i.e.*, during the first four stages. Local spread of disease is by means of conidia during period of high humidity, heavy dews and intermittent showers. Rain wash or splash plays a significant role in the dispersal of conidia as against the usual wind dispersal in conidial pathogens. However, during the later stages of the disease, the conidia are also forcibly discharged into air currents.

Ascospores are considered more important than conidia in the spread of the disease as the fungus produce only scanty conidia (Gauhl, 1994; Stover, 1980). However Jacome *et al.* (1991) opined that conidia are also capable of causing significant extent of disease and are especially important during dry periods. Spread of disease over larger distances within and outside the plantation is usually by ascospores during intense rains. These spores are formed in pseudothecia produced in mature lesions on older or dead leaves. Rainfall washes the ascospores out of the air and allows them to get dislodged from the foliage and to get deposited on the ground. Ascospores are disseminated by wind, however their long distance dispersal is restricted to few 100 km because of the vulnerability to ultraviolet radiation (Parnell *et al.*, 1998).

Conidia and ascospores require different temperature optimum for germination and germ tube elongation. Temperature above 21°C is favourable for inoculum production. The perithecium production is found to decline at temperatures below 21°C even when the rainy season prevailed (Stover, 1968). The production of asexual fruiting bodies and conidium during the rain free periods dependent on the presence of 10-12 hrs of dew.

Disease Cycle

Pathogen survives as perithecia in mature lesions on old and dead leaves. Ascospores produced are dispersed by wind to long distances or wind splashed rains to short distances within the plantation. The spores get deposited on the lower leaf surface during the unfurling of a new leaf. As a result, most of the infections are noticed on the abaxial surface of the leaves. Based upon the availability of high humidity and water on the leaves, the ascospores germinate profusely after a short period of incubation. Temperature of 26-28°C has been found optimum for germ tube elongation (Stover, 1980). Younger leaves are more susceptible than older ones. The incubation period varied with cultivar *i.e.*, in tolerant hybrids it took about 26 days for the symptom to develop as compared to 22.4 days in the susceptible cultivar, Grande Naine (Romero and Sutton, 1997). Under favourable weather condition, the disease takes lesser time of about 13-14 days to develop in the susceptible variety. The fungal hyphae, after penetration, colonise the host tissues inter cellularly for about 7 days to develop symptoms of the disease depending upon the favourable weather conditions and susceptibility of the cultivars.

The latent period or the time taken by the fungus to produce lesions with mature pseudothecium and ascospores varies according to the weather conditions, susceptibility of the host and intensity of infection. The latent period varied in a susceptible cultivar from 25 days during the rainy season to 70 days during dry season. Visible symptoms usually appear 15- 20 days after infection. The rapid spread of sigatoka is mediated by ascospore production, which in turn is dependent on warm and humid weather. Dry weather with very little dew formation during night is unfavourable for disease development. This disease is of common occurrence in ill drained soils and in shaded areas which cause increase in humidity within the plantation. Similarly, close planting, heavy weed population, profuse sucker production etc. favour increased build up of disease, but partial shading as in coconut gardens do not favour disease development. Frequently irrigated fields are more prone to this disease. The ramified mycelium generated through the germination of ascospores and further colonization emerge through the stomata. These hyphae penetrate the host through fresh stomata or produce conidiophores and conidia of the anamorph, *Psuedocercospora fijiensis*. The conidia are wind disseminated and help in the secondary spread of the disease.

Management

Black Sigatoka is a difficult and expensive disease to control. It causes rapid destruction of foliage that affects the photosynthesis and there by fruit yield. This necessitates the management of the disease through repeated application of fungicides. However, it is necessary to resort to other methods of disease management taking into consideration environmental and human health problems.

Cultural

Cultural methods such as removal of infected leaves, good drainage, and sufficient spacing play a significant role in reducing the conditions favourable for disease development. The reduction of inoculum level in the farms is an important cultural practice for managing the disease. The removal of the severely infected tissues helps to reduce the inoculum load. This practice is commonly known as deleafing or detipping or surgery according to the tissue removed. (Carlier *et al.*, 2000). By simply removing infected leaves (deleafing) and placing them on the ground, the efficacy of ascospore discharge can be significantly reduced. Since the infected leaf lying on the ground take about 22 weeks to get completely degraded, it is advisable to accelerate the decomposition process by addition of nitrogenous materials like urea. Quick decomposition helps to reduce further the available inoculum in the field.

Relative humidity in the field should be reduced to bring down the disease. Opening up the drainage channels can remove the running surface and sub-surface water.

Cultural management techniques such as wider plant spacing, better drainage of both water and air, intensified weed management and removal of severely diseased leaves or their portions from plants can also be used to contain the development and spread of the disease.

Biological

Research into developing biological control of black sigatoka is challenging because of the availability of effective chemical control and due to the polycyclic nature of the disease. The presence of plants of all ages and unfurling of highly susceptible leaves every 6-12 days makes the presence of the pathogen year–round and this makes implementation of biological control strategies difficult. However, due to the development of fungicide resistant strains of *M. fijiensis* and owing to the increasing need for environmentally friendly strategies emphasis has been laid on development of biocontrol of black sigatoka. Various epiphytic bacteria such as *Pseudomonas*, *Bacillus*, and *Serratia* spp. have been isolated and evaluated for their efficacy in suppressing the sigatoka. Effective isolates were selected based on the ability to secrete cell wall degrading enzymes such as chitinases and glucanases (Gonzalez *et al.*, 1996). Jimenez *et al.* (1987) obtained an isolate of *Pseudomonas* capable of suppressing the pathogen and performing better than chlorothalonil during *in vivo* screening studies under greenhouse condition.

However, *Pseudomonas* could not establish well in the phylloplane. A species of *Bacillus* with greater potential to colonize foliage and to tied over unfavourable environmental conditions also provided very little control on a commercial scale. *Serratia marcescens* showed good antagonism towards *M. fijiensis* and out yielded the control given by propiconazole in a greenhouse trial. (Jimenez *et al.*, 1987). Under field condition the effect of these isolates was similar to that of application of fungicides *i.e.*, alternation of propiconzole, tridemorph and mancozeb (Gonzalez *et al.*, 1996).

Though attempts on biological control have met with little success under field condition, further research on this strategy is to be strengthened in view of the frequent development of resistance to commonly used fungicides.

Resistant Cultivars

Use of resistant cultivars is really the only practical means of controlling black sigatoka for the small-scale and subsistence farmer since fungicides are generally unaffordable. Unfortunately, while some resistant cultivars of plantains and bananas are available, they are often unacceptable to local tastes. Resistance to *M. fijiensis* is found in some edible clones, wild diploids and some tetraploids (Meredith and Laurence, 1970). In Hawaii, all the AAA and AAB edible cultivars were found to be highly susceptible and the AAB plantains are moderately resistant whereas the ABB plantains are highly resistant. The diploid clones were found to be highly resistant to both *M. fijiensis* and *M. musicola*. However, clones with moderate levels of resistance to yellow sigatoka show high susceptibility that necessitate frequent spraying to avoid severe defoliation. (Vakili, 1968). High level of resistance has been located in wild diploids like *Musa acuminata* subsp. *burmannica*, sub sp. *malaccensis* and sub sp. *siamea* and in diploid cultivars such as 'Paka' and Pisang lilin. The reaction of banana towards BLSD may be categorised into two types: compatible and incompatible. Compatible interactions are characterized by full and rapid development of the disease from stage 1 to necrosis and sporulation. In such cases, the plant is deprived of almost all the functional leaves. In incompatible interactions, the development of the symptom is masked and there is hardly any sporulation. The defense reactions of the host swing into action immediately after the penetration of the fungus into the stomata.

The International Musa Testing Programme (IMTP) organized by INIBAP undertakes testing of hybrids at selected sites around the world. Three tetraploids *i.e.*, FHIA 01,–02 and–03 were identified to have resistance to BLSD. This resistance is being utilized for incorporation into edible types in order to breed new resistant banana cultivars. However, breeding for disease resistance is particularly difficult with banana plants. Commercial bananas are autotriploids (AAA), *i.e.*, they have three copies of the chromosome set instead of the two copies that the wild diploid species possess. While this extra chromosome set imparts favorable commercial characteristics such as seedlessness (unlike diploid wild species), as well as larger plant and fruit size, this high level of sterility is a major obstacle to banana breeders. Furthermore, the generation time (from seed to seed) for bananas is a long three years.

Chemical

Fungicidal management of BLSD is a regular practice being followed in commercial plantations. The use of fungicides for disease control started as early as 1930. It is generally achieved with the alternation of protectant fungicides such as mancozeb or chlorothalonil and systemic fungicides like benzimidazoles, triazoles, morpholines and strobilurins which may be applied either in oil or oil-water emulsion.

Oils

Petroleum oils have been used for control of sigatoka since late 1950s in the French Antilles (Guyot and Cuille, 1954). Oils were initially used as adjuvants, spreaders and stickers, later their fungistatic properties were recognized. Oils help in enhancing the penetration of systemic fungicides into leaves, and reduce the germination of conidium, germ tube elongation and infection process. They also retard the growth and development of the pathogen within the host (Stover, 1990). The rate of application of oil is based on the severity of the disease, weather conditions and the presence or absence of fungicide resistant population of the pathogen. High rate of application of petroleum oils can lead to phytotoxicity especially under dry conditions.

Heavy and light paraffinic oils were tested for their efficacy in BLSD suppression (Stover, 1990). Many of these were phytotoxic and hence not recommended for further use. Less phytotoxic and refined forms of naphthalene and paraffinic oils with unsulphonated residues of less than 90 per cent aromatic content of 12 per cent or less and distillation ranges between 346°C and 354 °C at 760 mm Hg are found suitable for application. However, in cases of reduced disease incidence the application of these compounds have been found to reduce yield by 5- 10 per cent (Israeli *et al.*, 1993).

Protectants

The protectant fungicides like mancozeb have been in use from 1970s for BLSD management (Stover, 1990). The oil based formulations caused phytotoxicity. Two new protectant fungicides *i.e.*, a flowable formulation of mancozeb (Dithane F) and chlorothalonil (Bravo 500) were introduced to overcome this defect. However, these formulations fail to perform well under periods of high rainfall and disease pressure, but were effective during dry periods, when the oil based formulation caused phytotoxicity. The new protectants either in combination or in alternation with systemics are now being used in management programmes.

Systemic Fungicide

The introduction of systemic fungicides provided better control of black sigatoka as compared to protectant fungicides. These fungicides were able to inhibit the development of symptoms even after their appearance. Benomyl was introduced for the disease management in early 1970s. The first steroid biosynthesis inhibitor belonging to morpholine group, tridemorph (Calixin) was subsequently introduced for combating this disease. Tridemoph has limited ability to penetrate banana leaves (Cronshaw and Akers, 1990). The first triazole fungicide (Tilt) was put to use for BLSD management in 1984 (Stover, 1990). There are now a large number of triazole fungicides with similar mode of action like bitertanol (Baycor), cyproconazole (Alto), difenoconazole (Score), flusilazole (Punch), hexaconzole (Contaf, Anvil), metconazole (Caramba) and tebuconazole (Folicur) have been registered for use for this purpose. Some of these compounds are translocated faster than others when applied on banana leaves (Guzman and Romero, 1996). Two strobilurin fungicides registered for BLSD management are azoxystrobin and trifloxystrobin. These systemic fungicides can be applied in oil water emulsion or in oil alone (Carlier *et.al.* 2000). The germination of ascospores is strongly inhibited by strobilurins. Two novel fungicides with distinct mode of action from the existing fungicides are piremethanil, an anilinopirimidine fungicide (Milling and Daniells, 1996) and spiroxamine, a spiroketalamine fungicide (Kramer *et al.*, 1999). Systemic resistance inducing compounds have also been tried for management of black sigatoka. Among this acibenzolar- S- methyl seems to be the most promising. The application of acibenzolar- S- methyl at intervals of 35 to 40 days at the rate of 40 g a.i./ha improved the natural resistance of banana cultivars and retarded the development of spots and streaks on leaves and resulted in considerable reduction in the severity of the disease (Madrigal *et al.*, 1998).

Fungicide Resistance

Development of resistance towards the effective systemic fungicides has resulted in reduction in their efficacy. Benomyl once an effective fungicide against black sigatoka was discontinued in plantations after 3-5 years of introduction. Similarly resistance to propiconazole became evident with reduced control using this fungicide in Central America in the 1990s. Much less is known about the resistance of *M. fijiensis* to new generation systemic fungicides like azoxystrobin. Maximum development of fungicidal resistance has been noticed for benzimidazole whereas moderate build up of resistance was observed in DMIs and strobilurins. Very low resistance has been indicated with contact fungicides such as chlorothalonil, copper and dithiocarbamates (Brent and Hollomon, 1998).

Forecasting

Early warning system for black sigatoka based on analysis of biological and climatic factors help to target fungicidal application to specific periods when disease severity is starting to increase and when environmental conditions are favourable for disease development. Application schedules in plantatations in several countries are now routinely determined based on disease forecast models (Stover, 1990).

Integrated Disease Management (IDM)

Integrated disease management programmes that help to reduce the incidence of severe epidemics include practices such as providing ample drainage, balanced nutrient management and nematode and weed control and proper inoculum management. Cultural practices such as deleafing followed by chemical treatment of diseased tissue for accelerating degradation as well as application of newer fungicides and biocontrol agents in the plantation in consonance with defense accentuating compounds are needed for developing holistic approach towards management of this dreaded disease affecting banana.

References

Brent, K.J. and Hollomon, D.W. (1998). *Fungicide Resistance: The Assessment of Risk.* FRAC Monograph No. 2. Global Crop Protection Federation. Brussels.

Carlier, J., Fouré, E., Gaul, F., Jones, D.R., Lepoivre, P., Mourichon, X., Pasberg-Gauhl, C., Romero, R.A. (2000). Fungal diseases of the foliage. In:. *Diseases of Banana, Abaca and Enset*, (Ed.) D.R. Jones. CABI Publishing, Wallingford, UK, p. 37–79.

Cronshaw, D.K. and Akers, A. (1990). Mode of action of tridimorph and sensitivity of *Mycosphaerella fijiensis*. In: *Sigatoka Leaf Spot Diseases of Bananas*, (Eds.) R.A. Fullerton and R.H. Stover. Proc. Int. Workshop, March 28–April 1, INIBAP. Montpellier, France, pp. 84–89.

Crous, P.W., Groenewald, J.Z., Aptroot, A., Braun, U., Mourichon, X. and Carlier, J. (2002). Integrating morphological and molecular data sets in *Mycosphaerella*, with specific reference to species occurring on *Musa*. In: *Proc. Int. Workshop Mycosphaerella Leaf Spot Dis. Bananas*, 2nd (In Press).

Foure, E. (1985). Black leaf streak diseases of Bananas and Plantains (*Mycosphaerella fijiensis* Morelet). Study of the symptoms and stages of the disease in Gabon. IRFA–CIRAD, Paris.

Fullerton, R.A. (1994). Sigatoka leaf diseases. In: *Compendium of Tropical Fruit Diseases*, (Eds). R.C. Ploetz, G.A. Zentmyer, W.T. Nishijinia, K.G. Rohrbach and H.D. Ohr. American Phytopathological Society, St. Paul, MN, pp.12–14.

Gauhl, F. (1994). Epidemiology and ecology of black Sigatoka (*Mycosphaerella fijiensis* Morelet) on plantain and banana (*Musa* spp.) in Costa Rica, Central America. *INIBAP*, Montpellier, France.

Gonzalez, R., Bustamante, E., Shannon, P., Okumotu, S. and Leandro, G. (1996). Seleccion de la Sigatoka negra (*Mycosphaerella fijiensis*) en banano. *Manejo Integrado Plagas*, 40: 6–11.

Guyot, H. and Cuille, J. (1954). Les formulesfongicides huileuses pourle traitement des bananeraies. *Fruits*, 9: 289–292.

Guzman, M. and Romero, R.A. (1996). Efecto de la applicacion der ceras y urea sobre la esporulacion de la Sigatoka negra. In: *Corporacion Bananera Nacional–Informe Annual* 1995. San Jose, Costarica pp. 63–65

Henderson, J.A., Pattemore, S.C., Porchun, H.L.,Hayden, S., Van Brunschot, K.R.E., Grice, R.A., Peterson, S.R.,Thomas-Hall, E.A. and Aitken, B. (2006). Black Sigatoka disease: new technologies to strengthen eradication strategies in Australia. *Australasian Plant Pathology*, 35: 2, 181

Israeli, Y., Shabi, E., Slabaugh, W.R. (1993). Effect of banana spray oil on banana yield in the absence of Sigatoka (*Mycosphaerella* sp.). *Scientia Horticulturae,* 56: 107–117.

Jacome, L.H., Schuh, W. and Stevenson, R.E. (1991). Effect of temperature and Relative humidity on germination and germ tube development of *Mycosphaerella fijiensis* var. *defformis. Phytopathology,* 81: 1480–1485.

Jacome, L.H. and Schuh, W. (1992). Effects of leaf wetness duration and temperature on development of black sigatoka disease of banana infected by *Mycosphaerella fijiensis* var. *defformis. Phytopathology,* 82: 515–520.

Jimenez, J.M., Galindo, J.J. and Ramirez, C. (1987). Estudios sobre combate biologico de *Mycosphaerella fijiensis* var. *defformis* mediante bacterias epifitas. In: *Proc. ACORBAT Meeting*, 7th. (Eds) J.J. Galindo and R. Jaramillo. Centro AgronomicoTropical de Investigacion y Ensenanza (CATIE), Turrialba, Costa Rica, pp. 105–109.

Kramer, W., Berg, D., Dutzmann, S., Gau, W., Stelzer, U. and Wesmuller, J. (1999). Chemistry, stereochemistry and biological properties of KWG 4168. *Pestic. Sci.*, 55: 610–614.

Leach, R. (1964). A new form of banana leaf spot in Fiji, black leaf streak. *Wild Crops*, 16: 60–64.

Lloyd, D. (2004). The Black Sigatoka threat. *Inside Bananas*. 1: 13–14.

Madrigal, A., Ruess, A. and Staehle-Csech, U. (1998). CGA 245704: A new plant activator to improve natural resistance of banana against black sigatoka (*Mycosphaerella fijiensis*). In: *Proc. ACORBAT Meeting* 8th, (Ed.) L.H. Arizago, CONABAN, Ecuador, pp. 266–274.

Marin, D.H., Romero, R.A., Guzman, M. and Sutton, T.B. (2003). Black sigatoka: An increasing threat to banana cultivation. *Plant Dis.*, 87: 208–222.

Mayorga, M.H. (1990). La Raya negra (*Mycosphaerella fijiensis* Morelet) del platano y del banano I. Ciclo de vida del patogeno bago las condiciones del Uraba. *Revista ICA.*, 25: 69–77.

Meredith, D.S. and Lawrence, J.S. (1969). Black leaf streak diseases of bananas (*Mycosphaerella fijiensis*): Symtoms of disease in Hawaii, and notes on the conidial state of the causal fungus. *Trans. Br. Mycol. Soc.*, 52: 459–476.

Meredith, D.S. and Lawrence, J.S. (1970). Black leaf streak diseases of bananas (*Mycosphaerella fijiensis*): Susceptibility of cultivars. *Trop. Agric.*, 47: 275–287.

Milling, R.J. and Daniells, A. (1996). Effect of pyrimethanil on the infection process and secretion of fungal cell wall degrading enzyme. In: *Modern Fungicides and Antifungal Compounds,* (Eds.). H. Lyr, P.E. Russel and H.D. Sisler. Intercept Andover, UK, pp.53–59.

Mourichon, X., Carlier, J. and Foure, E. (1997). Sigatoka leaf spot diseases: Black leaf streak disease (Black sigatoka) and Sigatoka disease (yellow sigatoka). Fact Sheet No. 8. *INIBAP.*

Mulder, J.L. and Stover, R.H. (1976). *Mycosphaerella* species causing banana leaf spot. *Trans. Br. Mycol. Soc.*, 67: 77–82.

Parnell, M., Bert, P.J.A. and Wilson, K. (1998). The influence of exposure to ultra violet radiation in simulated sun light on ascospores causing black sigatoka disease of banana and plantain. *Int. J. Biometeorol.*, 42: 22–27.

Ploetz, R. (2000). Black sigatoka. *Pesticide Outlook*, pp. 19–33.

Ploetz, R.C. and Mourichon, X. (1999) First report of black Sigatoka in Florida. *Plant Disease*, 83, 300.

Pons, N. (1987). Notes on *Mycosphaerella fijiensis* var. *difformis. Trans. Br. mycol Soc.*, 89: 120–124.

Rhodes, P.I. (1964). A new banana disease in Fiji. *Comm. Phytopathol. News*, 10:38–41

Romero, R.A. and Sutton, T.B. (1997) Sensitivity of *Mycosphaerella fijiensis*, causal agent of black Sigatoka of banana to propiconazole.*Phytopathology*, 87: 96–100.

Stover, R.H. (1968). Leaf spot of bananas *Mycosphaerella musicola*: Perithecia and sporodochia production in different climate. *Trop. Agric. Trin.*, 45: 1–12.

Stover, R.H. (1969). The *Mycosphaerella* species associated with banana leaf spots. *Trop. Agric. Trin.*, 46: 325–332.

Stover, R. (1980). Sigatoka leaf spot of banana and plantains. *Plant Dis.*, 64:750–755

Stover, R. (1986). Disease management strategies and the survival of the banana industry. *Annu. Rev. Phytopathol.*, 24:83–91.

Stover, R.H. (1990). Sigatoka leaf spots: Thirty years of changing control strategies: 1959–1989. In: *Sigatoka Leaf Spot Diseases of Banana* (Eds.) Fullerton, R.A. and R.H. Stover. INIBAP.Montpellier, France, pp. 66–74.

Vakili, N.G. (1968). Responses of *Musa acuminata* sp. and edible cultivars to infection of *Mycosphaerella musicola. Trop. Agric. Trin.*, 45: 13–22.

Plant Diseases Management for Sustainable Agriculture *Pages* **72–88**
Editor: **Dr. Shahid Ahamad**
Published by: **DAYA PUBLISHING HOUSE**

Chapter 7

Challenges in Tackling New Emerging Diseases of Maize

Meena Shekhar[1], Ram Dutta and Sangit Kumar[2]
[1]Senior Scientist, [2]Principal Scientist
Directorate of Maize Research, Pusa Campus, New Delhi – 110 012

Maize (*Zea mays* L.) occupies an important place in world agriculture due to its high potential of yield and greater demand. Among cereals, maize in India ranks third in yield per hectare, fourth in total production and 5th in total area. It is estimated that by the year 2020 demand for maize in developing countries will surpass demand for both wheat and rice. Rising income in much of developing world and the consequent growth in meat and poultry consumption has resulted in rapid increase in demand of maize for livestock and poultry feed. In India 35 per cent of the total produce directly consumed as a human food while the balance used in industry and poultry feed. The total production during the year was 14.14 million tones, from the estimated area of 7.49 million hectare. The productivity during the year was 1887 kg/ha (Anonymous 2006). The increase in the area has been reported mainly from the states of Gujarat, Bihar, U.P., A.P., M.P., Rajasthan, Chhatishgarh, Maharashtra and Tamil Nadu.

Biotic Stresses are Responsible for Lowering Yield

Among the factors adversely affecting productivity, ubiquitous incidence of diseases in the pre-harvest stage is prominent. Since there is practically no possibility of increasing maize area, the productivity can only be raised by providing seed replacement with improved cultivars, better agronomic practices and protection against diseases and pests.

Based on research efforts for the last three decades, under the aegis of All India Coordinated Maize Improvement Project, 16 out of 62 diseases adversely affecting this crop have been identified (Payak and Sharma 1985). Pythium stalk rot, Bacterial stalk rot and Peronosclerospora incited downy mildews posses the highest threat to reduce the maize yield. On a global basis, losses due to maize diseases have been estimated to the tune of 9.4 per cent annually. For the countries of Asia the figure is 12 per cent, while for the African countries the estimate is as high as 14 per cent (Cramer, 1967; James 1981). Even for the developed countries like USA, 12 per cent of the produce is lost due to diseases annually. For India the total loss in economic products of the crop due to diseases has been estimated to be the order of 13.2 per cent (Payak and Sharma, 1985).

Some Major Epidemics in India

1. Bacterial stalk rot (1959, 1969)
2. Brown stripe downy mildew (1962-1965)
3. Sorghum downy mildew (1968-1973)
4. Common rust (1973 onwards in rabi maize)
5. Turcicum leaf blight (1974)

Table 7.1: Estimation of Losses Due to Major Diseases of Maize in India

Diseases Groups	*% Loss*	*Annual Grain Loss (Tones)*
Seed and seedling blights	0.1	7,923.5
Downy mildews	2.1	166,395.5
Foliar diseases (Turcicum leaf blight, Maydis leaf blight, Phaeosphoria leaf spot, Rust, Brown stripe downy mildew)	5.0	396.180
Stalk rots, root rots, ear rots	5.0	396,180
Sheath blights, smuts, Viruses and others	1.0	79.236
Total	13.2	1,045,915

Sources: Payak and Sharma, 1985.

Changes in Diseases Spectrum Scenario

There is a major shift in disease pattern during the past years as the major diseases like TLB, MLB, BSDM and ESR are gradually becoming diseases of lesser economic importance now a days due to availability and use of sources of resistance in the newly developed hybrids and varieties. On the other hand the disease of minor importance became increasingly severe and assumed epidemic proportion in coming years. In early sixties the diseases like BLSB, Late wilt, Polysora Rust and Preharvest cob rots were considered as a disease of minor importance only, but due to changing the weather condition and pattern of disease spectrum these diseases are considered as major diseases not only in India but other countries of Asian region wherever maize is grown.

New Emerging Diseases in India

Banded Leaf and Sheath Blight

Banded leaf and sheath blight (BLSB) on maize incited by *Thanatephorus sasakii* (Shirai) Tu and Kimbro; anamorph *Rhizoctonia solani* f.sp. *sasakii* Exner, has become increasingly severe and economically important disease of maize during last two decades or so. For the first time was reported from Sri Lanka under the name 'Sclerotial' disease. Though it was considered a disease of minor importance till it appeared in an epidemic form in the foot hills region of Himalayas especially in the district of Mandi in Himachal Pradesh. In India, in early sixties the disease was little more than a curiosity and a disease of minor importance in the western and central Himalayan foothill region. However, it became increasingly severe and assumed epidemic proportions in the next two decades. Presently, the disease is considered as a major disease not only in India but also in several countries of Tropical Asia wherever maize is grown.

Distribution

On maize the disease was first recorded from Sri Lanka (Bertus, 1927) under the name 'sclerotial disease'. In recent years the disease outbreaks have occurred in more countries and have assumed epidemic dimensions. In India, the disease was first recorded from Tarai region of Uttar Pradesh in 1960 (Payak and Renfro 1966). Now it has become increasingly severe and assumed epidemic proportions in the next two decades. Presently, the disease is considered as a major disease in India wherever maize is grown in warm and humid conditions. It is known to be present in severe intensity in the states of Himachal Pradesh, Uttar Pradesh, Uttaranchal, Haryana, Punjab, Madhya Pradesh, Rajasthan, Bihar, West Bengal, Sikkim, Meghalaya, Assam and Orissa.

Economic Importance

The disease was earlier reported as a minor disease on maize (Payak and Renfro 1966). The importance of the disease was only realized in early 1970s when an epidemic occurred in warm and humid foot hills area, particularly in the Mandi district of Himachal Pradesh (Thakur *et al.*, 1973). The disease causes direct losses, resulting in premature death, stalk breakage and ear rot indirect losses by reducing the grain yield. In India, losses in grain yield have been estimated in the range of 23.9 to 31.9 per cent (Sangam Lal *et al.*, 1980). Singh and Sharma (1979) estimated 40.5 per cent loss in grain yield with 71 per cent disease index. However, the magnitude of grain loss may reach as high as 100 per cent if the ear rot phase of the disease predominates. Payak and Sharma (1985) have reported that annually at least 1 per cent of the total grain yield is reduced by BLSB in India.

Losses to the extent of 11-40 per cent were reported while evaluating 10 different varieties of maize (Singh and Sharma, 1976). Losses in grain yield showed a high positive correlation with premature death of plants and disease index. The disease caused drastic reduction in grain yield-to the tune of 97 per cent (Butchaiah, 1977) and exhibited a direct correlation with other yield parameters (Barua, 1979). Disease incidence varying from 2 to 80 per cent was observed affecting maize crop severely in Korea during previous two crop seasons (Kim *et al.*, 1992).

Symptoms and Signs

The disease generally appears at pre-flowering stage in 40-50 day old plants. The symptoms manifest on leaves, sheaths, stalk and ear. As the disease is soil borne it starts from the lowest leaf sheath or on leaves that are in contact with soil and travels up to the ear. The symptoms are more common on the leaf sheaths than on the laminate. The disease lesions are characterised by the presence of alternate bleached areas or zones that are initially water soaked and narrow, purple brown bands oriented perpendicular to long axis of leaves or leaf sheaths.

The hyphal masses on ears produce a caking effect and cement the husk leaves as well as styles (silk fibers) together. The mycelium develops above and between kernel rows. Sclerotia are found to be produced not only on sheaths, leaves and husk leaves but were also observed for the first time on glumes, in cupules, under the pericarp in caryopsis and silk fibres (Ahuja and Payak, 1982).

Figure 7.1: Crop Affected by Banded Leaf and Sheath Blight in Maize

Causal Organism

The disease banded leaf and sheath blight incited by *Rhizoctonia solani* f. sp. *Sasakii* Exner (*Thanetophorus sasakii* (Shirai) Tu and Kimbro), is found in most parts of the world and is capable of attacking a wide range of host plants including maize, causing seed decay, damping off, stem canker, root rots, aerial blight and seed decay. Detailed work on maize pathogen (Ahuja and Payak, 1978) suggests need for critically determining the correct taxonomy of the pathogen. The teleomorphic stage of maize pathogen has not been recorded in nature or *in vitro* so far, whereas it has been observed on rice (Anonymous, 1974). Pathogenicity tests revealed that *R. solani* AG-1-1A was the most strongly pathogenic to leaf sheaths and stalks of maize, followed by *R. solani* AG-2-2. Differences were observed in the susceptibility of four maize cultivars to the anastomosis groups of *R. solani* and *R. zeae* (Kim *et al.*, 1992).

Epidemiology

The rice pathogen has wide host range and infects plants belonging to over 32 families in 188 genera. *R. solani* f.sp. *sasakii* infects by artificial inoculations a number of crop plants belonging to families Gramineae, Papilionaceae and Solanaceae (Ahuja and Payak, 1988; Barua and Lal, 1981).Maize has also been found to be infected by isolates from rice, sugarcane, arrow root and some grasses (Ahuja and Payak, 1985; Haque, 1975). Cross inoculation tests indicated that *rice* and turmeric isolates did not differ either in pathogenicity or morphological characters (Lal *et al.*, 1980).

The primary source of inoculum are the sclerotia in soil or in infected host debris, and the active mycelium on the other grass hosts that grow in the vicinity of maize plant in the field. The viability of sclerotia is greatly influenced, to a large extent, by environmental and soil conditions. The germinated sclerotia come in touch with lower leaf sheath or drooping leaves contacting soil. The infection progresses upward in bands killing the tissues in advance. The infection extends to distal half of the leaf lamina exhibiting buff-white mycelial growth in patches on the affected parts within 48-72 h of infection. The primary inoculum becomes more active during the period of pre-flowering stage (30 to 40-day old plants).

Secondary spread is due to contact of healthy plants with infected leaf or sheath or vice-versa. The infection continues to mid-dough stage infecting the entire plant including the ear except tassel. These seeds are not considered to be source of inoculum and may not play a major role in severe disease outbreaks. If cob infection occurs at the late stage, the germination of seed is adversely affected resulting into seed rot and seedling blight. The sclerotia formed on the infected plant organs fall down on soil surface get, buried in soil with infected host debris and serve as a source of primary inoculum for maize and other host crops in field.

Environmental Factors

The disease is favored by warm and humid climate (Quazi and Iqbal, 1970). The optimum temperature for *in vitro* growth of the pathogen is 30°C and the highest level of disease is induced when RH is in the range of 90-100 per cent. Below 70 per cent the disease development was negligible. Maximum relative humidity and rainfall largely influence the spread of the disease. Studies on the meteorological factors in relation to disease progress revealed that optimum temperatures (near about 28°C) were essential for infection, disease development and subsequent spread of the pathogen. Rainfall over 100 mm in the first two weeks of inoculation favored early infection and disease development..

Positive correlation was noticed between rainfall and disease progress (Pandey, 1992). This led to obtain partial regression coefficients to derive a simulation equation based on meteorological factors for the prediction of disease. However, it needs more data to standardize the model for practical utility.

Disease Management

Through Chemical

- ☆ Fungicides like carbendazim, benodanil, validamycin, topsin M, Rhizolex etc. were effective in reducing disease severity under field conditions (Ahuja and Payak, 1986; Sharma and Rai, 1998).
- ☆ Benodanil (150 p.g/ml a.i), Validamycin (Ahuja and Payak, 1988) and Carbendazim (Sherwood, 1970) reduced disease by 33.8, 33.5 and 23.7 per cent, respectively.
- ☆ Bavistin 50 WP Benomyl 50 WP and Brestan gave 87, 82 and 77 per cent disease control, respectively and resulted in significant increase in grain yield (Sinha, 1992). Three foliar applications of Propiconazole at 10-day intervals exhibited reduction in disease severity and resulted in higher grain yields (Anonymous, 1995; Sharma, 1996).

Through Culture Practices

- ☆ Inter-cropping of maize with legumes specially with soybean effectively reduced the activity of the pathogen in soil (Lehman *et al.*, 1976).
- ☆ Maintaining the plant population at an appropriate level and applying cattle compost before planting decreased the disease level and its subsequent spread in field (Lee *et al.*, 1989).
- ☆ Mechanical stripping of the second and third leaf-sheath from the ground level at the age of 35-40 day old crop or knee high stage which is effective in checking disease development further up (Sharma and Hembram 1990).

Through Biocontrol Agents

Biological control of disease offers the preferred mode of environmental friendly control of disease in maize including banded leaf and sheath blight. Application of *Pseudomonas fluorescens* has been reported to control the disease in the field conditions besides improving the plant growth. The biocontrol agent used, exhibited the production of volatile ammonia and HCN under *in vitro* conditions. Sivakumar *et al.* (2000) demonstrated a very effective control by seed treatment of peat based formulation @ 16 g/kg, or soil application @ 2.5 kg/ha, or spraying liquid formulation twice @ 5 g/litre of water. Manisha *et al.* (2002) also observed that combined seed treatment and foliar treatment with *Pseudomonas fluorescens* from maize rhizosphere was most effective in giving 30 per cent reduction in disease incidence of banded leaf and sheath blight. Similarly, Meena *et al.*, 2003 demonstrated that among *Trichoderma, T. harzianum* has shown good promise against the pathogen both *in vitro* and *in vivo* conditions.

Disease Resistance

A variety of maize materials including–inbred lines, single crosses, double crosses, double top crosses etc., were evaluated in the Indian maize programme. Among the inbred lines C (coordinated) M (maize) 104 (yellow-flint), and CM 300 (white flint) were determined to be the elite lines in terms of resistance. In China, 844-1, SZ-I, 884-2, MS08, 8S4-1, WP-4, WP-5, H138, H158, KI 1414 and several other materials were rated as resistant during 1994-95 crop season. Maize varieties Jinguok, Suweon 83, Suweon 87, Suweon 89, P.3055, P.3160, DK 689 and XCG 51 showed high tolerance levels to BLSB in Korea (Lee *et al.*, 1989).

A number of CML lines from CIMMYT and other materials were evaluated in India and China and many lines were identified having reasonable level of resistance. Line CML-1 in particular was found to be elite one both at Delhi and Pantnagar in India. Inbred lines CM 104, CM 103, CM 300, CM

105, P 217407, CM 600 and hybrid VL 43 have been found resistant under field and laboratory evaluations (Ahuja and Payak, 1984). Of these, CM 104 and 105 have been reported earlier to be resistant source to banded leaf and sheath blight of maize. A total of 3000 genotypes have been screened. In India under ICAR-CIMMYT Collaborative Activity, out of them 50 materials have shown some level of tolerance to BLSB and these materials are being advanced to upgrade their level of resistance by adopting cyclic breeding approach.

Polysora Rust of Maize

Distribution

Among the rust diseases in maize Polysora rust or tropical rust or southern corn rust (*Puccinia polysora* Underw) is an important disease in tropical areas. It occurs in many parts of the world, and it is a recent introduction in the peninsular India in 1991 (Karnataka and Tamil Nadu during rainy season and Coastal districts of Andhra Pradesh during winter) on certain maize cultivars in Mysore distric (Payak, 1992). In recent years, the incidence of *P. polysora* has taken a heavy toll in majority cultivars grown in Karnataka namely Mysore, Mandya, Hassan, Kolar, part of Coorg, Shimoga and Chitradurga districts (Anon.2003) Polysora rust is a warm weather disease favoured by wet weather infection and disease development at 12- 27°C temperature (Hollier and King 1985, Melching 1983). Rain drizzle or even heavy dews allow disease formation.

Symptoms and Causal Organism

Pustules are small circular and orange in colour. They are evenly spread over the surface on the midribs, ear husks and tassels may be elongated or irregular in shape. With heavy infection leaf may die prematurely and defoliation starts from the base towards upward. The disease resembles common rust. The pustule appear on leaf are light cinnamon brown, circular to oval 0.2-2.0 mm. long densely scattered on the upper leaf. The uredospores are yellowish to golden, mostly ellipsoid or ovoid with minutely echinulate, 1-1.5 µm thick wall and sizes are 20-29, x 29-40µm. Pustule development on lower surface is more as compared to upper surface. The telia are circular to elongate, 0.2-0.5 mm in diameter dark chocolate brown to black and remain covered by the epidermis longer than the common rust. Telia often appear in circles around the uredial pustules. The teliospores are dark brown, smooth angularly ellipsoid or oblong, rounded at both the ends and 18-27 x 29-41 µm. They are brittle, usually two celled constricted at the septum and born on the short 10-30 µm persistant brownish pedicels one fourth length of the spore. Infected maize leaves become chlorotic and dry. It differs from common rust (*P. sorghi*) in pustule size, shape and colour however the most pronounced variation is that it kills the host unlike *P. sorghi.* Raid *et al.* (1988) found that even if infection comes as late as after anthesis, losses occur and the pathogen may cause heavy losses when the environmental conditions are conducive for disease development. However, Zummo (1988) reported that heavy losses may occur due to this disease if infection occurs an early stages.

Economic Importance

The pathogen has the potential of being destructive if infection comes after anthesis (Raid *et al.*, 1988). Melching (1975) reported yield loss up to 37 per cent due to *P. polysora* and rated this disease as the most destructive among the three rusts of maize. Ardon *et al.* (1988) recorded yield loss up to the tune of 45 per cent especially on late planted maize however, Frederekson *et al.*,(1990) reported 60 per cent loss in grain yield. Severe losses due to this disease can occur especially if infection occurs early. The presence of Polysora rust in the peninsular India, particularly in states like Karnataka, could have an adverse impact on maize production. It is also expected that the spread of the disease to new

Figure 7.2: Symptoms of Polysora Rust of Maize

locations could pose a major threat to maize cultivation, as this is considered to be the most destructive amongst the rust of maize.

Epidemiology

Polysora rust is a warm weather disease favoured by wet weather infection and disease development at temperatures (27°C) and high relative humidity. Disease develops rapidly in warm weather and uredospores comprise both primary and secondary inoculum (Bailey *et al.*, 1987). Teliospores of *P. polysora* are rare and are not known to be germinated therefore they are unimportant in disease cycle. The uredospores constitute both primary and secondary inoculum and are carried to maize plants by wind or on infected materials. No alternate host for this disease has been found.

Disease Management

Highly resistant genotypes have smaller uredosori than the moderately resistant or susceptible ones. Resistant lines normally show small chlorotic or necrotic flecks with no sporulation. Maize lines with small pustules surrounded by chlorotic or necrotic zone were rated as resistant while well developed pustules were considered susceptible (Hulbert *et al.*, 1990). Ullstrup (1965) reported a dominant gene of resistance and designated it as Rpp9. A major gene on chr. 10S accounted for 83 per cent of phenotypic variation for polysora rust resistance in a mapping population based on cross between tropical x temperate lines, with the tropical inbreds as a sources of resistance to polysora rust. RFLP markers flanking this locus have been identified (Holland *et al.*, 1998). Another study reported in a different genetic background, a QRL (Quantitative resistant locus) to *P. polysora* on chr. 2 and SSR markers closely linked to this locus (Brunellii *et al.*, 2002). Brewbaker (2005) demonstrated that synthetic MIRSYN3 based on 19 highly resistant inbreds showed resistance through 6 cycles of recurrent selection. Jines *et al.* (2006) prepared a multiple interval mapping model, including four QTL, accounted for 88 per cent of the variation among average disease ratings. A major QTL located on the short arm

Figure 7.3: Fusarium Ear Rot

Figure 7.4: Aspergillus Ear Rot

of chromosome 10, explained 83 per cent of the phenotypic variation, with the NC300 allele carrying the resistance. They also stated that the maturity and Polysora rust rating were slightly correlated, but QTL for the two traits did not co-localize. These studies clearly indicate that a few major loci undergrid general resistance to polysora rust in tropical maize germplasm. A net work project is going on at Directorate of Maize Research on gene pyramiding on Polysora rust and further studies are in progress. Based on the work carried out the management strategies for the Polysora rust is given below:

- ☆ A fungicide application is useful when pustules first appear on the leaves.

- ✰ Three sprays of Dithane M-45 beginning from first appearance of symptoms at 15 days interval can minimize the disease.
- ✰ Use resistant varieties–The sources of resistance are; NAI 112, SKV 18, SKV 21, NAH 2049 (Nithyashree) a resistant hybrid to Polysora rust, TLB and SDM has been released in Karnataka.

Cob Rots

The safety and integrity of the food supply are the paramount importance and are among the drivers of safe grain storage. In present scenario, the trade among the developing countries is expanding; hence quality of the produce is becoming a major concern. Demand for maize in both national and international market is high. Its uses are continuously increasing globally as important livestock, poultry feed and by industrial sectors. On the other hand the loss of grains is increasing with the increase of food production due to poor and improper storage as well as the pre harvest fungi associated with maize. These fungi are responsible for microbial spoilage of the grains and production of mycotoxins in pre- and post harvest stages.

Symptoms and Epidemiology

Fungi associated with the cob rots are–*Aspergillus flavus, Fusarium species, Gibberella zeae, Penicillium* etc. These field fungi infect the cobs either immediately or before harvest. The presence of mycotoxin traditionally regarded as indication of poor storage condition, but they may present in grains before coming to storage.

Fusarium Ear Rot

Fusarium ear rot caused by *F. moniliforme* occurs when harvest is delayed. The symptom of this disease is pink to reddish brown discoloration on the kernels and later on it spreads on whole ear. As the disease progresses, infected kernel becomes covered with a powdery/cottony pink mild growth. Kernels infected late in the season develop whitish streaks on the grains. Kernels become infected in several ways. The most common pathways is infection via silk channel. Air born spores present on residues can land on corn silks when it turns dark brown. Green silks are relatively resistant. Infection follows some form of injury, bird damage, feeding of corn borers. Disease development and spread one favoured by dry warm weather. Under certain conditions that are stressful for maize plant as the fungus becomes pathogenic and causes disease. Stalk rot, ear rot, kernel rot can occur in infected tissue although many times infected tissue have no symptoms.

Aspergillus Ear Rot

This rot caused by *Aspergillus* spp. and infection often follows drought stress and damage done by corn, earworms and corn borer and other insects. A tan sooty black, greenish or greenish yellow mold grows on and between kernels. The damage is most common at or near the tip of the ear; silk infection is favoured by high night and day temperature. Some times the susceptible maize genotypes got infected with *Aspergillus* sp. at the time of silking. The infection goes through silk to the maize seed/grains but remained latent and cause decay at the time of maturity in the form of cob rot/ear rot, sometimes the grain remain infected but does not show any symptoms even at the time of maturity. Corn planted and harvested late and grown under nitrogen stress more commonly contains aflatoxins prior to harvest than corn grown under good management practices and supplied with adequate nitrogen.

Maize grown in cooler areas usually contains low amount of mycotoxin (Deko *et al.*, 1994).

Strategies for Prevention of Mycotoxin in Maize

A number of different management strategies can be employed at the farm level to alleviate the development of mycotoxins in feeds and consequently reduce the adverse effects. It has been determined that the retention time of feed at the farm is highly correlated with both the frequency and concentration of aflatoxin; thus, more frequent delivery of feed with shorter time lag is one means of reducing problems with aflatoxicosis.

Prevention through pre-harvest management is the best method for controlling microbial spoilage of stored grain/seeds but when contamination does occur, the hazards associated with the toxin must be managed through post-harvest procedures, when the grains are to be used as a food and feed.

Pre-harvest Management

Prevention the growth of moulds on storage grains through pre-harvest management is the first step in minimizing the microbial spoilage of stored grains. Association between microbial decay and inadequate storage facilities cannot be ignored. Studies have revealed that some seeds are contaminated with moulds in the field itself. It is impossible to control, when these infected seeds are exposed to environmental factors as well as stored in poor storage conditions. Once the crop becomes infected under field conditions, the fungal growth will continue during post-harvest stages and storage. Therefore, the pre-harvest management is focused on controlling the main factors that are responsible to enhance the mycotoxins production in maize.

Some Common Management Strategies

Insect Infestation

It has been reported that the incidence of *Aspergillus flavus* is usually higher in insect damaged kernels and these damaged kernels are the routes for infection for mycotoxins producing fungi. Control of the growth of these fungi will subsequently minimize the fungal contamination.Discarding the damaged grains before storage helps in reducing the microbial colonization and subsequent formation of mycotoxins during storage.

Crop Rotation

It is an important factor in the spread of inoculum of grain molds. Adequate rotation of crops will provide minimum chance of spread of inoculum of these molds. Therefore, crop rotation will prove an aid to the prevention of mycotoxin contamination; such as maize-soybean rotation yielded a less extensive outbreak of fusarium than maize-maize planting operations.

Management of Crop Residues

Inoculum potential is a prerequisite for grain molds infection. Soil type and conditions, as well as availability of viable spores are important factors in mycotoxin production. At the time of crop harvesting, some residues remain on the field. This residue provides the source of inoculum for the next crop. Proper management of these crop residues would help to avoid this problem.

Avoidance of Environmental Condition that Favour Infection in the Field

To minimize the inoculum load, the adequate efforts should be made to avoid extreme conditions of either drought or excessive moisture. It has been observed by many workers that the drought stress followed high moisture condition is ideal for the development *Fusarium moniliforme* and subsequently production of fumonisin. Whenever such weather prevails it can be assumed that some degree of mycotoxin contamination will occur and accordingly the management strategies should be explored.

Development of Resistant/Tolerant Plant Varieties

The search for naturally resistant maize genotypes had not been successful till today. However, the extensive research on development of plant varieties and identifying the sources of resistance against mycotoxin producing molds is being corrected out. Based on screening of maize genotype mycotoxin producing fungi and level of mycotoxin conducted at DMR Priya sweet corn and Navjot showed some degree of tolerance up to 6 months stored period, whereas Shaktiman–1, found to be most promising by showing negative reaction for Aflatoxin in absence of *A. flavus* in cob samples even in 9 months storage period. In case of healthy maize seed samples, Shaktiman-1, Shaktiman-4 and KMH-1701 were found most promising by showing negative reaction for Aflatoxins even after 9 months stored period. In one year stored healthy grain samples, the most promising genotypes identified were Shaktiman–4 with a level of toxin (AFB_1) is 0.03 ppb followed by KMH–1701 (0.04 ppb); HQPM-1 (0.05 ppb); and QPM–2–136 (0.06 ppb) whereas the highly toxic samples was RP–4 (50.44) at moisture per cent ranged from 12 to 11.1 per cent in samples collected from Delhi. The maximum level of toxin observed in sample Mon- 4 with a level of toxin is 62.42 ppb at moisture 12.6 per cent obtained from Hyderabad. Shaktiman-1 also showed minimum (0.46 per cent) disease index of *A. flavus* in field conditions which again proving that this genotype is promising one.

Harvesting Management

During harvesting, three factors are very important to control such as:

1. Timeliness harvesting
2. Clean up of harvested crop
3. Drying of harvested crops.

These practices are essential for preventing microbial decay during storage. Timing of harvesting is greatly influences the microbial growth. Harvesting should be done as soon as the crop is fully grown. It has been observed that the crops left on the field for longer periods of time many present have higher level of toxin contamination. Adequate drying is also essential to prevent fungal proliferation during storage.

Post-harvest Management

We should ensure the proper grain storage conditions to minimize the potential for mold growth and microbial decay. The grain storage facilities should be monitored regularly to detect grain mold development. Keep moisture in stored grain below 12-13 per cent. Screening of grains before storage should be done to remove trash cracks and shriveled kernels to minimize the growth of molds. Mould development during storage causes loss of germination in seed grain, discolouration and change in fatty acid profile and other quality parameters. Mould development may also encourage mite and insect infestation (Wicklow, 1995).

Precautionary measurements must be done on newly harvested grains so that losses in quality and quantity could be avoided. A recommended and proven method is to reduce the moisture content as early as possible. This can be accomplished by drying and subsequently controlling the relative humidity and temperature during storage. In addition, chemical application prior to storage and during storage will be administered.

Drying

Field drying of most grain has been an accepted practice since commercial farming began; sun and wind are the primary drying agents for the most of farmers' community worldwide and developing countries in particular. After harvest, the moisture content of the produce must be dried to 12-14 per cent for safe storage with minimal deterioration.

Sun Drying

Sun drying is probably the oldest and most common way of reducing the moisture content of grains. It utilizes solar energy from the sun directly. In sunny days, the drying process will take 2 to 3 days. Mechanical drying is done by hot air blown by mechanical driers it consumes high power. Solar energy is tapped by the use of a collector and converted to a more convenient form of energy and either direct use or store it for later delivery to the point of use. In solar driers solar energy heats the air.

Corn at harvesting stage has high moisture content and should be mechanically dried to reduce its moisture content as early as possible to prevent grain quality deterioration by fungal invasion. The field shelled high moisture yellow corns could be dried in 5.5 m (18 ft.) diameter low temperature drying bins. One bin had two solar collector units, the second bin is used for ambient air only. Microflora activity can be minimized by shorter drying time and by the mixing of lower and higher moisture corn during the transfer to other storage bins.

Effect of Organic and Inorganic Acids on the Storage of Maize Grains for the Management of Mycotoxin Contamination

Some new promising non-toxic products like Propcorn, Nilspor and MycoCurb have been found effective are being further investigated for management of storage fungi. Propcorn, a product of the BP chemicals of England is based on feed grade propionic acid which prevents deterioration, loss of dry matter, nutrition loss and removes the associated health hazards to grain handlers, livestock and end-user, treated seeds could be stored upto 12 months without proper storage. Another chemical Nilspor, a propionic acid based chemical having higher acid content is more effective but having corrosive property. Kemin Industries of US produced MycoCurb a mould inhibitor which is non-corrosive, non-toxic and non-volatile. A trial was conducted to study the effect of these organic and inorganic acids and salts on different maize grain type at four months periodic interval at DMR, New Delhi. These salts showed encouraging results and further study is in progress.

Effect of Biocontrol Agents Isolated from Maize Grains

Some experiments conducted in the laboratory at DMR to find out the effect of biocontrol agents like strains of *A. niger*, *T. harzianum* and non-toxic strains of *A. flavus* isolated from stored maize grains and tested against *F. moniliforme* and *A. flavus*. It was observed that these bioagents could suppress the growth of *F. moniliforme* and *A. flavus*. and inhibit the production of mycotoxins. Biological control offers several advantages over chemical control like there is no pollution problems and health hazards as:

1. It is more stable without any problem of resistance development in pathogen
2. Biological control agents are not phytotoxic
3. It may have the growth promoting effect and
4. It causes little disturbance in ecological balance.

Prevention of mycotoxins in maize encompasses prevention of toxin biosynthesis and metabolism in the field or in storage. Mycotoxin detoxification refers to post-harvest treatments to remove, destroy or reduce the toxic effects. Failure to prevent mycotoxin formation in the field or in storage will inevitably lead to increased health risks and economic loss. However, successful monitoring should prevent mycotoxins from becoming a significant source of increased health risks. Grain spoilage and mycotoxin contamination of grain is not always a result of a poor storage as some mycotoxin particularly fumonisin may already be in the grain when they are brought in storage. For scientific storage, drying of food grains to safe moisture level is important. An integrated approach is the preferred strategy for managing mycotoxin contamination. The pre–or post-harvest strategy that is most appropriate will depend on the climatic conditions of that particular year. Understanding the environmental factors that promote infection, growth and toxin production is the first step towards an effective plan to minimize mycotoxin in food and feed. *Fusarium moniliforme* is the most common pathogen of maize. In maize, *F. moniliforme* is a seed-borne endophyte without symptoms, so its elimination is difficult. Tillage practices, crop rotation, weed control, late-season rainfall, wind and pest vectors can all affect the amount and source of the fungal inoculum that maintains the disease cycle in maize. Many of the new and exciting pre-harvest prevention strategies that are being explored involve production of genetically engineered resistant maize. Biocontrols using non-toxigenic biocompetitive agents are also a potentially useful strategy in maize. Post-harvest prevention of mycotoxin production is primarily dependent on good management practices before and after harvest. Detoxification strategies that are physical, chemical or microbiological can detoxify maize by destroying, modifying or absorbing the mycotoxins so as to reduce or eliminate the toxic effects.

In View the Work Carried Out the Package of Practices to Minimize Grains Molds and Mycotoxin Contamination

- ☆ Use recommended crop production practices.
- ☆ Plant early.
- ☆ Timely irrigation to reduce drought stress especially during flowering.
- ☆ Adopt practices to minimize the insect damage.
- ☆ Harvest crop in time.
- ☆ Minimize mechanical damage.
- ☆ Dry and store the grains properly at 13 per cent or less moisture.
- ☆ Use organic salts *i.e.* Sodium Carbonate, Sodium Bicarbonate, Propionic acid etc.
- ☆ Clean, segregate and destroy contaminated grains.
- ☆ Keep storage facilities clean.

References

Ahuja, S.C. and Payak, M.M. (1982). Symptoms and signs of banded leaf and sheath blight of maize. *Phytoparasitica,* 10: 41–49.

Ahuja, S.C. and Payak, M.M. (1984). Screening of maize germplasm for resistance to banded leaf and sheath blight. *India J. Agric. Sci.,* 54: 1019–1022.

Ahuja, S.C. and Payak, M.M. (1985). Comparative biology, pathology and karyology of rice and maize isolates of *R. solani* f.sp. *sasakii. Int. Rice Res. Newlt.,* 10: 5–6.

Ahuja, S.C. and Payak, M.M. (1988). Banded leaf and sheath blight of maize. In: *Perspectives in Mycology and Plant Pathology,* (Eds.) Agnihotri, V.P., Sarbhoy, A.K. and D. Kumar. Malhotra Publishing House, New Delhi, India, pp. 186.

Anonymous. (1974). *Plant Pathology, Annual Report*, 1973, Los Banos, Philippines, 105–106, 266 pp.

Anonymous. (2003). *Forty-sixth Ann. Prog. Rep.* of *Kharif* 2003. All India Coordinated Maize Improvement Project, IARI, New Delhi, pp. 1–62.

Anonymous. (1995). Maize at Pantnagar, Annual Report: Maize Pathology, AICMIP, pp 1–2.

Baicu, T., Sesan, T. and Gogoasa, C. (1989). Effectiveness of some *Viride* mutants tolerant to fungicides in biological control of some phytopathogenic fungi. *Probleme-de-Protectj, Planrelor*, 17: 205–218.

Bailey, B.A., Schuh, W., Frederiksen, R.A., Bockholt, A.J. and Smith, J.D. (1987) Identification of slow rusting resistance to *P. polysora* in maize inbreds and single cross. *Pl. Diseases*, 71: 518–521.

Barua, P. (1979). Studies on some aspect of banded sclerotial disease of maize caused by *Rhiznctoni solani* f. sp. sasakii. *MSc. Thesis*, G.B. Pant Univ. of Agric. and Tech., Pantnagar, 56 pp.

Barua, P. and Lal, S. (1981). Host range of *Rhizoctonia solani* f.sp. *sasakii*, the incitant of banded sclerotial disease of maize. *Indian Phytopath.*, 35: 494–496.

Bertus, L. (1927). A sclerotial disease of maize (*Zea mays* L.) due to *Rhizoctonia solani* Kuhn. *Year Book, Deptt. Agric.*, Ceylon, pp. 44–48.

Brunellii, K.R., Silva, H.P. and Camargo L.A. (2002). Mapping of resistant gene to *Puccinia polysora* in maize. *Fitopathol. Bras.*, 27: 2.

Butchaiah, K. (1977). Studies on banded sclerotial disease of maize. *MSc. Thesis.* G.B. Pant Univ. of Agric. and Tech., Pantnagar, 65 pp.

Cramer, H.H. (1967). *Plant Protection and World Crop Production*. Bayer, Leverkusen, Germany.

Deko, M.B., Rapior, S., Visconti, A. and Schjoth, J.E. (1995). Incidence and level of fumonisin contamination of maize genotypes grown in Europe and Africa. *Journal of Agriculture and Food Chemistry*, 43: 429–434.

Frederekson, R.A., Craig, J. and Smith, J.D. (1990). Characterization of slow rusting component in maize inbreds and single crosses *Phytopathology*, 80: 1069.

Holland, J.B., D.V. Uhr, D. Jeffers and Goodman, M.M. (1998). Inheritance of resistance to southern corn rust in tropical-by-corn-belt maize populations. *Theor. Appl. Genet.*, 96: 232–241.

Hollier, C.A. and King, S.B. (1985). Effect of dew period and temperature on infection of seedling maize plants by *Puccinia polysora. Plant Disease,* 69: 219–220.

Hulbert, S.H. and Bennetzen, J.C. (1990). Genetic fine Structure of a complex rust resistance locus in maize. *Phytopathology*, 80: 1007.

James, L. Brewbaker and Aleksander D. Josue. (2005). Breeding and genetics of polysora rust resistance in maize. *The ASA–CSSA–SSSA International Annual Meetings* November 6–10.

James, W.C. (1981). Estimated losses in crops from plant pathogens. In: *Handbook of Pest Management in Agriculture, Vol. I,* (Ed.) D. Pimentel. CRC Press, Boca Raton, Florida, pp 79–84.

Jines, M.P., Balint-Kurti P., Robertson-Hoyt, L.A., Molnar, T., Holland, J.B. and Goodman, M.M. (2007). Mapping resistance to Southern rust in a tropical by temperate maize recombinant inbred topcross population. *TAG Theoretical and Applied Genetics*, 114: 4.

Kim, W.G., Cho, W.D. and Lee, Y.H. (1992). Anastomosis groups and pathogenicity of isolates:I of *Rhizoctonia solani* and *Rhizoctonia zeae* from corn. Role of fungi as frontiers of biosciences: *Proceedings of the Asian Mycological Symposium in Commemoration of the 20th Anniversary of Korean Society of Mycology*. October 1–4, Seoul, Korea, pp. 147–149.

Lee, S.B., Kim, J.G., Han, M.S. and Han, H.J. (1989). Studies on the occurrence of corn sheath blight caused by *Rhizoctonia solani*. Influence of growth season and cultural environment on the occurrence of sheath blight disease. *J. Korean Soc. Grassland Sci.*, 9: 174–178.

Lehman, P.S., Machado, C.C., Tarrago, M.T. and Correa, C.F. (1976). Evaluation of recommended soybean, maize and sorghum cultivars and soybean lines for resistance to dead patch. *Phytopathologica*, 2: 47–49.

Meena, R.L. Rathore, R.S. and Mathur, K. (2003). Efficacy of biocontrol agents against *Rhizoctonia solani* f. sp. *Sassakii* causing banded leaf and sheath blight of maize. *J. Mycol. Pl. Pathol.*, 33(20): 310–312.

Melching, J.S. (1975). Corn Rust, Type, races and destructive potential. In: *Proc. 30th Annu. Conn. and Sorghum. Res. Conf.*, p. 90–115.

Pandey, M.K. (1992). Studies on the banded leaf and sheath blight of maize. *MSc. Thesis*. G.B. Pant Univ. of Agric. and Tech., Pantnagar, 79 pp.

Payak, M.M. and Renfro, B.L. (1966). Diseases of maize new to India. *Indian Phytopath. Soc. Bull.*, 3: 14–18.

Quazi, M.Z. and Iqbal, F.C. (1970). Effects of temperature on the growth of *Rhizoctonia solani. K. Agric. Pakistan*, 21: 447–451.

Raid, R.N., Pennypocker, S.P. and Stevenson, R.E. (1988). Characterization of *Puccinia polysora* epidemics in Pennsylvania and Maryland. *Phytopathology*, 78(5): 579–585.

Sangam L., P. Barauh and Butchaiah, K. (1980). Assessment of yield losses in maize cultivars due to banded sclerotial diseases. *Indian Phytopathol.*, 33: 440–443.

Sharma, G. (1996). Integrated management of banded leaf and sheath blight caused by *Rhizoctonia solani* Kuhn. *M.Sc. Thesis*. G.B. Pant Univ. of Agric. and Tech., Pantnagar, 90 pp.

Sharma, R.C. and Hembram, D. (1990). Leaf stripping: A method to control banded leaf and sheath blight of maize. *Curr. Sci.*, 59: 745–746.

Sharma, R.C. and Rai, S.N. (1998). Chemical control of banded leaf and sheath blight of maize. *Indian Phytopathol.*

Sherwood, R.T. (1970). Physiology of *Rhizoctonia solani*. In: *Rhizoctonia solani: Biology and Pathology*, (Ed.) J.R. Parmeter, Jr. Univ. Calif. Press, Berkeley, pp. 69–92.

Sinha, P.P. (1992). Chemical control of banded sclerotial disease of maize. *Orissa J. Agril. Res.*, 5: 54–57.

Shivkumar, G., Sharma, R.C. and Rai, S.N. (2002). Biocontrol of banded leaf and sheath blight of maize by peat based *Pseudomonas fluorescens* formulation. *Indian Phytopathol.*, 53(2): 190–192.

Srivastava, D.N. and Rao, V.R. (1964). Pythium stalk rot of corn in India. *Curr. Sci.*, 33: 119–120.

Thakur, M.S., Sharrna, S.L. and Munjal, R.L. (1973). Correlation studies between incidence of banded sclerotial disease and ear yield in maize. *Indian J. Mycol. Plant Pathol.*, 3: 180–181.

Ullstrup, A.J. (1965). Inheritance and linkage of a gene determining resistance in maize to an American race of *Puccinia polysora*. *Phytopathology*, 55: 425–428.

White, D. J. (1999). *Fungal Stalk Rots: Compendium of Corn Diseases* 3rd Edition, (Ed.) D.G. White. APS Press, St. Paul, M.N.

Wicklow, D.T. (1995). The mycology of stored grains: An ecological perspective. In: *Stored Grain Ecosystem,* (Eds.) Jayas, D.S., White, N.D.G. and W.G. Muir. Marcel Dekker, New York, pp. 197–249.

Zummo, N. (1988). Component contributing to partial resistance in maize to *Puccinia polysora. Plant Dis.,* 27: 157–160.

Plant Diseases Management for Sustainable Agriculture *Pages 89–96*
Editor: **Dr. Shahid Ahamad**
Published by: **DAYA PUBLISHING HOUSE**

Chapter 8

Integrated Disease Management on Rapeseed-Mustard

Shahid Ahamad[1], Pratibha Sharma[2] and Brajesh Kumar[3]
[1]*RARS, SKUAST–J, Rajouri – 185 131*
[2]*Division of Plant Pathology, IARI, New Delhi – 110 012*
[3]*SDS, Post Graduate College, Chandeshwar, Azamgarh, U.P.*

India is a major oilseed growing country covering more than 7 million hectare. Oilseeds holds a key position in Indian economy and contributed 4 per cent of India's GDP and 10 per cent of value of all Agricultural commodities during last decade (Anonymous, 1997).

Among nine oilseed crops, rapeseed and mustard is the second most important group of oilseed crops producing one fifth of world's rapeseed and mustard. These crops account for 30 per cent of the total oilseeds production and 13 per cent of the country's gross cropped area. There has been phenomenal growth in rapeseed and mustard production and productivity during the last 12 years. But still there is a big gap to reach the achievable productivity levels. The lack of judicial disease management programme is considered to be one of the most important factor in limiting the productivity of oil seed crops. The total loss of crop in India is estimated about 18-25 per cent : out of which 26 per cent accounts due to plant diseases. In case of alternaria blight of rapeseed and mustard, it was found that the loss in seed yield ranges from 24.5-71.5 per cent depending upon the severity of pod infection (Singh and Bhowmik, 1981). Yield loss of about 10-70 per cent have been reported in *Brassica juncea* due to Alternaria blight and white rust diseases (Kolte, 1985).

Diseases of Rapeseed-Mustard and their IDM Approaches

More than twenty diseases are known to attack the rapeseed–mustard in India but only five diseases namely alternaria blight, white rust, downey mildew, powdery mildew and sclerotinia stem rot (Table 8.1) are considered to be of major consequences because of their wide distribution and heavy yield losses (Degenhardt *et al*,1974). The typical symptom, epidemiology, mode of perpetuation and their IDM strategies was discussed in the following chapter.

Table 8.1: List of Major Fungal Diseases of Rapeseed-mustard

Name of Diseases	*Causal Organisms*
Alternaria blight	*Alternaria brassicae, A. brassicicola, A. parhani, A. alternata*
White rust/blister	*Albugo candida*
Downy mildew	*Perenospora parasitica*
Powdery mildew	*Erysiphe cruciferarum*
Slcerotinia stem rot	*Sclerotinia sclerotiorum*

Pathogen

The four *Alternaria* species *viz. Alternaria brassicae, A. brassicicola, A. raphani* and *A. alternata* can be easily distinguished by microscopic observation (Khalid *et al.*, 2004). The details of their morphology are given in Table 8.2 and Figures 8.1–8.3.

Table 8.2: Comparative Differentiation in Four *Alternaria* spp. Occurring on Rapeseed and Mustard

Morphological Characters: *A. alternata, A. brassicicola, A. brassicae, A. raphani*

Conidiophores Olive brown, erect, Pale mild olivaceous Mid-pale grayish, Olivaceous

simple, septate brown, simple, simple, erect, brown, simple

erect or curved, swellen at base, or occasionally

septate geniculate, septate branched, septate

Length 24.5-68.5mm 25-75mm 15-150mm 25-80mm

Width 3.50-7.25mm 1.5-7.5mm 5-10mm 3.5-7.5mm

Conidia

Conidia in chains up to 20 or more up to 20 or more Solitary or occasionally up to 3-4

Up to 4

Shape Polymorphic Cylindrical Obclavate, rostrate Obclavate or ellipsoidal

Colour Dark brown Pale to dark brown Olive or grayish Golden brown

olive to olivaceous

Cross septa 3-10 2-10 5-16 3-7

Longitudinal Septa 0-8 0-6 0-8 0-5

Length 8.50-53.50mm 20-120mm 40-150mm 20-60mm

Width 5.30-21.50mm 7-25mm 15-30mm 10-30mm

Beak

Length 1.0-45.5mm Non-exist, stunt 10-130mm 2.5-25mm

Width 1.5-6.5mm–3-8mm 2.0-6.5mm

Cross septa 0-3–1-7 0-3

Alternaria Blight

It is the most destructive and wide spread disease of rapseed–mustard and is caused by four species of *Alternaria viz., A. Brassicae* (berk) Sace., *A. Brassicicola* (Schwein) Wiltsh., *A. raphani* Groves and Solko and *A.alternata* (Fr.) Keissl.

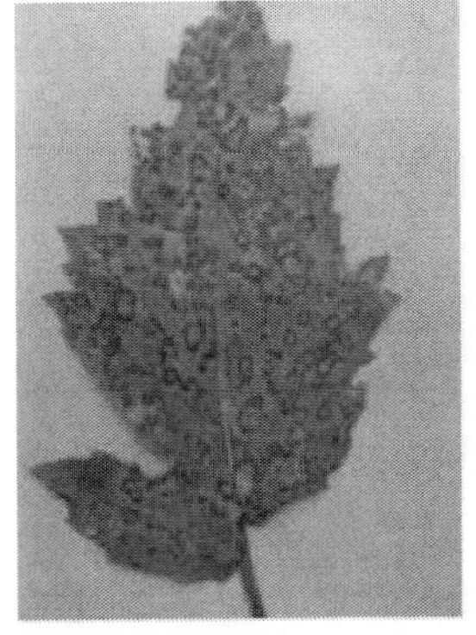
Brassica juncea

Brassica campestris var. toria

Alternaria infestation on stem

Brassica napus

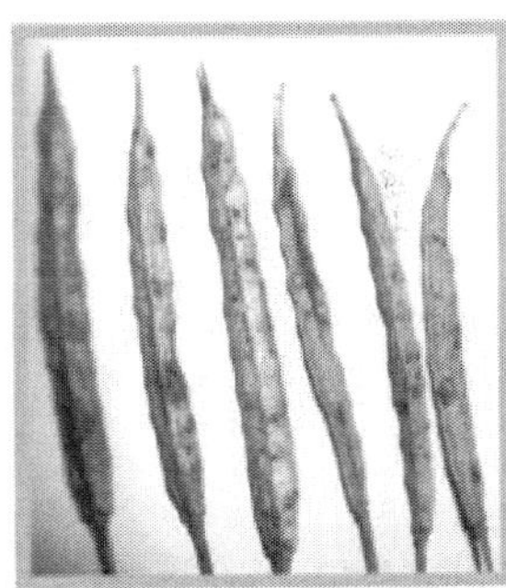
Alternaria spots on siliquae

Figure 8.1: Symptoms on *Brassica* spp.

Figure 8.2: Symptoms on Stems and Siliquae

Symptoms

The initial infection on lower leaves starts as minute brown to blackish lesions which later spread rapidly to the upper leaves, stems and silique. Lesions on leaves due to *A. brassicicola* are sooty black, velvety, whereas those caused by *A. brassicae* are grey, dense and those caused by *A. raphani* are small with raised margins and are surrounded by yellow translucent halo (Verma and Saharan, 1994). The lesions on stem are of linear in shape. Siliquae may show sunken, dark brown to black circular lesions. The severely infected siliquae contain small and shriveled seeds.

Epidemiology

A temperature range of 15-20° C, relative humidity of 70-90 per cent intermitten winter rains and wind velocity around 2-5 km per h has been reported to be conducive to alternaria blight development (Verma and Saharan, 1994). A minimum period of 4 h leaf wetness and reduced light intensity favours the lesion development. Closed spacing (30x15), higher dose of nitrogen (80 kg N/ha) and frequent irrigations are factors favoring the severity of the blight incidence (Saharan, 1991)

Perpetuation

The pathogens over winters on diseased crop residues seed and weed hosts.

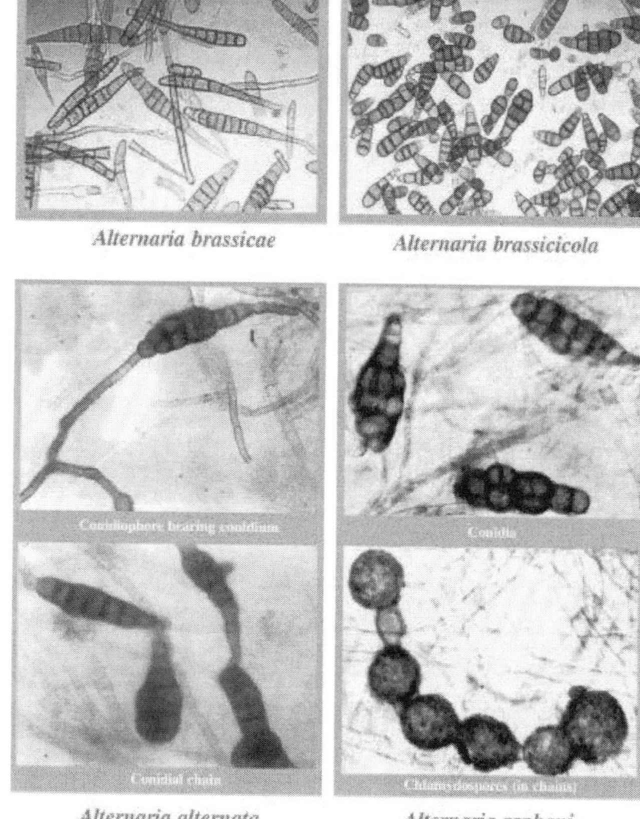

Figure 8.3: Morphological Characters of *Alternaria* spp.

White Rust

The white rust (caused by *Albugo candida*) is a major problem of mustard (*Brassica juncea*). It is estimated that the floral infection can cause losses from 23-54.5 per cent (Saharan *et al.*, 1984). The mixed infection of white rust and downey mildew is becoming a serious problem causing 37-47 per cent less pods with 17-32 per cent reduction in grain yield (Bains and Jhooty, 1979).

Symptoms

The initial symptom appears as small, localized, isolated white raised pustules or blisters on leaves, stem, petiole and floral parts. Later the pustules coalesce to form larger patches. Matured pustules rupture to release a mass of white powder of spores. In some cases pustule are surrounded by yellow hallow.The mixed infection of white rust and downy mildew is conspicuous at floral stage of crop growth where white rust produces pustules and downy mildew produces a coating of fine white powdery mass of spores.

Epidemiology

The optimum temperature for germination is about 10°C. The oospore germinates at 10-20°C. The placement of oospores in soil is an imporant factor in determining the time of infection. Placement of oospore at depth of 7.5 cm causes early infection while deeper (15cm) and shallow (on the surface) placement delays the infection in the host (Kumar *et al.*, 1995). The mixed infection of white rust and downey mildew is favoured by temperature between 13.8-14.8°C and rainfall more than 91.9 mm (Bains and Jooty, 1979).

Perpetuation

The pathogen perpetuates through oospore formed on diseased plant debris lying in the soil. The oospore can survive more than 21 years (Verma and Petrie, 1975). Perennial weeds host may also act as source of primary inoculam.

Downey Mildew

Downey mildew caused by *Peronospora parasitica* (Pers. Ex. Fr.), occurs in almost all the brassica growing areas of the country causing upto 75 per cent seedling deaths at cotyledonary infection. In case of mixed infection with white rust the pod yields losses is estimated about 37-47 per cent (Bains and Jooty, 1979).

Symptom

The symptom exhibits as yellow, irregular spots on the upper surface of the leaves and creamy white to light brown growth on the dorsal surface, opposite to spots. Severely diseased leaves dry up, shrivel and tear off easily. Stem swelling and malformation of floral parts are quite often found which is the result of mixed infection of downy mildew and white rust.

Epidemiology

The optimum temperature for formation and germination of conidia are 8-10° C and 8-12° C respectively. Severe infection and rapid disease development takes place when plants are exposed for 4-6 h leaf wetness at 20° C and 6-8 h at 24° C.

Perpetuation

Pathogen perpetuates in the soil through oospores which are formed in abundance in the malformed tissues of the infected plants. Seed borne inoculums may also serve as source of primary infection. Wild host and self grown plants are also reported to act as source of primary inoculums.

Powdery Mildew

Symptoms

The disease appears in the form of dirty white circular patches on both sides of lower leaves. Severely diseased plants appears as dusted with white grancular, chalk like powdery mass. Later the infected leaves, stem and siliquae show minute, circular black or dark brown spherical, scattered or concentrated bodies of perithecia (Saharan and Kausik, 1981).

Epidemiology

The temperature favouring germination of conidia, germ tube growth and appressorium formation is 20-25° C. It is reported that mean temperature of 16-28° C, mean relative humidity of less than 60 per cent and low or no rain fall are most congenial weather factors favouring epidemic development (Sahara and Sheoran, 1988).

Perpetuation

The pathogen survives in the form of perithecia on the diseased plant parts. Off season plants and weed hosts may also survive as source of primary inoculums.

Sclerotinia Stem Rot

Symptoms

Rotting of stem near the soil region. In severe conditions the whole plant topple down and dry off.

Epidemiology

Disease is favoured by humid weather.

Perpetuation

The pathogen can survice in the soil for several years as sclerotia.

Integrated Disease Management Approaches

Destruction of Diseased Debris

1. Diseased debris should be burnt after the harvesting of the crop.
2. Deep summer ploughing during the month of May–June to destroy the diseased debris.
3. Crop rotation with non cruciferous crops for atleast 3 years in case of severe infection.

Use of Resistant Varieties

1. *Alternaria blight*: The varieties PAB 93334, PAB 9511, RC 781, HC 9003, PBC 9221, PHR 1 and PHR 2 are found to be moderately resistant (Anonymous, 1997 and 1998).
2. *White rust and downy mildew*: The available resistant varieties are PWR 17, JYM 1, JYM 7 to 13, PWR 9538, PWR 9541, JMMWR 93-37, JMMWR 93-38, RH 8113, RH 9322, RH 8814, VSL 6, WDR 9601, PCR 10 and TERI (OE) (Anonymous, 1996,1997, 1998).
3. *Sclerotinia stem rot*: The resistant varieties for sclerotinia stem rot includes PCR 10, PRO AGRO- HO-1, RGN 8006 and RW 8401 (Anonymous, 1997 and 1998).
4. Use of bold seeded true certified seed.

Adoption of Proper Cultural Practices

1. Depending upon the weather condition and diseased severity prevailing at particular area, sowing of the crop can be delayed or advanced.
2. Plant spacing should be appropriate to maintain proper plant population.
3. First irrigation should be given 40-45 days after sowing.
4. Field should be inspected at regular intervals. Off–typed plants, diseased plant, alternate host and weed host should be roughed out.
5. Judicious application of fertilizer at proper time of crop growth phase.
6. In case of root- rot diseases, the movement of soil from diseased field to healthy field should be restricted.

Need Based Use of Chemical

1. *Alternaria blight*: Three sprays of Mancozeb (indofil M-45 0.25 per cent) at 60,80 and 100 days after sowing were found to be most effective in reducing the intensity of alternaria blight (anonymous, 1997).
2. *White rust*: Spray with Difolatan @ 0.2 per cent as soon as disease symptoms appears on the plant. Epidemic development can be checked by spraying Ridomil 0.25 per cent at 60 days followed by two sprays with Roveral 0.2 per cent at 80 and 100 days respectively.
3. *Downy mildew*: Seed treatment with Apron 35 @ 6g/kg seeds followed by Ist spray of Ridomil 0.25 per cent at 60 DAS and 2nd and 3rd spray of Roveral 0.2 per cent at 80 and 100 DAS (Anonymous, 1998). Spray with Zineb 0.2 per cent or Karathane 0.1 per cent for 2 to 3 times at 10 days intervals with the appearance of disease. Spray the crop with Ridomil MZ-72 @ 1.8-2.5 kg/ha in 750-1000 L of water against downy mildew in Toria as the disease starts appearing.
4. *Powdery mildew*: Spraying of Dinocap (Karathane 0.05 per cent) at 60,80 and 100 DAS was found most effective in minimizing the disease intensity (Anonymous, 1998).
5. *Sclerotinia stem rot*: Seed treatment with Bavistin @ 2.5g/kg when stem rot is a problem. Spraying of Roveral (0.2 per cent) or Bavistin (0.05 per cent) or Captan (0.02 per cent)were found quite effective (Anonymous, 1998).
6. Metasystox or Rogor can also be mixed with fungicidal solution @ 800-1000ml/ha for the control of aphids and other insects.

With the introduction of improved production technologies in the crop cultivation plant disease problem too have increased. Today it has become a challenge for farmers and Scientists to combat with the disease problem. If prior planning and suitable disease control strategies are adopted most of the disease can be managed so as to keep them below the economic threshold level. A crop may be affected by several diseases and it is not essential that a particular disease control method will be equally effective against all the diseases. Therefore, it is necessary to know the causal agent, mode of dispersal and disease cycle so as to develop such an strategy involving minimum activity and expenditure imparting complete or partial control of several disease simultaneously. Such a multi-disease management system can reduce the farmer expenditure and ease them in a application. It is essential for the success of IDM that it should be adopted in wide spread magnitude covering entire cropping areas. Since disease propagules spreads through wind, water, soil, insect- vectors etc. Therefore, if IDM is applied at community level it will be most effective with good economic returns.

Future Thrusts

Disease of rapeseed and mustard causes great destruction to the crop creating major economic loss. In spite of sufficient attention given to these diseases, no suitable method of its management has evolved so far. The only way to overcome the disease menace, to some extent, is the use of resistant varieties and crop rotation or intercropping. The discovery of the perfect state of the pathogen has provided substantial knowledge about the pathogens and has opened up new avenue for research on the disease and its casual organ ism. However, more efforts should be made to study:

1. Crop rotation, crop sequences and intercropping for natural suppression of the diseases.
2. Antagonism and biological control,
3. Survival of pathogen and factors adversely affecting the survival, and
4. Developing resistant varieties.

References

Anonymous (1996). *Annual Progress Report 1996*. All India Coordinated Research Project on Rapeseed–Mustard ICAR, Sewar, Bharatpur, Rajasthan, India.

Anonymous (1997). *Annual Progress Report 1997*. All India Coordinated Research Project on Rapeseed–Mustard ICAR, Sewar, Bharatpur, Rajasthan, India.

Anonymous (1998). *Annual Progress Report 1998*. All India Coordinated Research Project on Rapeseed–Mustard ICAR, Sewar, Bharatpur, Rajasthan, India.

Banis, S.S. and Jhooty, J.S. (1979). Mixed infection of *Albugo candida* and *Peronospora parasitica* on *Brassica juncea* inflorescence and their control. *Indian Phytopath.*, 32: 268–271.

Degenhardt, K.J., Kondra, J.P. and Skoropad, W.P. (1974). Effect of Alternaria black spot on yield, oil content and protein content of rapeseed. *Canadian J. Plant Sci.*

Kadian, A.K. and Saharan, G.S. (1983). Symptomatology, host range and assessment of yield losses due to alternaria brassicae infection in rapeseed and mustard. *Indian J. Mycol. Pl. Pathol.*, 13(3): 319–323.

Kolte (1985). *Diseases of Annual Edible Oilseed Crops*. CRC, Boca Raton, Florida, Vol. II, 135p.

Kumar, V., Kaushik, C.D. and Gupta, P.P. (1995). Role of various factors in development of white rust disease of rapeseed-mustard. *Indian J. Mycol. Pl. Pathol.*, 25: 145–148.

Saharan, G.S. (1991). Assessment of losses, epidemiology and management of black spot disease of rapeseed- mustard. In: *Proc. GCIRC 8th int. Rapeseed Cong.*, Saskatoon, Canada, 2: 465–470

Saharan, G.S and Kausik, J.C. (1981). Occurrence and epidemiology of powdery mildew of *Brassicas. Indian Phytopath.*, 34: 54–57

Saharan, G.S., Kausik, C.D.,Gupta, P.P.and Tripathi, N.N.(1984). Assessment of losses and control of white rust of mustard. *Indian Phytopath.* (Abstr.), 14: 11.

Saharan, G.S. and Sheoran, B.S. (1988). Persistance and efficiency of some common fungicides against Alternaria brassicae (Berk). Sace., the causal agent of leaf blight of rapeseed and mustard. In: *3rd Int. Symp. Indian Phytopath. Soc.*, December 14–18, 1981.

Verma, P.R. and Petire, G.A. (1975). Germination of oospores of *Albugo canadia. Canadian J. Bot.*, 53: 836–842.

Verma, R.P. and Saharan, G.S. (1994). Monograph on Alternaria disease of crucifers research. *Branch Agri. Food. Canada, Saskatoon Research Centre, Tech. Bull.*, 6E: 162.

Plant Diseases Management for Sustainable Agriculture *Pages* **97–109**
Editor: **Dr. Shahid Ahamad**
Published by: **DAYA PUBLISHING HOUSE**

Chapter 9

Fungal Diseases of Apple and their Management

***D. Gupta*[1] *and P.K. Sharma*[2]**

[1]*Department of Plant Pathology, Dr. Y.S. Parmar University of Horticulture and Forestry, Solan (H.P.)*
[2]*Regional Horticultural Research Station, Bhaderwah, Distt. Doda, SKUAST, Jammu*

Introduction

Apple (*Malus domestica* Borkh) is a major fruit crop of temperate regions of the world and is being grown successfully in the Indian Himalaya up to 3000m above mean sea level. Though, the apple production has increased, but the productivity per unit area is still quite low. Diseases are one of the factors contributing towards low productivity. With intensive cultivation, this fruit crop has been attacked by a number of diseases. Among various diseases affecting apple, the economically important ones caused by fungi are scab, powdery mildew, Marssonina blotch, white root rot, collar rot etc. Apple scab and Marssonina blotch are most destructive due to mid season defoliation. The diseases mentioned above have been managed well by following a combination of different methods like cultural practices, use of biological control agents and use of need based chemicals.

Apple Scab

Apple scab is a destructive disease which makes the trees unproductive and weak by mid season defoliation. The fruit lose their market value due to their deformed, scabby and knotted appearance. In India, the disease was first recorded in Kashmir Valley in 1930(Nath, 1930) and by 1973 it had finally firmly established in most of the apple plantations of the state. During 1977, scab was also detected in Himachal Pradesh (Gupta, 1978) and in 1983; it appeared in severe epiphytotic form resulting in huge crop loss (Gupta, 1989). The disease is present in all the apple growing regions of the country.

Symptoms

The most striking symptoms appear on leaves and fruits. However, symptoms are sometimes also observed on petioles, fruit stalk and blossoms. New and tender leaves in spring and summer are most susceptible to infection than the mature ones. Symptoms develop primarily on the blossom cluster leaves because these are the first tissue to emerge from the buds in the spring. Once the cluster leaves have unfolded and terminal leaves begin to develop, infections become evident on the upper surface of the leaves. Light brown to olive green spots which turn mousy black with age appear on the leaf lamina. At first, the margins of lesion are feathery and indefinite, but later on they become distinct. Severe spotting results in distortion and dwarfing of leaves and causes premature yellowing and defoliation. Lesions on young fruits initially resemble those on leaves, but later turn dark brown to black and corky. Severe early infection results in misshapen, knotty fruits. Infections that occur just before harvest may be symptom less at picking, yet develop into storage scab lesions after harvest.

Causal Organism

The apple scab is caused by a fungus *Venturia inaequalis* (Cke.)Wint.,with the imperfect stage as *Spilocea pomi* Fr. The fungus grows sub–cuticularly on host with little or no deleterious effect on the epidermal cells. *V. inaequalis* produces ascocarps (pseudothecia) in a stoma in over wintered leaves or fruit on the orchard floor. The pseudothecia are dark brown to black, spherical and have short beak and a distinct ostiole. Asci about 50 to 100 per pseudothecium, are fasciculate, cylindrical, short stipulate, and each contain 8 spores. Ascospores are yellowish green to tan and unequally two celled, with the upper cell shorter, wider than the lower cell.

Disease Cycle

Primary Cycle

In winter, the scab fungus over winters in infected leaves fallen on ground. As the temperature rises above 10°C in late January and February, the fungus enters sexual stage and produces microscopic, black pimple like structures called pseudothecia or perithecia embedded in leaf tissue. Each pseudothecium contains 100 to 150 microscopic scab infections on new growth. Good snow coverage and alternating periods of wetness and dryness in late winter and early spring favour the development of pseudothecia. Prolonged dry winter does not allow normal development of pseudothecia. A few ascospores are mature at the time of bud break (green tip), and maturity progresses slowly until about the tight cluster stage of blossom development. After this time, the percentage of mature spores begins to increase rapidly whenever temperatures are favourable for tree growth. Most ascospores have matured by the end of bloom. Once mature, they are forcibly ejected from the pseudothecia upon wetting of leaves on the orchard floor and may be then wind dispersed with in the orchard. Air currents carry them to the emerging tissues where infection takes place. Maximum ascospore discharge occurs within 30 minutes of wetting and complete discharge requires about 2 hours. Maturation and discharge of ascospores usually lasts 3 to 9 weeks. Germination begins as soon as a spore lands on new, green leaves, flower sepals or fruits, provided a water drop or film of moisture is present. The spore on germination produces a germ tube with appresorium at the tip, and a peg like structure emerging out of appresorium punctures the cuticle of plant tissue to start primary infection. Usually 9 to 17 days are required from inoculation to appearance of the olive green, velvety lesions.

Secondary Cycle

Large number of secondary spores (conidia) is formed with in primary scab lesions. Conidia are washed from lesions by rain and are spread to other tissues, where under appropriate environmental conditions, they cause secondary infection. The spread of the disease continues with each shower of rain through out the summer and in early autumn. Conidia production begins before lesions are apparently visible to the naked eye. Conidia development takes place at temperatures ranging from 4°C to 28°C and maximum number of conidia is produced between 60 to 100 per cent relative humidity. As such, high humidity and moderate temperature favours conidia production in scab lesion. Wetting is not required for spore formation. Many secondary cycles of the conidial occur during the growing season, and scab lesion produces thousands (50,000 to 1,00,000) of conidia. Conidia are then the principal source of disease build up during the summer and early autumn. Atmospheric humidity unto 90 per cent and temperature around 16 to 20 per cent favour conidial production.

Epidemiology

Weather is the most important factor in the development of apple scab, which is particularly severe in areas of high rainfall and relative humidity.

Development of Pseudothecia in Over Wintering Leaves

V. inaequalis perennates on the fallen leaves in the saprophytic phase. This phase starts with entry of the fungus hyphae into the leaf tissues. The mycelium from the subcuticular stroma proliferate into the palisade and mesophyll cells of the leaf and form a network of dark brown hyphae from which pseudothecia develop (MacHardy and Gadoury, 1985). Moisture, either rainfall or dew (Louw, 1951) is essential for the initiation and early development of pseudothecia which are initiated better in leaves kept alternately wet and dry. Besides moisture, temperature has a major influence on the growth of pseudothecium, pseudothecial number and rate of ascospore maturation (James and Sutton,1982).The thermal requirement of pseudothecia for different growth stages are different, 10°C for early development and 15 to 20°C for ascospore maturation. After the completion of dormant period, the asci and ascospores start developing within the ascus. Pseudothecia developed at low temperature of 6°C produce more asci than those at 12°C. In India, it was observed that lower temperature of 4°C was less favorable than 8°C for ascus development in the over wintered leaves but better than 15 and 20°C.

Discharge and Dissemination of Ascospores

Adequate moisture is essential for the discharge of ripe ascospores from the pseudothecia. In India, Gupta and Lele (1980) observed discharge of ascospores only under wet conditions. Wetting of pseudothecia just for 4 seconds by a thin film of water is enough for the release of a few spores, while such a process may start after 45 to 120 seconds of wetting if the pseudothecia are covered with a thick film of water (Brook, 1969). Humid and warmer conditions prior to wetting, result in the discharge of most of the mature ascospores, while during the night, few or no ascospores are discharged irrespective of wetting (Brook, 1969). The best conditions for ascospore discharge is a thin film of water over pseudothecia allowing asci to extend through ostiole and releasing the ascospores. Maturation and release of ascospores usually last for 3 to 11 weeks.

Climate and Infection

Ascospores after falling on susceptible host tissue, germinate under suitable temperature and moisture conditions. The conditions of temperature and leaf surface wetness duration for light, moderate and heavy scab infection have been well defined (Mills, 1944). It is referred to as "Mills period or table". Keitt and Jones (1926) were the first to investigate the relation of temperature and leaf wetness period to infection of apple leaves by ascospores under laboratory and green house conditions.

Ascospores on germination produce a germ tube, which produces an appressorium at the tip. A peg like structure emerging out of the appressorium punctures the cuticle of plant tissue to start primary infection. The fungus then grows beneath the cuticle and produces olive brown lesions within 2 to 4 weeks. Conidia produced on these lesions are mostly washed by rain or sometimes get detached by wind (Sutton *et al.*, 1981; Gupta *et al.*, 1984). The most critical period for the development of epidemics is from the time the buds start swelling till 2 to 4 weeks after petal fall, provided the weather remains wet.

Disease Prediction in Relation to Weather

Prediction of apple scab is mainly based on weather conditions *i.e.* Mills infection period, in addition to these host susceptibility and inoculum availability helps in forecasting of infection prior to the appearance of scab symptoms in spring. This can be used by growers to adopt suitable control measures.

Infection Period

In 1944, W. D. Mills determined the temperature and leaf wetting requirement for light, moderate and heavy infection by ascospores and conidia in the orchard and developed a static model called "Mills table". The interpretation of this table revealed that the minimum leaf wetness required for ascospores infection was 9 hours and for conidial infection 5.9 hours and symptom expression would take place within 9 days at optimum temperature range of 18.2 to 23.8 °C

Management

Three different approaches (1) eradication of primary inoculum by interrupting the life cycle of the scab pathogen; (2) protection against primary and secondary infection and; (3) resistance breeding have been applied for management of apple scab. In order to interrupt the life cycle of the pathogen to reduce the extent of primary inoculum in the spring season, orchard sanitation/or chemical spray can be employed for quick decomposition of leaves harboring the scab fungus. A spray of urea (5 per cent) in autumn shortly before leaf fall helps in decomposition of fallen leaves and prevents pseudothecial formation.

The intensity of apple scab can be substantially reduced by timely application of fungicides that are protective, curative or eradicants (Gupta, 1979). A protective spray programme being adopted in apple orchards in India include 6 to 7 spray of non systemic and systemic fungicides during the growing season and a single application of urea in autumn (Gupta, 1985). In curative spray programme, the fungicides are applied only after the occurrence of an infection period determined either by visual observations or predictive devices. Organic fungicides, which either belongs to dithiocarbamates or heterocyclic nitrogenous compounds do not provide curative activity against scab fungus. Among the benzimidazole compounds, benomyl may owe its activity as a curative fungicide to the fact that it is absorbed and translocated in the fruits. In recent years, benzimidazoles have been considered as the most effective curative fungicides in the control of apple scab in most countries.

Most of the EBI fungicides have excellent post infection curative activity against scab. These fungicides, particularly bitertanol, myclobutanil, prochloraz, penconazole and hexaconazole prevent the formation of visible lesions when applied 3 days after inoculation. Interruption in carrying on sprays at regular intervals in inclement weather may result in the establishment of scab fungus on foliage and fruits. To avoid this, fungicides having high level of eradicant or anti sporulant activity are applied. Benzimidazoles have this post symptom activity by almost completely inhibiting the germination of conidia in sporulating lesion. In India, scab resistant cultivars *i.e.*Prima, Priscilla, Sir

Prize, Jonafree, Redfree, Coop-12, Macfree, Liberty and Freedom were introduced in Himachal Pradesh (Sharma *et al.*, 1988).

Marssonina Blotch of Apple Large Dark Brown Blotches

The disease causes severe defoliation in apple plants, thereby, affecting fruit size, color and quality adversely. Development of fruit blotches on the produce makes it unsaleable in the market. It also affects tree vigor and fruit set in the following season. Its manifestation has been observed on all the commercial cultivars grown in India.

Symptoms

Disease appears both on leaves and fruits. Dark green patches develop on upper surface of leaves giving rise to 5 to 10 mm diameter brown leaf spots which turn dark brown in due course. Leaf spots initially occur on mature leaves present on lower portion of the plant. Small black acervuli are visible on the leaf surface. Various small and large lesions coalesce and the surrounding areas turn yellow. The yellowing advances towards petiole end and induce mid season defoliation. These symptoms are followed by severe leaf shedding in mid summer. In severely affected orchards, only fruits can be seen hanging on the defoliated trees. The mid season defoliation may be responsible for poor growth and development of fruiting buds and result into low productivity in ensuing years. Delicious cultivars have been found to be more prone to this disease.

Fruit infection is also not uncommon. Brown spots appear on the surface of the fruit. The spots are initially circular (4 to 5 mm in diameter) and become oval, depressed and dark brown with age and almost black at harvesting time. Small black acervuli are visible in the lesion. All the commercial cultivars *viz.* Red delicious, Royal delicious, Rich-a-red, Golden delicious and Red gold were found highly susceptible to this disease (Sharma and Gautam, 1997).

Causal Organism

The disease is caused by a fungus *Marssonina coronaria* (Ell. and Davis). The fungus produces asexual fruiting bodies called acervuli which are 100 to 200 µm in diameter, subcuticular and irregularly rupture the cuticle. Conidia are borne on small clavate conidiophores and are single septate, constricted at the septum, hyaline guttulate and 20 to 24 × 6Ì5 to 8Ì5 µm in size. Spermatia are produced in acervuli, solely or mixed with conidia and are rod shaped to elliptical, continuous, hyaline and 4 to 6 × 1 to 2 µm in size. *Diplocarpon rosae* is the perfect stage of the fungus (Harada *et al.*, 1974).

Disease Cycle

The fungus over winters in the form of apothecia on fallen leaves. Primary infections have been reported to be initiated by ascospores produced in apothecia on over wintered apple leaves (Harada *et al.*, 1974; Takahashi and Sawamura, 1990). Mature ascospore discharge begins just before bloom stage of bud development and the ascospores are released over a fairly long period in spring and summer months. Rain is necessarily required for spore release and infection. The secondary infection occurs more frequently through conidia. Air temperature, light intensity and soil moisture are the predisposing factors to trigger the disease outbreak, however, severity index increases with the maturity of leaves. Disease is severe when a cool, rainy period of 4 to 5 days proceeds several hot sunny days in the orchard.

Management

Disease control can be achieved through sanitation, adequate pruning and proper use of fungicides. As the disease is known to perennate in the infected leaves on the orchard floor, removal of over wintering leaf litter may reduce primary inoculum level. Proper pruning allows adequate air circulation in the tree canopy, thereby, modifying the microclimate and reduce the disease level. The orchard area should be kept free from weeds and bushes to avoid excessive humid conditions in the rainy season. Protective sprays of mancozeb (0.3 per cent), carbendazim (0.05 per cent), dodine (0.075 per cent), propineb (0.3 per cent), benomyl (0.05 per cent), thiophanate methyl (0.05 per cent) and fluquincoazole (0.02 per cent) provided complete control of the disease. However, hexaconazole (0.03 per cent) and bitertanol (0.05 per cent) are less effective.

Powdery Mildew

It is a major foliage disease of apple and prevalent in all apple growing countries of the world. It was first noticed on apple seedlings in Iowa, USA in 1871 (Bessey, 1871).

Symptoms

The disease infects leaves, blossom, twigs and fruits of most of the commercial cultivars. The symptoms first appear on lower surface of leaf as small whitish patches of fungal structures. The mycelium and powdery coating of spores soon become evident on the entire leaf. Infected leaves tend to crinkle, curl or roll upward along the edges giving them a narrow appearance. Severe infection leads to defoliation of mildewed shoot with subsequent curtailment of its growth and a reduction in the number of potential fruiting buds.

Infected terminal shoots are stunted with shortened internodes or show dieback symptoms with silvery appearance. Shoot infections are the typical result of over wintering infections of the fungus in the buds. When the terminals push in early spring, the fungus grows along with the new succulent growth and infects this tissue immediately. In some varieties, the tree shows stunted growth and typical russeting. In nurseries, the fungus is more destructive and spreads to all developing leaves leading to severe defoliation along with stunting of terminal growth.

Causal Organism

The disease is caused by *Podosphaera leucotricha*. It is an ascomycetous heterothallic fungus. Conidia are ellipsoidal, truncate and hyaline. They are produced in long chains of thin, effused, amphigenous mycelium. Perithecia are subglobose, dark brown and have apical and basal appendages.

Disease Cycle and Epidemiology

P. leucotricha is an obligate parasite and over winters in the form of mycelium in infected buds or as cleistothecia on the surface of infected twigs. Infected terminal and first three lateral buds serve as a principal source of inoculum (Stephan, 1991). Mycelia in infected buds produce conidia to initiate primary infections and these infections can be found as early as the tight cluster stage. There is often an abundance of susceptible tissue for the conidia to infect because infected buds break later than healthy buds. Conidia are disseminated by wind and can infect young leaf tissue, blossom and fruit. Conidial dispersal is low during night than day and is interrupted by heavy prolonged rains. Conidial germination takes place at temperature ranging between 5 to 30 °C with maximum at 20 °C and high relative humidity approaching 100 per cent favoured maximum germination (Sharma and Gupta, 1991). Dry weather favoured the rapid development of disease and wet conditions appear to restrict it.

Management

Disease management practices include the removal of dormant shoot tip and silvered area which causes reduction in primary inoculum, use of fungicides (Gupta and Verma, 1981), host resistance and biological control. In chemical control, first attempt to reduce powdery mildew infection in nurseries was made by sulfur dusting. Application of different fungicides including dinocap, carbendazim, benomyl, bitertanol and triforine has been recommended in India (Sharma *et al.*, 1992). Recently, the use of ergosterol biosynthesis inhibitor fungicides such as triadimefon, penconazole, hexaconazole, fenarimol, triforine, difenconazole and propiconazole was found effective.

Various crab cultivars namely, renetka, yantaka, dalgoe, karchemo and apple cultivars maharaja, chunth and golden Chinese have been found resistant to the disease in India (Verma and Gupta,1988). In biological control, attempts have been made to use the hyper parasite like *Cicnibolus casatii* and coccinelid insects like *Halyzia hayzeri* and *Vibidea duodecimguttata*.

White Root Rot

It is a disease of roots which remain covered with white fungal mycelial growth particularly in rainy season resulting in death of plants. This soil borne fungus has a very wide host range of around 173 plant species comprising of fruit plants, vegetables, forest trees and field crops. In apple, the disease was observed for the first time from Norwich in 1913 (Salman and Wormald, 1913). In India, the disease was first observed in Uttar Pradesh hills by Singh in 1939 (Singh, 1943).

Symptoms

The disease symptoms occur on lateral roots and the effects are noticed on above ground parts. The earliest above ground symptoms of the disease is bronzing of the leaves, reduction in their size and a stunted tree growth. Affected trees often persist for 2 to 3 years depending upon the severity of infection in roots. Root rot affected trees are usually associated with a heavy blossom and fruiting next year, however, in succeeding years, few leaves emerge and many immature fruits color early and fail to reach maturity. There is dieback/drying of twigs and branches. The under ground roots are completely rotten. The fibrous roots are attacked first and are completely devoured and infection spreads to main root through secondary root system. The lateral roots turn dark brown and the surface of roots may be covered with white cottony growth of the fungus. During rainy season white mycelial mat can be seen in the soil adhering to the roots.

Causal Organism

White root rot is caused by *Rosellinia necatrix* (Anamorph: *Dematophora necatrix* Hartig). The perfect stage of the fungus is not known to occur in India. Diagnostic feature of *D. necatrix* is the pear shaped swellings at each septum of the fungus hyphae. The fungus produces scattered black, microsclerotia measuring 130×98 µm and does not infect xylem tissues of apple roots, but bark is completely damaged.

Disease Cycle and Epidemiology

The fungus survives either in the form of mycelium or sclerotia in the infected roots. Infection of new roots takes place by the fungal mycelium present in the soil on debris by the contact of new plant roots with old dead roots. The fine roots are first to be infected by direct penetration. The fungus produces pectinolytic and cellulolytic enzymes which helps in the degradation of host tissues in advance and also toxins which help in killing the bark. With the ingress of the pathogen, it reaches the main root and ultimately complete root system is devoured. The mycelium spreads from upper side to lower side of the roots or from infected to healthy roots through infested soil and rain water. The disease severity is positively correlated with moderate temperature (15 to 25°C) and high moisture. The disease is most serious in water logged soils. Soil pH also seems to have a strong relationship with the survival of the pathogen. It flourishes best in 6 to 6.5 pH range.

Management

The measurement of soil borne diseases is difficult because of deep seated infection. White root rot of apple is best controlled by following preventive as well as curative measures consisting of cultural, biological, chemical methods and resistant rootstock. The cultural practices recommended for the control of white root rot are digging isolation trenches, improving the drainage of orchard by not allowing water to stagnate in the tree basins and removal of rotten roots followed by application of disinfectant paste. Soil solarization has been found to be effective in reducing the fungus inoculum in soil (Freeman *et al.*, 1986). The acidic soils should be amended by applying lime for some years (Jain, 1961).

The success of controlling soil borne diseases lies in the fact that the fungicide should reach the point of infection in the soil. The systemic fungicides such as benomyl, aureofungin and carbendazim are widely used against root rot. The fungicide should be applied through deep holes (15 to 20 cm) made in the basin of the tree at 30 cm distance from each other. Before the application of chemical, the tree basin is brought to a soil moisture level of 30 to 40 per cent. The treatment is done 3 to 4 times every year during rainy season. Sousa (1985) observed preventive as well as curative effects of benomyl and carbendazim and reported that treatment of diseased plants in winter further improved control. Some degree of resistance against the pathogen has been observed in the rootstocks namely MM109, M16 and MM104. Use of various antagonists in the management of root rot has been explored and few antagonists such as *Trichoderma viridae*, *T. harzianum* along with soil solarization have been found to protect the plants from root rot infection.

Collar Rot

The disease was reported for the first time by Baines (1939) on Cox orange variety of apple. It causes extensive losses and sometimes results in death of trees.

Symptoms

The above ground symptoms of collar rot are often confused with white root rot. The disease starts from the collar region and spreads mostly to the under ground parts. The above ground stem portion is also in highly susceptible scion cultivars. Bark at the soil level becomes slimy and start rotting which ultimately results in cankered areas. The wounds are irregular in outline, but usually roughly oval which extends in lateral and vertical directions, often resulting in girdling of the tree. Trees attacked by this disease may be recognized at a distance by the lighter foliage. The attacked trees have chlorotic foliage which appears only on the branches above the cankered area, while remainder of the tree may appear normal and may bear a crop for several years.

Causal Organism

The disease is caused by *Phytophthora cactorum*, a soil inhabiting fungi. *P. cactorum* is semi aquatic, soil borne, homothallic fungus with a complex life cycle consisting of an asexual phase in which motile zoospores are produced from sporangia and a sexual phase resulting in the formation of oospores. The sporangia are papillate, borne on sympodially branched sporangiophores. The sporangia are 50 µm in length and produce about 40 to 90 zoospores. The antheridia are paragymous. Oogonia are mostly 25 to 32 µm in diameter.

Disease Cycle and Epidemiology

The fungus is known to survive in the orchard soil and a soil temperature of 12 to 20°C and soil pH 5 to 6 is best for its spread. The fungus survives as chlamydospores in plant debris or soil may colonize fallen apple fruit on the apple floor. Majority of infections of the stem occur through mycelia penetration near the ground line, but zoospore infection is also quite common since these are liberated in soil. Fungus usually produces oospores which serve as the source of primary inoculum. Host penetration occurs through unwounded bark, cracks, wounds and burns on collar region. The fungus grows both inter and intracellularly through the bark tissues. Soil temperature of 20 to 25 °C along with high soil moisture is favorable for sporangia production, but oospores are produced at low moisture level at the same temperature. Flooded soil predisposes collar region to infection and colonization of bark tissues is greatest between pink bud stage of tree phenology and beginning of shoot elongation (Gupta, 1996).

Management

The pathogen gets entry through the graft union, therefore, the graft point must be kept about 30 cm from soil level. General orchard management practices such as improved drainage around the tree base, removal of crop refuge and avoiding injury to collar region are helpful in restricting the disease spread (Gupta and Rana, 1996).

The disease can also be controlled by drenching with mancozeb(0.3 per cent) or blitox/fytolan (0.5 to 1 per cent) in 30 cm radius around the tree trunk (Rana and Gupta, 1981). Systemic fungicide like ridomil MZ is also effective both as soil drench and paint (Gupta, 1990). Rootstocks such as M2, M4. MM110, MM114 and crab apple have been reported to possess high degree of resistance against *P. cactorum* under Indian conditions. Utkhede and Smith (1990) identified the isolates of *Trichoderma* and *Gliocladium* with potential for biocontrol of *P. cactorum*.

Seedling Blight

This disease is found in nurseries and 2 to 3 year old seedlings are attacked. In India, the disease was first reported on apple seedlings from Mysore state(Anonymous, 1939) and in Himachal Pradesh by Jain(1962). Sclerotium diseases occur primarily in warmer climates.

Symptoms

Upon infection of roots, the leaves of affected plants start wilting and show a characteristic reddish or grayish purplish discoloration. Ultimately blighting of foliage of infected seedlings is the common symptom. In the vicinity of dead seedlings, the fungus produces numerous small roundish sclerotia of uniform size that are white when immature, becoming dark brown on maturity.

Causal Organism

The disease is caused by *Sclerotium rolfsii* Sacc. The fungus grows on a wide range of media and produce round mustard seed size, tan to dark brown sclerotia. The asexual spores are not produced. *Aethalium rolfsii* is the sclerotial stage of S. *rolfsii*. The sclerotia have a long but variable survival period and germinate eruptively to infect plant directly.

Disease Cycle and Epidemiology

The fungus over winters mainly as sclerotia and serves as a source of primary inoculum in the nursery. The sclerotia germinate better in light and well aerated soils and attack tissues directly. Once established in plants, the fungus advances and produces mycelium and sclerotia quite rapidly, especially at high temperature. A soil temperature of 30 to 33 °C, pH 6 and soil moisture of 38 per cent or above have been found to be most suitable for disease development (Bhardwaj and Agarwala, 1986).

Management

Control of *Sclerotium* diseases is difficult. Crop rotation, cultural practices such as deep ploughing to bury sclerotia in surface debris and application of fungicides to the soil is recommended. Affected seedlings in the nursery beds are treated with thiram and aureofungin. Water logging in nursery beds should be avoided as the disease is aggravated by high soil moisture. The nursery site must be rotated every 4 to 5 years.

Sooty Blotch and Fly Speck

Sooty blotch and fly speck are two of the most common diseases of apple in many temperate areas of the world. The two diseases have been mentioned in horticultural literature since the early 1800s, but it was not until 1920 that A. S. Colby determined that they were caused by two distinct fungi. Sooty blotch and fly speck cause a discoloration or blemish of near-mature fruit. The discoloration is superficial, and while neither disease actually damages the fruit, the presence of disease reduces the grade and market value of the fruit. Both diseases are most common during years with a cool, wet spring, rains in late summer and low temperatures in early fall.

Symptoms

Sooty blotch manifests as brown or olive green colored spots. These spots are smudgy looking in appearance and manifest anywhere on the surface of the apple fruit. Brown to dull black, sooty blotches with an indefinite outline form on the fruit surface. Blotches may be l/4 inch in diameter or larger. The blotches may coalesce to cover practically the entire fruit. The sooty blotch fungus is restricted to the outer surface of the fruit, and in many cases the blotches can be easily rubbed off.

Fly speck usually appears as anywhere from ten to thirty clusters of raised spots which can appear anywhere on the surface of the apple fruit. The variable cluster of raised spots are black and tends to be very shiny and reflective compared to the rest of the surface of the apple fruit.

Causal Organism

Fly speck in apples is caused by the *Microthyriella rubi* fungus, while sooty blotch is caused by the *Gloeodes pomigena*. Both *Gleodes pomigena* and *Microthyriella rubi* are more often found together infecting apple trees rather than found individually.

Disease Cycle

Both fly speck and sooty blotch over winter on apple tree twigs, branches and mummies, or fallen fruits that have mummified. The fungi remain dormant throughout the winter until springtime when wet weather erupts in conjunction with mild temperatures. Sooty Blotch results from the presence of hundreds of minute, dark fungal fruiting bodies (pycnidia) that are interconnected by a mass of loose, interwoven dark hyphae. In spring, pycnidia on wild plants produce large numbers of spores (conidia) that ooze out and collect in a gelatinous mass. The conidia are then spread by water splash or wind blown mists into orchards from late May or early June until fall. The fungus first affects apple twigs, and then secondary colonies are initiated on the fruit. Cool, humid weather (18°C) is essential for disease development. The disease does not develop when temperatures reach 30 °C. When May and June are cool and moist and are followed by a hot July and August, sooty blotch symptoms often do not appear for two to three months. The disease is absent or rare when hot, dry weather prevails until close to harvest time. The disease is most severe when cool, rainy weather in the spring is coupled with late summer rains and low temperatures in early fall. Under ideal conditions, the incubation period from the time the fungus reaches the fruit to the appearance of symptoms may be as short as five days. In the orchard, the incubation period usually lasts three to four weeks on fruits that are 42 to 45 days old.

Primary infections of flyspeck are initiated by airborne spores, discharged during rainy periods by the over wintering stage of the fungus from just before bloom to one to two months after bloom. The optimum temperature for the germination of the primary spores is 20°C, although they will germinate in temperatures from 16-28°C. High humidity is important for the development of flyspeck. Under optimal conditions, colonies become visible in 10-12 days after infection. In the orchard, at least three weeks of incubation are usually needed before symptoms become apparent. Secondary spread of flyspeck is by airborne spores, which are produced abundantly on the fruit. The optimum conditions for secondary spore production is a temperature of approximately 17°C and relative humidity greater than 96 per cent.

Management

Infected apples are edible, despite being unsightly. To remove the smudges, and spots, rub the apple fruits vigorously either up and down on a cloth or in a circular motion. Select an orchard site that always has full sunlight, good air circulation, and good soil (water) drainage. Prune trees annually to an open center for maximum air circulation. Both diseases are most prevalent in the damp, low shaded areas of the orchard. Any practice that opens up the trees to greater air movement and promotes faster drying greatly aids in control. Remove or destroy nearby wild or neglected apple trees. Especially in commercial plantings, fungicide sprays are important for controlling these diseases.

References

Anonymous. (1939). Some diseases of apple. *Mysore Agric. Col.*, pp. 25–33.

Baines, R.C. (1939). Phytophthora trunk canker or collar rot of apple trees. *J. Agric. Res.*, 59: 159–184.

Bessey, C.W. (1871). On injurious fungi: the blight (Erysiphae). *Iowa State College of Agric. Bienn. Rep.*, pp. 185–204.

Bhardwaj, L.N. and Agarwala, R.K. (1986). Effect of some important edaphic factors on the incidence of seedling blight of apple caused by *Sclerotium rolfsii* Sacc. In: *Advances in Research on Temperate Fruits,* (Eds.) Chadha, T.R., Bhutani, V.P. and Kaul, J.L.. UHF, Nauni, pp. 403–406.

Brook, P.J. (1969). Effects of light, temperature and moisture on the release of ascospores by *Venturia inaequalis. N.Z. Agric. Res.*, 12: 194–198.

Colby, A.S. (1920). Sooty blotch of pomaceous fruits. *Trans. Ill. Acad. Sci.*, 13: 139–179.

Freeman, S., Sztejnberg, A., Chet, I. and Katan, J. (1986). Solar and biological control of white root rot disease in apple orchards caused by soil borne fungus *Rosellinia nectarix. Harradeh,* 66: 1608–1613.

Gupta, G.K. (1978). Present status of apple scab in Himachal Pradesh and strategy for its control. *Pesticides*, 12: 13–14.

Gupta, G. K. (1979). Role of on season, post harvest and preleaf fall sprays in the control of apple scab. *Indian J. Mycol. Pl. Pathol.*, 9: 141–149.

Gupta, G.K. 1985. Recent trends in forecasting and control of apple scab. *Pesticides*, 19: 19–31.

Gupta, G.K. (1989). Apple scab: Economics and technology. *Pesticides*, 23: 33–39.

Gupta, G.K. (1990). Major temperate fruit diseases in India and their management strategies. *Plant Dis. Res.*, 5: 136–142.

Gupta, G.K. (1996). Fungal diseases of temperate fruits in India. In: *Advances in Diseases of Fruit Crops in India*, (Eds.) S.J. Singh. Kalyani publishers, Ludhiana, 492 pp.

Gupta, G.K. and Lele, V.C. (1980). Morphology, physiology and epidemiology of the apple scab fungus, *V. inaequalis* in Kashmir Valley. *Indian J. Agric. Sci.*, 50: 51–60.

Gupta, G.K. and Verma, K.D. (1981). Influence of cultural and chemical practices on the incidence of powdery mildew of apple. *Indian J. Mycol. Pl. Pathol.* 10: 60 (Abst.).

Gupta, G.K., Verma, K.D. and Pal, J. (1984). Some observations on the prevalence and severity of apple scab in Himachal Pradesh during the year 1978 to 1983. *Indian J. Mycol. Plant Pathol.*, 14: 12–14.

Gupta, V.K. and Rana, K.S. (1996). Effect of soil factors and amendments on the viability of *Phytophthora cactorum* sporangia and oospores in apple orchard soil. *J. Tree Sci.*, 1: 64–70.

Harada, Y., Sawamura, K. and Kono, K. (1974). *Diplocarpon mali* sp. nov, the perfect stage of apple blotch fungus *Marssonina cronata. Ann. Phytopathol. Soc. Japnnzdling*, 40: 412–418.

Jain, S.S. (1962). Studies on apple blight *Sclerotium rolfsii* Sacc. In: *Himachal Pradesh Proc. Indian Sci. Cong.*, 3: 247.

Jain, S.S. (1961). Root and collar rot diseases in Himachal Pradesh. *Himachal Hortic.*, 2: 19–23.

James, D.R. and Sutton, T.B. (1982). A model for predicting ascospore maturation of *Venturia inaequalis. Phytopathology* 72: 1081–1085.

Keitt, G.W. and Jones, L.R. (1926). Studies of the epidemiology and control of apple scab. *Wisc. Agric. Exp. Stn. Res. Bull.*, 73: 104.

Louw, A.J. (1951). Studies on the influence of environmental factors on the over wintering and epiphytology of apple scab in the winter rainfall area of Cape Province. *S. Afr. Dep. Agric. Sci. Bull.*, 310: 40.

MacHardy, W.E. and Gadoury, D.M. (1985). Forecasting the seasonal maturation of ascospores of *Venturia inaequalis. Phytopathology*, 75: 381–385.

Mills, W.D. (1944). Efficient use of sulfer dusts and sprays during rain to control apple scab. *Cornell Univ. Ext. Bull.*, 630: 4.

Nath, P. (1930). Studies on the diseases of apples in Northern India. II. A short note on apple scab due to *Fusicladium dendriticium. J. Indian Bot. Sci.*, 14: 121–124.

Rana, K.S. and Gupta, V.K. (1981). Cause and control of collar rot of apple in India. *Indian Phytopath.*, 34: 17–19.

Salman, E.S. and Wormald, H. (1913). Report on economic mycology. *J. South East Agric. Coll.*, 22: 453.

Sharma, J.N. and Gautam, D.R.(1997). Studies on premature leaf fall in apple: A new problem. *Indian J. Plant Prot.*, 25: 8–12.

Sharma, R.L., Kumar, J. and Ram, U. (1988). Performance of scab resistant cultivars of apple in Kullu Valley, Himachal Pradesh. *J. Tree Sci.*, 7: 45–49.

Sharma, S.K. and Gupta, V.K. (1991). Effect of temperature and relative humidity on conidial germination and germ tube length of *Podosphaera leucotricha* on glass slide. *Pl. Dis. Res.*, 6: 91–95.

Sharma, S.K., Gupta, S.K. and Kaith, D.S. (1992). Efficacy of funicides in relation to number of sprays against powdery mildew of apple. *Indian J. Mycol. Pl. Pathol.*, 22: 267–269.

Singh, U.B. 1943. Control of fruit diseases in Kumaon. *Indian Farm.*, 4: 411–412.

Smith, V.L., Wilcox, W.F. and Harman, G.E.(1990). Potential for biological control of phytophthora root and crown rot of apple by *Trichoderma* and *Gliocladium* spp. *Phytopathology*, 80: 880–885.

Sousa, A.J.T.D. (1985). Control of *Rosellinia necatrix* causal agent of white root rot, susceptibility of several species and chemical control. *European J. Plant Pathol.*, 15: 323–332.

Stephan, S. (1991). Studies on primary and secondary shoot infection with apple powdery mildew. *Arch. Phytopathol. Pflanzenschutz*, 27: 181–192.

Sutton, T.B., James, J.R. and Nardacci, J.F.(1981). Evaluation of a New York ascospore maturity model for *Venturia inaequalis* in Northern Carolina. *Phytopathology*, 71: 1030–1032.

Takahashi, S. and Sawamura, K.(1990). Marssonina blotch. In: *Compendium of Apple and Pear Diseases*, (Eds.) Jones, A.L. and H.S. Aldwinckle. APS Press, St. Paul Minnesota, pp. 33.

Utkhede, R.S. and Smith, E.M.(1990). Field resistance of apple rootstocks to *Phytoththora cactorum* infection. *J. Hort. Sci.*, 69: 467–472.

Verma, K.D. and Gupta, G.K.(1988). Field reaction of apple germplasm to powdery mildew. *Indian J. Agric. Sci.*, 58: 233–234.

Plant Diseases Management for Sustainable Agriculture *Pages* **110–120**
Editor: **Dr. Shahid Ahamad**
Published by: **DAYA PUBLISHING HOUSE**

Chapter 10

Diseases of Walnut and their Management

J.N. Srivastava[1], P.K. Sharma[2], A. Mishra[3] and Shahid Ahamad[4]

[1]*Regional Horticulture Research Station, Bhaderwah, Distt. Doda, SKUAST-Jammu*
[2]*ICAR Research Complex for North Eastern Hill Region, Manipur Centre, Imphal*
[3]*Advance Centre for Plant Virology, IARI, New Delhi*
[4]*RARS, SKUAST-J, Rajouri – 185 131*

The Persian or English walnut (*Juglans regia* L.) is one of the most important temperate nut fruits of India. It is grown in Himalayan region of Jammu and Kashmir, Uttarakhand and Himachal Pradesh. It is not only consumed on large scale within the country, but also earns foreign exchange. The walnut has a tremendous export potential (Singh *et al.*, 2001). It is grown in marginal areas at altitudes 4000-7000 feet above sea level (Razdan and Puttoo, 2005). The good quality, thin-shelled cultivars are propagated vegetatively (grafted/Budded) for planting, instead of seedlings. Seedling tree come to bearing 10-15 years after planting and do not produce nuts of uniform size and shape. Whereas vegetatively propagated trees produce thin-shelled nuts of uniform shape and size, come to early bearing *i.e.* 4-5 years after planting, and have an export potential (Singh *et al.*, 2001).

Walnut is prone to several diseases that harm its growth, quality and yield of the nuts. Plant diseases in walnut tree involve almost all plant parts, resulting in both direct and indirect losses. In this chapter, important diseases of walnut and various approached for their effective management are discussed.

Fungal Diseases

Anthracnose/Leaf Blotch

Anthracnose also known as 'walnut leaf blotch' and is caused by the fungus *Gnomonia leptostyla* (imperfect stage = *Marssonina juglandis*). It is a very serious disease of walnut. The disease has been reported from Kashmir valley and other walnut cultivated areas.

Symptoms

Pathogen attacks the leaves, young shoots, and fruits (Nuts). On the leaves, it produces roundish, reddish-brown spots that develop in to large blotches involving a major part of lamina, that have grayish brown center and are up to ½ inch in diameter. Leaf show chlorosis and premature leaf fall. (Pathak, 1980).

On the young shoots, the lesions appears as oval to irregularly circular sunken, light grayish brown, dead areas, which have dark, reddish-browm margins (Berry,1961). On the fruits (Nuts), this disease is most serious and destructive and causes depressed, circular to irregularly dead spots in the husk. They are not quite so large (small >2mm) as on leaves. Very young nuts which are infected do not develop normally and drop to ground prematurely. Pathogen attack on fruits results in premature, defoliation and poorly filled nuts.The disease may reduce the yield in quality and quantity (Belisario, 1992).

Disease Development

The anthracnose fungus forms black specks (perithecia) during the winter. Perithecial formation is favored by temperature between 45° and 50°F (7° to 10°C). Two compatible strains (+and–) are required to form the sexual, spore-bearing perithecia. Within each perithecium, many sac-like structures called asci developed, each containing eight ascospores. In wet spring weather, the ascospores are forcibly discharged from the perithecium and are carried by air currents to walnut leaves where infection occurs. Rain is important in spore discharge and dissemination. The optimum temperature for ascospore discharge is 50°F (10°C). (Belisario, 1992). Ascospore discharge begins an hour after the start of a wet period and is greatest within 4 to 6 hours, and ends after 20 to 30 hours. Wind and rain splash carry the spores to the new green leaves. Symptoms of anthracnose can be seen as early as late May.

Black acervuli form in lesions on leaves especially on the underside (lower surface) of the leaflet near the veins resulting from spore infections. Acervuli production is greatest at 65°F (18°C). The acervuli produce large numbers of microscopic, asexual spores (conidia). These spores give rise to secondly infections, which occur from the late June until the fall. Rain splash is required for the spores to spread. Free moisture must remain on the leaf surface for more than 6 hours for infection to occur. The optimum temperature for infection is 70° F (21°C). Little infection occurs at temperatures above 81° to 86° F (27° to 30° C) and below 50° to 59° F (10° to 15°C). Symptoms develop 14 to 21 days after inoculation. Spores can remain alive on the leaf surface for at least two weeks, even if free moisture is not present (Berry, 1961).

The anthracnose fungus over winters in fallen leaves as perithecia. More recently, perithecia embedded in the leaf stems have been suggested as being important because of the resistance of these structures to decay. The fungus may also over winter as spores on walnut twigs.

Trees affected by the anthracnose for one or two years are not usually damaged enough to noticeably affect growth. Trees affected for several successive years, however may be stunted in growth and weakened.

Management

- ☆ Collect and burn fallen leaves in autumn.
- ☆ The first spray of zineb or captan (0.01 per cent) is necessary to control the fungus which is required to be followed by another two or three spray at an interval of a fortnight (Rathore, 1999).

- ☆ Spraying of benomyl (0.1 per cent), dodine (0.15), chlorothalonil (0.4 per cent), maneb (0.1 per cent) give effective control of disease. Benomyl is more effective than dodine (Berry, 1977).
- ☆ Spraying mixture of Urea and Bordeaux mixture has also given excellent results (Reznikova, 1971).
- ☆ Fertilization with nitrogen may help reduce the harmful effects like delayed premature defoliation due to disease (Neely 1981).
- ☆ Spray Bordeaux mixture (4:4:50) or copper oxychloride @ 2.5 g/L of water at leaf unfolding stage and then after two weak and at full leaf stage (Saha, 2002).
- ☆ Spray the tree with copper oxychloride (300 g/100 L water) in late spring.Repeat 1-2 spray at fortnightly intervals.

Downy Leaf Spot

Downy leaf spot, also known as yellow-leaf blotch, white mold, microstroma white mold, or white leaf spot, is caused by *Microstroma juglandis.* Downey leaf spot occurs in all regions of world, where walnut is grown. (Burde and Lucy, 1988). After anthracnose, this disease is most common in walnut cultivated areas. It is a serious disease in Kashmir (Puttoo, 1976).

Symptoms

The disease is more unsightly than damaging. This disease does not kill the leaf and is not known to cause defoliation. The effect to the walnut tree is minimal, probably resulting in a minor reduction in photosynthesis. The fungus causes a yellow blotching/discoloration on the upper leaf surface that ultimately becomes brown and dry. On the lower surface, the whitish growth of fungus can be seen.This growth is usually confined to the area between veins (Sinclair *et al.*, 1987).

Disease Development

Whitish growth of downy leaf spot fungus is composed of small white sporodochia which burst through the epidermis and produce large number of spores (conidia). These spores are responsible for spread of disease (Cook, 1975). The disease cycle of *Microstroma juglandis* is not completely known, some scientist believe that the fungus over winters as stromata in fallen leaves. Inflectious spores are released from the remains of fallen leaves in the spring to cause new infections (Sinclair *et al.*, 1987).

Management

- ☆ Fallen leaves should be gathered and composted or burned.
- ☆ Nitrogen fertilization helps minimize the harmful effect of *Microstroma juglandis.* Proper maintenance of soil fertility based on a soil test is important. (Sinclair *et al.*, 1987; Nancy, 1987).
- ☆ Downy leaf spot can be controlled by spraying twice with systemic fungicide bavistin 50 WP (0.03 per cent) in june at an interval of 15 days (Puttoo, 1976).

Mycosphaerella Leaf Spot

Mycosphaerella leaf spot is caused by fungus *Mycosphaerella juglandis* (asexual state *Cylindrosporium juglandis*). It is important in young walnut plantations, where it causes premature defoliation, thus reducing growth and nut production.

Symptoms

Mycosphaerella juglandis causes leaf spot that are angular, reaching a maximum size of 4mm in diameter. By mid summer, as lesions increase from secondary infections, affected walnut trees begin to look chlorotic or yellowish. Closer examination reveals leaf scorch, vein pattern necrosis and lesion fleck symptoms. By late summer/mid August severely diseased leaves fall, especially in dry weather (Nancy, 1987). Minor symptoms are as mentioned below.

- *Leaf scorch*: Look for killing of portions of leaves, particularly at the tip of leaflets.
- *Vein pattern necrosis*: Look for necrotic flecking among major leaf veins.
- *Leaf spotting*: Look for angular sided necrotic flecks up to 4 mm in diameter.

Disease Development

From July to September pycnidia bearing conidiophores are produced on undersurfaces on leaf spot lesions. Pycnidia become apparent about two weeks after lesions are first noticed. Pycnidia bearing spores are readily detected by the presence of white mosses of oozing spores. In the absence of spore mosses, pycnida are difficult to detect with out a hand lens. They can be made more apparent by wetting the leaf to cause them to swell (Kenneth and Linda, 2006).

Management

- Fallen leaves should be gathered and composted or burned.
- Apply herbicides to reduce weed and grass around seedlings. This reduces the high humidity around seedlings, which is conducive to leaf infection (Nancy, 1987)
- Nitrogen fertilization helps minimize the harmful effect of pathogen. Proper maintenance of soil fertility based on soil test should de done (Sinclair *et al.*, 1987).

Phytophthora Root and Crown Rot

Phytophthora root and crown rot in walnut is caused by the fungus *Phytophthora* spp. *viz. Phytophthora cactorum*, (Crandall and Harley, 1938), *P. cinnamomi* (Crandall *et al.*, 1945), *P. citricola* (Ploetz and Green, 1977), *P. citrophthora* and *P. megasperma*. Cool, wet or water logged soil are favorable for disease. Period of 24 hours or more of water saturated soil favour phytophthora infection. Good water drainage reduces the risk of root and crown rot. Root stocks vary in susceptibility to different *Phytophthora*. and none are resistant to all species of fungus (Ramos, 1998)

Symptoms

A number of *Phytophthora* spp. may cause root or crown rot or both in walnut tree. The first indication of infection in the above ground part of the tree is usually a somewhat stunted appearance accompanied by sparse, yellowing green foliage and poor twig growth. Common symptoms are chronic infections, usually of the roots, causing reduction in growth and early senescence and leaf fall. Terminal shoots may die and foliage may be sparse. These trees may be unthrifty for several years before succumbing to the disease. Root/crown rot is characterized by decayed bark at the base of infected trees. Decayed bark area are soft or spongy and usually are confined to the bark. Black exudates from the bark may be noticeable. When the bark is stripped away, the cambium is brown with a water soaked, zonate appearance between white healthy tissue and brown diseased tissue. Once crown rot girdles the tree, it will collapse and die soon after the first warm weather of spring. Phytophthora infection typically kills young trees because their root systems and crown area are small compared to those of mature trees (Ramos, 1998). The tree appears drought–stricken in midsummer, the turn yellow, wither or drop prematurely, the twig and smaller branches die back.

Disease Development

Phytophthora fungi produce spores capable of moving long distance in the soil. Nutrients exuded from roots of host plants attract the motile spores. After reaching the surface of root or crown rot area, the spore is inactive for a brief period before it releases enzyme that dissolves epidermal cell walls. At this point, it rapidly invades the roots and under continuing cool conditions, adjoining portion of root system is invaded until entire root system is colonized. Depending on *Phytophthora* species involved, a number of spore types are produced. These spores, some of which develop in the roots, can survive for one or more year under conditions varying from a moist, cool soil (Zentmeyer and Mircetich, 1966) to a dry warm soil (Hoppe, 1966). The germination of these dormant spores requires favorable soil conditions plus the presence of substances exuded by host root or crown area.

Management

- ✰ Plant orchards/nursery on well drained soil (Green, 1975).
- ✰ Avoid excessive irrigation that saturates the soil or frequently wets the trees crowns
- ✰ In soil infested with *Phytophthora* spp, use the paradox rootstock. Paradox is significantly more resistant to many species of *Phytophthora*. (Ramos, 1998).
- ✰ Disease can be controlled, when the lesions are still small or localized by cutting out the infected areas. The wounds should be disinfected with bichloride of mercury 1- 1000 (1 part bichloride of Mercury, 250 part of denatured alcohol, 750 part of water)and then covered with Bordeaux–linseed oil paint or some other good wound dressing that contains a fungicide (Miller, 2005).
- ✰ Fungicide based on metalaxyl (ridomil 5 per cent) applied at the rate of 100g/tree in the root zone reduce disease incidence. (Ramos, 1998).

Armillaria Root Rot

Armillaria root rot on walnut is caused by *Armillaria mellea* fungus, whose fruiting bodies consist of honey colored or light brown "toad stools" or "mushroom", which often appear in cluster about the base of infected areas in month of October-November. These fungus fruiting bodies can be seen at the root surface of infected trees. They develop depending on the weather and the extent of decay (Miller, 2005).

Armillaria (*Armillariella*) *mellea* is soil borne fungus. "*Armillaria*" in used loosely to refer to a group of about 20 species of *Armillaria* which cause root rot.

Symptoms

Symptoms of root rot on root crown and basal part of plants generally appear as yellowing or browning of leaves, stunting of growth, a general decline in the vigor of plant and twig, branch and main stem die back. Such a decline usually occurs over several years but may appear to progress very quickly as the tree show advanced symptoms of decline and death. Small plants die quickly after the first symptoms appear, but large or old age trees survive for a number of years. Infected trees exude resin, gum or fermenting watery liquid from the lower trunk (Schlesinger and Funk, 1977).

White or creamy white, paper-thick, fan-shaped sheets of *Armillaria* mycelium can be seen growing over the water-soaked sap wood when exposed. The *Armillaria* fungi have a strong mushroom odor. By the time a tree wilts and dies, the trunk is usually encircled by the fungus.

The death of only a few branches can result from the killing of one or several main lateral roots. After the plant dies, rhizomorphs (dark brown to black, slender, root like "shoestrings" with a white interior) develop beneath the bark. The rhizomorphs are 1 to 3 millimeters in diameter, round or flattened and branched, and they consist of hyphal strands bundled together and enclosed with in suberized cells of infected roots. Rhizomorphs may be absent if conditions are unfavorable for their development (Sinclair *et al.*, 1987).

Disease Development

The fungus produces fruiting structure, commonly called toadstools, if conditions are favorable. The toadstools produce spores, which germinate and produce the vegetable part of the fungus (mycelium).The fungus, first develop as a saprophyte that feeds on rotting roots, dead tree stumps. On decaying vegetation in the soil, rhizomorphs push their way through the soil and gain access to the plant or root through wounds made during cultivation or by borers, mice etc. They may also penetrate through the epidermis directly by means of a germ tube and infect the living tissue. The pathogen spreads from one spot or one tree by fungus rhizomorphs. (Miller, 2005). *Armillaria* can pass from tree to tree via root grafts. Roots of trees under stress are most easily infected. *Armillaria* is generally inhibited at soil temperature above 79°F (26°C).

Management

- ☆ Planting material brought to an area should be disease free.
- ☆ Be careful not to introduce *Armillaria* in to an established orchard via diseased root pieces that may be transported on equipment or in surface water.
- ☆ In soil infested with *Armillaria,* use paradox root stock. Paradox rootstock is more tolerant of *Armillaria* root rot.
- ☆ It is possible to kill the fungus in the soil by treatment with powerful disinfectants such as carbon disulfide and choloropicrin. Carbon disulfide is preferable as it is less dangerous to use and not so costly. The method used in applying carbon disulfide is to inject 2 ounces by weight (3/5 ounce liquid measure) of the material in holes 8 inches deep and spaced on an equilateral triangle 18 inches apart. The holes must be plugged immediately after application for the gas fumes to be effective. All infected trees should be taken out before the soil is treated (Miller, 2005).
- ☆ If the disease is discovered while the lesion are still small and localized, it is sometimes possible to save the tree by cutting out all the infected areas and painting the wounds with a good disinfected such as alcoholic solution of bichloride of mercury (1 part bichloride of mercury, 250 parts of denatured alcohol, 750 parts of water). (Miller, 2005).

Bacterial Diseases

Walnut Blight

Walnut blight, also known as bacterial blight, is caused by *Xanthomonas compestris* pv. *juglandis* (Adhikari *et al.*, 1998). Persian walnuts are more susceptible than black walnuts to bacterial blight.

Bacterial blight of walnut is of greater economic importance than all other diseases of walnut together. Its prevalence and destructiveness vary from year to year.

Symptoms

Blight attacks leaves, buds, catkins, nut and succulent new shoots. The first symptoms on the young leaves appear in spring as small, reddish brown spots at the margins or in the tissue between the veins. On the stem, the disease causes black, slightly depressed spots, which often girdle the shoots and cause them to die-back. Young leaf, catkins and buds turn dark brown or black and die. These spots enlarge and become sunken.

Walnut blight can be destructive when it attacks nuts. Small, round, elevated black spots appear on nuts. These spots soon become sunken. The young diseased nuts fall. In older ones, the bacteria invade the shell and the kernel, making them black and the kernel becomes unfit for food (Amani, 1977).

Disease Development

The bacterium that causes walnut blight overwinters primarily in dormant buds and to a lesser extent in canker on twigs of the previous year growth (Miller 2005). The bacteria are disseminated mainly by raindrops or carried by insect to new leaves, shoots and catkins. The bacteria enter the tissues of the current growth through pores (stomata) of plant parts. Infected pollen grain from diseased catkins also sometimes distributes the bacteria (Mulrean and Schroth., 1982).

Management

- ✰ Bacteria over winter in dormant buds, making the control difficult. It is, however, checked by timely spray of copper containing chemicals and Bordeaux mixture. The first spray is applied on the opening of first female flower which should be followed by subsequent sprays at 7-14 days interval according to weather conditions. The most effective rate of metallic copper is 4.5 kg/ha (Teviotdale *et al.*, 1985).
- ✰ A summer oil emulsion, at the rate of 1 pint of oil emulsion to every 100 gallons of Bordeaux mixture (4-2-100) should be added and applied at pre-bloom stage to reduce the severity of leaf injury. One to three applications of tetra copper calcium oxychloride at ten days interval, at the rate of 2 pounds in 100 gallons of water reduces the incidence of disease (Miller, 2005).
- ✰ Spray yellow cuprous oxide (yellow cuprocide) at the rate of 1 pound in 100 gallons of water at pre-bloom stage (Olsen *et al.*, 1976).

Crown Gall

Crown gall is caused by a gram-negative soil bacteria *Agrobacterium tumefaciens* (Smith and Townsend, 1907) that services in soil and gall tissues. Bacteria enter primarily through wounds. Crown gall is most damaging to trees that are 1 to 8 year old. Paradox roots stock is especially susceptible (Ogawa and English, 1991). Crown gall caused by *A. tumefaciens* in walnut tree was reported in Himachal Pradesh, in 1994 by Jindal and Dwivedi.

Symptoms

Crown gall symptoms appear as soft over growths or swelling on collar or root crown portion, but occasionally on the trunk, branches and lateral roots. Galls or tumors start increasing in size, the surface becomes convoluted, dark brown or black and roughened due to the death of peripheral cells.

They may be on only one side of the tree or they may extend around the circumference of the crown or collar region (Miller 2005).

Disease Development

The bacteria perpetuate in infested soil, where they can live as a saprophyte for several years. When susceptible host plants are raised in such sites, the bacteria enter the crown region through the fresh wounds caused by pruning, grafting, and mechanical injury from cultural practices, chewing insect, or by the emergence of lateral roots.

After the bacterium enters through a wound, a small piece of its DNA is transferred in to the trees DNA. The "foreign" DNA transforms normal cells in the wounded area in to tumor cells. Once transformed, tumor cell proliferate automatically. The result is a gall, a disorganized mass of hyperplastic and hypertrophic tissue (Zambryski, 1992).

Due to the invasion of tumors by secondary microbes decay of peripheral cells occurs. The bacteria from these galls may escape to contaminate the surrounding soil or healthy roots. The bacterium can also be spread by splashing rain, irrigation water, tools, wind, insects and parts use for propagation (Smith and Townsend, 1907).

Management

- ✰ Use disease-free nursery stock.
- ✰ Avoid planting in sites with a recent history of crown gall.
- ✰ Remove and destroy badly affected plants.
- ✰ Crown gall can be controlled by surgery (in early stage of infection only) or by painting the galls with a chemical mixture consisting 20 per cent sodium dinitro cresylate (elgetol) and 80 percent methyl alcohol (methanol) (Miller 2005).
- ✰ Biological control by *Agrobacterium radiobacter* strain K84 is preventive only. Latent infection (symptom less) and existing galls are not controlled. Control of crown gall is achieved by dipping root system of nursery seedlings or rootstocks up to crow region in *Agrobacterium radiobacter* strain K84 suspension (Kerr, 1980). Bacterial suspension can also be applied as spray or soil drench. This biocontrol agent occupies the infection sites and produces a bacteriocin called agrocin 84 which is inhibitory to the pathogen. After continuous use of strain k84, *A. tumefaciens* develops resistance to this biocontrol agent through the transfer of agrocin 84 resistant gene form K84 strain to the pathogen. At present, a genetically engineered strain K-1026 of *A. radiobacter* which lack the ability to transfer its agrocin 84 resistant gene to pathogenic *A. tumefaciens* strain is available (Jones and Kerr, 1989).

Deep Bark Canker

Deep bark canker (DBC) is caused by the bacterium *Brenneria rubrifaciens* (previously known as *Erwinia rubifaciens*). The bacterium is capable of infecting English walnut cultivars and is characterized by late onset of symptoms in trees greater than 15 year old. The bacteria infect throughout the year except winter (Lopez *et al.*, 1994).

Symptoms

Deep and large longitudinal cracks in the bark of trunk and scaffolds are the external symptoms of canker (Rathore, 1999). A reddish brown to dark brown substance oozes from these cankers from late spring through early fall, giving them a "bleeding" appearance. Internally dark brown to black streak of varying width extend through the inner bark and may run for many feet up and down on trunk and scaffolds of affected tress(Lelliott and Stead, 1987).

Disease Development

Bacterium *Brenneria rubrifaciens* survives the winter in crackered tissue affected trees or dried exudates on the surface. Bacteria become active in late spring and begin to ooze from cankers together with plant sap. Pathogen enters the tree most often through wounds caused by equipment. High temperature is favourable for disease development.

Management

- ☆ Tree vigor should be maintained through culture practices.
- ☆ Bark eating insects must be controlled.
- ☆ The injuries from mechanical shakers during harvesting needs to be avoided.

Viral Diseases

Walnut Blackline

Blackline disease is by *Cherry leaf roll Nepovirus*. *Cherry leaf roll virus* (CLRV) was first described in 1955 by Cropley as causing a disease of sweet cheery (*Prunus avium*) in England. Since then, it has been shown to exhibit a wide natural host range of herbaceous and woody plants including English walnut (*Juglans regia L.*) and sweet cheery (Jones, 1986). On several species of Juglans, this virus causes a disease called walnut backline.

Symptoms

The first symptoms on walnut are tree poor growth of shoot, yellowing and drooping of leaves and premature leaf fall. Death usually follows in 3 to 5 year after that first symptoms are noted in the top of tree. The decline and eventual death of tree is caused by breakdown or separation of the conductive tissue at the graft union (Miller, 2005).

The external evidence of blackline disease on walnut tree is the appearance of a dark line between the stock and grafted tissue ('scion'). In other words, symptoms of walnut blackline disease is a gradual girdling of cambium and phloem tissue by a narrow, black, necrotic strip at the rootstock scion union, resulting in decline and death of the English walnut scion. (Jones, 1986).

Epidemiology

Cherry leaf roll virus has a bipartite single strand positive sense RNA and belongs to the genus *Nepovirus*. CLRV is naturally transmitted through seed, pollen, (Bandte and Bittner, 2001), mechanical inoculation, grafting and nematodes (Jones, 1986) but, unlike the majority of other members of this genus, CLRV is not transmitted by soil borne nematodes (Wang *et al.*, 2002)

Management

No satisfactory method for the control of black-line is known. Inarching the black walnut suckers coming from the rootstock into the Persian top has been attempted. Persian seedling walnuts have been planted around the trees and attempts made to inarch them into the trunks of affected trees, but the result have not been very successful. Seedlings of the franquette and other varieties of Persian walnut are used as rootstocks instead of hindus black walnut by certain nurseryman in order to avoid the disease. This should result in control as the disorder has so far never been found in franquettes grafted on Persian walnut rootstocks. (Miller, 2005).

References

Adhikari, R.S., Bora, S.S. and Singh, S.B. (1988). *Xanthonomas campsetris* pv. *juglandis:* A new report from India. *Current Science,* 57: 728.

Amani, B.O. (1977). Bacterial blight of walnut in Iran. *Iranian Journal of Plant Pathology,* 13:15–23.

Bandte, M. and Bittner, C. (2001). Occurrence, tranamission and diagnosis of cherry leaf roll nepovirus: A literature revieu. *Pflanzenschutzberiche,* 59: 1–9.

Belisario, A. (1992). Phytopathological problems of walnut. *Informatore Agraio.,* 48: 51–53.

Berry, F.H. (1977). Control of walnut anthracnose with fungicide in black walnut plantation. *Pl. Dis. Reptr.,* 61: 378–379.

Berry, F.H. (1961). Etiology and control of walnut anthracnose. *Md. Agric. Exp. Stn. Bull.,* A–113: 22.

Burde, E.and Lucy (Ed) (1988). Walnut Notes. St. Paul MN. U.S. Department of Agriculture, Forest service, North Central Experiment Station, p. 501–502.

Cook, A.A. (1975). *Diseases of Tropical Subtropical Fruits and Nuts.* Hafner Press, New York, 317 pp.

Crandall, B.S. and Harley, C. (1938). *Phytophthora cactorum* associated with seedling disease in forest nurseries. *Phytopathology,* 28: 358–360.

Crandall, B.S., Gravatt, G.F. and Ryan, M.M. (1945). Root disease of Castanea species and some coniferous and broad leaf nursery stocks, caused by *Phytophthora cinnamomi. Phytopathology,* 35: 162–180.

Cropley, R. (1960). Cheery leaf roll virus. *Appl. Biol.,* 49: 524–229.

Green, R.J. Jr. (1975). Phytophthora root rot of black walnut seedlings. In: *Forest Nursery Diseases in United States,* Agric. Hanlb. 470. Washington. D.C. U.S. Department of Agriculture, pp.19–22.

Hoppe, P.E. (1966). *Pythium* species still viable after 12 year in air dry muck soil. *Phytopathology,* 56: 1411.

Jindal, K.K. and Dwivedi, M.P. (1994). Occurrence of crown gall on walnut in India. *Plant Disease Res.,* 9: 67–68.

Jones, A.T. (1986). Cherry roll virus. CA/AAB description of plant viruses No. 306. Commonwealth Mycological Institute, Kew, England.

Jones, D.A. and Kerr, A. (1989). *Agrobacterium radiobacter* Strain 1026, a genetically engineered derivative of strain K 84 for biological control of crown gall. *Plant Disease,* 73:15–18.

Kenneth, J. Kessler and Linda, B.H. Swanson. (2006). Walnut: Mycosphaerella leaf spot notes. U.S. Department of Agriculture, Forest Service. North Central Experimental Station.

Kerr, A. (1980). Biological control of crown gall through Agrocin 84. *Plant Dis.,* 64: 25–30.

Lelliott, R.A. and Stead, D.E. (1987). *Methods for Diagnosis of Bacterial Disease of Plants.* Blackwell Scientific Publication, Oxford, U.K.

Lopez, M.M., Marti, R., Morente, C., Orelian, N., Nonot, T. and Aleta, N. (1994). Phytopathogenic bacteria identified in walnut in Spain. *Fuera de Seria.* Instituto Valenciano de Investigaciones Agrarias (IVIA), Moncada, Valencia, Spain, 2: 307–314.

Miller, Paul W. (2005). Felberts and Persian walnuts. In: *Diseases of Fruits and Nuts*, (Ed.) Alefred Stefeured. Biotech Books, Delhi, pp. 320.

Murlean, E.N. and Schroth, M.M. (1982). Ecology of *Xanthomonas campestris* pv *juglandis* on Persian(English) walnut. *Phytopathology*, 72: 434–438.

Nancy, R.P. (1987). Fungal leaf spot of walnut. RPD–600. Department of Crop Science. University of Illinois, Urbana.

Neely, D. (1981). Application of nitrogen fertilizer to control anthracnose of black walnut. *Plant Disease*, 65: 580–581.

Ogawa, J.M. and English, H. (1991). *Diseases of Temperate Zone Tree Fruit and Nut Crops*. Pub. 3345. Univ. Calif. Div. Agric. Natural Resources, Oaklank. U.S.A. 461pp.

Olsen, W.H., Moller, W.J., Fitch, L.B. and Jeter, R.B. (1976). Walnut blight control. *California Agriculture* 30: 10–13.

Pathak, V.N. (1980). *Disease of Fruit Crops*. Oxford and IBH Publishing Co.

Ploetz, R.C. and Green, R.J. (1977). The root rot of black walnut seedling caused by *Phytophthora citricola*. In: *Proceeding of Indiana Academy of Science*, 87: 105–112.

Puttoo, B.L. (1976). Fungicidal control of downy leaf spot of walnut in Kashmir. *Pesticide*, 10: 21–22.

Ramos, D.E. (1998). *Walnut Production Manual*. University of California Publication, 3373: 320.

Rathore, D.S. (1999). Walnuts. In: *Temperate Fruit Horticulture*, (Eds.) S.K. Mitra and D.S. Rathore. Allied Publishers, Calcutta, pp. 406–408.

Razdan, V.K. and Puttoo, B.L. (2005). Diseases of nut fruits. In: *Diseases of Fruits and Vegetables and their Management*, (Ed.) T.S. Thind. Kalyani Publisher, Ludhiana (Punjab), pp. 474.

Reznikova, L.M. (1971). Urea against brown spot of walnut. *Zashchita Rastenii*, 10: 23.

Saha, L.R. (2002). *A Handbook of Plant Diseases*. Kalyani Publisher, Ludhiana (Punjab), pp. 327

Schlesinger, R.C. and Funk, D.T. (1977). Manager's handbook of black walnut. Gen. Tech. Rep. NC–38. St Paul, M.N., U.S. Department of Agriculture. Forest Service North Central Forest Experiment Station. 22pp.

Sinclair, W.A., Lyon, H.H. and Johnson, W.T. (1987). *Disease of Trees and Shrubs*. Comstock Publishing Associates, Cornell University Press, 574 pp.

Singh, S.S., Prasad, D. and Lal, H. (2001). Status and management of pest and diseases. In: *Crop Protection: Management Strategies*, (Ed.) D. Prasad. Daya Publishing House, Delhi, pp. 677.

Smith, E.F. and Townsend, C.O. (1907). A plant tumour of bacterial origin. *Science*, 25: 671–673.

Teviotdale, B.L., Schroth, M.N. and Mulrean, E.N. (1985). *Walnut Orchard Management*, (Ed.) D.E. Ramos, Div. Agn. Not. Res. Univ. of Calif. Publ., 21810: 153–157.

Wang, S., Gergerich, R.C., Wickize, S.L. and Kim, K.S. (2002). Localization of transmissible and non transmissible viruses in the vector nematode *Xiphinema americana*. *Phytopathology*, 92: 646–653.

Zambryski, P.C. (1992). Chronicles from the Agrobacterium plant cell DNA transfer story. *Annu. Rev. Plant Phsio. Plant Mol. Biol.*, 43: 465–490.

Zentmeyer, C.A. and Mircetich, S.M. (1966). Saprophytism and persistence in soil by *Phytophthora cinnamomi*. *Phytopathology*, 56: 710–712.

Plant Diseases Management for Sustainable Agriculture *Pages* **121–135**
Editor: **Dr. Shahid Ahamad**
Published by: **DAYA PUBLISHING HOUSE**

Chapter 11

Occurrence and Management of Die-Back in Chillies: An Overview

Jameel Akhtar, H.C. Lal and Atul Kumar

Department of Plant Pathology, Birsa Agricultural University, Ranchi – 834 006 (Jharkhand)

The genus *Capsicum* includes various cultivated species, of which *C. annum* L. is most important. It comprises of both sweet (Bell pepper) and pungent (chilli) fruits of numerous shape and size. The most notable feature of capsicum is flavour and is an excellent source of vitamin A and C besides medicinal uses. Pungent pepper are used as condiment; mild as salad and pickles; bell shaped (non pungent) for cooking and paprikas or Spanish pimiento for colouring and flavouring of culinary preparations.

The native home of chillies is considered to be tropical America especially Brazil, where it is still found growing in a wild state. In India, its introduction is believed to be through the Portuguese in 17th century. At present chillies are indispensable and common ingredient of Indian dietary. It is used throughout the country as a condiment.

The capsicum is vulnerable to various parasitic diseases and they are responsible for taking away a big chunk of production. Among the fungal diseases, anthracnose/die-back/fruit-rot of chillies, is an important disease causing serious losses in field, transit, transport and storage (Singh, 1987). It is present throughout India but is reported to be more common and aggressive in Assam, Bihar, A.P. and U.P. in India (Singh, 1987).

Economic Importance

Fruits showing no symptoms or mild symptoms of the disease when stored under warm humid conditions subsequently are rendered unsuitable for drying or canning (Walker, 1952). Chowdhary (1957) reported 12-32 per cent yield loss in Assam. Smith and Crossan (1958) had observed 48 per cent infection on chillies seeds due to attack of *C. capsici* while Rai and Chohan (1966) reported approximately 29 per cent mortality in chilies. Bansal and Grover (1969) while studying the impact of disease in Punjab and Haryana, reported 10-60 per cent yield loss. Hiremath and Govindu (1973) observed that fruit- rot due to *Colletotrichum capsici* is the most important factor of yield loss during marketing. Even during the post-harvest period, the disease is reported to cause 10-15 per cent loss in yield (Khan, 1978). Ascorbic acid, an important component in fruits, is reported to decrease gradually due to fruit rot (Adisa, 1983). Singh (1987) and Muthukrishnan *et al.* (1988) have also observed development of symptoms on healthy looking fruits during transit and storage. Datar (1995) also observed loss of 8-27 per cent during transit, transport and storage.

Symptomatology

The pathogen infects growing plants and causes shoot necrosis or die-back symptoms (Figure 11.1). The twigs are attacked from tip backward (Verma and Sharma, 1999). The lesions initially appear as water soaked and brown which later on turn grayish white or straw coloured. Within the lesions and necrotic areas a large number of pin-headed black bodies which are asexual fruiting bodies (acervuli) are found scattered (Singh, 2002).

Ripe fruit rot phase of the disease occurs when chilli fruits just start maturing (turning red). However, anthracnose symptoms do occur frequently on well developed matured green fruits as well. Initial development of symptom on fruits starts first after the formation of depressed lesions of variable sizes *i.e.* from mere dots to those covering most of the fruit surface areas (Figure 11.2). In anthracnose caused by *C. capsici*, infected fruits appear black (Smith and Crossan, 1958). Fully developed lesions are sunken and appear dark red to light tan with varying amount of visible dark acervuli of the fungus. The pathogen also colonizes frozen green fruits because of leakage of carbohydrates, proteins and electrolytes.

Colletotrichum coccodes has been reported to cause anthracnose of capsicum seedlings. The initial spots that develop become sunken, coalescing to form large irregular lesions and finally showing blight symptom. Severe infection may lead to defoliation and lodging of the plants (Oh *et al.*, 1988).

Park *et al.* (1989) observed that susceptibility of red fruit is due to higher amount of total carbohydrate, reducing sugars and total amino acids compared to their amounts in green fruits. Sucrose is the main carbohydrate in green fruits while glucose and fructose are dominant in red fruit.

The lower portion of infected fruit skin is invariably found covered with minute black, spherical elevated stromatic masses or sclerotia of the fungus. In advanced stages, the seeds are covered with felt of white mycelium in which are embedded a few black to grey green stromatic bodies. Infected seeds turn rusty in color. (Choudhary 1957; Smith and Crossan 1958; Rai and Chohan, 1966; Akhtar and Singh, 2007).

The Causal Organism

The disease is caused by *Colletotrichum capsici* (Syd.) Butler and Bisby though number of other *Colletotrichum* species such as *C. gloeosporiodes* (Verma, 1973), *C. acutalum* (Kaur and Singh, 1990) and *C. coccodes* (Yu *et al.*, 1987and Oh *et al.*, 1988) have been found associated with the disease in different geographical areas. *Colletotrichum dematium, C. graminicola* and *C. atramentarium* have also been reported to cause this disease on chillies in India (Verma, 1973). *Gloeosporium piperatum* has been observed to cause anthracnose in Argentina (Laguna and Docampo, 1971).

Figure 11.1: Plants Showing Die-back Symptoms

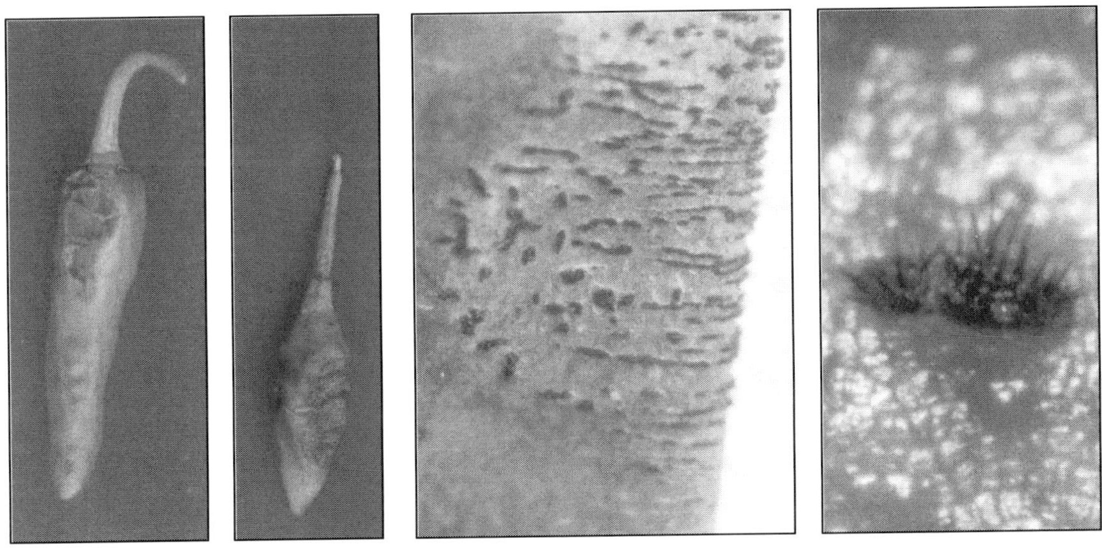

Figure 11.2: Fruit-rot and Anthracnose Symptoms on Chilli

Colletotrichum capsici (Syd.) Bulter and Bisby was first described as *Vermicularia capsici* by H. Sydown (1913) from pepper fruit collected by W. Mc Rae in India during 1912. Taxonomic position of the pathogen as proposed by Ainsworth and Bisby (1995), is given below.

Kingdom: Fungi

Phylum: Ascomycota

Class: Coelomycetes

Order: Melanconiales

Family: Melanconiaceae

No definite perfect stage is known so far although a *Glomerella cingulata* (Stonem) Spauld and Schrenk) is suspected as the telomorphic stage (Singh, 1987).

On potato dextrose agar, the colonies of the fungus appear initially white but subsequently turn grey. The aerial mycelium network forms light to dark grey felt (Figure 11.3). Sometimes, it is zonated due to acervuli that are formed subsequently (Akhtar, 2007). On the host, the septate mycelium is both inter and intracellular. The saucer shaped acervuli are formed on fruits, leaves and stems with closely

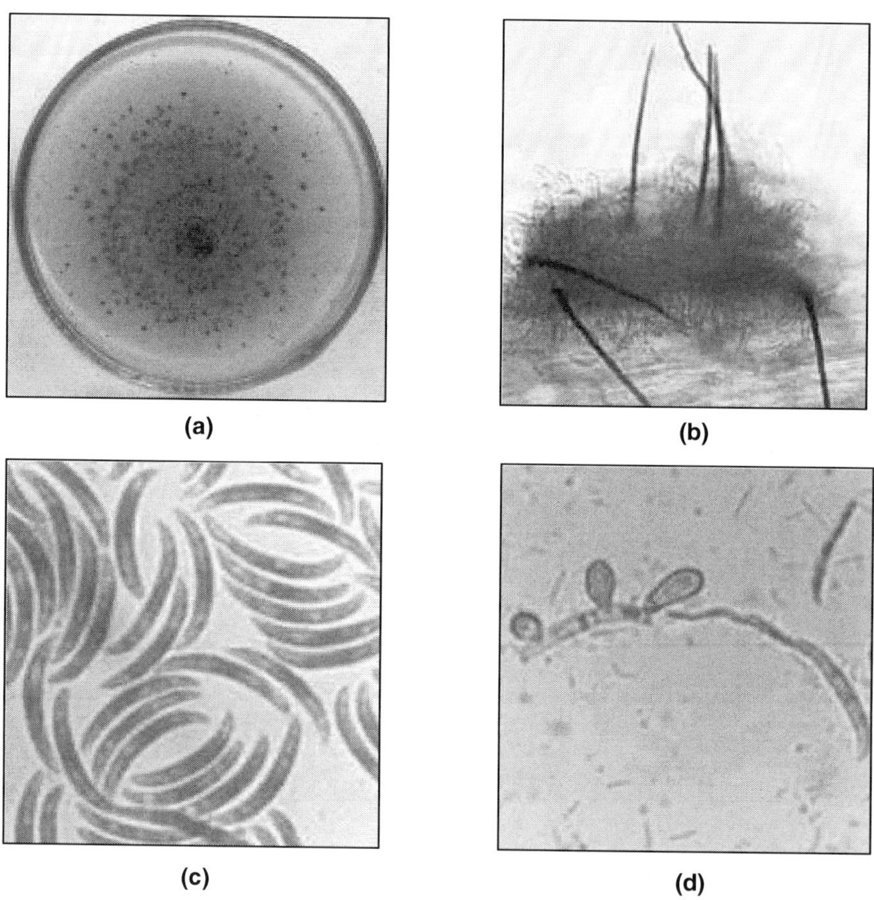

Figure 11.3: Morphology of *C. capsici*
(a) Dark grey zonated colony; (b) Saucer shaped acervuli with setae; (c) Falcate conidia; and (d) Germination of conidia with appressoria

packed stalks which are almost rounded and sometimes elongated. They measure 350 µm in diameter (Verma and Sharma, 1999). Initially the acervuli are subepidermal but later disrupt the host surface to dispose off conidia. Abundant setae that develop are rigid, hard, brown to black, one to five septate, tapering at apex but hardly swollen at the base. They measure upto 250 µm in length 5.8 µm in width. (Singh, 1983; Verma and Sharma, 1999, Akhtar and Singh, 2007).

Abundant conidia are produced. They are falcate to fusifom with acute apices and narrow truncated base. They are one celled, hyaline and uninucleate. The conidiophores are faintly brown, cylindrical and either septate or aseptate and phialidic. These conidia measure 16-30 (18-23) x 2.5-4 (3.5-4.0) µm in size (Smith and Crossan, 1958; Singh, 1987; Verma and Sharma, 1999 and Akhtar and Singh, 2007). They form pale buff to salmon coloured droplets in culture. Often at the time of germination, the conidia become one septate and produce germtube. The development of appresorium is fast as they are formed immediately after spore germination. The appressoria are medium brown clavate to circular and measure 6-25 x 4-11.5 µm in size (Das Gupta, 1981; Singh; 1987 and Roberts and Snow, 1990).

Chowdhary (1957) studied the effect of temperature and humidity and hydrogen ion concentration on growth of *C. capsici in vitro*. Best growth of the pathogen occurred at 28°C, relative humidity 92.0 per cent and pH of medium in between 5.0 and 6.0. Mishra and Mahmood (1960) studied the factors affecting the growth. They observed that among synthetic media mycelial growth was most profuse in Richard's medium, where as sporulation was maximum in Czapek's medium. Among non-systemic media, PDA was best, both for mycelial growth and sporulation. The size of conidia and setae was maximum when the fungus was grown on this medium. Maximum spore germination took place in neutral solution and in 1 per cent sucrose solution. However, leaf extract and leaf decoction media were most suitable substrates for the growth and sporulation. They also reported that the fungus grew over a wide range of pH, ranging from 3.0 to 9.0, the optimum pH around 5.0. The formation of acervuli was maximum in alkaline medium than in acidic medium.

Malhotra and Chaturvedi (1974) studied the effects of hydrogen ion concentration and different carbon, nitrogen and sulphur sources on growth and sporulation of *C. truncatum* isolated from diseased pods of soybean. Optimum pH for growth and sporulation was 5.5. Sucrose was the best carbon source for growth. Nitrates in general were comparatively better nitrogen sources than ammonium compounds. Glutamic acid was best nitrogen source for growth. Best growth and excellent sporulation of the organism was obtained on $MgSO_4.7H_2O$. Palarpawar *et al.* (1997) observed that of the twelve different nitrogen sources, individually incorporated in basal media, peptone promoted the greatest mycelial growth of *C. capsici*. The response of *C. curcumae* to different nitrogen sources varied greatly. It was concluded that the nitrogen requirements of *C. capsici* and *C. curcumae* are almost quite different. Chacko *et al.* (1978) recorded differences in growth of two isolates of *C. truncatum* on three synthetic and four non-synthetic media.

According to Wong *et al.* (1983) oat meal agar was the best medium for growth of *C. truncatum* while Czapeck Dox and potato dextrose agar was best for sporulation. Sporulation was highest under 12 hr alternating light/dark treatment. Optimum temperature of growth and sporulation was 25°C.

Adebitan (1994) tested six media for the growth of *C. truncatum* from cowpea. The best medium observed for growth was cowpea stem decoction agar, while PDA was best for sporulation. 28°C being optimum for growth and 25°C for sporulation.

Sook *et al.* (1995), observed colonies of *C. truncatum* on culture media. In general, the function of conidia and setae on water agar leaf medium was better than on potato dextrose agar medium.

Akhtar and Singh (2007) observed colonies of *C. capsici* on five culture media. In general, the growth of mycelium, production of conidia and setae on potato dextrose medium was better than on other media. Five monoconidial isolates of *C. capsici* differed in producing acervuli and spores on different culture media. This signifies the fact that different culture media do show differences in supporting growth and devlopment.

Three mycelial isolates of *C. lindemuthianum* differed in their ability to produce conidia on different culture media (Mendes *et al*, 1996). Pria *et al.* (1997) reported that different culture media do show differences in supporting growth and sporulation in *C. lidemuthianum*.

Minor differences in morphology and virulence occur between and within the strains but the species is generally regarded as unspecialized pathogen and the ability of isolates to adapt to new hosts has been demonstrated (Singh, 1987). Dasgupta (1988) working on variability of the spores among the isolates of *C. capsici*, found no consistently observable differences, among the 12 isolates of *C. capsici* isolated from betelevine. The measurement of setae and distance between two septa of setae did show some variations. However, all the isolates showed variation in their ability to cause infection, on stem and leaf. Roberts *et al.* (1990) studied four isolates each of *C. capsici* and *C. indicum*, from different geographical regions and found significant variability in the conidial dimensions between both the species and also between/within the isolates.

Akhtar and Singh (2007) have also reported significant variations in the acervular density, conidial dimensions and germination among the five isolates of *C. capsici* derived from different geographical locations of Uttarakhand.

Verma (1973) in a detailed investigation at Jabalpur, India observed that four species of the fungus namely *C. dematium, C. gloeospoiroides, C. graminicola* and *C. atramentarium* cause seedling collar rot, leaf spot, dieback and fruit rot of chilli. The comparative study on the virulence of eight isolates belonging to four species further revealed that *C. atramentarium* was more virulent on seedling; *C. gloeosporioides* on stem and *C. dematium* on fruit and *C. graminicola* was intermediate in virulence. The five varieties (Pandurna, Yellow Long, S.G. 5 Anand, S.2 Anand and Sweet pepper) tested against the eight isolates showed that all the varieties were highly susceptible to *C. atramentarium* and *C. gloespoiroides*. The varieties S.G.5 Anand and Sweet pepper were susceptible to all the species. However, infectivity on Pandurna, Yellow Long and S. 2 Anand varied from isolate to isolate, demonstrating physiological specialization.

In another experiment, the fruits of 4 chilli cultivars *viz.*; Kandhari, Pant chilli-1, Pusa Jwala and G4 were inoculated with five isolates of *C. capsici* (KG, S2, BIL, SIT and DIN) to observe differences in virulence and host reactions. The difference in the amount of disease produced on inoculated fruits by the pathogen isolates was significantly different from each other. Isolates S2 and BIL were found most aggressive while isolate KG was least aggressive. Similarly, among the cultivars, Pant chilli-1 was least susceptible (Singh, 2002). Bansal and Grover (1969), Das Gupta (1988) and Jayalakshmi and Seetharaman (1995) have also studied differences in virulence of the isolates of *C. capsici* and differences, in host reaction.

Host Range

Since, the first description of *C. capsici* in 1913 by H. Sydow, it has been reported to infect at least 176 host genera (Sutton, 1980). In addition to chillies, the pathogen has been reported to cause infection on, gram, cotton, eggplant, jute, tomato, betelevine, mung, soybean, castor and *Curcuma zeodaria*, (Tripathi and Beniwal, 1978; Chang and Chi, 1993; Sinha and Singh, 1993; Khan, 1994 and Pring *et al.*,

1995). In India, anthracnose of betelevine caused by *C. capsici* is of great economic importance (Singh and Shankar, 1971 and Maiti and Sen, 1979).

Disease Cycle

The pathogen over winters in crop debris and seeds of infected fruit. Mycelium invades the inner surface of seed through seed coat and colonizes the outer layer of endosperm (Sanchote and Juangbhanich, 1984). Seed borne infection affected germination and early seedling growth but was not found responsible for development of fruit-rot and die-back at later stage of the plant growth (Perane and Joi, 1988). Podaganur and Naik (1991) reported externally seed borne nature of *C. capsici*. Histopathological observations have proved pathogen transmission from endosperm to hypocotyls and radicals (Lee *et al.*, 1995).

The young or mature fruits wounded or weakened are infected by the pathogen (Grover and Bansal, 1970). Conidia present in a gelatinous matrix in acervuli on plant debris are disseminated by rain splashes. Conidia germinate readily and cause visible symptoms within 3 to 5 days depending upon climatic conditions.

The optimum temperature for germination of conidia varies from 28 to 30°C with 100.0 per cent relative humidity. The entry of infection peg through the cuticle is gained apparently by a cutin dissolving enzymes. Subsequently, the mycelium invades the fruit cavities and enters the newly formed seeds Figure 11.4. The disease is favoured by heavy dew and rains (Dastur, 1922; Kogan, 1977). Chowdhary (1957) found that the disease did not develop below 90.0 per cent relative humidity and greatest disease development occurs at 28°C. Sproulation and maximum spore germination occur at 30°C and 100.0 per cent relative humidity (Mishra and Mahmood, 1960; Astuti and Suhardi, 1986).

The optimum temperature for the growth of the fungus in culture was around 30.0°C. No growth occurred at 7.2°C, and only slight growth occurred at 35.5°C. The optimum relative humidity for the growth of the fungus at 30.0°C was found to be 82.5 per cent. Mycelial growth and sporulation were maximum when the pathogen was exposed to alternate light and darkness.

In an experiment pertaining to the effect of temperature on growth and sporulation of the *C. capsici* isolate, it was conclusively proved that a temperature of 28±1°C was most suitable and favourable for the growth, sporulation and germination and formation of appresoria of the fungus. The five isolates too differed from each other at every temperature studied. The germination and formation of appresoria even at 20±4°C too was observed. As already discussed, since temperature does affect growth and sporulation, there is nothing unusual, rather it is normal that conidial geminability and subsequently formation of appressoria of the isolates differed from each other (Singh, 2002).

The conidia germinated readily in tap water within 4 hours. The percentage of spore germination increased with increase of temperature being highest at 30.0°C. It was observed that 100.0 per cent atmospheric humidity was congenial for spore germinability and there was no effect of light.

Fruit pH also plays an important role in determining the infection by *C. capasici*. Since the green fruit pH lies at 5.5 it is unsuitable for infection by the fungus (Thirupathaiah and Subramaniam, 1978).

Management

Several workers (Chowdhary, 1957; Grover and Bansal, 1970; Narain and Panigrahi, 1971; Joshi and Wangikar, 1978; Thakur and Singh, 1987; Das Gupta, 1988; Perane and Joi, 1989; Acharya and

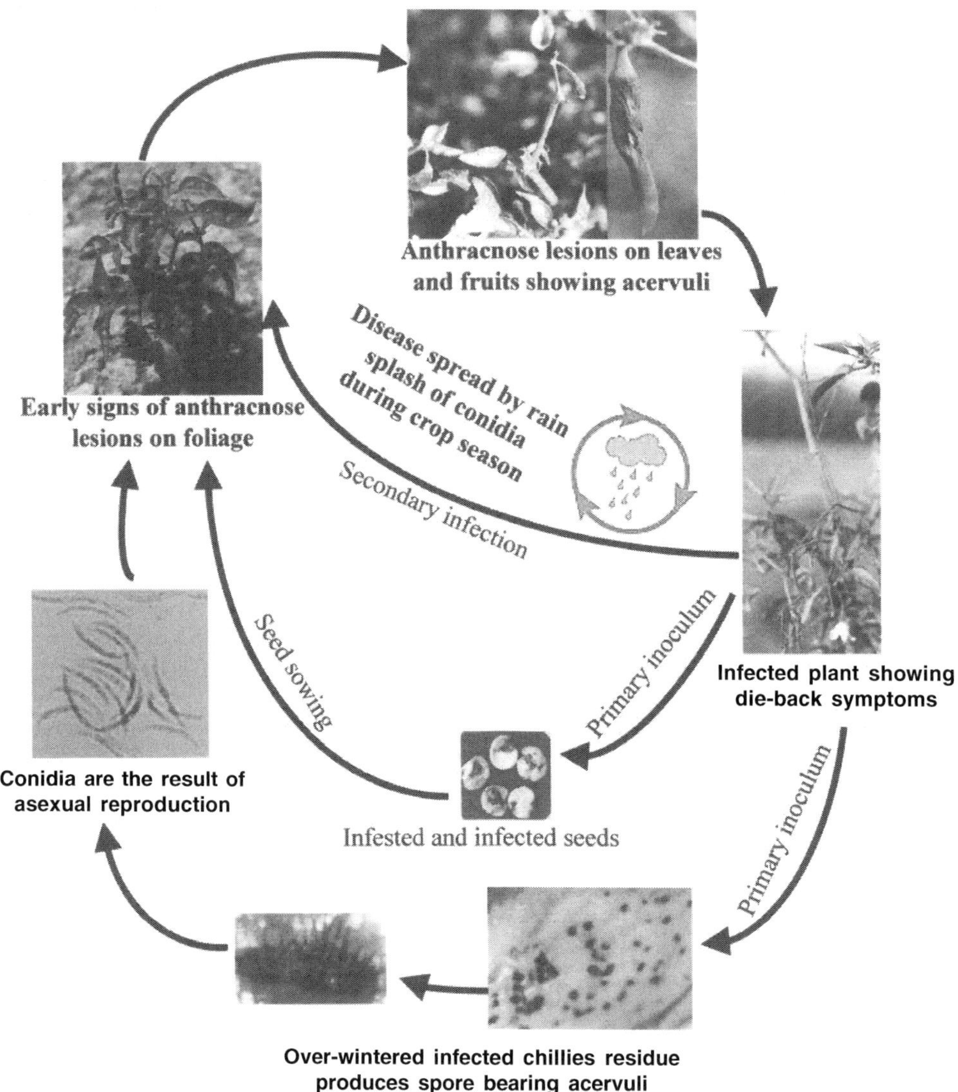

Figure 11.4: Disease Cycle of *Colletotrichum capsici* Causing Chillies Anthracnose

Das, 1995; Singh, 2002; Srinivas *et al.*; 2006 and Akhtar and Singh, 2007) have tested several systemic and non-systemic fungicides either in field or *in vitro* against *C. capsici*. Invariably, most of the fungicides tested by them resulted in significant control of both the disease and the pathogen.

Datar (1996), reported Bavistin followed by Dithane M-45 to be highly effective in checking mycelial growth of the fungus. Joshi and Wangikar (1978) have also made similar observation. Bavistin @ 200µg/ml completely inhibited mycelial growth of the fungus. Azad (1992) observed 10 per cent contamination on chilli seeds treated with Carbendazim while no contamination was observed with

Zineb and Tridemorph. However, Kakade and Khure (1989) did not find Bavistin (Carbendazim) effective against *C. capsici.* Ali *et al.* (1993) reported complete inhibition of mycelial growth of *C. gloesporoides* at 100 ppm concentration of Bavistin (Carbendazim), 200 ppm concentration of Tilt (Propiconazole) and 400 ppm concentration of Folicur (Tebuconazole).

Narain and Panigrahi (1971) studied effects of Plantvax and Vitavax on spore germinability. Both were highly effective and checked spore germination of *C. capsici.*

Goswami *et al.* (1996) reported best control of *C. catechu* causing leaf spot of betel nut by Knowin 50 WP (Carbendazim) followed by Tilt, Folicur, Alliete and Dithane M-45 at 1000 ppm concentration.

Freeman *et al.* (1997) reported effective dose causing 50 per cent inhibition (ED 50) of *C. acutatum* growth in culture media at 150 ppm concentration of Tilt. Effectiveness of Tilt in controlling *Colletrotrichum* spp. have been reported by several workers (Black and Smith, 1993; Read and Hide, 1995 and Koelsch *et al.*, 1995).

Foliar sprays of the chemicals such as Captan, Carbendazim, Benomyl, Nabam, Zineb, Capafol and Mancozeb at 10 to 15 days interval after transplanting have invariably been recommended by various workers (Thakur and Singh, 1973; Chakravarti and Kumar, 1975; Chauhan and Duhan, 1977; Sivaprakaram *et al.*, 1978; Sankaralingam *et al.*, 1983 and Mishra 1988). Seed dipping in Thiram has also been reported to control fruit rot effectively (Chakravarti and Kumar, 1975).

Datar (1996), in field trial observed and reported that Bavistin (Carbendazim) @ 1000 ppm to be most effective followed by Dithane M-45 (Mancozeb). However, in another field experiment, Ebenezar and Alice (1996) found the above two fungicides giving contrary results. It was Mancozeb which gave best control followed by Carbendazim. Maximum control of *C. capsici* infection in field by Bavistin is also reported several others (Maiti *et al.*, 1978; Khatua *et al.*, 1981; Gupta, 1988; Perane and Joi, 1989 and Biswas, 1992). Similarly, several workers (Natrajan and Subramaniam, 1976; Chauhan and Duhan, 1977; Kadu *et al.*, 1977; Joi and Sonone, 1980; Sitaramraju and Srinivas, 1985 and Ebenenzar and Narayanaswami, 1988) have reported Dithane M-45 (Mancozeb) a highly effective fungicide under field conditions.

Perane and Joi (1989) tried seed treatment as well as foliar sprays. Seed treatment with Carbendazim and spraying of Mancozeb reduced incidence of disease significantly. As seed protectant, Delesene M (Carbendazim + Mancozeb) and Benlate have also been reported effective (Alabi and Emechebe, 1990).

Adaptation of *C. capsici* to higher doses of fungicides, was reported by Kore and Apet (1989). *C. capsici* showed adaptation to 3000 ppm concentration of Dithane Z-78, 2500 ppm concentration of Dithane M-45 (Mancozeb), 1000 ppm concentration of Bavistin (Carbendazim), 2500 ppm concentration of Iprodoine, 1000 ppm concentration of Thiophante methyl, and 1000 ppm concentration of Captafol. Similar results have also reported by Mohanraj *et al.* (1972), Pan and Sen (1981) and Reddy *et al.* (1981).

Sinha (1990) reported best control of *C. capsici* in terms of cost benefit ratio by Dithane M-45 (Mancozeb), though Foltaf (Captafol) was found best. Similar observations were made by Acharya and Das (1995) is controlling anthracnose of betelevine.

In two years of field trials, Goswami *et al.* (1996) used @ of 2000ppm and found that disease severity was reduced by 45.83 per cent and the yield increased by 105.64kg/hectare over check in areca nut.

Barros *et al.* (1999) in field trial for the management of common bean anthractonse (*C. lindemuthianum*) and angular leaf spot (*Phaeoisariopsis griseola*) found to be Azoxystrobin, Tebuconazole

+ Fentinhydroxide and Tebuconazole (Folicur), highly effective. Lindner *et al.* (2000) studied the control of lupin anthracnose through seed treatment. They found that seed treatment with Solitar (active ingredients: Fludioxonil, Cyprodinil and Tebuconazole) @ 3 ml/kg, highly effective. Foliar application of Folicur (Tebuconazole) further increased disease control and yield.

Romer *et al.* (2000) have also reported that seed dressing with Folicur and other chemicals as spray fungicides controlled anthracnose of lupins satisfactorily.

Akhtar (2007) has reported that the isolates of *Colletotriclatm capsici* when exposed to different concentrations of five fungicides namely Dithane M-45, Bavistin, Vitavax, Tilt and Folicur, differences in sensitivity of the isolates to a particular fungicide and concentration were clearly evident. This signifies the fact that all the isolates were not equally sensitive to all the five fungicides tested. Folicur was highly toxic to the isolates as the reduction in growth was significant.

Jaylakshmi *et al.* (1998) have studied the potential of biocontrol agents against *Colletotrichum capsici* causing anthracnose/fruit-rot/die-back disease of chilli.

Chloroform extracts of non volatile antibiotics (NVAC) of *T. viride* added to the culture media inoculated with *C. capsici,* showed reduction in biomass and synthesis of RNA, DNA and protein (Rajathilangam and Kannabiran, 2001).

Five different cultivars of chilli seeds infected to a varied degree by *C. capsici* were treated with pure cultures of *Trichoderma harzianum* and *Pseudomonas fluorescens* and their formulations to evaluate per cent reduction of *C. capsici*, seed germination, vigour index and field emergence. It was observed that pure culture of *P. fluorescens* was more effective in reducing *C. capsici* than *T. harzianum* and untreated seeds. Formulations of *P. fluorescens* were also effective in reducing *C. capsici* infection and in increasing the seed germination, vigour index and field emergence followed by *T. harzianum* in comparision with control (Srinivas *et al.*, 2006).

The interaction between the five isolates namely S2, DIN, SIT, BIL and KG and a strain of *T. harzianum* indicated differences in sensitivity of the isolates to bioagent. While there was maximum inhibition of the isolate DIN, it was minimum in isolate BIL. There was non-significant inhibition in isolates S2 and SIT (Singh *et al.*, 2007).

References

Acharya, A. and Das, J.N. (1995). Control of Anthracnose of Betelvine by fungicidal chemicals. *Current Agricultural Research*, 8: 58–60.

Adebitan, S.A. 1994. Some factors influencing the growth and sporulation of *Colletotrichum truncatum* in culture media. *International Journal of Tropical Plant Diseases*, 12: 197–207.

Adira, V.A. (1983). Fruit rots of *Capsicum annum* L. and *C. frutescens* L. in Negiria. Part III. Effect of post harvest infection by two fusarial species on a few nutrients. *Nahrung*, 27: 669–674.

Akhtar, J. (2007). Effect of fungicides on growth and sensitivity of the isolates of *Colletotrichum capsici* causing fruit rot of chilli. *J. Pl.Dis.Sci.*, 2(1): 59–62.

Akhtar, J. and Singh, M.K. (2007). Studies on the variability in *Colletotrichum capsici* causing chilli anthracnose. *Indian Phytopath.*, 60 (1):63–67.

Alabi, O. and Emechebe, A.M. (1990). Field evaluation of seed treatment and foliar spray fungicides for the control of cowpea brown blotch induced by *Colletotrichum capsici. Samaru Journal of Agricultural Research*, 7: 151–158.

Ali, M.A., Ali, M., Huq, M. and Ahmed, M. (1993). In vitro studies on fungicides against *Colletotrichum gloeosporoides* (Penz.) sat-the die-back of tea. *Sri Lanka Journal of Tea Science*, 62: 25–31.

Astuti, E.B. and Suhardi (1986). The effect of storage temperature and fruit maturity on the incidence of anthracnose on chilli pepper. *Bull. Penelitian Hortik,* 13: 40–50.

Azad, P. (1992). Efficacy of certain fungi-toxicants against *Colletotrichum capsici* (Syd) Butter and Bisby, the incitant of ripe rot of chilli. *Journal of the Assam Science Society*, 34: 34–39.

Bansal, R.D. and Grover, R.K.(1969). Reaction of chilli (*Capsicum frutescens*) varieties of *Colletotrichum capsici J. Res.*, 6:345–348.

Barros, B.C., Castro, J.L. and Parici, F.R.A. (1999). Comparison of fungicides on control of bean diseases. *Arquivos do Instituto Biologico Sao Paulo*, 66: 53–58.

Biswas, A. (1992). Efficacy of fungicides in control of anthracnose disease of chilli in Sundarban region of West Bengal. *Journal of Mycopatological Research*, 30: 31–35.

Black, L.L. and Smith, B.J. (1993). *In vitro* activity of fungicides against *Colletotrichum fragariae. Acta Horticulutrae*, 348: 509–512.

Chacko, S., Khare, M.N. and Agarwal, S.C. (1978). Cellulolytic and pectinolytic enzyme production by *Colletotrichum dematium* f. sp. *Trurncatum* causing anthraconose of soybean.

Chakravarti, B.P. and Kumar, A.(1975). Control of seed borne infection of *Colletotrichum capsici* in chillies. *Curr. Res.*, 4: 172.

Chang, G.M. and Chi, P.K. (1993). Identification of falcate spores of *Colletotrichum* on medical plants cultured in Guangdong Province. *Acta Phytopathologica Sinica*, 23: 121–129.

Chauhan, M.S. and Duhan, J.C. (1977). Efficacy of some systemic and non systemic fungicidal compound to control anthracnose and ripe fruit-rot of chillies. *Pesticides*, 11: 17–18.

Chowdhary, S. (1957). Studies on the development and control of fruit rot of chillies. *Indian Phytopath.*, 10: 55–62.

Dasgupta, B. (1981). Sporulation and relative virulence among isolates of *Colletotrichum capsici* causing anthranose of betel vine. *Indian Phytopath.*, 34: 196–199.

Dasgupta, B. (1988). Variation in spore morphology among isolates of *Colletotrichum capsici* causing anthracnose of betel vine. *Journal of Plantation Crops*, 10: 65–66.

Dastur, J.F. (1922). Dieback of chillies in Bihar. *Mem. Dept. Agric.*, 11: 129–144.

Datar, V.V. (1995). Pathogenecity and effect of temperature on six fungi causing fruit rot of chilli. *Indian J. Mycol. Pl. Pathol.*, 25: 195–197.

Datar, V.V. (1996). Efficacy of growth regulators and fungitoxicants on fruit rot of chilli. *Indian J. Mycol. Plant Pathol.*, 26: 239–242.

Ebenezar, E.G. and Alice, D. 1996. Field evaluation of fungicides against fruit rot and die-back of chillies. *Indian Journal of Plant Protection*, 24: 50–52.

Ebenezar, E.G. and Narayanaswami, R. (1988). Effect of certain fungicides on the incidence of *C. capsici* in chilli varieties K-1 and K-2. *South Indian Horticulture*, 36: 205–206.

Freeman, S., Nizani, V. Dotan, S., Evaen, S. and Sando, T. (1997). Control of *Colletotrichum acutatum* in straw berry under laboratory greenhouse and field conditions. *Plant Diseases*, 81: 749–752.

Goswami, B.K., Dey, T. K., Kader, K.A., Ahmed, A., Neg, B.L. and Miraj, K.M. (1996). Efficacy and economics of fungicides against leaf spot (*Colletotrichum catechu*) disease of betel nut. *Bangladesh Journal of Plant Pathology*, 12: 1–13.

Grover, R.K. and Bansal, R.D. (1970). Seed borne nature of *Colletotrichum capsici* in chilli buds and its control by seed dressing fungicides. *Ind. Phytopath.*, 23: 664–668.

Gupta, B.D. (1988). Variation in spore morphology among isolates of *Colletotrichum capsici. Journal of Plantation Crops*, 16: 65–66.

Hiremath, P.C. and Govindu, H.C. (1973). Studies on market diseases of vegetables in Mysore state. *Mysore J. Agric Sci.*, 7: 428–435.

Jaylakshmi, C, Durairaj, P., Seetharaman, K. and Sivaprakasham, K. (1998). Biocontrol of fruit-rot and die-back of chilli using antagonistic microorganisrns. *Inilian Phytopath.*, 51: 180–183.

Joi, M.B. and Sonone, H.N. (1980). Chemical control of leaf curl, fruit rot and powdery mildew of chilli. *Journal Maharashtra Agril. Univ.*, 5: 219–222.

Joshi, C.G. and Wangikar,P.D. (1978). Efficacy of fungicides against *Colletotrichum capsici. Indian Phytopath.*, 31: 222–223.

Kadu, U.K., More, B.B. and Utikar, P.G. (1977). Laboratory and field evaluation of fungicides against *Colletotrichum capsici* (Syd.) Butler and Bisby the incitant of anthracnose of chillies. *Pesticides*, 11: 19–20.

Kakade, D.S. and Khure, N.N. (1989). A study on efficacy of some fungicides against selected phytopathogenic fungi. *Journal of Maharashtra Agricultural Universities*, 14: 141–142.

Kaur, S. and Singh, J. (1990). *Colletotrichum acutatum:* A new threat to chilli crop in Punjab. *Indian Phytopath.*, 43: 108–110.

Khan, G.M. (1978). Post harvest diseases of fruits and vegetables. *M.Sc. (Agric.) Thesis,* MAU, Parbhani Maharashtra, India.

Khan, M.K. (1994). New host records for some folicolous fungi from India. *Indian Phytopathology*, 47: 274–275.

Khatua, D.G., Das Gupta, B. and Sen, C. (1981). Foliage disease of betelvine caused by fungal pathogens, *Group Discussion on Improvement of Betelvine Cultivation*, NBRI, Lucknow, Feb. 27–28: 63–70.

Kogan, E.D. (1977). The pathogen of the anthracnose of fruits of solanaceous vegetable crops in Maldavia. *Referatavnye Zhurnal*, 8: 1317.

Kolesch, M.C., Cole, J.C. and Broembsen, S.L. (1995). Effectiveness of selected fungicides in controlling foliar diseases of Common periwinkle (*Vinea minor* L.) *Hort. Science*, 30: 554–557.

Kore, S.S. and Apet, K.T. (1989). Adaptation of *Colletotrichum capsici* to fungicides *J. Maharashtra Agric. Univ.*, 14: 103–104.

Laguna, I.G. and Docampo, D.M. (1971). Capsicum anthracnose in Carboda province. *Rev. Fac. Agron Vet. B. Aires*, 19: 47–52.

Lee, T.H., Chang, H.S., Lee, T.H. and Chung, H.S. (1995). Detection and transmission of seed borne *Colletotrichum gloeosporioides* in red papper, *Capsicum annuum. Seed Sci. Tehnol.*, 23: 533–541.

Lindner, K. Flath, K., Garba, V. Barteles, G., Breschewitz, B., Heidel, W., Hartleb, H., Bohlmann, J., Rittmann, B., Dittrich, R., Schmiechen, U. and Santhen, E.V. (2000). The effectiveness of chemical and physical seed treatment to control anthracnose in *Lupinus luteus*. In: *Lupin: An Ancient Crop for the New Millennium*, June 20–24, Germany, p. 57–59.

Maiti, S. and Sen, C. (1979). Fungal disease of betelvine. *P.A.N.S.*, 25: 150–157.

Maiti, S., Khatua, D.C. and Sen, C. (1978). Chemical control of two major foliage disease of betelvine. *Pesticides*, 12: 45–47.

Malhotra, R.K. and Chaturvedi, C. (1974). Physiological studies on *Colletotrichum trunctum* causing pod blight disease in soybean. *Frilsea*, 10: 4–5.

Mendes, M.C., Mendonea, H.A. and Demendonea H.A. (1996). Morphology, growth and conidial production of *Colletrotrichum lindemuthianum* in different culture of *Colletotrichum linedmuthionum* in different culture media. *Lieneia and Agro techologia*, 20: 475–479.

Mishra, A.P. and Mahmood, M. (1960). Factors affecting growth of *Colletotrichum capsici. Indian Phytopath.*, 13: 12–17.

Mishra, D. (1988). Fungicidal control of anthracnose and fruit rot of chilli. *Indian J. Agric. Sci.*, 58: 147–149.

Mohanraj, D.P., Vidhyasekaran and Kandaswamy (1972). An assessment of adaptation of *Gloeosporium ampelophagum* to fungicides. *Madras Agric. J.*, 54: 237–239.

Muthukrishnan, C.R., Thangarajk, T. and Chatterjee, R. (1988). Chilli and capsicum. In: *Vegetable Crops in India*, (Eds.) Bose and Som. Naya Prakash, Calcutta, India, p. 343–384.

Narain, A. and Panigrahi, C. (1971). Efficacy of some fungicidal compounds to control *Colletotrichum capsici in vitro* and *in vivo*. *Indian Phytopath.*, 24: 593–595.

Natrajan, S. and Subramanian, C.L. (1976). Fungicidal control of chilli anthracnose. *Madras Agril. J.*, 63: 423–424.

Oh, I.S., In: M.S., Woo, T.S., Lee S.K. and Yu, S.H. (1988). Anthracnose of pepper seedlings caused by *Colletotrichum coccodes*. *Korean J. Mycology*, 16: 151–156.

Palarpawar, M.Y. and Ghurde, V.R. (1997). Influence of different nitrogen sources on growth and sporulation of *Colletotrichum capsici* and *Colletotrichum curcumae*. *Indian J. Mycol. Plant Pathol.*, 27: 227–228.

Pan, S. and Sen, C. (1981). Adaptive ability of isolates of *Macrophomina phaseolina* towards four fungicides. *Indian Phytopath.*, 14: 346–354.

Park, W.M., Kim, S.H., Kim, S.H. and Ko, Y.H. (1989). Susceptibility of red pepper to *C. gloeosporoides* (penz) in relation to the ripening of fruits. *Korean J. Plant Pathol.*, 5: 260–270.

Perane, R.R. and Joi, M.B. (1988). Control of fruit rot and die back of chilli by seed treatment and spray. *J. Maharashtra Agric. Universities*, 14: 368.

Podaganur, G.M. and Naik, K.S. (1991). Mycoflora of chilli seeds from fruit rot affected and healthy fruit. *Curr. Res.*, 23: 183–184.

Pria, M.D., Bergamin Filtho, A. and Amarin, L. (1997). Evaluation of different culture media for sporulation of *Colletotrichum lindemuthianum, Phaesiariopsis griseola* and *Altenaria* sp. *Summa Phytopathologica*, 23: 181–183.

Pring, R.J., Nash, C., Zakaria, M. and Baiely, J.A. (1995). Infection process and host range of *Colletotrichum capsici*. *Physiological and Molecular Plant Pathology*, 46: 137–152.

Rai, I.S. and Chohan, J.S. (1966). Studies on variation and perpetuation of *Colletotrichum capsici* (Syd.) Butler and Bisby causing fruit rot of chillies in the Punjab. *J. Res.*, 2: 32–36.

Rajathilagam, R. and Kannabiran, B. (2001). Antagonistic effects of *Trichoderma viride* against anthracnose fungus *Colletotrichum capsici. Indian Phytopath.*, 54: 135–136.

Read, P.J. and Hide, G.A. (1995). Effects of fungicides on the growth and conidial germination of *Colletotrichum coccodes* and on the development of black dot disease of potatoes. *Annals of Applied Biology*, 126: 437–447.

Reddy, M.S., Pandu, S.R. and Rao, A.A. (1981). Development of resistance to *G. ampelophagum* and *C. capsici* to fungicides. *Proc. Indian Acad. Pl. Sci.*, 90: 535–539.

Roberts, R.G. and Snow, J.P. (1990). Morphological and pathological studies of *Colletotrichum capsici* and *C. indicum. Mycologia*, 82: 82–90.

Romer, P., Masutt, K., Rocha, J.P., Rocha, M.J.C. Santen, E.V., Wink, M. and Wieissman, S. (2000). Further trials to control anthracnose (*Colletotrichum capsici*) in white lupins (*Lupinus albus*) with chemicals. In: *Lupin: An Ancient Crop for the Mew Millennium* June, 20–24.

Sanchote, S. and Juangbhanich, P. (1984). Seed transmission of *Colletotrichum capsici* in pepper. *Kasetsart J. Nat. Sci.*, 18: 7–13.

Sankaralingam, A., Suthanthirapandian, I.R. and Ravikumar, V. (1983). Efficacy of certain fungicides against fruit rot of chillies. In: *Proceedings of the National Seminar on the Production Technology of Tomato and Chillies.* TNAV, Coimbatore, P. 165–166.

Singh, M.K. (2002). Studies on anthracnose of chillies and its incitant *Colletotrichum capsici* (Syd.) Butler and Bisby. *M.Sc. Ag. Thesis*, G.B. Pant University of Agriculture and Technology, India, 103. pp.

Singh, M.K., Akhtar, J., Kumar, A. and Abdul Khalid (2007). Sensitivity of different isolates of *Colletotrichum capsici* to *Trichoderma* spp. *Joumal of Eco-friendly Agriculture*, 2(1): 54–5.

Singh, R.A. and Shankar, G. (1971). Some parasitic fungi on *Piper bettle* L. in Varansi, U.P. *Mycopath. Mycol. Appl.*, 43: 109–115.

Singh, R.S. (1985). Use of *Epicoccum perpurescens* as an antagonist against *Microphomina phaxolina* and *Colletotrichum capsici. Indian Phytopath.*, 38: 258–262.

Singh, R.S. (1987). *Disease of Vegetable Crops*, 2nd Edition.Oxford and IBH Publishing Co., New Delhi, India, 362 pp.

Sinha, J.N. and Singh, A.P. (1993). *Ricinus communis*: A new host for *C. capsici. Indian Phytopathology*, 46: 97.

Sinha, P.P. (1990). Cost effective control of die-back and fruit rot of chillies. *Vegetable Science*, 17: 110–112.

Sitaramraju, K. and Srinivas, R.G. (1985). Effect of combined application of Dithane M–45 with different insecticides to control fruit rot and pest complex on chilli. *Indian J. Mycol Pl. Pathol.*, 15: 239–246.

Sivaprakasam, K., Jaganathan, K., Pi llayayarsamy and Anavaradhan, L.(1978). Control of dieback and fruit rot of chillies. *Vatika*, 1: 103–105.

Smith, R.W. and Crossan, D.F. (1958). The taxonomy, etiology and control of *Colletotrichum peperatum* (E and E) E and H and *Colletotrichum capsici* (Syd.) B. and B. *Plant Disease Reporter*, 42: 1099–1103.

Sook, H.K., Hyung, L.D., Han, K.S. and Lee, D.H. (1995). Identification and etiological characteristics of anthracnose fungi isolated from soybeans, Adzuki beans and mung beans. *Korean Journal of Plant Pathology*, 11: 30–38.

Srinivas, C., Niranjana, S.R., Praveen Kumar, L., Chandra Nayaka, S. and Shetty, H. S. (2006). Effect of fungicides and Bioagents against *Colletotrichum capsici* on seed quality of chilli. *Indian Phytopath.*, 59(1): 62–67.

Sutton (1980). *The Coelomycetes.* Commonwealth Mycological Institute, Ferrylane, Kew, Surrey, England.

Thakur, D.P. and Singh S.A. (1973). Efficacy of chemicals *in vitro* and *in vivo* in controlling fruit rot of chillies caused by *Colletotrichum capsici* (Syd.) Butler and Bisby. *Haryana J. Hort. Sci.*, 2: 128–133.

Thirupathaiah, V. and Subramanian, D. (1978). Production of pectic and cellutolytic enzymes by *Colletotrichum capsici* in the medium containing isolated cell walls of chilli fruits. *J. Indian Bot. Soc.*, 57: 114–116.

Tripathi, H.S. and Beniwal, S.P.S. (1978). Anthracnose of urdbean (*Vigna mungo* L.) caused by *C. capsici* (India). *Indian Phytopath.*, 30: 273–274.

Verma, L.R. and Sharma, R.C. (1999). *Diseases of Horticultural Crops: Vegetables, Ornamentals and Mushrooms* (Eds.). Indus Publishing Co., New Delhi, pp. 1460.

Verma, M.C. (1973). Comparative studies on virulence of isolates of four species of *Colletotrichum* parasitic on chillies. *Indian Phytopath.*, 24: 28–331.

Walker, J.C. (1952). *Disease of Vegetable Crops*. Mc-Graw Hill Book Co., Inc., N.Y., USA, pp. 529.

Wong, C.F.J., Nik, W.Z. and Lim, T.K. (1983). Studies on *Colletotrichum dematium f.sp. truncata* on soybean. *Pertanika*, 6: 28–33.

Yu, S.H., Park, J.S., Oh, I.S., Wu, I.S. and Mathur, S.B. (1987). *Colletotrichum coccodes* found in seeds of *Capsicum annuum* and pathogenicity of solanaceae plants. *Korena J. Mycol.*, 15: 183–186.

Plant Diseases Management for Sustainable Agriculture *Pages* **136–140**
Editor: **Dr. Shahid Ahamad**
Published by: **DAYA PUBLISHING HOUSE**

Chapter 12

Diseases of Saffron

P.K. Sharma[1], A. Vaid[2], J.N. Srivastava[3], D. Gupta[4] and Shahid Ahamad[5]

[1]*ICAR Research Complex for North Eastern Hill Region, Manipur Centre, Imphal*
[2]*Division of Plant Pathology, SKUAST, Jammu (J & K)*
[3]*Regional Horticulture Research Station, SKUAST, Jammu (J & K)*
[4]*Temperate Horticultural Research Station, Kotkhai, Shimla (H.P.)*
[5]*RARS, SKUAST-J, Rajouri*

Saffron (*Crocus sativus*) is a low growing perennial bulbous herb (15-35 cm in height) with narrow, linear radical leaves. It is a stemless plant belonging to Iridiacaae family. The stigmas along with orange colored partition of style designate saffron of commercial importance. Saffron is native to Southern Europe. The crop is known since the time of Greeks and the Hebrews. Its cultivation is confined to Pampore and Kishtwar in Jammu and Kashmir and in some pockets of Kinnaur, Lahaul and Spiti in Himachal Pradesh and Bharsar, Joshimath, Doonda and Chaubatia in Uttarakhand. The crop is subject to attack by fungal, bacterial and viral diseases which reduce its yield.

Corm Rot

A disease of crocus corms rot to prominence in the decade 1925-1935. Its pathogenicity was shown in 1928 (Drayton, 1928). A mysterious disease of the saffron crocus flourished in France until about 1898 when saffron ceased to be grown there. It was first described in 1785 (Bondaroy,1782). Thakur *et al.*(1992) identified *Macrophomina phasaeoli* as the cause of corm rot of saffron. They observed rotted corms of *C. sativus* during planting at Berwar- Kishtwar in August 1989 and 1990, with 30-40 per cent corms showing symptoms. This was the first report of corm rot of *C. sativus* in India. *F. oxysporum* f.sp. *gladioli* has been reported as the causal organism of corm rot by Shah and Srivastava (1984) and (Cappelli, 1994). Carta *et al.*(1982) reported the occurrence of *Macrophomina phaseolina* on saffron. Sud *et al.* (1999) reported *Fusarium solani* as the cause of corm rot in saffron in Kinnaur and

Lahaul and Spiti districts of Himachal Pradesh, where as Xu and Ge (1990) identified *Pseudomonas gladioli* as the causal organism of rot.

Symptoms

Infection begins either in the roots or at the base of the corms and it spreads through the lower parts of the latter, causing the tissues to become brown and rotten. Premature yellowing and dieing off of the foliage takes place. The decay proceeds upwards through the vascular tissues and becomes evident on the lower part of the corm as rings of blackish slightly sunken spots, each with a white centre. If a corm is cut through longitudinally, the rot appears as brownish black strands spreading from the base and following the course of the water conducting strands. The disease can be suspected from the appearance of outer covering scales. The rot gradually progresses in storage and ultimately affected corms are rendered worthless. Cappelli *et al.* (1991) reported *Penicillium corymbiferum* on saffron in Italy. In most severely affected areas, 40-50 per cent of corms were infected. Damping off, basal stem rot and drooping and wilting of shoots were observed. Francesconi (1974) observed the rotting of bulbs of *C. sativus* in the Abruzzese Appennines, Italy during 1972 when July-August was warm and damp. The rotting was caused by *Penicillium cyclopium*.

Control

The corms can not be cured once they are infected and the presence of these in the stock only serves to perpetuate the disease. Preventive measures supported by good crop husbandry leads to the maintenance of a healthy stock. At the time of uprooting of corms from the field, the rotted corms should not be allowed to remain in the field. They should be collected and burned. *Fusarium oxysporum* f. sp. *gladioli* on *C. sativus* was successfully controlled with difolatan (80 per cent captafol), bavistin (carbendazim) or benlate (benomyl) each at 0.2 per cent by dipping the corms for 20 minutes before planting in September (Shah and Srivastava, 1984). Cappelli *et al.* (1994) recommended the use of healthy reproductive material and chemical control using benzimidazoles for the prevention of *F. oxysporum* which is often, associated with *Penicillium corymbiferum* in saffron in Abruzzo, Italy. Sud *et al.* (1999) found bavistin (0.2 per cent) and tecto (0.2 per cent) to be highly effective in checking corm rot by dip or drench methods as evidenced by increased number of daughter corms produced and the yield of saffron, followed by difolatum and blitox respectively as dip method.

Smut

Smut is caused by *Urocystis gladiolicola*. W. G. Smith found the fungus in England in March 1876 and named it *Urocystis gladioli*. The fungus had been previously described as *Uredo gladioli*. Masse recorded it in 1908 at Kew.

Symptoms

Leaden colored, rounded, weal like swellings with yellow margins occur in the corm, stems and leaves of infected plants. The skin over these swellings splits and the dark brown powdery spore masses of the fungus are exposed. Sori may be found in leaves of plants grown from cormels. Spore mass is powdery and dark brown. Such leaves are malformed and the plant prematurely dies in the second year.

Control

Prolonged soaking before hot water treatment at 47°C for between 30 minutes to 1 hour kills mycelium and spores. Dry storage at 34°C for 6 weeks also gives good control.

Rust

Rust caused by *Uromyces croci* Pass attacks the leaves. The fungus is capable of producing an underground attack. Teleutospores in soil cause local infection of mother corms. Lesions on scales can penetrate deeply into the flesh and teleutosori may be produced at their centers within the corms. Mycelium may grow directly from the mother corm into the new developing corms, giving rise to systemic infections. Deep penetration of the corm tissues may be accompanied by darkening caused by the production of internal teleutosori [Boerema (1959); Boerema and van Kesteren (1965)]. Crocuses are also attacked by *Puccinia croci*.

Bacterial Scab

Scab caused by *Pseudomonas marginata* was first investigated in 1913 by McCulloch in America but no account appeared until 1921. In Europe, scab was studied by Fischer (1930) in Austria, by Rainio (1936) in Finland and Muntanola (1948) in South America.

Symptoms

Minute reddish brown specks appear on the leaves. The specks gradually enlarge and become circular or elongated dark brown spots with a slightly raised somewhat darker margin. The spots are few in number on the upper parts of the leaves and they generally remain small and do little or no damage. On the basal parts, where they are more numerous, the spots may coalesce to form extensive diseased areas in which the softer tissues become destroyed, while the veins are left more or less undamaged. The rot generally spreads inward and the whole of the foliage from soil level may become affected. A distinctive feature of the disease is copious gum like exudation from the corm lesions. Particles of soil are embedded in the mass of gum which also contains large numbers of bacteria. The disease is contracted from the soil or may be carried over the corm.

Young (1954) found an association between the amount of gladiolus scab and the prevalence of wire worms. Forsberg (1955) showed that an insecticidal soil treatment superimposed on fungicidal corm dips reduced scab index figures from about 93 per cent to about 4 per cent in one variety and 89 per cent to 55 per cent in another. It was found that corms free from white grub damage were also free from scab

Control

Thiram or formalin has been successfully used to control this disease. Nichols (1960) found that a proprietary mixture containing 60 per cent thiram and 15 per cent dieldrin dusted on to corms before planting gave very good control. Shumilenko and Sharanova (1963) reported good control of disease by dipping of corms in a 0.2 per cent thiram for 15 minutes. Pape (1958) found effective control for 30 minutes dip in 4 per cent formalin at 43-50°C. Milholland and Aycock (1965) reported that hot water treatment at 57°C alone gave good control.

Soft Rot

Severini (1913) described a soft rot of *Ixia maculate* observed in Italy. Bacteria was isolated from this host and named *Bacillus ixiae* Sever. This bacterium was capable of causing a rot of *Crocus sativus*. Mizusawa (1923) described a bacterial rot of roots and corms in the field, leading to yellowing and withering of foliage in Japan. The corms became dark brown and rotten and the rot was soft one, but under dry conditions the affected parts dried up and were then porous in texture. The original species having been named *Bacillus croci* Mizus, became *Erwinia croci* and is now included under *Erwinia carotovora* var. *carotovora*.

Viruses

Tobacco rattle virus causes stunting and necrotic flecks, streaks on leaves and flowers of infected crocuses in Netherlands (Slogteren, 1958). Cucumber mosaic virus causes leaf distortion, leaf chlorosis, flower streaking, and stunting in Germany and Sweden (Ryden, 1974). Russo *et al.* (1979) reported *Bean yellow mosaic virus* latently infecting *C. sativus*. The virus was identified on the basis of host range and symptoms, light and electron microscopy.

References

Boerema, G.H. (1959). *Versl. Meded. plZiektenk. Dienst* Wageningen, 13: 130–139.

Boerema, G.H. and Kesteren, H.A.Van. (1965). The underground attacks on *Crocus* and *Colchicum* by the rusts *Uromyces croci* and *Uromyces colchicii* respectively. *Neth. J. Pl. Path.*, 71: 136–144.

Bondaroy, F.De. (1782). Premiere memoire sur le safran. *Hist. Acad. Sci.* Paris, 89–105. Second memoire sur la maladie du safran connue sous le nom de nom de Tacon lb., 105–112.

Cappelli, C. (1994). Occurrence of *Fusarium oxysporum* f. sp. *gladioli* on saffron in Italy. *Phytopathologia Mediterranea*, 33(1): 93–94.

Cappelli, C., Buonaurio, R. and Polverari, A. (1991). Occurrence of *Penicillium corymbiferum* on saffron in Italy. *Plant Pathology*, 40(1): 148–149.

Cappelli, C., Donato, G. di. and Di-Donato, G. (1994). Attacks of *Fusarium oxysporum* Schlecht in crops of saffron in Abruzzo. *Informatore Agrario*, 50(25): 65–76.

Carta, C., Fiori, M. and and Franceschini, A. (1982). Charcoal rot of saffron (*Crocus sativus* L.) bulbs. *Studii-Sassaresi*, 3(29): 193–197.

Drayton, F.L. (1928). Examination of plant importations. *Rep. Dep. Agric. Ican. Botany*, 1929: 17–25.

Fischer, R.C. (1930). Phytopatholgische Mitteilungen. II. Uber die durch *Bacterium marginatum* verursachten Gladiolen-Krankheiten. *Zbl. Bakt.*, 11(81): 80–86.

Forsberg, J.L. (1955). The use of insecticides and corm and soil treatments for control of bacterial scab of gladiolus. *Pl. Dis. Reptr.*, 39: 106–114.

Francesconi, A. (1974). The rotting of bulbs of *Crocus sativus* L. by *Penicillium cyclopium*. *Annali-Botanica*, (1973), Publ. 1974, 32: 63–70.

Masse, G. (1908). The wild fauna and flora of the Royal Botanic Gardens, *Kew Bull. Add. Ser.*, 5: 223.

McCulloch, L. (1921). A bacterial disease of gladioli. *J. Agric. Res.*, 27: 225–230.

Milholland, R.D. and Aycock, R. (1965). Propagation of disease free gladiolus from hot water treated cormels in south-eastern North Carolina. *Bull.N. Carol. Agric. Exp. Stn.*, pp. 168.

Mizusawa, Y. (1923). A bacterial rot disease of saffrons. *Ann. Phytopath. Soc. Japan*, 1: 12.

Muntanola, M. (1948). La podredumbre del cuello del gladiolo. *Publ. Misc. Minist. Agric. Ser., A* 4: 41.

Nichols, L.P. (1960). Corm and soil treatment for the control of bacterial scab of gladiolus. *Pl. Dis. Reptr.*, 44: 417–418.

Pape, H. (1958). *Rep. 14th Int. Hort. Congress 1955*. The Hague, Netherlands, p. 1008–1018.

Rainio, A.J. (1936). Tutkimuksia gladiolus-Kasvien bakteeritaudeista (*Pseudomonas marginata* McCl, *Bacillus omnivorus* Hallja *B.variegatusi* Rainio nov.spec.) *ja niden* torjunnasts. *Vatt. Maatalouskoet. Julk*, 84: 102.

Russo, M., Martelli, G.P., Cresti, M. and Ciampolini, F. (1979). Bean yellow mosaic virus in saffron. *Phytopathologia Mediterranea*, 18: 189–191.

Ryden, K. (1974). Crocus breaking: A disease caused by an isometric virus. *Phytopath. Z.*, 80: 361–365.

Schumilenko, E.P.and Sharanova, A. (1963). Disinfection of flower planting material. *Zasch. Rasti. Moskva*, 8: 38.

Severini, G. (1913). Una bacteriosi dell' *Ixia maculate* e del *Gladiolus colvilli. Annali Bot.*, 2: 414–424.

Shah, A. and Srivastava, K.K. (1984). Control of corm rot of saffron. *Progressive Horticulture*, 16(1/2): 141–143.

Slogteren, D.H.M.Van. (1958). Ratelvirus als oorzaak van ziekten in bloembolgewassen de mogelijkheden de infectie door middle van grondontsmetting te bestrijden tijdschr. *PlZiekt*, 64: 452–462.

Smith, W.G. (1876). The gladiolus disease. *Monthly Microsc. J*, 16: 304–311.

Sud, A.K., Paul, Y.S. and Thakur, B.R. (1999). Corm rot of saffron and its management. *J. Mycol. Pl. Pathol.*, 29(3): 380–382.

Thakur, R.N., Singh, C. and Kaul, B.L. (1992). First report of corm rot in *Crocus sativus. Indian Phytopathology*, 45(2): 278.

Xu, C.X. and Ge, G.X. (1990). A preliminary study on corm rot of *Crocus sativus* L. *Acta Agriculturae Universitatis Zhejiangensis*, 16(2): 241–246.

Young, R.A. (1954). Fungicide-insecticide mixtures in pre-planting corm treatments for control of bacterial scab in gladiolus. *Pl. Dis. Reptr.*, 38: 55–56.

Plant Diseases Management for Sustainable Agriculture *Pages* **141–154**
Editor: **Dr. Shahid Ahamad**
Published by: **DAYA PUBLISHING HOUSE**

Chapter 13

Fruit-rot of Eggplant: A Threatening Disease

Jameel Akhtar[1], Abdul Khalid[2] and H.S. Chaube[2]

[1]*Department of Plant Pathology, Birsa Agricultural University, Ranchi – 834 006 (Jharkhand)*

[2]*Department of Plant Pathology, GBPUAT, Pantnagar – 263 145 (Uttaranchal)*

Phomopsis vexans causing pre and post-emergence damping-off of seedlings, stem canker, foliage blight and fruit-rot has probably been known since 1881, when it was described as a new pathogen on eggplant (*Solanum melongena* L.) in Italy (Spegazzini, 1881). Halsted (1892) was the first to describe the disease in New Jersey. It was later reported from Florida (Sherbakoff, 1915, 1916 and 1917; Gratz, 1942 and Harrison and Kelbert, 1944), Louisiana (Edgerton and Moreland, 1921), Barbados (Bourne, 1923), Bermuda (Ogilvie, 1924), New Jersey (Martin, 1928), Puerto Rico (Nolla, 1929), Philippines (Palo, 1936), Canada (Conners, 1939), Brazil (de Figueiredo and Pereira, 1944), Moritius (Orian, 1951; Felix *et al.*, 1965) and in India, Uppal *et al.* (1935). Thereafter, the occurrence of this disease as seedling blight, leaf spot, fruit-rot and stem canker has been reported from other states of India (Panwar *et al.*, 1970; Chowdhury and Hasija, 1979; Rangaswami, 1979; Akhtar and Chaube, 2003). In India it occurs in almost all states, including Punjab, Delhi, Uttar Pradesh, Uttarakhand, Madhya Pradesh, Maharashtra, Gujrat, of the country wherever the eggplant is cultivated.

Phomopsis blight and fruit-rot caused by *Phomopsis vexans* has been reported to be most serious and destructive disease of brinjal and the fruit-rot is distinctly the most destructive stage of disease which damages the fruit partially or completely in the field as well as during transit. Reduction in the fruit size and weight due to fruit-rot ultimately leads to reduced crop yield. Severe epiphytotics of disease have been recorded in the United States of America (Martin, 1928; Hyre, 1944). In Lauisiana, at least 50 per cent reduction in the yield was recorded in eggplant crop due to *P. vexans* infection in the field (Edgerton and Moreland, 1921). Later from Puerto Rico, Nolla (1929) also mentioned about 50 per cent loss in yield due to Phomopsis blight (*P. vexans*) in eggplant. According to de Figueirdo and

Pereira (1944) *P. vexans* caused such a devastating loss of brinjal crop that it rendered all the control measures ineffective. An epiphytotic caused by *P. vexans* in Mauritius was reported in 1964 (Felix *et al.*, 1965). The per cent yield losses due to fruit-rot ranged from 10 to 20 in Punjab and Delhi (Panwar *et al.*, 1970; Akhtar and Chaube, 2002 and 2003). The incidence of disease reduced the average yield of crop by 20 per cent in Jabalpur (Chowdhury and Hasija, 1979). Severe leaf blight and fruit-rot of brinjal was observed in Ugar, Belgaon district by Pawar and Patel (1957). Seed quality is adversely affected and infected seeds are reported to show poor germinabilty and reduced seed viability (Toole *et al.*, 1941; Porter, 1943; Pawar and Patel, 1957; Vishunavat and Kumar, 1993; Kaushal and Sugha, 1995; Akhtar, 2003). Thippeswami *et al.* (2005) reported poor germination and 30-50 per cent yield loss due to *Phomopsis vexans* and *Alternaria solani* in eggplant.

Symptoms

The disease symptoms produced by *P. vexans* have been described by various workers (Halsted, 1892; Harter, 1912; Edgerton and Moreland, I921; Nolla, 1929; Palo, 1936; Walker, I952; Pawar and Patel, I957; Lapis and Deangkinay, 1967; Panwar *et al.*, 1970; Kaushal, 1992 and Vishunavat, 1992; Akhtar *et al.*, 2005). In seedbeds, the disease causes "damping-off" of seedlings and "seedling blight", on stem and branches the disease manifests itself as elongated, blackish brown lesion–'the canker' (Figure 13.1).

When there is girdling of the stem due to infection, the shoot above the infected area shows necrosis, wilting and eventually dries-up. However, the disease is more prominent on leaves and fruits. On leaves it forms small almost circular, buffy-olive to cinnamon buff spots with irregular blackish margin. The older spots coalesce forming large irregular necrotic areas measuring from 3 to 16 mm in diameter and giving a blighted appearance to the plant. Affected leaves drop pre-maturely, and the blightened area is covered with numerous black minute pycnidial bodies (Figure 13.2).

Lesions on the petioles or at a lower part of midrib result in the death of the entire leaf. The diseased plants usually produce smaller leaves and the axillary buds are often killed.

On fruits, the disease appears as minute sunken grayish spots with brownish halo, which later enlarge and produce concentric rings with yellow and brown zones. The outermost rings get separated from the healthy fruit surface. These spots increase in size and form large rotten areas on which pycnidia develop, covering most of the rotten fruit surface. If the infection enters the fruits through the calyx, the whole fruit becomes mummified (Figure 13.3) due to dry rot (Pawar and Patel 1957; Panwar *et al.*, 1970; Kumar *et al.*, 1986; Akhtar *et al.*, 2005). Mature tissues exhibit hypertrophy and hyperplasia below the infected region, preventing further spread of the fungus (Divinagracia, 1968).

Etiology

This pathogen, and the disease is variously known as leaf blight, fruit-rot, leaf spot, stem blight and eggplant blight. Its cause have been studied by numerous investigators, among whom are Sherbakoff (1918), Edgerton and Moreland (1921), Nolla (1929), Palo (1936), Toole *et al.* (1941) and Howard and Desrosiers (1941), Uecker (1988).

The pathogen *Phomopsis vexans* is confined to the host eggplant (*Solanum melongena* L.) only (Ramanujam, 1966). Other genera, even of the family Solanaceae have not been found to be the host of the fungus even after repeated artificial inoculation (Howard and Desrosiers, 1941; Walker, 1952; Pawar and Patel 1957; Chupp and Sherf, 1960; Ramanujan, 1966).

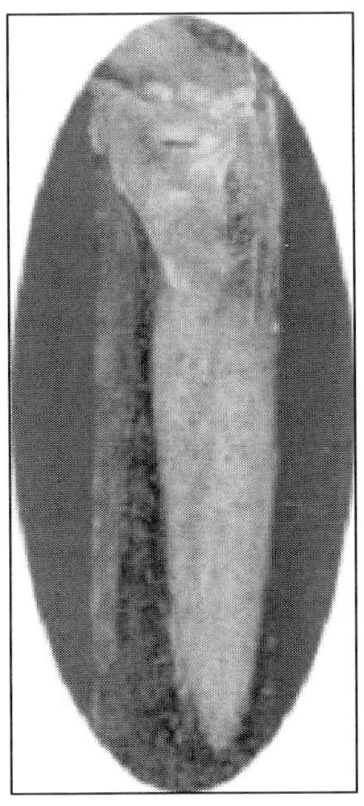

Figure 13.1: Infected Plants Showing Typical Symptoms of Stem Canker

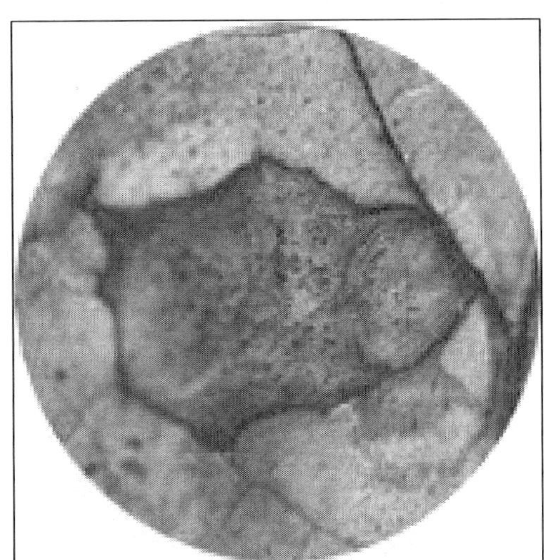

Figure 13.2: Leaf Showing Typical Blightening Area with Numerous Minute Pycnidia

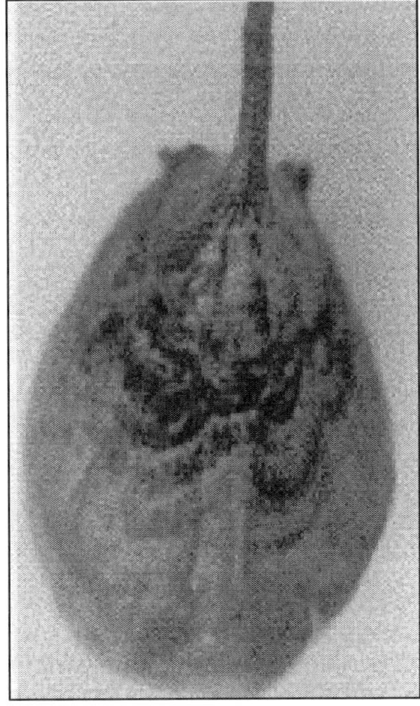

(a)

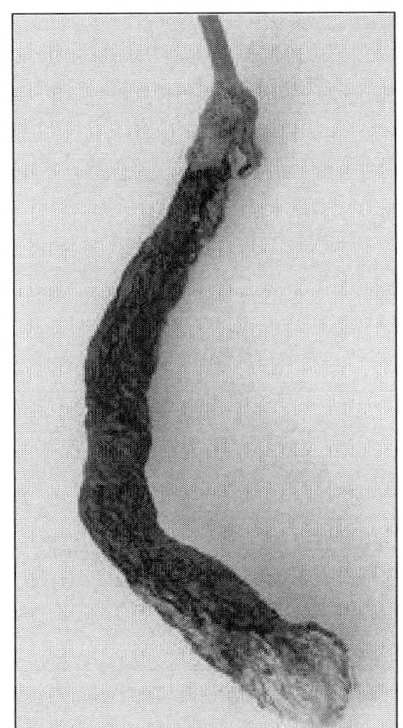

(b)

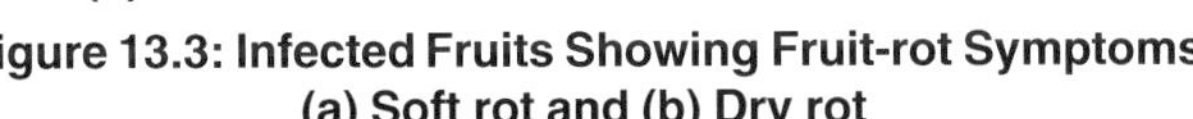

Figure 13.3: Infected Fruits Showing Fruit-rot Symptoms (a) Soft rot and (b) Dry rot

Spegazzini (1881) described a fungus occurring on foliage of eggplant as *Phyllosticta hortorum*. Later, the same fungus was reported from fruits and leaves of eggplants by Halsted (1892) in New Jersey and named as *Phoma solani*. The name *Phoma solani* assigned by Halsted was preoccupied by Cooke and Harkness for a fungus on another host, therefore, Saccardo and Sydow (1899) substituted *Phoma vexans* for the *Phoma solani*. Later Smith (1905) proposed a name *Ascochyta hortorum* instead of *Phyllosticta hortorum*. Voglino (1907) studied a fungus on eggplant in Italy and concluded that the fungus, previously described by Spegazzini (1881) as *Phyllosticta hortorum* is *Ascochyta hortorum*. Further investigations conducted by Harter (1912) indicated that *Phoma solani* and *Phyllosticta hortorum* were the same, and concluded that the genus to which these belong is neither *Phoma*, *Phyllosticta* nor *Ascochyta* but belongs to genus *Phomopsis* and proposed the name *Phomopsis vexans*.

Cultures of the pathogen may show different colony colour such as white, dirty white or blackish. Colony may be slightly irregular, regular or circular. Mycelium may be aerial, subaerial or appressed towards periphery. Texture of the fungus may be bushy tufts or compact. Zonation in the colony may be defined or undefined. Distribution of pycnidia in culture may be few and scattered, numerous and scattered around the point of inoculation or very few and scattered (Panwar and Chand, 1968, Akhtar and Chaube, 2006). The mycelium is hyaline and septate measuring 2.7 to 3.9 µm in diameter. The pycnidia may or may not contain beak. Pathogen produces spherical or flask-shaped and invariably black to brown coloured pycnidia (Figure 13.4a). The conidiophores (phialids) in the pycnidium are hyaline, simple or branched, some times septate, 10-16 µm long, and arise from innermost layer of the cells lining the pycnidial cavity. Alpha conidia are hyaline, single celled, subcylindrical to elliptical and 5.0-12.9 x 2.0-3.4 µm in size. Number of guttulae (oil drops) present in α conidia may vary from 1-3 (Figure 13.4b). Another form of conidia, the stylospores (β conidia), are either filiform or curved or sickle shaped, hyaline and septate and measure about 14.2-32.5 x 0.5-1.6 µm in size (Figure 13.4c). Unipolar germination of *a* conidia takes place. Bipolar germination too has been recorded (Figure 13.4c) (Akhtar, 2006b; Akhtar and Chaube, 2006). Stylospores do not germinate.

The perithecia under culture conditions are usually in clusters, 130-350 µm in diameter, beaked, carbonaceous, sinuate, with irregular ostiole. The beak is 80-500 µm long. Asci are clavate, sessile and contain eight spores, each measuring 28-44 x 5-12 µm in size. Ascospores are hyaline narrowly ellipsoid to bluntly fusoid with one septate and constricted at the septum. They measure 9-12 x 3-4.4 µm in size (Singh, 1987).

Several workers have recorded variation in the natural population of pathogen. Panwar and Chand (1968) grouped nine isolates of *Phomopsis vexans* into three strains based on their cultural and pathogenic behaviour. These strains/isolates differed in their colony colour, colony type, growth rate, type of spores and sporulation density.

Similarly, Islam and Pan (1990a) while studying the morphological characteristics of 16 isolates of *Phomopsis vexans* collected from diseased eggplant from various locations in India, found that isolates varied in colony diameter (69.00-94.33 mm), dry weight (436.66-642.66 mg), relative abundance of conidia production, colour, texture, zonation and arrangement of pycnidia in colonies and pathogenicity.

Akhtar and Chaube (2003 and 2006) while studying the morphological/cultural characters of thirty two isolates of *Phomopsis vexans* collected from different places of U. P. and Uttarakhand grouped into five strains based on their morphological/cultural and pathogenic behaviour. These strains/isolates differed in their colony colour, colony type, growth rate, type of spores and sporulation density.

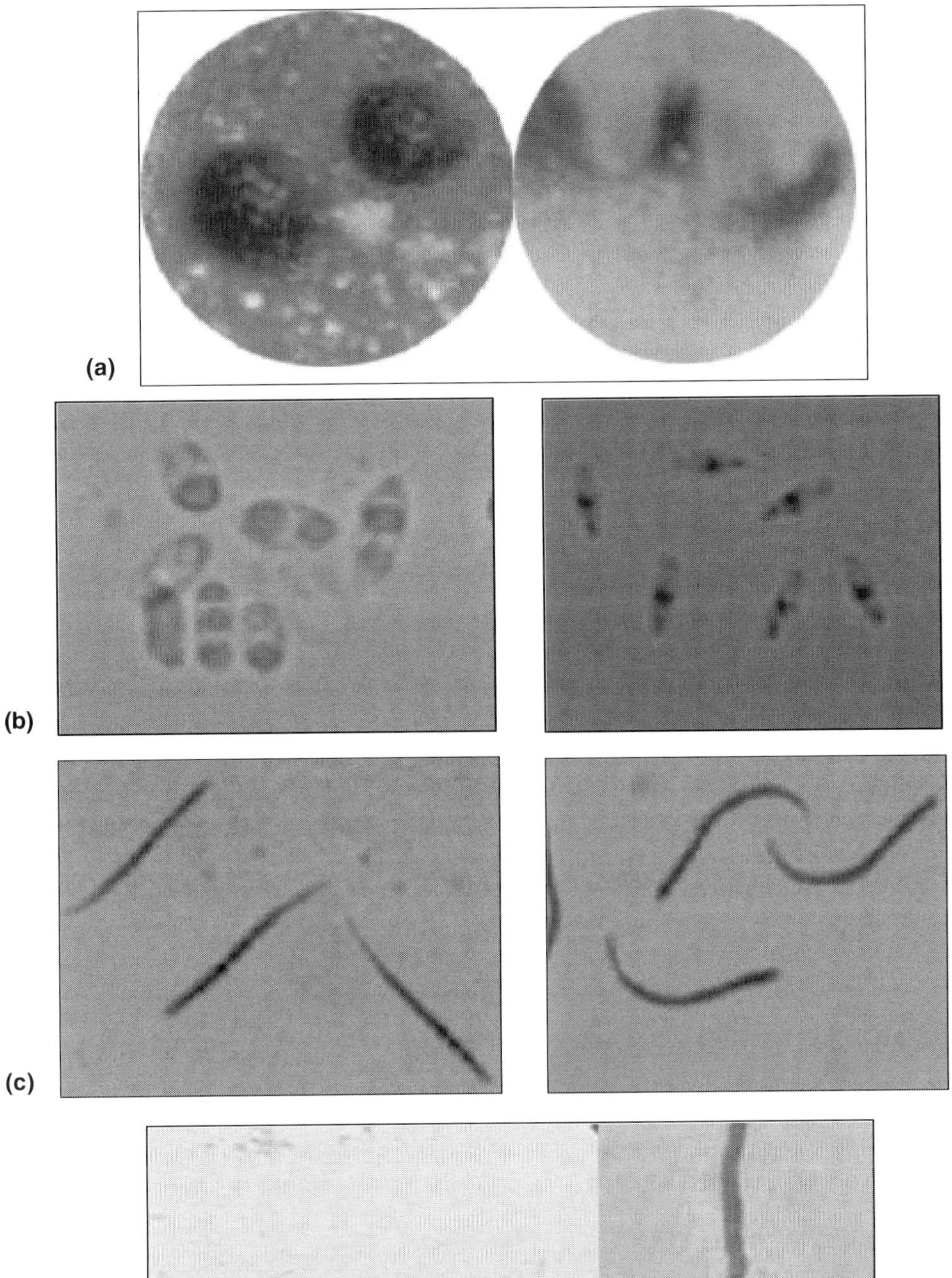

(a)

(b)

(c)

(d)

Figure 13.4: Pycnidial and Conidial Morphology of *P. vexans*
(a) Spherical and flask-shaped pycnidia; (b) Subcylindrical and elliptical α conidia; (c) Filiform and sickle shaped stylospores and (d) Unipolar and bipolar germination of α conidia

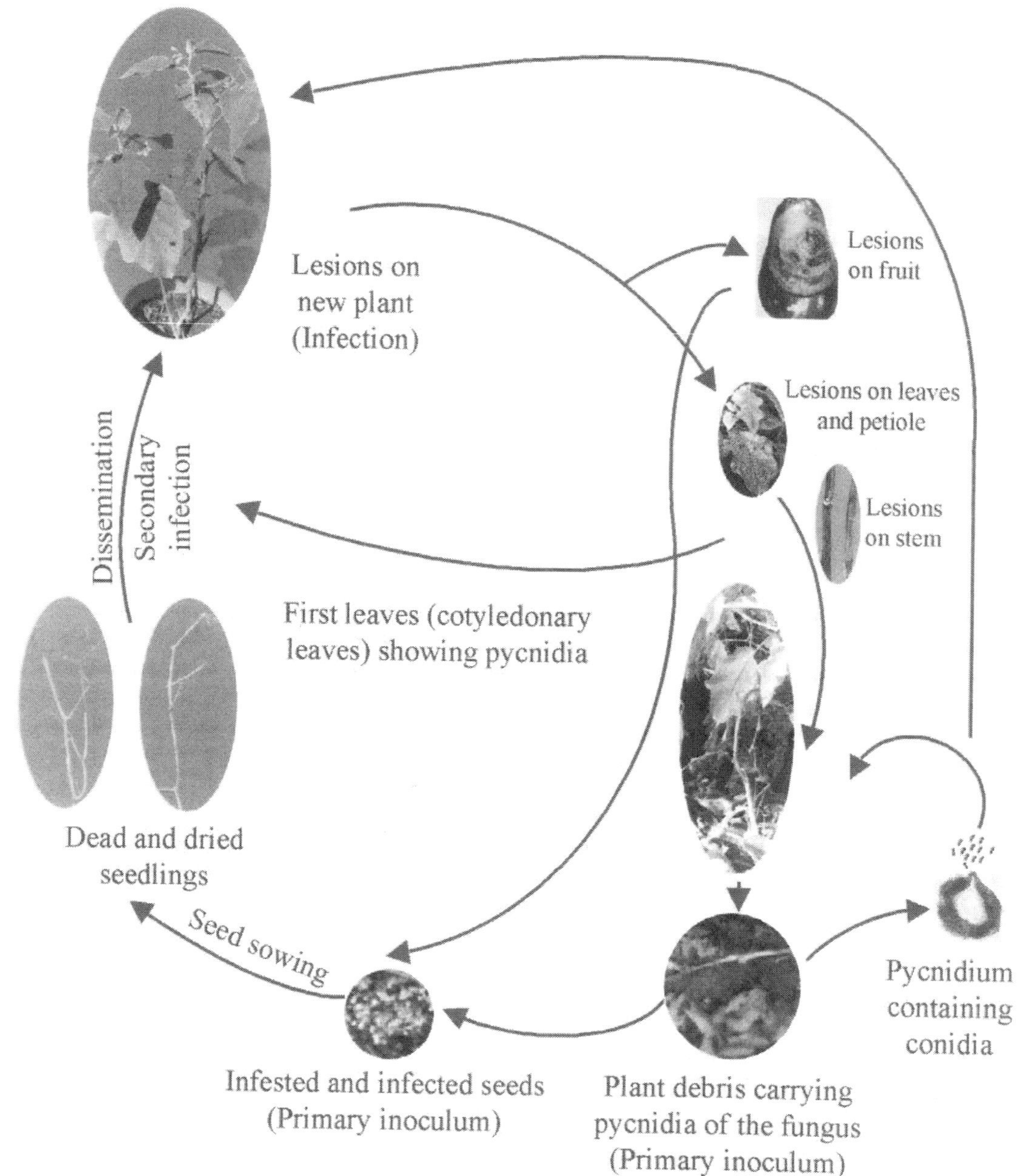

Figure 13.5: Disease Cycle of Fruit-Rot of Brinjal

Five isolatcs of *Phomopsis vexans* were evaluated against 25 genotypes of brinjal showed substantial differences in their aggressiveness. Disease severity ranged from moderately susceptible to highly susceptible and resistant to moderately resistant. Aggressiveness of 5 isolates of *P.vexans* on the basis of responses shown by 25 host genotypes revealed that mean aggressiveness response value in terms of incubation period was 0.57 (range 0.48–0.68) and in terms of lesion area, it was 0.56 (range 0.42–0.70). The mean response values for isolates against the genotypes ranged from 0.42 to 0.70 (Akhtar, 2006[c]).

Host-pathogen interactions in terms of parasitic fitness revealed differences in virulence/aggressiveness and the host reactions upon inoculation of 5 isolates of *P. vexans* on 12 brinjal genotypes. Isolate Pv 36 was the most aggressive followed by Pv 16 and isolate Pv 11 was the least aggressive. When evaluated against 12 genotypes of brinjal all the isolates of *P. vexans* showed differences in reactions. The difference in the amount of disease produced on fruits and leaves by the pathogen isolates was significant (Akhtar, 2007b). Noorj and Damme (1988) have also studied differences in mean pathogenecity within the population of the fungus *Phomopsis* sp. infecting *Plantago lanceolatct*.

Five monoconidial isolates of *P. vexans* collected from diverse locations varied considerably at morphological and cultural level when grown on nine different culture media. The colony colour ranged from white to dull white to blackish white to yellowish white to grayish with circular to irregular colony shape. Zonation in cultures of different isolates on different media varied from distinct to indistinct to absent. Colony growth rate ranged from slow to fast. Radial growth observed on culture media revealed PDA as best, while Richard's Agar as least supporting medium (Akhtar, 2007[a]).

Disease Cycle

The pathogen is seed transmissible (Nolla, 1929; Vishunavat and Kumar, 1993, 1994 and Pan and Acharya, 1995). Martin (1930, 1934) reported that the fungus *P. vexans* is seed-borne with infection occurring under the seed coats from badly decayed fruits. Several other workers have also reported *P. vexans* to be both internally as well as externally seed-borne (Edgerton and Moreland, 1921; Toole *et al.*, 1941; Porter, 1943; Suryanarayana *et al.*, 1962; Panwar *et al.*, 1970; Singh and Chakrabarti, 1982 and Thippeswami *et al.*, 2005).

Thippeswamy *et al.* (2005) found that the *P. vexans* was distributed in different seed parts. *P. vexans* ranged 0.0-2.0. per cent in seed coat, 0.0-4.0 per cent in cotyledons and 0.0-4.0 per cent in embryonic axis.

Akhtar (2003) reported that the isolates of the causal organism differed in their ability to cause seed infection. The different grades of infection caused differences in seed viability and germinability too.

Fungus also survives on diseased crop debris in off-season (Ogilvie, 1924; Panwar *et al.*, 1970). Infected seedlings bear pycnidia on first true leaf which serves as primary inoculum. The conidia are locally disseminated by raindrop splashes to other plants. Thus, the inoculum may be brought in contact with healthy seedlings either through the infected seedlings from the sick nursery or from diseased crop debris already present in field. Dissemination may be by tools and insects as well (Figure 13.5).

Predisposition

P. vexans requires hot and humid climate for infection and subsequently for disease development. A number of workers have studied the factors affecting growth and sporulation of the fungus *in-vitro* (Singh and Chand, 1986).

The optimum relative humidity for disease development is 55 per cent and above (Lapis and Deangkinay, 1967). Singh (1987) observed that the disease is favoured by wet weather with temperature around 20°C and temperature of 5, 10 and 40°C were unfavourable for disease development. The mode of infection of eggplant on artificial inoculation with conidia of *P. vexans* has been studied (Divinagracia, 1969). Temperature of 25°C and RH 55 per cent for 15 days are reported most congenial for fruit-rot of brinjal (Ahmed and Ahmed, 1987). Islam and Pan (1990b) observed good correlation between environmental factors and Phomopsis leaf blight of brinjal and revealed the significance of temperature, humidity, rainfall and number of rainy days on disease severity. Kumar and Sugha (2006) reported that temperature in the range of 21.4-24.5°C favoured the disease development optimum being 23-25°C. Akhtar (2006) reported that temperature below 20±1°C favours conversion of α conidia to β form and above 20±1°C, β conidia started to convert into α form.

Management

Phomopsis blight in eggplant is a serious disease in India as well as different countries of the world. Sayre (1924) advocated the use of copper lime dust in controlling fruit-rot of brinjal. Nolla (1929) suggested soil sterilization with 2.0 per cent formaldehyde solution at the rate of half gallon per square foot before raising the crop.

Several workers have tried to control seed-borne inoculum by hot water treatment. Martin (1930) recommended seed treatment with hot water (55°C for 30 min.) for the control of *P. vexans*. Felix *et al.* (1965) in Mauritius recommended that the seed-borne inoculum of *P. vexans* may be controlled by hot water seed treatment at 50°C for 30 min. Similarly, Singh (1987) has also reported that hot water seed treatment at 50°C for 15 min controlled the seed-borne inoculum of *P. vexans* of brinjal effectively. Seed treatment followed by four foliar sprays of carbendazim gave excellent control of fruit rot than copper oxychloride and mancozeb (Sinha, 1989). Porter (1943) obtained better and more effective control of leaf spot due to *P. vexans* with 10-90 Femate (Ferrid dimethyl dithiocarbamate) talc dust. de Figueiredo and Pareira (1944) recommended the application of Bordeaux mixture to growing plants for the control of Phomopsis blight in Brazil. Fresa (1963) found that captan and ziram were promising for the control of *Phomopsis vexans*. Bordeaux mixture and copper oxychloride at 0.25 per cent proved effective against *Phomopsis vexans* on brinjal (Panwar and Chand, 1969).

Significant improvement in seedling stand was observed by seed treatment with carbendazim, thiram, and carbendazim+thiram. However, carbendazim alone or in combination with thiram provided best control of pre and post-emergence damping-off (Kaushal and Sugha, 1995).

Vidhyasekaran *et al.* (1980) reported that bavistin is highly effective against fruit-rot of brinjal. They also reported that seed treatment with bavistin+thiram controlled most of the fungi effectively in the seedlings of brinjal.

Teo (1980) obtained that stem blight infection on eggplant was reduced by a mixture of copper oxychloride + benlate (benomyl) or copper oxychloride alone, a reduction of 32.73 per cent in the disease was reported by benomyl. Teo (1982, 1984) reported that spray of trimanzone, copper oxychloride, and captan reduced the disease incidence but benomyl was not effective. He also suggested that copper oxychloride alone or in combination with benomyl is most effective against the disease. Grewal and Jhooty (1987) in India obtained best results with dithane Z-78 at 0.2 per cent. Singh and Chakrabarti (1982) reported that sprays with difolatan and captan gave effective control against the disease but the fruits touching the ground were observed to rot irrespective of fungicide treatment.

Jacqua and Gerion (1988) in France recommended chlorothalonil (deconil) and a mixture of captafol and carbendazim and maneb + thiophenate methyl spray at 10 to 12 days intervals for control of disease. Bavistin @ 0.1 per cent was reported to provide best control of the disease with increased yield (Islam and Pan, 1989). Mohanty *et al.* (1994) tested five fungicides (blitox-50, bavistin, dithane M-45, ridomil MZ and difolaton) *in- vitro* against *P. vexans* and reported that all the five chemicals inhibited mycelial growth significantly at all the three concentrations (0.1, 0.2 and 0.3 per cent), however, bavistin caused 100 per cent inhibition. Three sprays of carbendazim alone or in combination with mancozeb and copper oxychloride showed better curative and protective action against the pathogen as well as superior control of fruit-rot of brinjal (Anonymous, 1991; Kaushal, 1992).

Akhtar, (2007c) reported that five monoconidial isolates of the fungus, *P. vexans*, when exposed to different concentrations of three fungicides namely, carboxin, carbendazim and mancozeb, employing two basic techniques–Poisoned Food and Spore germination, differences were significant in growth and inhibition of spore germination of the isolates. Differences in sensitivity of the isolates to a particular fungicide and concentration were clearly observed.

Kaushal and Sugha (1995) obtained that seed treatment with carbendazim alone or in combination with thiram gave best control of pre and post-emergence damping-off caused by *P. vexans*.

Das (1995) studied bioefficacy and economics of fungicidal control of Phomopsis leaf blight/ fruit-rot of brinjal in Orissa and reported that 3 foliar sprays with carbendazim (0.15 per cent) gave maximum protection against leaf blight and fruit-rot (76.97 and 74.62 per cent, respectively).

Thippeswamy *et al.* (2006) observed that among fungicides tested against *P. vexans* by standard blotter method mancozeb, carbendazim and captaf were superior in inhibiting seed borne inoculum.

Mohanty *et al.* (1995a) investigated aqueous leaf extract of five plants for allelopathic control of *P. vexans* and observed maximum fungal growth inhibition (93.75 per cent) with leaf extracts *of Allemanda cathertica.* Leaf extracts of *Catharanthus roseus, Polyalthia longifolia* and *Azadirachta indica* were found equally effective.

Homeopathic drugs from *Thuja* spp. and *Teucnum* spp. each at 2.0 per cent were reported to inhibit the growth of *P. vexans* as effective as fungicide bavistin and leaf extract of *Allemanda cathartica* were found to have excellent potential as a fungicide (Mohanty *et al.* (1995b; Panda *et al.*, 1996).

Host resistance is another strategy of disease management that is most economical and environment friendly. Decker (1951) bred and released two eggplant varieties, Florida Market and Florida Beauty, resistant to Phomopsis blight. Kalda *et al.* (1976) tested over 300 entries including *Solanum* spp, eggplant cultivars and interspecific F_1 hybrids for resistance to *P. vexans* and found that two *S. melongena* lines were resistant and F_1 hybrids showed variable reactions. Resistance to Phomopsis blight was observed to be recessive and governed by polygenes.

Regarding the sources of resistance, Ahmad and Ahmed (1987) reported that two wild spp. of *Solanum viz. S. gilo* and *S. sismbrifolium,* were immune to Phomopsis blight and may provide the plant breeders with a source of resistance.

References

Ahmad, Q. and Ahmad, Quaiser (1987). Source of resistance in brinjal to phomopsis fruit rot. *Indian Phytopath.,* 40:1, 98–99.

Akhtar, J. (2003). Studies on Phomoposis blight of brinjal with special reference to variability in *Phomopsis vexans* (Sacc and Syd.), the incitant of the disease. *Ph. D. Thesis,* G. B. Pant University of Agriculture and Technology, India, 156. pp.

Akhtar, J. (2006*a*). Intraspecific diversity in *Phomopsis uexans*, inciting Phomopsis blight of brinjal. *Journal of Research (BAU)*, 18 (1): 81–87.

Akhtar, J. (2006*b*). Variability in production, interconversion and germinability of α and β conidia in *Phomopsis vexans*. *J. Mycol Pl. Pathol.*, 36(2):193–197.

Akhtar, J. (2006*c*). Host-pathogen interaction with reference to aggressiveness of the isolates of *Phomopsis vexans* to different genotypes of the brinjal. *J. Pl. Dis. Sci.*, 1(2): 164–167.

Akhtar, J. (2007*a*). Effect of nutritional factors on growth and sporulation of *Phomopsis vexans* isolates. *J. Pl. Dis. Sci.*, 1(2): 129–131.

Akhtar, J. (2007*b*). Host-pathogen interaction and parasitic fitness of *Phomopsis vexans* to brinjal genotypes. *J. Mycol. Pl. Pathol.*, 37(1): 22–24.

Akhtar, J. (2007*c*). Bioefficacy of fungicides and sensitivity of the isolates of *Phomopsis vexans*. *Pantnagar Journal of Research*, 5(1):66–524.

Akhtar, J. and Chaube, H.S. (2002). Morphological and pathogenic variability of *Phomopsis vexans*, the incitant of fruit-rot of brinjal. *J. Mycol. Pl. Pathol.*, 32(2): 432.

Akhtar, J. and Chaube, H.S. (2003). Studies on Phomopsis blight of brinjal with special reference to variability in *Phomopsis vexans*, the incitant of the disease. *J. Mycol. Pl. Pathol.*, 33(3): 465.

Akhtar, J. and Chaube, H.S. (2005). Phomopsis blight of brinjal: A review. In: *Biotechnology: Plant Health Management*, (Eds.) Neeta Sharma and H.B. Singh. International Book Distributors, Lucknow, pp. 421–438.

Akhtar, J. and Chaube, H.S. (2006). Variability in Phomopsis blight pathogen [*Phomopsis vexans* (Sacc. and Syd.) Harter]. *Indian Phytopath.*, 59(4): 439–444.

Akhtar, J., Lal, H.C., Kumar, A., Ekka, S. and Chaube, H.S. (2005). Studies on symptomatology and severity of fruit–rot of brinjal on different genotypes/cultivars caused by different isolates of *Phomopsis vexans*. *Indian Phytopath.*, 58(3): 363.

Anonymous. (1991). *Annual Progress Report for 1990–91*. Department of Plant Pathology, HPKV, Palampur, 128 p.

Bourne, B.A. (1923). Report of the Assistant Director of Agriculture on the entomological and mycological work carried out during the period 1922–23. *Rev. Rept. Dept. Agric. Barbados*, 1922–23, pp. 7–9.

Chaudhury, S.R. and Hasija, S.K. (1979). Phytopathological studies on *Phomopsis vexans* causing soft rot of brinjal fruits. *Indian Phytopath.*, 32: 495–496.

Chupp, C. and Sherf, A.F. (1960). *Vegetable Disease and their Control*. Ronald Press Co., New York.

Conners, I.L. (1939). *Eighteenth Annual Report of the Canadian Plant Disease Survey*, 1938, 12: 112.

de Figueiredo, E.R. and Pereira, H.F. (1944). Uma doenca grave da Berinjela causada for *Phomopsis vexans* (A serious egg plant disease caused by *Phomopsis vexans*). *Biologico*, 10:349.

Das, S.R. (1995). Bioefficacy and economics of fungicidal control of Phomopsis leaf blight/fruit rot of brinjal in Orissa. *Orissa J. Hort.*, 23: 1–2.

Divinagracia, G.C. (1968). Pycnidial production by *Diaporthe vexans* (Sacc. and Syd.) Gratz. and the pathological history of infected stem and leaves of the eggplant. *Plant Dis. Abstract*, 29: 836–837.

Divinagracia, G.C. (1969). Some factors affecting pycnidial production of *P. vexans* in culture. *Philipp. J. Agric.*, 53: 173.

Edgerton, C.W. and Moreland, C.C. (1921). Egg plant blight. *Louisiana Agric Exp. Stat. Bull.*, 178: 1–44.

Felix, S., Lallmohamed, G.M. and Orienx, L. (1965). Eggplant wilt *Diaporthe vexans* (Saccardo and Sydow) Gartz in Mauritius. *Rev. Agric. Surc. Ile Maurice*, 44: 182–186.

Fresa, R. (1963). *Corticum solani* in egg plant. *Idia.* 184:1–12.

Gratz, L.O. (1942). The perfect stage of *P. vexans. Phytopathology*, 32: 540.

Grewal, R.K. and Jhooty, J.S. (1987). Control of Phomopsis fruit rot of egg plant with different fungicides. *Indian Phytopath.*, 40: 263–264.

Halsted, B.D. (1892). Some fungus diseases of the egg plant. *New Jursey Exp. Stat. Twelfth Ann. Rept. 1891*, pp. 277–283.

Harrison, A.L. and Kelbert, D.G.A. (1944). Late blight in Florida. *Plant Dis. Reptr.*, 28:116.

Harter, L.L. (1912). Fruit rot, leaf spot and stem blight of the egg plant caused by *Phomopsis vexans. J. Agric. Res.*, 5: 331–338.

Howard, F.L. and Desrosiers, R. (1941). Studies on the resistance of egg plant varieties to Phomopsis blight. *Proc. Amer. Soc. Hort. Sci.*, 39: 337.

Hyre, R.A. (1944). Diseases of solanaceous crops in Southern Florida. *Plant Dis. Reptr.*, 28: 117.

Islam, S.J. and. Pan, Sitansu. (1989). Chemical control of leaf blight and fruit rot of hrinjal caused by *Phomopsis vexans. Indian J. Mycol. Res.*, 27: 159–163.

Islam, S.J. and Pan, Sitansu. (1990a). Variabilities among isolates of *Phomopsis vexans. Environment and Ecology*, 8: 315–319.

Islam, S.J. and Pan, Sitansu. (1990b). Effect of humidity and temperature on Phomopsis fruit rot of brinjal *Solanum melongena. Environment and Ecology*, 8: 1309–1310.

Jacqua, G. and Gerion, A.L. (1988). *Corynespora crassicola* and *Phomopsis vexans* in aubergine crop. *Bulletin Agronomique Antilles–Guyana*, 7: 39.

Kalda, T.S., Swarup, V. and Choudhury, B. (1976). Studies on resistance to phomopsis blight in egg plant (*S. melongena* L.). *Vegetable Sci.*, 3: 65–70.

Kaushal, N. (1992). Studies on Phomopsis blight and fruit rot of brinjal. *M.Sc.(Ag.) Thesis*, HPKV, Palampur.

Kaushal, N. and Sugha, S.K. (1995). Role of *Phomopsis vexans* in damping-off of seedlings in eggplant and its control. *Indian J. Mycol. Pl. Pathol.*, 25(3): 189–191.

Kumar, A., Aulakh, K.S. and Grewal, R.K. (1986). Incidence of fungal fruit rots of brinjal in Punjab. *Indian Phytopath.*, 39: 482–485.

Kumar, S. and Sugha, S.K. (2006). Role of weather variables in the development of Phomopsis blight of brinjal. *Pl. Dis. Res.*, 21(1): 30–32.

Lapis, D.B. and Deangkinay, M.T. (1967). Factors affecting growth and sporulation of *Phomopsis vexans* Harter. *Philipp. J. Agric.*, 50: 276–288.

Martin, W.H. (1928). Report of the Department of Plant Pathology. *Ann. Rept. New Jersey Agric. Exp. Stat.* for the year ending June 30, 1927, 48: 205–238.

Martin, W. H. (1930). *Plant Pathology*. Ann. Rept. New Jersey Agric. Exp. Stat. for the year ending June. 30, 1929. 44: 235.

Martin, W. H. (1934). *Plant Pathology*. Ann. Reptr. New Jersey Agric. Exp. Stat. for 2 year period ending June 30, 1933, 57–66.

Mohanty, A.K., Panda, R.N., Sethi, P.N. and Kaur, A.K. (1994). Efficacy of certain fungicides in controlling Phomopsis vexans causing fruit-rot of brinjal. *Orissa J. Agric. Res.*, 7: 85–86.

Mohanty, A.K., Kar, A.K. and Sethi, P.N. (1995a). Efficacy of crude leaf extract of some selected plants in controlling brinjal blight and fruit rot pathogen, *Phomopsis vexans. Crop Res.*, 9(3): 447–448.

Mohanty, A.K., Panda, R.N., Kar, A.K. and Sethi, P.N. (1995b). Effect of certain homeopathic drugs on the growth of *Phomopsis vexans* causing fruit rot of brinjal. *Crop Res.*, 9(1): 165–167.

Nolla, J.A.B. (1929). The egg plant blight and fruit rot in Puerto Rico. *J. Dept. Agric. Puerto Rico*, 13: 35–37.

Ogilvie, L. (1924). Preliminary Report of the Plant Pathologists for the period Sept. 27th to Dec 31st, 1923. Report Board and Deptt. of Agriculture, Bermuda, p 28.

Orian, G. (1949). Division of Plant Pathol. Rep. Deptt. Agric. Mauritius. 1949. pp. 66–72, 1950, pp. 80–85.

Palo, M.A. (1936). The Phomopsis disease of egg plant and its control. *Phiippl. J. Agic.*, 7: 289–315.

Pan, S. and Acharya, S. (1995). Studies on seed borne nature of *Phomopsis vexans* (Sacc. and Syd.) Harter. *Indian Agriculturist*, 39: 193–198.

Panda, R.N., Tripathy, S.K., Kar, J. and Mohanty, A.K. (1996). Antifungal efficacy of homeopathic drugs and leaf extracts in brinjal. *Environment and Ecology*, 14(2): 292–294.

Panwar, N.S. and Chand, J.N. (1968). Cultural characters and pathogenecity of three isolates of *P. vexans* (Sacc. and Syd.) Harter, *Indian J. Microbiol.*, 8: 203–205.

Panwar, N.S. and Chand, J.N. (1969). *In vitro* effect of different fungicides on growth and sporulation of *Phomopsis vexans* (Sacc. and Syd.) Harter, the causal organism of fruit rot of brinjal. *Inc. Appl. Science*, 1: 41–42

Panwar, N.S., Chand, J.N., Singh, H. and Paracer, C.S. (1970). Phomopsis fruit rot of brinjal (*S. melongena L.*) in the Punjab. 1. Viability of the fungus and role of seeds in the disease development. *J. Res. P.A.U., Ludhiana*, 7: 641–643.

Pawar, V.H. and Patel, M.K. (1957). Phomopsis blight and fruit rot of brinjal. *Indian Phytopath.*, 10: 115–120.

Porter, R.P. (1943). Seed-borne inoculum of *Phomopsis vexans*, its extent and effects. *Plant Dis. Rept.*, 27: 167–169.

Ramanujam, K. (1966). Studies on Phomopsis blight and fruit rot of brinjal with special reference to its control. *M.Sc. (Ag.) Thesis*, Indian Agricultural Research Institute, New Delhi.

Rangaswami, G. (1979). *Diseases of Crop Plant in India*, 2nd Edn. Prentice Hall of India Pvt. Ltd., New Delhi, pp. 298–302.

Saccardo and Sydow. (1899). *Phomopsis vexans* Saccardo Syll. *Fungi*, 14: 889.

Sayre, C.B. (1924). Recent developments in sprayng and dusting vegetables. *Trans. Illinois State. Hort. Sci.*, 7: 360–367.

Sherbakoff, C.D. (1915). Report of the Assistant Plant Pathologist, *Flo. Agri. Expt. Sta. Ann. Rept*, p. 94–98.

Sherbakoff, C.D. (1916). Report of the Assistant Plant Pathologist, *Flo. Agri. Expt. Sta. Ann. Rept.*, p. 80–98.

Sherbakoff, C.D. (1917). Some important diseases of truck crops in Floridea. *Flo. Agri. Expt. Sta. Ann. Rept.*, p. 76–86.

Sherbakoff, C.D. (1918). Report of the Assistant Plant Pathologist. *Flo. Agri. Expt. Sta. Ann. Rept.*, p. 68–78.

Singh, D. and Chakrabarti, A.K. (1982). Chemical control of Phomopsis fruit rot of brinjal. *Indian Phytopath.*, 35: 314–315.

Singh, H. and Chand, J.N. (1986). Factors affecting growth and production of pycnidia by *Phomopsis vexans* causing fruit rot of brinjal. *J. Mycol. Pl. Pathol.*, 16 (2): 207–208.

Singh, R.S. (1987). *Diseases of Vegetable Crops*. Oxford and IBH Publishing Co. Pvt. Ltd., pp. 140–145.

Sinha, P.P. (1989). Control of Phomopsis blight and fruit rot of brinjal. *Indian J. Mycol. Pl. Pathol.*, 19: 320–321.

Smith, C.O. (1905). Study of the diseases of some truck crops of delware. *Delawre Agriculture Experiment Station Bulletin*, 70: 16.

Spegazzini, C. (1881). Novaddend ad Mycolgium Venetam, *Att. Soc. Crittog. Ital. Ann.*, 24: 42–71.

Suryanarayana, D., Ram Nath and Bhombe, B.B. (1962). Spotting seed-borne diseases of vegetables. *Indian Farming*, 12: 20–21.

Teo, C.H. (1980). Plant Pathology-Brinjal (*S. malongena*). In: *Annual Report of the Research Branch*, Dept. of Agriculture, Sarawak, for the year 1980, Kuching, Sarawak, Ministry of Agriculture and Community Development, 240 p.

Teo, C.H. (1982). Plant Patholgy-Brinjal (*S. melongena*). In: *Annual Report of the Research Branch*, Dept. of Agriculture, Sarawak, for the year 1980, Kuching, Sarawak, Ministry of Agriculture and Community Development, 240 p.

Teo, C.H. (1984). Plant Pathology-Brinjal (*S. malongena*). In: *Annual Report of the Research Branch*, Dept. of Agriculture, Sarawak, for the year 1982, 280 p.

Thippeswamiy, B., Krishnappa, M. and Chakravarthy, C.N. (2005). Location and transmission of *Phomopsis vexans, Alternaria solani* in brinjal. *Indian Phytopath.*, 58(4): 410–413.

Thippeswamy, B., Krishnappa, M., Chakarvathy, C.N., Sathisa, A.M., Jyothi, S.U. and Vasantha Kumar, K. (2006). Pathogenicity and management of *Phomopsis vexans* and *Alternaria solani*. *Indian Phytopath.*, 59(4): 259–261

Toole, E.H., Western, R.E. and Toole, V.K. (1941). The effect of fruit rot of egg plant on germination. *Proc. Americ. Soc. Hort. Sci.*, 38: 496–498.

Uecker, F.A. (1988). A world list of *Phomopsis vexans* with notes on nomenclature, morphology and biolgy. *Mycologia Memoir*, No. 17, Cramer Publisher, Burlin.

Uppal, B.N., Patel, M.K. and Kamat, M.N. (1935). *The Fungi of Bombay Bull.*, 178: 56.

Vidhyasekharan, P., Thulasidas, G. and Kandaswamy, T.K. (1980). Preservation of viability of brinjal seeds in storage. *Indian Phytopath.*, 33: 259–261.

Vishunavat, K. (1992). Phomopsis blight and fruit rot of eggplant. Plant diseases of international importance. In: *Diseases of Vegetables and Oil Seed Crops, Vol. II,* (Eds.) Chaube, H.S., Singh, U.S., Mukhopadhyay, A.N. and J. Kumar. Prentice Hall, Inc., Englewood Cliffs, New Jersey, U.S.A. pp. 325–342.

Vishunavat, K. and Kumar, S. (1993). Detection and transmission of seed-borne inoculum of *Phomopsis vexans* (Sacc. and Syd.) Harter and the effect of infection on seed quality in egg plant (*S. melongena* L.). *Seed Research,* 21: 66–71.

Voglino, P. (1907). Intorno ad un parasita dannoco al Solanum melongena, *Malpighia,* 21:353–363.

Walker, J.C. (1952). *Diseases of Vegetable Crops.* McGraw Hill Book Company, Inc., New York, Toronto, London, pp. 529.

Plant Diseases Management for Sustainable Agriculture *Pages* **155–164**
Editor: **Dr. Shahid Ahamad**
Published by: **DAYA PUBLISHING HOUSE**

Chapter 14

Fungal and Viral Diseases of Gladiolus

***Kumud Jarial*[1]*, Dharmesh Gupta*[2]*, P.K. Sharma*[3] *and U.R. Sangle*[3]**

[1]*Department of Mycology and Plant Pathology, Dr. YSPUHF, Solan*
[2]*Temperate Horticultural Research Station, Kotkhai, Shimla (H.P.)*
[3]*ICAR Research Complex for North Eastern Hill Region, Manipur Centre, Imphal*

Gladiolus, the sword lily commonly known as 'Glads' is frequently planted in home gardens or fields grown for cut flowers. The name gladiolus was originally coined by Pliny and Elder (AD 23–29) from Latin word *'gladius'* meaning a sword, on account of sword like shape of the foliage (Bose and Yadav, 1989). Its cultivation started towards the end of 16th century (Innes, 1985). In India, its cultivation dates back to nineteenth century.

It is a popular cut flower in the international cut flower trade. Its magnificent inflorescence with a variety of colors has made it attractive for use in herbaceous borders, beddings, pots and for cut flowers. In addition, to cut flower trade, gladiolus plants are used in many different ways. People residing at high altitude mountains relished the flowers of certain species of gladiolus. Flowers of certain species are used as salad after nipping off the anthers.

Gladiolus plants and corms are susceptible to diseases caused by fungi, bacteria and viruses. It is always best to avoid diseases whenever possible by starting with healthy stock, however, if problems do occur, one should know to identify the disease symptoms so that proper steps may be taken for their management. Important fungal and viral diseases of gladiolus will be discussed in this chapter.

Fungal Diseases

Fusarium Rot and Yellows

The disease results in yellows type of symptoms on plants and dry corm rot in storage. The disease is also called brown rot, dry rot or vascular wilt. The disease was first reported from California (Pryal, 1909). In India, the disease was reported by Singh (1969). The fungus invades vascular tissues of roots, corms and leaf bases and leads to rot of underground parts, yellowing and distortion of leaves, stem and flowers (Sohi, 2005).

Symptoms

A gradual yellowing of tips of oldest leaves is first noticed. The disease is recognized by inter veinal yellowing of leaf from tip downwards which in advance stages of infection turns the whole leaf to brown colour (Chandel and Bhardwaj, 2001). The spike including the florets may be stunted and faded in colour. The infected stem often gets crooked just below the florets and greener than the normal. The floret colour, size and shape get modified as a result of slight corm infection. On the advancement of disease, plant remains stunted, turns yellow and wilts (Chandel and Bhardwaj, 2001). Flowers may not develop while, the stalks are curved in S shape.

Corms may rot in ground or while in storage. Brown rot of corms develops in ground and begins in basal plate and core which further extends upwards into leaf bases via vascular strands.Oval, sunken spots on corms surface are brown and may have concentric rings.In field, the rot is soft and accompanied with brown wiry roots. The centre of the corm turns black and rot completely. Lesions on corms are reddish brown with well defined margins, round to oval and somewhat depressed. This rot produces hard sunken mummified corms.Dry rot develops in storage especially when corms are not properly dried or cured before storage and is confined often to corm base (Chandel, 1999). Lesions on corms vary in colour from very light brown to tan. The spots are sunken and wrinkled in concentric rings (Salinger, 1987).

Symptoms on root are visible as discrete brown lesion and result in turning of roots into wiry thread like structures (Chandel and Bhardwaj, 2001).

Causal Organism

The disease is caused by *Fusarium oxysporum* f.sp *gladioli.* Other species associated with the disease are *F. solani, F. subglutinans* and *F. heterospermum* (Sarbhoy and Agarwal, 1983; Georgieva and Peikova, 1976; Hornok, 1982).

Disease Cycle and Epidemiology

The disease is either soil borne or latent infection in the corms of previous year becomes the primary source of inoculum (Chandel, 1999). The fungus is carried practically in all stocks of corms and cormels as latent infection (Roebroeck *et al.*, 1990). Additionally, the disease is disseminated through infested soil, contaminated water and leaf hoppers (Albouy, 1966). Optimum temperature for disease development in field is 26–28°C with maximum being 34–36°C and minimum 10°C. In storage, maximum rotting occurs at 7–10°C (Sohi, 2005). Use of high nitrogen fertilizers tends to increase corm rot development.Virulence of pathogen is enhanced by certain micronutrients and ammonia nitrogen (Woltz and Jones, 1968; Groen, 1990), low pH, warm soil with 27–33°C and accumulation of CO_2 in soil, corms and storage air (Magie, 1971).

Control Measures

Management includes removal of infected plants and corms with obvious decay. Follow good harvesting and storage practices. Use disease free corms for plantations (Chandel and Bhardwaj, 2001). Rotate cropping site every year and 2-3 years crop rotation is beneficial for crop. Proper curing of corms at 29°C for a week after harvest and storage in a single layer keeps it under check (Sohi, 2005). The cormels floating in the water before hot water treatment should also be discarded and removal of corm scales (cataphylls) before planting is recommended by various workers (Magie and Poe, 1972;

Katan, 1980). Plant disease free corms in clean soil. Hot water treatment of infected corms at 52°C and cormels at 55°C for 30 minutes (Chandel and Bhardwaj, 2001) eliminates the fungus. Soil pH should be maintained from 6.6 to 7.0 and 80 to 90 per cent nitrogen fertilizer should be used in the form of nitrate (Groen, 1990). Adequate aeration in storage, raised plots, improved water drainage and incorporation of organic mulches in soil helps in preventing the corm rot (Chandel, 1999). Fumigate soil with formalin before plantation. Drench soil with a solution of carbendazim (0.1 per cent), thiram (0.3 per cent), benomyl (0.1 per cent) or captan (0.3 per cent) at disease appearance (Chandel and Bhardwaj, 2001). Bavistin followed by Dithane M-45 have been reported to be effective for management of disease (Shakir *et al.*, 1998). Pre planting or post harvest dusting of corms with 10 to 20 per cent benomyl or 10 to 20 per cent TBZ is very effective (Bose and Yadav, 1989).

Botrytis Blight, Neck rot and Grey Mould

This disease causes severe damage to the flowers and corms throughout the world (Chandel, 1999). The disease was first reported from Holland (Timmermans, 1941).

Symptoms

The disease starts by producing small brown or grey spots of variable shape and size that may have reddish margins appear on both sides of the leaf. These spots are often covered with grey masses of spores. Similar spots on stem become soft and rot in wet weather. With the progression of disease, the tips of leaves and stem may turn yellow and die. Infection of flowers usually begins near the edges of the petals and appears as water soaked spots. Small, clear spots in petals become brown. The spots enlarge rapidly, and the flowers become slimy, brown and begin to droop. When infection is at or near the soil level, it may progress inwards through leaf bases and causes basal stem and corm rot. Vascular bundles may show discolouration and outer leaves turn yellow and die. Under severe conditions, corms also get infected where small, circular, reddish brown sunken spots appear with a cavity formation beneath the spots. The infection later extends to the vascular tissue leading to the core and finally the whole core gets infected which later produces a soft, spongy rot in the corm. A white mouldy growth and hard structures (sclerotia) appears at spongy and dead rotten tissue and surface. Corm decay can continue in storage too (Bose and Yadav, 1989; Chandel, 1999; Chandel and Bhardwaj, 2001).

Causal Organism

Botrytis gladiorum, B. gladioli and B. cinerea.

Disease Cycle and Epidemiology

The fungus survives in old plant parts or debris left in the soil. During cool and moist weather and after frost injury the disease becomes infectious, destroying the crop at maturity. Pathogen grows best at 25°C, pH 7.0 and continuous darkness. Secondary infection takes place through air borne spores on petals and buds during dew and rainfall. Storage of corms at 17 °C leads to high incidence of storage corm rot (Tariq *et al.*, 1993; Bose and Yadav, 1989; Chandel, 1999; Chandel and Bhardwaj, 2001).

Control Measures

Follow proper cultural practices. Disease free corms should be used for planting and dense planting should be avoided. Plant debris is to be destroyed and weeding as well as rouging of diseased plants is required from time to time. Treat corms with hot water (52°C) added with benomyl (0.1 per cent) and chlorothalonil (0.2 per cent) for 30 minutes. Sprays of dithiocarbamates including mancozeb (0.25 per cent), zineb (0.25 per cent) captafol (0.2 per cent), or benomyl (0.05 per cent) is recommended immediately after the appearance of symptoms in the field. Fungicides like chlorothalonil, iprodione and vinclozolin have also been reported effective against the disease. Rotation of fungicides is also required to avoid the development of resistance in the pathogen. Harvest corms in dry weather and store as early as possible. Additionally, corms should be dried at 25 to 28 °C before storage (Holloman and Young, 1951; Magie, 1961; Nagler, 1965; Pirone, 1978; Salinger, 1987; Chandel, 1999; Chandel and Bhardwaj, 2001).

Dry Rot

It is also called neck rot and root rot. The disease is wide spread attacking gladiolus plants in the field and corms in storage (Bose and Yadav, 1989). The first report of disease came in 1928 (Drayton, 1928) and a complete account of the disease was described by Gloud (1958) and Magie (1973).

Symptoms

The symptoms in field appear as premature yellowing of plants ultimately leading to their death. The plants are infected in groups as the disease spreads from one infected plant. Small plants derived from cormels are killed soon after emergence just as in case of damping off. The leaves slowly turn brown from tip downwards and at bases they decay causing neck rot. Under such conditions too, the corms though diseased remain attached to the stems firmly. The pathogen is destructive on young cormels where it destroys or girdles the plant at soil line. The stem develops lesions, leaf bases and cataphylls get shredded and covered with numerous round and black sclerotia. Symptoms on corms develop as lesions in ring form as small, black, more or less superficial spots which can cause collar rot. Lesions are mostly at the nodes and on the tips of corms. Root rotting results in (Magie, 1957) short flower stems. An extensive dry rot is apparent in storage which turns the corms to hard, shriveled mummies. When an infected corm is cut in half, dark streaks can be seen radiating out from the core to the surface of corm (Salinger, 1987; Chandel, 1999; Chandel and Bhardwaj, 2001).

Causal Organism

Stromatinia gladioli. Earlier the pathogen was named as *Sclerotinia gladioli* (Bose and Yadav, 1989).

Disease Cycle and Epidemiology

The fungus is soil borne. Additionally, the infection originates from infected corms and sclerotia. The pathogen can survive in corms and soil for 10 years or more (Bose and Yadav, 1989; Chandel, 1999). The infection first takes place at cataphylls and leaf bases. Neck rot phase is favored by moist soil (Beute, 1971). Corm rot phase is common in cool climates where soil temperature is below 20°C. High doses of nitrogen and low doses of potassium lead to increase in disease levels (Gloud and Russel, 1961).

Control Measures

Maintain healthy planting stock and follow 4 to 6 years crop rotation (Chandel, 1999). Hot water treatment has also been reported effective for disease management (Roistacher, 1951). Plant corms in soil that has already been pasteurized with chemical fumigants like methyl bromide–chloropicrin combination or steam. Before planting the corms, dip them in dichloran (0.3 per cent) for 20 minutes and spray with same chemical in furrow before planting (Chandel and Bhardwaj, 2001). If infection still persists, plants should be sprayed with carbendazim (0.05 per cent) or vinclozolin and soil drenched with thiram (0.3 per cent). Give hot water added with carbendazim (0.1 per cent) treatment for 30 minutes to corms. Store corms at 4 to 10 °C after submerging in hot water (53°C) for 30 minutes (Chandel and Bhardwaj, 2001).

Curvularia Leaf Spot, Blight and Corm Rot

The disease is serious and was first reported by Magie (1948) from Florida, USA.

Symptoms

The disease is characterized by firm, sunken black and corky lesions. The symptoms of disease appear on leaves as well as corms. Aerial parts are more prone to infection causing unmarketable spikes. First symptoms on leaves appear as severe yellowing and browning of leaf tips, initially with circular to oval spots which in favourable conditions become irregular and extend lengthwise along the leaf blades. The spots start as a point which later on coalesce to form bigger spots. These spots in later stages turn tan brown and are surrounded by darker reddish border. Soon the central area is filled with black spores. In later stages, the entire leaf is invaded causing distortion and breaking of leaves. The spots develop severely on stems and corms too. Distorted flowers with brown oval spots on petals and young leaves are seen in abundance. These water soaked spots are colourless which later turn brown to black after spore production. Such infected plants are girdled near soil line resulting in their death. The symptoms on corms develop as small to medium sized, dark brown, rounded, elongated or irregularly shaped lesions surrounded by reddish brown ring. Brown vascular strands extend from the core through the flesh to the surface of corm (Bose and Yadav, 1989; Chandel and Bhardwaj, 2001).

Causal Organism

Curvularia trifoli f.sp. *gladioli* and *C. lunata* (Magie, 1948).

Disease Cycle and Epidemiology

The disease is soil borne and pathogen survives in soil for three or more years. Infected plants or soil which are moist, provide a good source of inoculum. Disease is more serious during warm (18 to 32°C) and humid weathers (Bose and Yadav, 1989).

Control Measures

Plant disease free corms. Plant corms in pasteurized soil. Avoid injury in the corms during storage and handling. Destroy the weed and provide better sanitation in the field. Dip the corms in mancozeb or zineb @ 0.25 per cent for 30 minutes before sowing. Spray zineb or chlorothalonil @ 0.2 per cent or mancozeb @ 0.25 per cent at 10 to 15 days interval. Hot water treatment (52°C) of corm with carbendazim (0.1 per cent) for 30 minutes is effective in eradicating the pathogen from planting stock (Bose and Yadav, 1989; Chandel and Bhardwaj, 2001).

Storage Rots

During storage, the corms are severely affected due to attack of certain fungi. Two types of rots appear in storage: Dry corm rot and soft rot.

Symptoms

The corm rot is introduced only through mechanical injury. The decay is corky with sunken lesions which are reddish brown and get covered with green mould of fungus with time. The infection originates where the husks are attached to corms. This causes series of infection rings which eventually carries the disease upwards and downwards. Numerous light brown minute sclerotia appear on corms which remains vital for several months. The infection on corms initially occurs on tunic scars and spreads rapidly. Such corms pose a problem when corms are piled in heaps without treatment (Pirone, 1978; Bose and Yadav, 1989; Chandel and Bhardwaj, 2001).

Soft rot occurs when corms are injured by artificial heat or sun scald after lifting and during curing and result in watery rot (Bose and Yadav, 1989).

Causal Organism

Dry corm rot: *Penicillium gladioli*

Soft rot: *Rhizopus arrhizus*

Disease Cycle and Epidemiology

Dry corm rot is perpetuated by fungal sclerotia, while previously infected corms are responsible for the spread of soft rot (Bose and Yadav, 1989; Chandel and Bhardwaj, 2001).

Control Measures

Prevent corm injury during handling and avoid damp storage conditions. Discard infected corms. Dip corms in Bordeaux mixture (4:4:50) or carbendazim (0.1 per cent) before planting (Bose and Yadav, 1989; Chandel and Bhardwaj, 2001).

Other fungal diseases of gladiolus include leaf spots caused by *Alternaria fasciculate, A. tenuis, Septoria gladioli, Stemphyllium botryosum* (Bose and Yadav, 1989).

Viral and Phytoplasmal Diseases

Gladiolus plants may be infected by one or more viruses or by strains of a particular virus. Identification of the virus, based on symptoms alone, is difficult because the gladiolus varieties often react differently to the same virus. The most common virus infections in gladiolus plantings are those caused by the *Cucumber mosaic virus* or the *Bean yellow mosaic virus*, both of which are found worldwide. These viruses can cause extensive losses in the quality and quantity of flowers and corms (Anonymous, 1983;).

Cucumber Mosaic Virus (White Break Mosaic)

Symptoms

Infection of the gladiolus plant by one or more strains of the *Cucumber mosaic virus* produces conspicuous blotches in the flowers that are white, light gray, or yellowish. Hence, the common name of "white break mosaic". Flower discoloration may be accompanied by crinkling, shrinking, or flower deformations. Affected flowers open slowly and imperfectly and fade early. Coarseness, twisting or a reduction in the size of the florets may also occur. Flower bracts often become severely yellowed and wither early. Foliar symptoms usually do not appear during the season in which the plants are infected. During the following season, however, conspicuous symptoms appear in the leaves of some gladiolus varieties. Other varieties never show leaf symptoms. The typical foliar symptoms consist of small, chlorotic squares between two leaf veins. The spots may be gray, yellow, brown, or reddish. When the symptoms are severe, the plant may have an overall yellow appearance (Anonymous, 1983; Moran, 1996).

Transmission and Distribution

Cucumber mosaic virus has a very wide host range that includes such garden vegetables as bean, carrot, celery, cucumber, melon, onion, pepper, potato, spinach, squash, and tomato. Susceptible garden flowers include calendula, China-aster, columbine, delphinium, geranium, lilies, lobelia, lupines, periwinkle, petunia, primulas, violas, zinnia, and many others. The virus can be transmitted to gladiolus plants by the feeding of aphids, as a result of planting infected corms, and mechanically by contaminated tools (Anonymous, 1983).

Bean Yellow Mosaic Virus (Mild Mosaic)

Symptoms

Leaves and flower stems often develop an angular, light and dark green mottling in the early summer when the gladiolus plants are growing rapidly and the temperatures are mild. The symptoms tend to be masked when high temperatures occur. Infected flowers have faint, often inconspicuous, pencil-stripe break patterns that are usually lighter in color than the normal flower (Anonymous, 1983).

Transmission and Distribution

Bean yellow mosaic virus infects about 80 species of wild and cultivated legumes, including the garden bean and pea, soybean, clover (alsike, crimson, white sweet), lespedeza, and sweet pea, as well as a few species that are not legumes. The virus can be transmitted readily by the feeding of aphids, or mechanically by contaminated tools. Plantings of gladiolus adjacent to bean, pea, or clover fields may be heavily inoculated with the virus as a result of insect migrations (Anonymous, 1983).

Tobacco Ringspot Virus (Ringspot)

Symptoms

Leaves on affected plants develop necrotic or chlorotic ringspot patterns. The flowers are not affected. Mild strains of the virus may cause mild chlorotic mottling in gladiolus (Anonymous, 1983).

Transmission and Distribution

The *Tobacco ringspot virus* has a wide range of over 250 species in 54 plant families. This virus infects many herbaceous and woody plants. The *Tobacco ringspot virus* can be transmitted by a dagger nematode (*Xiphinema americana*), although other vectors such as thrips, mites, grasshoppers, and the tobacco flea beetle have been reported as transmitting agents to other hosts. A low rate of mechanical transmission (3 per cent) by cutting tools has been shown for gladiolus plants (Anonymous, 1983).

Tobacco Rattle Virus (Notched leaf)

Symptoms

Gladiolus plants infected with the *Tobacco rattle virus* showed severely distorted growth. Such plants usually have notches along the margin or along the veins of one or more of the leaves. These markings are often combined with chlorotic or brown, dead (necrotic) stripes and spots or a crumpling of the tissue between the veins. The stems and buds of the flowers may stop growing and may also show deformities and necrosis (Anonymous, 1983).

Transmission and Distribution

Tobacco rattle virus has a host range of over 350 plant species and is transmitted by soil-inhabiting nematodes of the genus *Trichodorus*. Stubby-root nematodes prefer light and sandy or peaty soils and can transmit the virus either in the adult or the larval stages. The nematodes retain the capability of transmitting the virus for several weeks after feeding on infected plants. Transmission by contaminated tools is also possible with some virus strains (Anonymous, 1983).

Tomato Ring Spot Virus (Stunt)

Symptoms

Gladiolus plants infected with the virus are stunted and have short flower spikes that often contain only 4 or 5 florets. Leaves on affected plants are usually smaller, stiffer, and more erect than those on normal plants.

Transmission and Distribution

The *Tobacco ringspot virus* is transmitted by the same dagger nematode that vectors the *Tobacco ringspot virus*. The adult stage and all three larval stages are capable of transmission. The virus has a wide host range of woody as well as herbaceous hosts and can be found where the nematode is prevalent in the soil. The mechanical transmission of the virus to other host plants has been reported, although this has not been successful with gladiolus plants.

Grassy Top (Aster Yellows Phytoplasma)

Symptoms

Current season infection results in early maturity, small corms and arrested root development. Next year, the corms produce numerous thin, weak shoots leading to grassy top. Flowers produced by grassy top plants are green.

Transmission and Distribution

Disease is transmitted by leaf hoppers. Host range is broad including some weeds.

Control of Virus Diseases (Anonymous, 1983)

1. Plant only certified, disease-free corms, preferably CVI (culture-virus-indexed) corms (incorrectly called bulbs) from a reputable nursery.
2. Remove and destroy (burn or compost) all infected plants when symptoms first become evident. Some diseased plants may not show symptoms until the season after infection has occurred. Since the viruses overseason in the corms, plants known to be infected should be dug up and all parts destroyed.
3. Eradicate all nearby weeds (for example, groundcherry, milkweed, pokeweed, pigweed, clover, henbane, jimsonweed, wild cucumber, bryony, shepherd's-purse, goosefoot, thistle, chickweed, nettle, deadnettle, and sowthistle) that may serve as virus reservoirs or may harbor insects that move onto gladiolus plants later in the season.
4. Avoid transferring viruses from infected to healthy plants during the harvesting of flowers and corms by disinfecting tools and other utensils in a 70 per cent solution of denatured alcohol between cuts.

5. Where feasible, place aluminum strips between the rows of gladiolus plants to repel aphids. This should help reduce the prevalence of the viruses transmitted by these insects. The aluminum strips also help to keep down weeds.
6. Avoid growing gladiolus near bean, clover, cucumber, melon, or tomato.
7. Apply a nematicide, such as EDB or ethylene dibromide, or mixed dichloropropenes (according to label directions) to nematode-infested soil, where treatment would be economically feasible. Soil treatment with a nematicide cannot completely prevent the transmission of the *Tobacco rattle virus* to gladiolus when viruliferous nematodes occur at great depths, as in many sandy soils with a low water table.

References

Albouy, J. (1966). Le problems des' germes fins' du glaieul. *Ann. Epiphyt.*, 17: 81–91.

Anonymous. (1983). Gladiolus viruses. *Report on Plant Disease, Univ. Illinois Extension*. RPD No. 612.

Beute, M.K. (1971). Factors affecting Stromatinia root rot of gladiolus. *Phytopathology*, 61: 1329–1331.

Bose, T.K. and Yadav, L.P. (1989). *Commercial Flowers*. Naya Prokash Publishers, Calcutta, 874 p.

Chandel, S. (1999). Fungal and bacterial diseases of bulbous ornamentals. In: *Diseases of Horticultural Crops: Vegetables, Ornamentals and Mushrooms*, (Eds.) L.R. Verma and R.C. Sharma. Indus Publishing Co., New Delhi, pp. 502–531.

Chandel, S. and Bhardwaj, L.N. (2001). *Flower Diseases: A Practical Guide to Farmers and Students.* Department of Mycology and Plant Pathology, Dr. Y. S. Parmar University of Horticulture and Forestry, Nauni, Solan, 78 p.

Drayton, F.L. (1928). Examinations of plant importations. *Rep. Dep. Agric. Can.* (*Bot.*). 14: 32.

Georgieva, M. and Peikova, E. (1976). Fusarium and Penicillium rots of stored gladiolus bulbs and their control. *B'lgarski Plodove, Zalenchusti I Konservi*, 8: 22–29.

Gloud, C.J. (1958). The dry rot disease of gladiolus. *Plant Dis. Rep.*, 42: 1011–1024.

Gloud, C.J. and Russel, T.S. (1961). Relation of previous nutrition of gladiolus to infection by *stromatinia gladioli. Phytopathology*, 51: 482–486.

Groen, N.P.A. (1990). Measurement before application is important. Nitrogen encourages Fusarium infection in gladioli. *Bloembollencultuur*, 101: 35.

Holloman, A. and Young, R.A. (1951). Evaluation of fungicides for control of leaf spot disease of gladiolus caused by *Botrytis gladiolorum* Timmermans. *Plant Dis. Rep.*, 38: 456–458.

Hornok, L. (1982). Fusarium infections of gladiolus corms. *Novenyvedelem*, 18: 319–322.

Innes, C. (1985). *The World of Iridaceae: A Comprehensive Record*. Holly Gate International Ltd., England.

Katan, J. (1980). Solar pasteurization of soil for disease control: Status and prospects. *Plant Dis. Rep.*, 64: 450–454.

Magie, R.O. (1948). *Curvularia* spot, a new disease of gladiolus. *Plant Dis. Reptr.*, 32: 11–13.

Magie, R.O. (1957). Soil fumigation in controlling gladiolus Stromatinia disease. *Proc. State Hortic. Soc.*, 70: 373–379.

Magie, R.O. (1961). Control of *Curvularia* and *Botrytis* disease of gladiolus. *Ann. Rep. Agric. Exp. Stn.*, pp. 302–303.

Magie, R.O. (1971). Carbon dioxide treatment of gladiolus corms reveals latent Fusarium infection. *Plant Dis. Rep.*, 55: 340–341.

Magie, R.O. (1973). Gladiolus Stromatinia disease: Symptoms, epidemiology and control. *Am. Gladiolus Counc. Gladio Grams*, 11: 1–5.

Magie, R.O. and Poe, S.L. (1972). Disease and pest associates of gladioli. In: *The World of Gladiolus* by Noreen Koenig (ed.). N. Am. Gladiolus Council, pp. 155–181.

Nagler, M. (1965). Experiments on the chemical control of corm rot of gladiolus. *Acta Bot. Hortic. Bucur*, p. 177–184.

Pirone, P.P. (1978). *Diseases and Pests of Ornamental Plants.* John Wiley and Sons, New York, 566p.

Pryal, W.S. (1909). Disease of gladioli. *Rural New Yorker*, 68: 1009.

Roebrock, E.J.A., Groen, N.P.A. and Mes, J.J. (1990). Detection of latent *Fusarium oxysporum* in gladiolus corms. *Acta Hortic.*, 266: 469–473.

Roistacher, C.N. (1951). Hot water treatment of gladiolus cormels. *Phytopatholgy*, 41: 94–96.

Salinger, J.P.(1987). *Commercial Flower Growing*. Butterworths Horticultural Books, New Zealand, 269 p.

Sarbhoy, A.K. and Agarwal, D.K. (1983). Two new diseases of ornamental plants: *Fusarium* rot of gladiolus and *Mammalaria* species. *Curr. Sci.*, 52: 821–822.

Shakir, A.S., Haq, E. and Ayub, M. (1998). Studies on pathogenicity and eradication of some fungal diseases of gladiolus in Pakistan. *Pakistan J. Biol. Sci.*, 1(1): 23–26.

Singh, R.N. (1969). Vascular disease of *Fusarium oxysporum* f. sp. *gladioli* in India. *Indian Phytopath.*, 22: 402–403.

Sohi, H.S. (2005). Important disease problems of ornamental plants and their control. In: *The Fungi: Diversity and Conservation in India*, (Eds.) J.S. Dargan, N.S. Atri and G.S. Dhingra. Bishen Singh Mahendra Pal Singh, Dehra Dun, pp. 253–263.

Tariq, M., Mirza, J.H. and Shakir, A.S. (1993). Physiological studies of *Botrytis gladiorum* and its *in vitro* sensitivity to fungicides. *Pak. J. Phytopathol.*, 5: 89–92.

Timmermans, A.S. (1941).The *Botrytis* rot of gladioli caused by *Botrytis gladiolorum Meded. Inst. Phytopath. Lab. Bloembol. Lisse.*, 67: 32.

Woltz, S.S. and Jones, J.P. (1968). Micronutrient effects on the *in vitro* growth and pathogenicity of *Fusarium oxysporum* f. sp. *lycopersici. Phytopathology*, 58: 336–338.

Plant Diseases Management for Sustainable Agriculture *Pages* **165–182**
Editor: **Dr. Shahid Ahamad**
Published by: **DAYA PUBLISHING HOUSE**

Chapter 15
Fungal Diseases of Stone Fruits and their Management

D. Gupta[1], P.K. Sharma[2] and Kumud Verma[3]

[1]Temperate Horticultural Research Station, Kotkhai, Shimla – 171 202 (H.P.)
[2]ICAR Research Complex for North Eastern Hill Region, Manipur Centre, Imphal
[3]Department of Mycology and Plant Pathology, UHF, Nauni, Solan (H.P.)

Stone fruits include the fruit trees within the genus *Prunus*. The most commonly produced commercial stone fruits are peaches, plums, apricots, cherries, nectarine, almonds and prunes. These fruits are characterized as having a hard stony pit (seed) surrounded by a fleshy fruit. These are grown in sub-temperate and temperate regions of the world, however, in India, these are grown in and near the Himalayas and in very small acreage in south. These fruits are attacked by fungi, bacteria and viruses. Among these, fungi cause the maximum damage throughout the world and have been studied in detail. Among fungal diseases, the most important ones are peach leaf curl, brown rot, rust, powdery mildew, leaf spots, peach scab, canker and die-back and root rot. The diseases can be successfully managed by the combined use of cultural, chemical, biological measures along with use of resistant cultivars. Since many of the diseases attacking stone fruits are common and thus have been discussed together under the host on which they are of great importance.

Brown Rot

Brown rot is a common and destructive disease of peach and other stone fruits (plum, nectarine, apricot and cherry). The brown rot fungus may attack blossoms, fruits, spurs and small branches. The disease is most important on fruits just before ripening, during and after harvest.Under favorable conditions for disease development; the entire crop can be completely rotted on the tree. Peaches not kept in cold storage may be rotted in two to three days by the fungus.

Symptoms

The symptoms of brown rot are very similar on all stone fruits. Symptoms first appear in the spring as the blossoms open. Diseased flowers wilt, turn brown and may be covered with masses of brownish-gray spores. The diseased flowers usually remain attached into the summer. The fungus spreads through the peduncle and reach branches resulting in twig blight. Stem cankers develop usually from blighted twigs or fruit spurs.

Young fruits are normally resistant (Mari *et al.*, 2004), but become infected through wounds. As fruits mature they become more susceptible to attack, even in the absence of wounds. Fruit infection appears as soft brown spots which rapidly expand and produce a tan powdery mass of conidia. The entire fruit rots in a few hours. Rotted fruit either fall to the ground or persist as mummy on the tree (Sharma and Kaul, 1988).Under wet and humid conditions, ash gray tuffs of fungus develop over the surface of the lesions. These structures called sporodochia, produce conidia important in spread of the disease. *Monilinia laxa* causes blossom wilt of apricot and almond in north India (Gupta and Byrde, 1988).

The fungus may move from diseased blossoms or fruit into the spur. The fungus may then invade and cause diseased areas (cankers) on the twigs below. Succulent shoots are sometimes infected by direct penetration near their tip. A canker may form encircling the twig, causing death of the twig beyond the canker (twig blight).

Causal Organism

Brown rot is caused by the fungus *Monilinia fructicola*. The mycelium produces chains of conidia on hyphal branches arranged in groups or tuffs (sporodochia).The sexual stage (apothecium) originates from pseudosclerotia formed in mummified fruit partly or wholly buried in soil or debris.

Etiology

The brown rot fungus survives the winter in mummified fruits and in twig and branch cankers produced the preceding years. Both sources may produce spores that can infect blossoms and young shoots. At about blossom time, a mummified fruit that has fallen on the ground produces up to 20 or more small, tan, cuplike structures on slender stalks that are called apothecia. As an apothecium matures, it becomes thicker and the cup opens to a bowl-like disc 1/8 to1/2 inch in diameter across the spore-containing sacs (asci).At this stage, the slightest disturbance of air movement will cause an apothecium to forcibly discharge millions of spores. These spores (ascospores) are carried by wind to the open or unopened blossoms and young shoots. If a film of water is present for 5 hours or longer, the spores can germinate and penetrate the plant. Infected blossoms soon wilt and tan-gray tufts, composed of masses of another type of spore (conidia),develop on the blossoms. If the infected blossoms do not drop off, the fungus soon grows through the pedicel to the twig and forms a canker.

Masses of conidia are soon produced on the newly cankered twig surface during moist periods. These spores are easily detached and like the ascospores are mainly wind borne.They are also splashed by rain or carried by insects to the growing fruit. Brown rot conidia can germinate and infect at temperatures of 32 to 90°F.

Following spring and summer rainy periods, mummified fruit still hanging in the tree become covered with masses of conidia that may result in blossom blight or fruit rot. Mummies hanging on the tree do not produce ascospores. Although the flesh of young fruit is very susceptible to brown rot infection, the fruit has such a tough skin that the germ tubes of the summer spores do not normally penetrate. For this reason, young uninjured fruits are fairly safe from infection. However, any type of injury to the fruit will provide entry points for brown rot spores. Insect and hail wounds, fruit cracking, limb rubs, twig punctures and a variety of picking and packing injuries greatly increase the losses due to brown rot. The fungus causes the maceration of infected tissues by secreting pectolytic enzymes in invaded fruits. The fungus may survive for two or more years on the mummified fruits fallen on the ground.

Epidemiology

Warm, wet, humid weather is particularly favorable for brown rot. At 20°C, 3-5 hrs.of wetting may lead to significant infection. With 24 hrs. of wetting, severe blossom infection may result regardless of temperature. Optimum temperature for blossom infections of peach and cherry is 25°C (Biggs and Northover, 1988).

Control

1. Sanitation is very important in controlling brown rot. All dropped and rotted fruit should be picked up and destroyed promptly. At the same time, remove all mummies from the trees. Prune out all cankers during the dormant season. Overripe or rotting fruit in the packing shed should be removed and destroyed at once.
2. Control of insects that feed on fruit is essential because wounding will increase the incidence of brown rot. Special care should be taken during harvesting and packing to prevent puncturing or bruising of the fruit.
3. Remove wild or neglected stone fruit trees that serve as reservoir for the disease. Orchards abandoned after a destructive frost can have a tremendous buildup of brown rot on the few remaining fruits.
4. The use of fungicides is an important part of the disease management program for brown rot (Mancini and Cotroneo,1996).The disease can be best controlled by spraying two to four times with an effective fungicide from the time the blossom bud show pink until petal fall. Benomyl, thiophanate methyl, iprodione, vinclozolin, captan, thiabendazole, dicloran and chlorothalonil are the fungicides used for brown rot control [Byrde and Willetts (1977); Osorio *et al.* (1993); Ruegg and Siegfried (1993); Balardin *et al.* (1994); Andrade and Matos (1996); Celiker *et al.* (2003)].

To control brown rot in ripening fruit, the fungicides are applied to the trees a few weeks before harvest and continue until just before harvest.

Post harvest control of brown rot begins with a good protective fungicide program that prevents build up of the disease in the orchard. Harvest needs to be closely supervised to avoid fruit injury due to careless handling. Post harvest brown rot can be reduced by dipping fruit in carbendazim, thiabendazole, benomyl and thiophanate-methyl solution (Sharma, 1994).

Peach Leaf Curl

Leaf curl is a serious disease that occurs on peach, nectarine and related ornamental plants. The disease is prevalent in all the peach growing areas of the world. In India, it is prevalent in Himachal Pradesh, Jammu and Kashmir, Assam, Bihar and Uttaranchal (Jain, 1962, Ferraris, 1928 and Sydow and Butler, 1916). The disease can be severe during cool, wet springs that follow mild winters. The disease causes early defoliation that weakens the trees, making them more susceptible to other diseases and to winter injury. Subsequent refoliation uses energy reserves which would otherwise be used to produce fruit. Weakened trees produce fewer fruit the following season. When the fruit is infected, it is likely to drop before maturity.

Symptoms

As the name implies, the disease is expressed as distorted or curled leaf tissue.These puckered, thickened leaves often have a characteristic reddish-purple colour where the distortion occurs. These symptoms may begin early in the season as the leaves are unfolding. Later the colour turns to a powdery gray on the upper leaf surface as the fungus begins to produce spores. As the season progresses and the weather turns warmer and drier, the leaves turn yellow and dry and finally drop from the tree.

Infected twigs are stunted, distorted, thickened and may die. Leaves that grow from such twigs are also stunted and distorted. Infected twigs tend to die easily. Fruit infections are not common but appear as wrinkled or swollen reddish lesion which lacks the normal peach fuzz.

Causal Organism

Peach leaf curl is caused by the fungus *Taphrina deformans*. Asci of the fungus break through the cuticle of the distorted leaf. They appear as a powdery gray like area on the thickened leaf. The mature asci on being ruptured release ovoid ascospores. During period of spring rain, the ascospores produce large number of blastospores or bud conidia by budding.

Etiology

The fungus survives the winter as ascospores or thick walled conidia on bark and buds. Infection occurs very early in the growing season. During cool, wet spring weather the conidia infect new leaves as they emerge from the buds. Host plant tissues are susceptible for only a short period. As the tissues mature they become resistant. The fungus produces another type of spore (ascospore) on the upper surface of the diseased leaves. During wet weather, ascospores produce additional conidia by budding. These conidia are carried to other parts of the tree by rain and wind, where they will over winter until the next spring.

Epidemiology

Leaf curl is damaging when the weather is cool and wet. Low temperatures are thought to retard maturation of leaf tissue, thus prolonging the time infection may occur. The fungus can penetrate young leaves readily at temperatures between 10 and 21.1° C but only weakly below 7.2° C. Heavy rain is necessary for infection (Rossi *et al.*, 2005).

Control

Peach leaf curl is not difficult to control. A single fungicide application made in the fall after leaves have dropped or in spring before bud swell will control the disease (Sohi *et al.*, 1968). Once the fungus enters the leaf, the disease can not be controlled. Fungicides such as copper oxychloride, carbendazim [Sharma *et al.* (1987, 1988a)] Thakur *et al.* (1991); Bhardwaj and Ved Ram (1995); Pandey and Tripathi (1998) or lime sulphur (Bose and Sindhan,1972) are effective. Other fungicides effective against peach leaf curl are ziram, captafol, thiram, dodine, captan and or chlorothalonil (Rana and Jain, 1992; Brunell and Ponti, 1993; Take and Wood, 1994; Spada *et al.*, 2005).The fungicides kill the spores on twig surfaces. For either the spring or the fall spray to be effective, application must be thorough. Complete coverage of the twigs, branches and trunk is essential.

Peach cultivars Red Heaven, Favorita, Fertilia-1,Eixan-1,Modeline Pouyet, Flacara X HB9-35,Flacara X Miorita and J.H.Hale X Sunbeam are either tolerant or resistant to the disease(Coroianee and Ivascu,1981;Ritchie and Warner,1981). Summer grand was most susceptible and 5745 and Mme Guillou was least susceptible to peach leaf curl (Warloop, 2005).

Spraying does not always control the disease. These failures may be due to:

1. Too-late application of fungicide.
2. Lack of the thoroughness of fungicide coverage.
3. An effective choice of fungicide.
4. Weather conditions at the time of spraying or immediately following spraying.

Trees that have been defoliated by leaf curl should be given the best of care and culture in order that they may overcome the effects of the disease.

Powdery Mildew

Symptoms

Powdery mildew of stone fruits is a disease of leaves, twigs and fruits. On the leaves the disease appears first as blister like areas that soon become covered with grayish white patches of fungus growth. Infection causes leaf distortion, curling and premature leaf fall. White patches of fungus appear on green shoots, which become narrow, strap like and distorted. New shoots are shorter than normal. Buds and flowers may also be directly attacked. In such cases buds either fail to open or open improperly while flowers are discolored, stunted and eventually dry up. Infected fruits first show circular spots which may spread over the whole fruits. The fungus mycelium and its spores produce the white spots. Later the mycelium sloughs off and leaves a rusty-coloured patch. About the time of pit hardening, the skin of the fruit under the spot turns pinkish. Eventually the skin becomes leathery or hard, turns brown and may crack.

Causal Organism

Different species such as *Podosphaera clandestina, P.leucotricha, P.oxycanthe, P. tridactyla* and *Sphaerotheca pannosa* cause powdery mildew in cherry, peach, apricot, plum and peach. The disease is very serious in nurseries where seedling stocks infected early in the growing season remain stunted. Sharma (1985), Kaul (1967) and Pandotra *et al.* (1968) reported occurrence of powdery mildew on almond and plum due to *Phyllactinia* and *Uncinula.* Powdery mildew fungi generally do not require moist conditions to establish and grow and normally do well in warm climates.

Etiology

All powdery mildew fungi require living plant tissue to grow. The fungus survives from one season to the next as mycelium in buds. Cleistothecia are only rarely found on peach and apricot. Primary inoculum is therefore, represented by conidia produced by the over wintering mycelium and sometimes by ascospores. Spores, which are the primary means of dispersal, make up the bulk of the powdery growth of powdery mildew and are produced in chains. Conidia and ascospores are carried by wind to new host/green tissues where they germinate, each producing a germ tube which produces a fine hypha. This hypha penetrates the host directly through the cuticle and when it reaches the cell lumen of an epidermal cell it forms a globose haustorium by which the fungus obtains its nutrients. The aerial mycelium on the other hand produces short conidiophores each bearing a chain of egg shaped conidia. These conidia are dispersed by wind and infect expanding leaves, young shoots and fruit until they are 2-3 cm in diameter.The optimum temperature for conidia germination lies between 21-27° C (Weinhold, 1961).

Management

Successful, sustained powdery mildew management is a season long proposition. Early season disease control is essential for successful season-long disease management.The disease can be effectively controlled by three sprays of wettable sulphur or carbendazim, thiophanate methyl before opening of blooms, at petal fall and two weeks later (Abelentsev *et al.*,1979, Anonymous,1995). New fungicides *viz.* myclobutanil, fenarimol,flusilazole, triademefon, captafol and tebuconazole provided good control of powdery mildew in stone fruits (Torre Almaraz and Ceballos Silva,1990,Dong *et al.*,1991,Jones *et al.*,1993 and Huang *et al.*,1995).Prune trees to promote good air circulation and light penetration.

Coryneum Blight of Stone Fruits

Coryneum blight or "Shot-Hole" is a fungal disease affecting peach, apricot, nectarine, sweet cherry and almond. The disease was first reported in India by Munjal and Kulshrestha (1968) on peach, apricot and almond and by Puttoo and Rajdan (1991) from Jammu and Kashmir on cherry.

Symptoms

Bud and twig infections are the most common cause of damage in peach. Twig infections result in cankers which often surround leaf or flower buds. The cankers are purple or brown in colour and slightly sunken. They enlarge and coalesce to girdle a branch or twig and kill it. Gumming is common around infected buds and cankers.

Leaf infections appear as small tan spots on young leaves,which later turn brown with purple borders.These spots eventually drop out of the leaf, resulting in "Shot-Hole" symptoms.

Fruit infections occur infrequently on peach, but occur on susceptible varieties of apricot. Infection consists of small reddish brown spots with light centers on immature fruits. On apricot fruits the spots become rough and hard with dry, corky areas in the fruit underneath the spots. The spots may grow together and produce large, irregular, leathery patches with cracking and gumming. Such fruit are easily invaded by soft rot fungi and develops soft rot.

Causal Organism

The disease is caused by a fungus *Wilsonomyces carpophilus* (Adaskaveg *et al.*, 1990). The fungus produces black dot like fruiting bodies on the necrotic cankers and twigs. These fruiting bodies produce numerous conidia which are thick walled, ellipsoidal or fusiform, with three to five transverse, double walled septa and infect buds and twigs of trees in winter with the onset of rains.

Disease Development

The fungus over winters in blighted buds and twig cankers. Spores formed in the buds and cankers are spread by fall and spring rains and may infect leaf scars produced when leaves drop in the autumn. Spores are also moved by rain to leaf or flower buds and remain under the bud scales until ready to germinate. In the spring, these spores infect the swollen buds and result in death of the bud or cause blossom blight or "Shot-Hole" symptoms in the leaves. Cankers on new wood can girdle and kill developing buds and twigs. Spores produced on twig cankers in the spring infect susceptible plant tissue during periods of spring rains.

Control

Infection on fruits can be reduced through orchard sanitation and protective sprays. Sanitation is important in reducing inoculum in the orchard, especially in wet years. During spring pruning, all diseased wood should be removed and destroyed. Two sprays with captan 50WP (0.3 per cent) first at leaf fall stage in autumn and second at bud swell stage in spring provide effective control of the disease. Fall applications of copper fungicides prevent the formation of new twig cankers but will not remove existing cankers.

Rust

Stone fruits are affected by this disease and is of worldwide occurrence but is severe only in warmer areas. If not controlled, a severe infection can cause premature leaf fall. The effects of premature loss of leaf may be greater than the loss of the immediate crop. The leafless trees must live on stored food reserves for the rest of the season, depleting food reserves for bud development. This can cause weak bud development and below normal crop yield in the following season. The disease was first reported from Australia in 1890 (Dunegan, 1938). Later on, it was reported from China, Japan, Brazil, Yugoslavia (Mijuskovic, 1992), New Zealand and is now widely distributed in Europe and North America also. In India, the disease was reported by Butler and Bisby in 1931. The disease has been observed in a serious form on almonds and plum (Sharma *et al.*, 1988b; Bhardwaj and Shyam, 1986; Kaul, 1971; Waraich and Khatri, 1977; Singh *et al.*, 1976; Sharma *et al.*, 1987).

Symptoms

Leaves

On peach, pale yellow spots appear on both surfaces of leaves which later change into bright yellow islands prominent on the upper surface. The underside of the leaves develops rusty brown spots. Severe infection leads to defoliation.

Shoots

Peach shoots may also be attacked, resulting in small dead patches where the bark splits on one and two year old shoots.

Fruits

Fruit infection occurs only when conditions are favorable for the fungus to cause infection at a relatively late stage of fruit maturity. Water soaked dark green sunken spots appear on fruits which later turn yellow.

Causal Organism

Tranzschelia discolor is a fungal pathogen that attacks plants of the genus *Prunus*, including almond, apricot, cherry, peach, nectarine, plum and prune. It is most common, however, on almond, peach and prune. Hawksworth *et al.* (1983) have described seven species of *Tranzschelia* genus. However, in India only *T.discolor* (Kaul, 1971; Waraich and Khatri, 1977) and *T. prunespinosae* (Waraich and Khatri, 1977; Bhardwaj and Shyam, 1986) have been recorded. The fungus can be separated by special forms, or formae specialis (f.sp.), based on the host where it is found (Bolkan *et* al.,1985). These forms are *T.discolor* f.sp.*persicae* on peach, *T.discolor* f.sp. *dulcis* on almond and *T. discolor* f.sp.*domestica* on prune. Although cross infection of forms among *Prunus* spp. occurs (eg. the pathogen from almond infecting peach), symptom expression is usually delayed, lesions are fewer in number and fungal sporulation is reduced.

Etiology

The fungus over winters as mycelia in fruit wood infections from the previous summer and fall. In the spring, these infections most likely become the twig cankers that are the source of primary inoculum each year. Urediospores from twig cankers then infect nearby leaves (Jafar, 1959), more spores are produced in the leaf lesions (secondary inoculum), and under favorable environmental conditions, the disease becomes epidemic due to repeated infections of leaves and fruit. Urediospores are disseminated by wind. Rainfall and higher wind velocity result in increased numbers of airborne spores. Because urediospores germinate over a wide temperature range of 5 to 30°C with an optimal temperature of 10 to 25°C, the availability of viable spores (inoculum) and wetness are major factors for determining infection periods. Leaf and twig infections can occur over a wide range of wetness periods (12 to 36 hrs.) and temperatures (15 to 25°C) (Goldworthy and Smith, 1931).Epidemics are favored by rainy period in summer when temperature is around 20-30°C which is most favorable for disease development (Simeone *et al.*, 1985).

Control

The following sanitation measures are designed to break the disease cycle in single orchards. They can also reduce the severity of rust by delaying epidemics. If an area is free of disease at the beginning of a new season, it will take rust much longer to invade from distant orchards or other sources.

1. While pruning peach trees, remove all diseased wood and any diseased leaves and burn them.
2. During winter, remove and destroy all dead leaves from within the framework of orchard trees. Removing and destroying dead, infected leaves is preferable to dropping them on the orchard floor. The rust spores can survive on dead infected leaves on the ground as it does within the tree framework.
3. Some stone fruit species, such as almonds and some varieties, tend to retain some living leaves through the winter. Remove these leaves if possible, as they may be a source of rust infection in the new season.
4. Where rust has been a problem in previous seasons and the crop is being grown in conditions conducive to disease infection, a full fungicide spray program to control the disease should be undertaken. Michailides and Ogawa (1986) found wettable sulphur and mancozeb effective in reducing the disease in sprayed trees. New EBI fungicides such as propiconazole, RH-2161 and RH-3866 (myclobutanil) also showed strong curative activity against rust infection in plum leaves. Foliar spray (25-50 ppm) of these fungicides applied during latent period up to 7 days after inoculation completely suppressed the disease. Shabi *et al.* (1990) showed that rust incidence was low in plants treated with captan, maneb, dithianon, bitertanol and systhane.

Disease control programme in Australia include the use of fungicides and integrated disease management (Penrose,1998).Ram *et al.*(2000) reported sprays of SAN-619 (cycloconazole) or contaf+kavach very effective in reducing the incidence of apricot rust. Corazza *et al.*(1984) studied the reaction of about 25 cvs. under natural conditions and found Brusca, Fracasso, Frasso and San Castrese to be moderately resistant. Ram *et al.* (2000) found the cultivars of apricot namely Fragmore Early, Kaisha and Shakarpara resistant to rust under Indian conditions.

Simeone *et al.* (1984) showed Japanese plums to be less susceptible than European plums and Santa Rosa plums to be immune to rust disease. No attempts have been made to control this disease with biological methods, but few reports of parasitism of the pathogen with other organisms exist. In India, Kaul (1967) showed that *Derluca filum* parasitizes rust fungi in Kashmir valley.

Peach Scab

Peach scab can be extremely damaging to the trees primarily in the warmer, wet peach growing areas. The disease is known to occur on nectarines, apricot and cherry as well. Bagghee and Singh (1960) and Baruah *et al.* (1980) reported this disease on peach from India.

Symptoms

The disease symptoms occurs on twigs and leaves (Matthee and Thomas, 1981), but the most notable symptoms of peach scab occur on the fruit, where small, greenish, circular spots gradually enlarge and deepen in colour to black as spore production begins. In case of severe infection, the individual spots merge and form a uniform, dark olivaceous, velvety blotch. Fruit lesions are most common on the shoulders of the fruit but can occur anywhere on the surface. Where numerous, they often coalesce and may lead to cracking of the skin as the fruit enlarges, allowing rot organisms to enter. On twigs, fungus produces dark brown oval lesions of 3-6 mm diameter. The over wintering twig lesions are clearly visible during the early season as small, grayish, more or less circular,slightly sunken lesions on the previous season's shoot growth. The under surface of the leaf shows angular, pale green areas which gradually turn dark brown. Dark brown long and narrow lesions are formed on the midrib. If infection is severe, premature defoliation may occur.

Causal Organism

The disease is caused by *Venturia carpophila*, the imperfect stage is known as *Cladosporium carpophilum.*

Disease Cycle

The fungus over winters in twig lesions (Ian and Schern, 2003). In the spring, spores are produced on the lesion surface (primary inoculum) and are washed or splashed primarily by wind blown rain to fruit, leaves or new growing twigs. On peaches, fruits are first susceptible starting 5-7 days after shuck split. Periods of rainfall with temperatures of 65-75° F are optimal conditions for infection.

Although fruit remains susceptible until harvest, the four week period following start of split is the most critical period for infection to occur. The time from infection until lesions are visible may be 4-6 weeks.

Control

All varieties are susceptible to scab, with some being more severely affected than others. Generally, scab is more severe the first year the tree bears fruit. This results from the large numbers of twig lesions that develop during the first two growing seasons. Trees should be properly pruned to allow good air circulation, thus promoting rapid drying of fruit, twigs and leaves. Low lying planting sites should be avoided. Fungicide sprays applied well in time can control the scab. Fungicides such as captan and wettable sulphur provide adequate control of the disease if applications are properly timed (Scherm and Savelle, 2001).Huang *et al.*(1993) found that bitertanol spray with summer oil week before bloom followed by bitertanol spray twice at 15-20 days interval reduced scab incidence. Chlorothalonil was found highly effective against scab (Cragnolini *et al., 2005*). Pandey and Tripathi (1998) found blitox and bavistin effective against peach scab.

Canker and Die-back

Silver Leaf

Silver leaf disease is caused by a fungus *Chondrostereum purpureum*. It is a destructive disease of stone fruit trees but can affect other tree fruits also. The fungus is a wound parasite and enters through pruning cuts. Toxins produced by the fungus affect leaves and on some hosts it may kill branches or entire trees. The leaves develop a silvery sheen, caused by air in the tissues and shoots die off progressively as the disease works back along branch. As an infected branch thickens it will show a purplish stain in the centre when cut across. Sometimes the bark of the tree, especially in older one, may bear flat, purplish fungal growths, usually on dead wood and this must be removed for burning. The risk of infection and disease is enhanced if trees are pruned during late winter and early spring, when concentrations of nutrients in sap, such as nitrogen and carbohydrates are highest. Branches which become heavily laden with fruit must be supported to prevent breakages that can let in disease. Application of fungicidal paste/paints should be done immediately after pruning on the cut ends surfaces. Application of an antagonist (*Trichoderma viridae*) spores on the fresh pruning wounds provide an effective control of this disease. Other promising antagonists are *Truncatella* spp.and *Gliocladium* spp. (Dubos and Ricard, 1974; Corke, 1978).Pruned cut of the trees when sprayed immediately with 0.8 per cent captafol provides effective control. A foliar application of iron compounds has been shown to reduce the disease incidence (Patil, 1981).

Leucostoma or Perennial Canker

Cytospora or *Leucostoma* canker is most serious on peach and other stone fruits. The physical appearance of *Leucostoma* canker depends on the part of the tree infected.

Symptoms

Infections of small twigs appear as sunken discolored areas, often with light and dark concentric circles of dead tissue, usually around buds killed by winter cold or leaf scars from previous year foliage. Nodal infections are noticed 2 to 4 weeks after bud break. Infected tissues darken with time and amber gum may ooze from the infected tissue unless the twig is killed entirely. One year old shoots that develop in the center of the tree are especially susceptible to nodal infections. Branch cankers that result from such infections will have dead twigs or twig stubs at the center of the canker. Cankers are elliptical, parallel to the long axis of the stem. Usually the first external symptom of such cankers is the copious production of amber colored gum. Gum production is a natural host response to irritation, but gum production due to infection by *Leucostoma* spp. is excessive. As the canker ages, the gum becomes dark brown and the infected bark dries and cracks open exposing blackened tissue beneath. *Leucostoma* canker can occur on weakened branches without the production of gum, especially if winter injury precedes infection. Beginning in late spring and continuing through summer, the trees grow rapidly and resist further penetration of the fungus into healthy tissues. During this time, the tree may form a callus ring around the canker, but the fungus usually invades this tissue again in late fall or early spring when the tree is dormant and cannot actively resist penetration. The yearly alternation of callus production and canker extension produces a canker with concentric callus rings.

Branches and twig infection may result in foliar symptoms during the growing season. Leaves on an infected branch often turn yellow, droop and may wilt and die. Dead twigs and branches are usually covered with a multitude of pin-head sized black structures erupting through the dead bark, which contain the asexual reproductive structures of the pathogen.

Causal Organism

The fungi which cause *Leucostoma* canker on stone fruits, *Leucostoma cincta* and *L. persoonii* are ascomycetes in the order Diaporthales, family Valsaceae. The imperfect stages of these fungi, *Leucostoma cincta* [=*Cytospora cincta*] and *L. leucostoma* [=*Cytospora leucostoma*], respectively are commonly encountered in the field.

Epidemiology

Leucostoma pathogen can invade the host only if the host is predisposed to infection through mechanical wounding or freeze injury. In nature peach and nectarine orchards where *L.cinctum* is the primary pathogen, many new infections appear at the nodes on one year old shoots. The pathogen enters the node through leaf scars or dead buds during dormant season. Small twigs killed by *L. cinctum* are pathways by which the pathogen can reach older limbs and initiate cankers that may then result in death of large portions of the tree. Many new infections also take place at pruning wounds and the fungus most often found at these sites is *L. persoonii.* At temperatures from 14 to 20°C, *L. persoonii* initiate necrosis, however, lesion expansion is limited by host defense reactions.

The primary inoculum for new infections is conidia. The role of ascospores in the disease cycle remains to be determined. Conidia are most abundant under the cool, moist conditions of late fall and early spring but are present throughout the year, depending upon rainfall. There are significant correlations between spore capture for *L. persoonii* and *L. cinctum* and the number of hours that temperatures are between 10 and 15°C, the duration of wetness and the length of time that relative humidity is above 90 per cent. Short range dispersal of conidia indicates rain or wind driven rain as dispersal agents. Localized splash dispersal may also contribute. Ascospores are discharged forcibly into the air following rain showers at any time of the year when rainfall, but they are most abundant in late spring and early autumn. Spread of *L. persoonii* to a limited scale may occur also via infested pruning tools and possibly by the shot hole borer.

Disease Management

All attempts to control peach canker must take place within the framework of an integrated crop management strategy. All phases of orchard management from establishment of new plantings to care of bearing orchards are important. Management of canker is based on preventive measures designed to decrease winter injury and insect damage, promote optimum plant health and facilitate rapid wound healing. Pruning should be well planned each year so that large cuts, which heal more slowly, will not be needed. When pruning side branches from larger limbs, the cut should be made just beyond the ridge of thickened bark where the smaller branch joins the larger limb. Prune to open the center of trees to light penetration because shaded branches are weakened and more susceptible to winter injury and *Leucostoma* infection.

Cankers should be removed from the tree and burned, buried or moved out of the orchard. Cankers on trunks and large limbs can be removed surgically in mid summer when trees heal most rapidly. Surgery should be performed in dry weather with a forecast of dry conditions for at least three days. Disinfect cutting tools between cuts with alcohol or bleach solution. There are no fungicides registered specifically for control of *Leucostoma* spp. but spraying with benomyl immediately after pruning and before it rains provides some protection (Biggs, 1989; Jones and Aldwinkle, 1990).

Eutypa Die-back

The disease primarily occurs on apricots, but can infect other stone fruits also.

Symptoms

Often the first symptom of the disease appears in mid-late summer on infected branches. The fungus blocks the supply of water up the branch and so there is rapid wilting of leaves in hot weather when the tree has a high moisture demand or is under stress. These dead leaves remain attached to the tree. In advance stages, it produces soft rot symptoms in xylem and makes these branches brittle (English and Davis, 1978). If present, fruit often remains small and ripens before fruit on healthy branches.

On the branches, the infected areas become brown and dry and can shrink away from healthy tissue, causing the bark to split and to exude large amounts of gum. When an infected branch is cut, the internal cross section is discolored brown-black in small flecks or in segments.

Causal Organism

Eutypa die-back of apricot is caused by *Eutypa armeniacae* and now designated as *Eutypa lata.*

Etiology

The fungus enters the plant through fresh pruning and other wounds during wet weather. Spores for new infection develop on diseased dead wood and are released during rain.

Control

The main control for *Eutypa* die back in dry regions is hygiene. Control measures are more important in the wetter region. Spores are released only when the fruiting bodies of the fungus are wet and the most abundant releases accompany rainfall in autumn and spring. Control measures are aimed at reducing the likelihood of spores reaching unprotected wound surfaces. Two essential elements of control are:

1. Keeping pruning wounds to a minimum and protecting them.
2. Burning/pruning of all possible hosts.

Winter pruning of apricot trees should be carried at early in winter when spore production is lowest; but no time of year is safe if wounds are left unprotected. Where large cuts are needed, immediately apply a fungicide to the cut surface.

Careful pruning should be used to remove any diseased branches. This is the only way to eradicate *Eutypa* dieback from a infected tree. If removal of infected branches is not carried out, the fungus will grow down the xylem tissue of the branch at a rate of 10-20 cm per year until the limb is killed. When the infection reaches the trunk, the tree will eventually die. When a diseased tree is cut down, ensure no part of the stump is allowed to remain above soil level. The roots are not a source of infection.

Armillaria Root Rot

Armillaria (or shoestring root rot) is a fungus that rots the roots of many different trees and plants and is widely distributed throughout the stone fruit growing areas. It is found infecting over 700 species of plants. This root rot is most common where old orchards have been replanted or planted on new land cleared of forest trees, particularly oaks.

Symptoms

Affected trees exhibit poor growth, smaller yellowish leaves, dieback of twigs and branches and gradual or sudden death of the tree. White mycelial fan shaped strands of the fungus on the roots is a distinguishing feature of the disease. The fungus over winters as mycelium or rhizomorphs in the diseased trees or in decaying roots.

Causal Organism

The disease is caused by a fungus *Armillaria mellea*. Another closely related fungus causing root rot similar to shoe string root rot is *Armillaria tabascens*.

Control

There is no cure for severely infected plants. They should be removed along with the stump. *Armillaria* can survive for many years on dead stumps and roots. Slightly affected plants should have soil removed from the root crown and allowed to be air dried. Avoid watering the base of the trunk; keep this area as dry as possible. Other control measures include digging a trench (1.8m deep) around infected trees and kept open for several years to prevent growth of rhizomorphs to adjacent trees and fumigation of the infected area to destroy the fungus in soil before *Armillaria* killed trees can be replaced.

Several examples of integrated control of *Armillaria* have been reported. *Trichoderma viride* can replace *Armillaria* in roots after fumigation between tree crops. Unfortunately such biological fumigants are unsuitable for use on established trees as they are too phytotoxic (Bliss,1951). Mycophagous nematodes have also been implicated in the biological control of *Armillaria*. Mycophagous nematodes greatly reduced mortality of ponderosa pine seedling inoculated with *Armillaria*.

Phytophthora Root and Crown Rot

Phytophthora root and crown rots are serious soil borne diseases which for a long time have caused considerable losses in stone fruit orchards. A number of *Phytophthora* species are notorious for being the cause of root and crown rots in stone fruit trees.It causes significant losses of production by killing both young and mature trees.

Symptoms

On stone fruits, pale amber, cloudy drops of gum exude from the bark near the base of infected trees. The gum darkens with age until it is almost black, while new drops exude further up and around the trunk as the infection progresses. The disease produces cankers at or below the ground line in the root crown area. Canker may extend up to the point of origin into the root system and up to the trunk to the bud union and occasionally into scaffold branches. Girdling of the entire crown region may lead to death of the tree. The bark has a characteristic sickly-sweet odor when removed and the inner bark and cambium in the lesion are banded or mottled with shades of cream and brown.

Symptoms in the foliage are secondary and are rarely obvious until girdling of the trunk is well advanced or a major part of the root system is invaded. They vary with the time of girdling or root destruction. Stunted leaves, yellow discolouration, cessation of shoot growth, wilting and dieback are common symptoms. Affected trees sometimes bear fruits which ripen early.

Causal Organism

In peaches and cherries, the disease may be caused by different species of *Phytophthora viz. P. cactorum,P. cambivora,P. cinnamomi,P. cryptogea, P. megasperma, P. citricola* and *P. dreschleri* (Taylor and Washington,1984).

Disease Cycle

These fungi exist in soil as resting spores. The resting spores germinate when the soil is wet and warm to form sac-like sporangia which release many zoospores into the soil water. The zoospores swim to the underground plant parts and infect them. The cycle is completed when the fungus growing in the plant produces more resting spores, which are released when the dead tissue rots. Infections of root at a distance from the trunk are less likely to result in trunk rot than are direct infections of the root crown region. The pathogen survives in the soil for several years. Pathogen population is directly influenced by soil moisture and temperature and a soil temperature of 12-20°C with pH 5-6 is best for survival of the fungal propagules (Rana and Gupta, 1984).

Control

The disease can be best controlled by superficial heating of the localized lesions with the help of blow lamp or scarifying cankered lesion upto healthy tissues followed by application of chaubatia paste or copper oxychloride paint and irrigation of 30 cm radius around tree trunk with 0.3 per cent mancozeb or 0.5-1 per cent blitox/fytolan/blue copper (Agarwala, 1970; Rana and Gupta, 1981 and 1983). Soil application of systemic fungicides *i.e.* metalaxyl and soil or foliar application of alliete reduced *P. cactorum* incidence in peach (Taylor and Washington, 1984).

Beside chemical control, soil water management is important in reducing the incidence of the disease. Infection is favoured if the amount of water applied exceeds losses through transpiration, evaporation and drainage.

Hilling up the top soil along the tree line and shallow planting on the hill improves surface drainage and expose less of the susceptible trunk tissue to the infected soil. Avoid movement of the soil and infected plant material from diseased to healthy areas. Sod culture is preferred to clean cultivation for this. It is advisable to keep graft union at least 30 cm above soil level. Precautions must be taken to apply benomyl, carbendazim/thiophanate methyl in crown rot affected trees as it is bound to increase the severity of the disease (Byrde and Jordan, 1977).

References

Abelentsev, V.T., Savhenko, V.T., Vishnevskaya, A.M. and Shelkovnikov, V.F. (1979). Fungicides against diseases of stone fruits. *Zashchita Rastenii*, 5: 33.

Adaskaveg, J.E., Ogawa, J.M. and Butler, E.E. (1990). Morphology and ontogeny of conidia in *Wilsonomyces carpophilus*, gen. nov. and comb. nov., causal pathogen of shot hole disease of *Prunus* species. *Mycotaxon*, 37: 275–290.

Agarwala, R.K. (1970). Relative importance of the control methods of *Phytophthora cactorum* collar rot disease of apple trees. In: *Plant Disease Problems. Proc. I*[st] *Int. Symp. Pl. Path.*, Indian Phytopathological Society, IARI, New Delhi, p. 632–638.

Andrade, E.R.D.E. and Matos, C.S. (1996). Chemical control of *Monilinia fructicola* on postharvest peach fruits. *Fitopatologia Brasileria*, 21: 301–303.

Anonymous. (1995). *Package of Practices for Fruit Crops*. Dr. Y.S. Parmar University of Hort. and Forestry, Nauni, Solan, 186 pp.

Bagghee, K.D. and Singh, V. (1960). *Indian Forest Records, New Series*, 10: 199–348.

Balardin, R.S., Balardin, C.R.R. and Chaves, L.C. (1994). Fungicides and doses efficiency on *Monilia fructicola* (Wint.) post-harvest control in peach. *Ciencia Rural*, 24: 15–17.

Baruah, A., Baruah, S.N. and Bora, K.N. (1980). A fungal disease of peach plant collected from Gauhati University Campus, Assam (*Venturia carpophilla*). *Science and Culure*, 46: 264.

Bhardwaj, L.N. and Shyam, K.R. (1986). Occurrence of rust on Japanese plum. *Himachal Journal of Agricultural Research*, 12: 62.

Bhardwaj, S.S. and Ved Ram (1995). Incidence of apricot leaf curl and its management. In: *Integrated Disease Management and Plant Health*, (Eds.) Gupta, V.K. and R.C. Sharma. Scientific Publishers, Jodhpur, pp. 227–279.

Biggs, A.R. (1989). Integrated approach to controlling *Leucostoma* canker of peach in *Ontario. Plant Disease*, 73: 869–874.

Biggs, A.R. and Northover, J. (1988). Influence of temperature and wetness duration on infection of peach and sweet cherry fruits by *Monilinia fructicola. Phytopathology*, 78: 1352–1356.

Bliss, D.E. (1951). The destruction of *Armillaria mellea* in citrus soils. *Phytopathology*, 41: 665–683.

Bolkan, H.A., Ogawa, J.M., Michailides, T.J. and Kable, P.F. (1985). Physiological specialization in *Transzchelia discolor. Plant Disease*, 69: 485–486.

Bose, S.K. and Sindhan, G.S. (1972). Peach leaf curl (*Taphrina deformans*) in Kumaon and its control. *Progressive Horticulture*, 4: 5–8.

Brunell, A. and Ponti, I. (1993). Observations of epidemiology and control measures against leaf curl on peach. *Bulletin OILB/SROP*, 16: 64–67.

Butler, E.J. and Bisby, G.R. (1931). *The Fungi of India*. Empirical Council of Agricultural Research (India). Science Monograph, 1.

Byrde, R.J.W. and Jordan, V.W.L. (1977). In: *Systemic Fungicides* (Ed.) R.W. Marsh. Longman, London and New York, 171 pp.

Byrde, R.J.W. and Willetts (1977). *The Brown Rot Fungi of Fruit: Their Biology and Control*. Pergaman Press, New York, 171 pp.

Celiker, N.M., Ibis, M.A., Demir, S.T. and Kaya, U. (2003). Investigations on chemical control of brown rot on cherry fruits. *Bitki-Koruma-Bulteni*, 43(14): 125–133.

Corazza, L., Simeone, A.M. and Ialongo, M.T. (1984). Relation of some apricot cultivars of rust (*Tranzschelia pruni-spinosae* Persoon) at a location on the latium coast. *Informatore Fitopatologico*, 34: 45–48.

Corke, A.T.K. (1978). Microbial antagonists affecting tree diseases. *Annals of Applied Biology*, 89: 89–93.

Coroianee, A. and Ivascu, A. (1981). Investigations on the behavior of some peach cultivars and hybrids to attack by *Taphrina deformans. Lucr. Stiinti*, B24: 59–63.

Cragnolini, C.T., Novo, R.J., March, G.J., Conles, M.Y. and Balzarini, M. (2005). Fungicide application timings and efficacy in the control of peach scab. *Agriscientia* 22(1/2): 37–45.

Dong, C.O., Kim, S.B., Jang, H.L., Cho, M.D. and Lee, E.K.(1991). Studies on the ecological characteristics and control methods of peach powdery mildew. *Journal of the Korean Society for Horticultural Science*, 32: 191–198.

Dubos, B. and Ricard, J.L.(1974). Curative treatment of peach trees against silver leaf disease (*Stereum purpureum*) with *Trichoderma viride* preparations. *Plant Disease Reporter*, 58: 147–150.

Dunegan, J.C. (1938). The rust of stone fruit. *Phytopathology*, 28: 411–426.

English, H. and Davis, J.R.(1978). *Eutypa armeniacae* in apricot. Pathogenesis and induction of xylem soft rot. *Hilgardia*, 46: 193–204.

Ferraris, T. (1928). Agricultura ef fitopathologianel Kashmir.*Curiamo de plante*, 6: 61–86.

Goldworthy, M.C. and Smith, R.E. (1931). Studies on a rust of clingstone peaches in California. *Phytopathology*, 21: 133–168.

Gupta, G.K. and Byrde, R.J.W. (1988). *Monilinia laxa* associated with blossom wilt of apricot and almond in Himachal Pradesh, India. *Plant Pathology*, 37: 591–593.

Hawksworth, D.L., Sutton, B.C. and Ainsworth, G.C.(1983). *Ainsworth and Bisby's Dictionary of the Fungi, 7th edm.* Comm. Mycol. Inst., Kew Surrey.

Huang, J.W., Chen, J.C., Chung, W.C. and Yang, S.H. (1995). Chemical control of powdery mildew on Japanese apricot. *Journal of Agriculture and Forestry*, 44: 13–18.

Huang, J.W., Yong, H.C. and Leu, L.S. (1993). Studies on control of Japanese apricot scab. *Plant Pathology Bulletin*, 2: 7–11.

Ian, Z. and Scherm, H. (2003). Moisture sources in relation to conidial dissemination and infection by *Cladosporium carpophilum* within peach canopies. *Phytopathology*, 93(12): 1581–1586.

Jafar, H. (1959). Current Research and Investigations. *Orchard, NZ*, 32: 17–19.

Jain, S.S. (1962). Tests of fungicides for control of peach leaf curl (*Taphrina deformans*). *Himachal Horticulture*, 3: 11–15.

Jones, A.L. and Aldwinkle, H.S. (Eds.) (1990). *Compendium of Apple and Pear Diseases*. APS Press, St. Paul, Minnesota.

Jones, A.L., Ehret, G.R.,Garcia, S.M., Kesner, C.D. and Klein, W.M. (1993). Control of cherry leaf spot and powdery mildew on sour cherry with alternate side applications of fenarimol, myclobutanil and tebuconazole. *Plant Dis.*, 77: 703–706.

Kaul, T.N. (1971). Rust of almond a new record. In: *Proceedings of Second International Symposium on Plant Pathology*, New Delhi, Indian Phytopathological Society, p.159.

Kaul, T.N. (1967). Diseases of stone fruits in Kashmir. *Horticulturist*, 2: 52–58.

Mancini, G. and Cotroneo, A. (1996). Defense strategies against the principal fungal diseases on stone fruits in Piedmont. *Informatore-Agrario-Supplemento*, 52(15): 4–9.

Mari, M., Casalini, L. and Pratella, G.C. (2004). Brown rot of stone fruits: unripe fruits are less susceptible. *Rivista-di-Frutticoltura-e-di-Ortofloricoltura*, 66(3): 60–62.

Matthee, F.N. and Thomas, A.C. (1981). Occurrence of peach scab. *Deciduous Fruit Grower*, 31(11): 464–465.

Michailides, T.J. and Ogawa, J.M. (1986). Chemical control of prune leaf rust. *Plant Disease*, 70: 307–309.

Mijuskovic, M. (1992). Rusts of stone fruits in Montenegro. *Crnogorska-Akademi Ja-Nauka-i-Umjetnosti-Glasnik-Odjeljenja-Prirodnikh-Nauka*, p. 145–189.

Munjal, R.L. and Kulshreshta, D.D. (1968). Some *Dematiaceae hyphomycetes* from India. *Indian Phytopathology*, 21: 309–314.

Osorio, J.M., Adaskaveg, J.E. and Ogawa, J.M. (1993). Comparative efficacy and systemic activity of iprodione and the experimental anilide E-0858 for control of brown rot on peach fruit. *Plant Disease*, 77: 1140–1143.

Pandey, J.C. and Tripathi, K.C. (1998). Peach scab in U.P. hills and its control. *Progressive Horticulture*, 30(3/4): 214–216.

Pandotra, V.N., Kachroo, J.L. and Sastry, K.S.M. (1968). Six powdery mildew from Jammu and Kashmir state. *Proceeding of Indian Academy of Sciences*, 67B: 119–124.

Patil, J. (1981). *Cultural Practices and Infectious Crop Diseases*. Springer Verlag, Berlin, 243 pp.

Penrose, L.J. (1998). Disease control in stone fruits in Australia. *Pesticide Outlook*, 9(2): 13–17.

Puttoo, B.L. and Razdan, V.K. (1991). Stigmina blight of cherry: A new record from India. *Plant Disease*, 6: 60.

Ram, V., Sharma, Usha and Bhardwaj, L.N. (2000). Management of apricot leaf rust. *Plant Disease Research*, 15: 113–115.

Rana, B.S. and Jain, V.B. (1992). Screening of peach varieties against leaf curl disease and its control. 22: 87 (Abst.)

Rana, K.S. and Gupta, V.K. (1984). Effect of fungicides on the viability of *Phytophthora cactorum* propagules in soil. *Phytopath Z.*, 110: 245–250.

Rana, K.S. and Gupta, V.K. (1981). Cause and control of collar rot of apple in India. *Indian Phytopathology*, 34: 17–19.

Rana, K.S. and Gupta, V.K. (1983). *In vitro* and *In vivo* efficacy of systemic and protectants fungicides against *Phytophthora cactorum*. *Indian Journal of Mycology and Plant Pathology*, 13: 272–276.

Riffle, J.W. (1973). Effect of two mycophagous nematodes on *Armillaria mellea* root rot of *Pinus ponderosa* seedlings. *Plant Disease Reporter*, 57: 355–357.

Ritchie, D.F. and Werner, D.J. (1981). Susceptibility and inheritance of susceptibility to peach leaf curl in peach and nectarine cultivars. *Plant Disease*, 65: 731–734.

Rossi, V., Botognesi, M., Giosue, S., Mazzini, F., Ponti, T. and Spada, G. (2005). Biology and epidemiology of the agent of leaf curl in peach. *Informatore Agrario*, 61(10): 5 9–67.

Ruegg, J. and Siegfried, W. (1993). Integrated fruit production: news on blossom and fruit Monilia. *Schweizerische zeitschrift fiir obst-und weibau*, 129: 620–624.

Scherm, H. and Savelle, A.T. (2001). Control of peach scab with reduced midseason fungicide programs. *Plant Disease*, 85(7): 706–712.

Shabi, E., Elisha, S., Birger, R. and Singer, G. (1990). Almond diseases and their control. *Alow Hanolea*, 42: 445–450.

Sharma, R.C., Jindal, K.K. and Gupta, V.K. (1987). First report of rust caused by *Tranzschelia discolor* f.sp. *dulcis* on almond in India. *Plant Disease*, 71(9): 850.

Sharma, R.C., Jindal, K.K. and Gupta, V.K. (1988b). Reaction of some almond cultivars to rust. *Plant Disease Research*, 4: 80–81.

Sharma, R.L. and Kaul, J.L. (1988). Occurrence of brown rot on stone fruits in Himachal Pradesh. *Plant Disease Research.*, : 46–47.

Sharma, R.L. and Kaul, J.L. (1989). Effect of post-harvest fungicides treatments on brown rot of apricot. *Plant Disease Research*, 4: 54–58.

Sharma, A.K. (1985). A new species of *Phyllactinia* (Erysiphaceae) from India. *Trans. British Mycological*, 85: 756–759.

Sharma, R.C., Gupta, V.K. and Garg, R.C. (1988a). Dormant applications of fungicides for the management of peach leaf curl. *Indian Phytopath.*, 41: 259–260.

Sharma, R.L. (1994). Eficacy of post-harvest fungicidal treatments on brown rot of peach. *Indian Journal of Mycology and Plant Pathology*, 24: 60–61.

Simeone, A.M., Ialonga, M.T. and Corazza, L. (1985). Reaction of some plum varieties to rust in a locality of Latium coast. *Informatore Fitopatologico*, 35: 37–41.

Simeone, A.M., Ialonga, M.T. and Corazza, L. (1984). The reaction of certain plum cultivars to rust in Latium Littoral region. *Annual Dell's Institute sperimenta perla Fruitticoltura*, 15: 39–49.

Singh, G., Chauhan, J.S., Bhatt, A.S. and Malthi, C.S. (1976). Evaluation of germplasm collections of almond rust in Punjab. *Indian Journal of Mycology and Plant Pathology*, 6: 31.

Sohi, H.S., Sharma, S.L., Verma, B.R. and Shyam, H.S. (1968). Chemical control of peach leaf curl. *Indian Journal of Horticulture*, 25: 48–51.

Spada, G., Ceredi, G., Brunelli, A., Gianati, P., Berardi, R., Mazzini, F., Rossi, R. and Ponti, T. (2005). Old and new fungicides to protect against peach leaf curl. *Informatore Agrario*, 61(10): 68–72.

Sydow, H. and Butler, E.J. (1916). Fungi Indiae Orientalis Part IV. *Annals of Mycology*, 14: 177–220.

Take, K.G. and Wood, P.N. (1994). Field evaluation of fungicides for control of peach leaf curl (*Taphrina deformans*). In: "*Proceedings of the Forty Seventh New Zealand Plant Protection Conference*", Waitongi Hotel, New Zealand, 9–11 August, Rotoura, New Zealand. New Zealand Plant Protection Society, p. 289–293.

Taylor, P.A. and Washington, W.S. (1984). Curative treatments for *Phytophthora cactorum* in peach trees using metalaxyl and phosetyl-Al. *Australian Journal of Plant Pathology*, 13: 31–34.

Thakur, V.S., Gupta, V.K. and Garg, R.C. (1991). Management of fungal diseases of stone fruits with fungicides in Himachal Pradesh. *Indian Journal of Agricultural Sciences*, 61(9): 677–681.

Torre Almaraz, R., De La and Ceballos Silva, A.P. (1990). Chemical control of peach disease at Acajete,Pucbla. *Revista Mexicana de Fitopathologia*, 8: 181–190.

Waraich, K.S. and Khatri, H.L. (1977). Occurrence of pink disease and rust on plum in India. *Indian Journal of Mycology and Plant Pathology*, 7: 202–206.

Warloop, F. (2005). Organic peach.Selection of peach varieties tolerant to peach leaf curl and adapted to organic agriculture. *Arboriculture-Fruitiere*, 598: 25–28.

Weinhold, A.R. (1961). Temperature and moisture requirements for germination of conidia of *Sphaerotheca pannosa*. *Phytopathology*, 51: 699–703.

Plant Diseases Management for Sustainable Agriculture *Pages* **183–194**
Editor: **Dr. Shahid Ahamad**
Published by: **DAYA PUBLISHING HOUSE**

Chapter 16

Banded Leaf and Sheath Blight of Maize Incited by *Rhizoctonia solani* f. sp. *sasakii* and its Management

Ashraf Ali Khan and P.P. Thirumalaisamy

DMR, IARI Campus, Pusa, New Delhi – 110 012

Introduction

Globally, maize is a leading cereal cultivated over an area of 140 million hectare with the production over 600 million tones and average productivity of 4 tonnes/hectare. A major shift in global cereal demand is realized and it is postulated that, demand for maize in developing world alone will increase from 282 mt in 1995 to 504 mt in 2020 (Zaidi and Singh, 2006). Plant diseases represent important preventable hazards to crop production. In case of maize, diseases are responsible for lowering grain yield considerably. On a global basis losses due to maize diseases have been estimated to the tune of 9.4 per cent annually. For the countries of Asia the figure is 12 per cent, while for African countries the estimate is as high as 14 per cent (Cramer, 1967; James, 1981). Even for the developed countries like U.S.A., 12 per cent of the produce is lost due to diseases, annually. For India a per cent loss of 13.2 has been estimated (Payak and Sharma, 1985).

Among various diseases of maize, banded leaf and sheath blight incited by the fungus *Rhizoctonia solani* f. sp. *sasakii* Exner [= *Thanetophorus sasakii* (Shirai) Tu and Kimbro], has attained the status of the most economically important disease. The disease was first reported from Sri Lanka (Bertus, 1927) under the name 'Sclerotial' disease. Subsequently, it was reported from Malaysia as 'Banded sheath rot', in the Philippines as 'Banded sclerotial disease' and as 'summer sheath blight' in Japan. In India, in early sixties the disease was little more than a curiosity and a disease of minor importance in the western and central Himalayan foot hill region. However, it became increasingly severe and assumed epidemic proportions in the following two decades. Presently the disease is considered as a major disease not only in India but in several countries of Tropical Asia, wherever maize is grown (Sharma *et al.*, 1993).

Economic Importance

The banded leaf and sheath blight of maize has been found to cause extensive losses not only in the total productivity of maize but also has been observed to cause considerable reduction in percent yield of high yielding varieties, both released entries and also those in the pipeline. India and South China have generated information on the impact of the disease on reduction of yield. The disease causes direct losses, resulting in premature death, stalk breakage and ear rot, indirect losses by reducing the grain yield. In India, loss in grain yield has been estimated in the range of 23.9 to 31.9 per cent in ten cultivars (Sangam Lal *et al.*, 1980). Singh and Sharma (1969) estimated 40.5 per cent loss in grain yield with 71 per cent disease index. In Guangxi province in South China, yield losses of 87.5 and 57.8 per cent have been determined under natural conditions in the hybrids Luyu 13 and Guiding planted at Bao Qiao and Chen Xiang counties (Zhang *et al.*, unpublished). However, the magnitude of grain loss may reach as high as 100 per cent if the ear rot phase of the disease predominates. In India, out of the estimated 13.2 per cent annual loss in grain yield due to its, sheath blight contributes 1 per cent (Payak and Sharma, 1985). Sumner and Minton (1989) observed that yields were reduced upto 47 per cent in soil infested with the high inoculum level compared with noninfested soil.

Distribution

The disease on maize was first recorded in Sri Lanka as 'Sclerotial disease'. In recent years, disease outbreaks have occurred in many more countries and have assumed epidemic dimensions. Besides India and Sri Lanka, it has been found to occur in various hot humid tropical maize growing areas in Bhutan, Indonesia, Cambodia, Bangladesh, Pakistan, Nepal, Myanmar, Thailand, Laos, Vietnam, Philippines, China, Taiwan, Malaysia, Korea and Japan. Except these there are also reports of its occurrence in Sierra Leone, Ivory Coast, Nigeria, England, and Venezuela. In 1988, the disease was also reported on field corn in Arkansas, USA (Hirrel *et al.*, 1988).

Symptoms and Signs

The disease generally appears at pre-flowering stage in 40-50 day old plants. The symptoms manifest on leaves, sheaths, stalk and ear. As the disease is soil borne, it starts from the lowest leaf sheath or on leaves which are in contact with soil and travels up to the ear. Symptoms are more common on the leaf sheath than on the laminae. The disease lesions are characterized by the presence of alternate bleached areas or zones, which are initially water soaked and narrow, purple brown bends oriented perpendicular to the long axis of leaves or leaf sheaths (Figure 16.1). Sharma *et al.* (2004) reported that the symptoms were irregular, water-soaked, straw-coloured lesions on leaf bases and sheaths. The lesions enlarged rapidly resulting in discolored areas alternating with dark bands, apparent on lower leaves after 7 to 8 days.

Bertus (1927) recorded the disease symptoms on leaf sheaths as small purplish brown or greenish olive brown large continuous patches, on stalks immediately below the diseased leaf sheath and on the stem at the base of female inflorescence as pale olive brown lesions. He also reported the presence of sclerotia on diseased parts and restricted development of flowers, withering and outright death of diseased plants. The ear rot phase may be distinguished into three types (*a*) in case of infection occurring prior to ear emergence, the development of ear is completely suppressed or at best it remains rudimentary with bleaching of entire husk leaves, (*b*) if the pathogen reaches the ear shoot after ear

Water Soaking

Spreading Upward

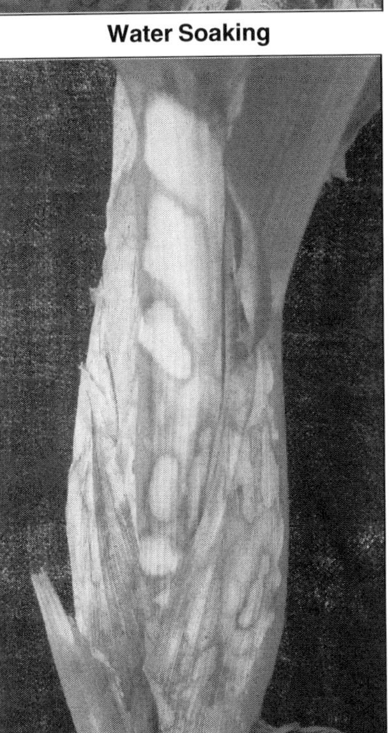

Symptoms on Cob

Severely Affected Plant

Figure 16.1: BLSB Symptoms

emergence, the stalk fibers at the tip darkened become caked up and turn into a hardened lump leading to poor filling, (*c*) if infection occurs after grain formation, the kernels become light in weight, chaffy and lusterless.

Ahuja and Payak (1982) documented in depth symptoms and signs of the disease in addition to those described above. Such symptoms included stalk lesions or rind spotting, stalk breakage, clumping and caking of style (silk fiber), horse-shoe shaped lesion with banding on caryopsis which is considered most important for diagnosis of the disease in pathological tests of the seed, as kernel rots incited by other fungi do not produce blemishes in field or during storage.

Two types of mycelia are found on the host–the straight non-infectious cottony, runner-type of hyphae and multi-branched lobate-type of infection hypae. The former type may be present on the unaffected parts of the maize plants. The later type develops aggregation of non-infectious hyphae on the tissues, which have turned necrotic, they sometimes give rise to sclerotia. The hyaline mycelium is especially prominent on the abaxial surface of sheaths and husk leaves. A raised and spreading type of mycelium occurs on both surfaces of leaves and its greater concentration has been observed on unaffected parts adjacent to the diseased ones. The disease is noteworthy in that, infection appears some distance away from the main center of growth infection cushions develop further away. Mycelia as well as infectious cushion turn buff accompanied by the development of alternating bands and blotches.

The hyphal masses on ears produce a caking effect and the husk leaves as well as styles (silk fibers) together. The mycelium develops above and between kernel rows. Irregularly shaped sclerotia, roundish to convex in frontal view (curled up inside), give a pastular appearance to the affected parts which have already turned necrotic. Sclerotia are found to be produced not only on sheaths, leaves and husk leaves but also for the first time on glumes, in couples, under the pericarp in caryopsis and silk fibers (Ahuja and Payak, 1982).

The Pathogen

The genus *Rhizoctonia* was erected in the year 1815 by De Candolle to accommodate the non-sporulating root pathogen *R. crocorum* D.C. ex. Fr. It was described under the name *Rhizoctonia solani* which was earlier reported about 100 years ago by the German scientist Julius Kühn on diseased potato tubers. Sheath blight pathogen has been considered distinct from *R. solani* complex on the basis of cultural characters, symptomatology, pathogenicity and compatibility. Comparative studies on mycelial and sclerotial characters of basidial (teleomorph) and anamorph of *Corticium solani*, *C. microsclerotia* and *C. sasakii* led to the conclusion that these did not differ significantly as distinct species and considered as *forma specialis* of *Pellicularia filamentosa* (Pat.) Rogers, a synonym of *T. cucumeris* (Frank) Donk. The telomorph stage has not been recorded on maize. Hence its description is not provided here.

Hyphae hyaline to sub-hyaline at first (Figure 16.2), brownish later, separate with dolipore septum and septum near origin of branch, 3-11mm in diameter, multi-nucleate (3-16 but 4-8 per cell with a mean of 6.3-6.6 and a mode value of 6), moniloid cells present but clamp connections absent. The mycelium is branched at right angle and slight constriction observed at the branching. Sclerotia are not differentiated into rind and medulla (primary rind type), variable in shape and size but mostly flat and discoid and irregularly curled.

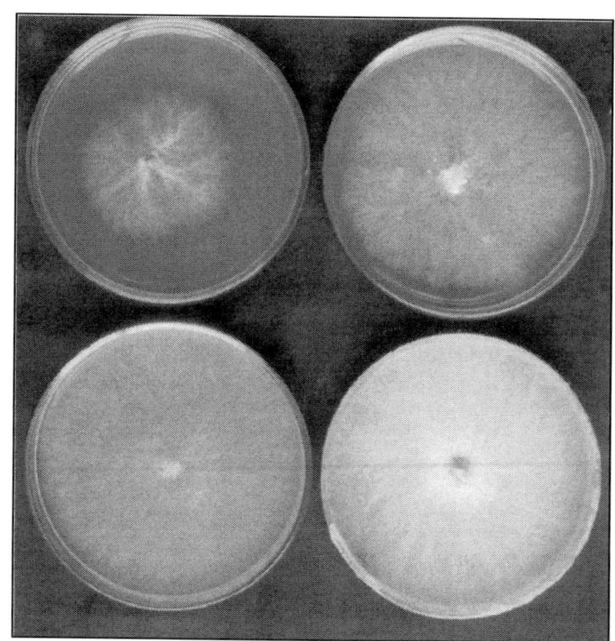

Figure 16.2: Mycelium of *Rhizoctonia solani* f. sp. *sasakii*

Systemic Position

Sclerotial Stage

Sub division: Deuteromycotina

Form-Class: Deuteromycetes

Form-Sub class: Agonomycetales

Genus: *Rhizoctonia*

Species: *solani* (Kuhn) (Alexopoulus *et al.*, 1996)

Basidial Stage

Sub division: Basidiomycotina

Class: Basidiomycetes

Sub-Class: Holobasidomycitidae (Hymenomycetes)

Order: Tulasnellales

Family: Ceratobasidiaceae

Genus: *Thanetephorus*

Species: *cucumeris* (Frank) Donk (Alexopoulus *et al.*, 1996)

Mycelium

Young mycelium of pathogen is silvery, becoming yellow and brown with maturity, 1-2μ broad and infrequently septate. Three types of mycelium are seen (Dasgupta *et al.*, 1992).

Runner Mycelium

Straight, creeping trophic, non-infection hyphae or some times thick and flattened.

Lobate Mycelium

Branching out from the former as short, swollen, much branched, and single or multiple lobate appresoria and penetration peg, which later becomes intracellular all through the lesions. The lobate state associated with the infection process may be counterpart of the sporogenous moniloid state.

Moniloid Mycelium

Forming the sclerotia but also formed on the walls of culture wares.

Sclerotia

The sclerotia of *R. solani* which are initially whitish to creamish, turning salmon-red later. The sclerotia appear as convex or orbicular bodies in frontal view but wrinkled and curled up types become common in dry weather. The sclerotia are brown to black composed of clusters of melanin encrusted, thick walled cells, rich in nutrients, formed by repeated branching from short, thick, lateral hyphae, when produced on plant parts it is difficult to separate the sclerotia from their surrounding embedded mycelia (Gangopadhyay and Chakraborty, 1982). Five phases of sclerotial morphogensis have been recognized in *R. solani*. (1) repeated hyphal branching resulting in short, thick, lateral moniloid hyphae; (2) hyphal aggregation as clusters of thick walled cells and network formation, (3) formation of sclerotial initial, (4) formation of whitish immature sclerotia and (5) maturation and pigmentation by melanin encrustation. Freshly formed sclerotia are dense and sink in water but with age they become buoyant and the cell wall becomes thicker (0.51 µm).

Basidia and Basidiospores

The basidia are borne on small perfectly symmetrical cymes. They are barrel-shaped, ob-pyriform or clavate and 9-15 x 5-12 (mostly 14 x 9) microns. They bear 4 sterigmata measure 13 microns in length and 3 microns broad at the base. Basidiospores are ellipsoid or oblong-ellipsoid, flattened on one side, hyaline, truncate, and 5-14.5 (9) x 4–8 (average 9 x 5.5) microns.

Host Range

The pathogen has a wide host range, which infects plants belonging to 32 families and 188 genera. Under artificial inoculations *R. solani* f. sp *sasakii* infects a number of crops belonging to families, Poaeae, Papilionaceae, Solanaceae and *Paspalum scrobiculatum, Penicum americanum, Coix lachryma jobi, Echinochloa frumentacea, Zea mays, Z. mexicana, Oryza sativa, Saccharum officiniarium* and *Sorghum bicolor.* Maize also has been found to be infected by isolates from rice, sugarcane, arrowroot and some grasses.

Characters for Identification

Rhizoctonia solani Kühn is morphologically characterized by the following features: pale to brown colour of mycelium, branching near the distal septum in young growing hyphae, presence of a constriction and formation of a septum in branch near the point of origin, absence of clamp connections, sclerotia of undifferentiated texture, young multinucleate hyphal cells with a prominent septal pore apparatus and rapid growth rate. *R. solani* f. sp. *sasakii*, the mycelial state of *T. sasakii* can not be easily differentiated from *R. solani*, the mycelial state of *T. cucumeris* (Frank) Donk on the base of mycelial and sclerotial characters.

Also, it should be noted that there has been considerable confusion regarding classification of species of *Rhizoctonia* and its perfect state. In Japan, much attention has been paid to differentiate between *sasakii* type and other related fungi. Watanabe and Matsuda (1966) divided 200 isolates into six groups on the bases of cultural characteristics, pathogenic or saprobic nature. The *sasakii* type belonged to first group 1A. Ogoshi (1987) examined hypal anastomosis of the clones isolated from various plants and fields soil including representative strains of Watanabe and Matsuda. All the clones were classified into five anastomosis groups and the sasakii type belonged to the first group AG1. Group 1A (of Wantanabe and Matsuda) and AG1 (of Ogoshi) have characteristics in common.

Disease Cycle

The primary source of inoculum are the sclerotia in soil or in infected host debris, and the active mycelium on the other grass hosts that grow in the vicinity of maize plant in the field. In undistributed soil, sclerotia are present up to 2 cm depth. Population of sclerotia in ploughed field is higher at 12 cm depth in soil have better buoyancy and germinability. Sclerotia, which survive on plant debris often, come up on the soil surface during field preparation and other operations. They come in contact with newly planted seedlings/plants and cause infection.

Secondary spread is due to contact of healthy plants with infected leaves/sheaths or vice-versa. The infection continues to mid-dough stage covering the entire plant including the ear but not the tassel. Seeds are not considered to be a source of inoculum and may not play a major role in severe disease outbreaks. If ear infection occurs at the late stage, germination of the seed is adversely affected resulting in seed rot and seedling blight. The sclerotia formed on infected plant organs on the soil surface together with infected host debris serve as a source of primary inoculum for maize and other host crops.

Secondary spread by basidiospore has not been observed in maize crop. Though the pathogen produces distinct horse-shoe shaped lesions on seed, the seed inoculum is of little importance in areas where disease is already known to occur. The pathogen is soil-borne and also survives on a large number of cultivated and wild plants.

Epidemiology

The optimum temperature for *in vitro* growth of the pathogen is 30°C and highest level of the disease is induced when the relative humidity (RH) is in the range of 90-100 per cent. At RH 70 per cent or below, disease development is negligible or absent. High relative humidity and rainfall largely influence the spread of the disease. Studies on the meteorological factors in relation to disease development revealed that an optimum temperature about 28°C was essential for infection and disease development to a large extent. High relative humidity (88 to 90 per cent) in the first two weeks of infection favoured rapid disease progress but RH below 70 per cent checked the development and spread of the disease. Similarly, rainfall over 100 mm in the first two weeks of infection favoured early infection and disease development.

Disease Management

The genetic bottleneck has prompted development of alternative methodology related towards the arresting of proliferation and spreads of disease through agronomic practices. Management of disease through chemical, biological and cultural practices are still the main stay for minimizing the devastation due to BLSB in maize growing areas of South and South East Asia. Information regarding on some of these aspects is summarized below:

Genetic Resistance

The genetic variability for resistance to BLSB has been found to be limited which puts limitation on a resistance breeding programme. Screening for BLSB resistance is being practiced in national programs in India, Guangxi province in South China, Indonesia and Philippines. A variety of maize materials including–inbred lines, single crosses, double crosses, and double top crosses etc. were evaluated in Indian maize programme. Among the inbred lines CM 104 (Co-ordinates Maize–yellow flint) and CM 300 (white flint) were determined to be the elite lines in terms of resistance. In Guangxi, China, B 44-1, SZ-1, B 84-2, MS 08, BS-1, WP-4, WP-5, H 138, H 158, KI 1414 and several other materials were rated as resistant during 1994-95 crop season. Maize varieties Jinguok, Suweon 83, Suweon 87, Suweon 89, P 3055, P 3160, DK 689 and XCG 51 showed high tolerance levels to BLSB in Korea (Lee *et al.*, 1989). However, performance of these materials in other countries of the region needs to be seen before they are mobilized in any breeding programme. Selection of disease resistance has been activated through collaborative project introduced by CIMMYT and national programs, in Asian region. It is expected that through these coordinated activities, resistant germplasm will be made available to other national programme in the Asian region for utilization in the development of disease resistant cultivars. In fact, a very good beginning has been made in this direction. A large number of CML (CIMMYT Maize Lines) and other materials were evaluated in India and China and many lines were identified having reasonable level of resistance. Both in early maturing and late maturing groups few lines have been identified to be tolerant at locations in China, India and Indonesia. Some of these are; CA 00106, CA 00102, CA 03147, CA 03106, CA 003134, CL 02846, CA 00396, CA141518, CA 00310, F-

2-32-B-B, P 42c9MH17-26-1-B-2, EWDMR-G-7-HS-(S1-B)-9B-1-B-B-B, HKH-1140, PRD-340, PAC-79001, NMH-9858, FH-3097, FH-3133, NECH-01, F-7001 and BIO-81009 and POB 502c3 F2 10-8-1-1-B-B, (Balla *et al.*, 2000; Singh and Saxena 2005; Meena, 2004; Sharma *et al.*, 2000). Among 45 inbred maize lines evaluated against BLSB, the resistance of R15 was higher than that of R 09. Of the 45 inbred lines tested, none was immune to the disease. CML 270 was found to be highly resistant (Yang *et al.*, 2003).

Cultural

Sharma and Hembram (1990) have worked out a simple and economical method to control this disease. The methodology comprises of mechanical stripping of the second and third leaf sheaths from the ground level at the age of 35-40 days old crop or knee high stage which is effective in checking disease development further up. This method also was found to be effective in minimizing the yield loss. In Japan, also resistance to this disease has been observed after the fall of the lower sheath (Kato and Inoue, 1995). Qing *et al.* (1994) demonstrated that removing infected maize leaves at the whorl stage gave significantly better control than spraying with Jinggangmycin. Brushing Jingganmycin liquid on infected plant parts after leaf removal resulted in a disease control of up to 100 per cent.

Bio-control

In recent years there has been considerable popular and scientific resistance to use of pesticides/ fungicides for the control of diseases and pests in food crop plant in the world over. It, therefore, becomes all the more important that alternative environmental friendly methods should be discovered to alleviate the ill effects of chemical control agents in food chain. Biological control of diseases offers the preferred mode of environmental friendly control of diseases in maize including banded leaf and sheath blight. Several antagonists such as species of *Trichoderma, Gliocladium virens, Bacillus subtilis* and *Pseudomonas fluorescens* have been tried by different workers. However, the encouraging results have been obtained by the use of *P. fluorescens.* Shivakumar *et al.* (2000) demonstrated a very effective control by seed treatment of the peat based formulation @16g/kg, or soil application @ 2.5 kg/ha, or spraying liquid formulation twice @ 5 g/litre of water. The increase in phenolic content was also observed in leaf sheaths of plants raised from *P. fluorescens*-treated seeds when inoculated with *R. solani.* Peroxidase (PO), polyphenol oxidase [catechol oxidase] (PPO) and phenylalanine ammonia-lyase (PAL) activities increased when leaf sheaths were inoculated with the pathogen, while plants raised from *P. fluorescens*-treated seeds showed higher activity. However, the leaves from *P. fluorescens*-treated seeds did not show any further increase in the PO and PPO activities after inoculation with *R. solani.* The bacterized seed with *P. fluorescens* accumulated high phenolic compounds and high PO, PPO and PAL activities, which may play a role in the plant defense mechanism against the pathogen (Sivakumar and Sharma, 2003). The combined seed and foliar treatment with Fluorescent pseudomonads from maize rhizoplane was most effective in giving 30 per cent reduction in disease incidence of banded leaf and sheath blight. Similarly, among several spp. of *Trichoderma, T. harzianum* has shown good promise against this pathogen both *in vivo* and *in vitro* conditions (Meena *et al.*, 2003; Khan and Sinha, 2005). In Venezuela, a very effective control of banded leaf and sheath blight pathogen using *T. harzianum* has been demonstrated (Personal Communication). Spraying of Talc+ CMC based formulations of bioagents *viz., T. harzianum, T. virens,* and *T. viride* before disease appearance are quite effective against Sheath Blight (Singh and Sinha, 2005; Khan and Sinha, 2006; Khan and Sinha, 2007).

Chemical

Several fungicides like carbendazim, bengard, thiophanate-methyl, iprobenfos, captan, quintozene, mancozeb + thiophenate-methyl, copperoxychloride and thiram have been found effective in inhibiting the growth of pathogen under *in vitro* conditions. Under field conditions also, all the fungicides except thiram were effective in reducing disease severity (Ahuja and Payak, 1986.; Sharma and Rai, 1999; Puzari *et al.*, 1998). In China, the antibiotic Jinggangmycin (Validamycin) gave satisfactory control in field conditions (Jiang *et al.*, 1991). In India, the formulation of Validamycin has shown good control against banded leaf and sheath blight pathogen (Batsa, 2003). Foliar sprays of Contaf @ 1ml/litre significantly reduce infection of *R. solani* (Khan and Sinha, 2007).

Integrated Disease Management

Economic control of plant diseases is rarely achieved by a single method. It must be supported by different control measures. An integrated pest management system involves simultaneous or sequential use of an optimal combination of various means of disease control eco-friendly. Biological control is of particular interest as a component of IPM and can best be exploited within the framework of an IPM system (Klassen, 1982). Species of *Trichoderma* and *Gliocladium* have distinct advantage as biocontrol agents, as they are compatible with most of the agricultural practices and hence, can be successfully utilized as a part of total crop management practices or broadly as a component of agro-ecosystem management strategies. *Gliocladium* spp. and *Trichoderma* spp. have been integrated with cultural practices, physical soil treatment (solarization), fungicide and disease resistant varieties for the management of different plant diseases.

Combining the antagonist *T. harzianum* with soil solarization in *R. solani* infested soil improved disease control and delayed inoculum build up (Elad *et al.*, 1980, Chet *et al.*, 1982). One of the greatest advantage in using *Trichoderma* and *Gliocladium* is that these two antagonists are insensitive to many fungicides of common use for controlling plant diseases (Mukhopadhyay *et al.*, 1992). Gogoi and Ali (2005) found that *T. harzianum* was more compitable with captan, as it inhibited the growth of *R. solani* more than *T. viride* and *G. virens*. Single or combination doses of iprodione (Rovral 50) and carbendazim (Bavistin 50 WP), both at 0.5, 1.0, 2.0, or 4.0 ppm, and the compatibility of these fungicides with *T. harzianum* and *G. virens* were tested. Single and combine applications of the fungicides were highly effective in controlling the diseases. The plant pathogen was more tolerant to the fungicides than the biological control agents. Single dose of carbendazim exibited better results than iprodione and the doses of carbendazim (Sitansu *et al.*, 2001). Many positive results obtained with fungal biocontrol agents and organic amendments in various crops suggest possibility of combining these two systems for effective crop protection. The beneficial effects of soil amendments in plant disease management lie in their ability to suppress pathogen populations through the stimulation of soil saprophytes and also by toxic substances produced during their decomposition. Soil amended with organic amendments such as rice chalf, mustard cake, sawdust, neem cake, composed coir with and farmyard manure significantly reduced the seedling infections, decrease severity (Kanniyan and Prasad, 1981; Meena and Muthusamy, 1999). Baby and Manibhushanrao (1993) reported that there was significant decrease in ShB severity when two fungal biocontrol agents (*G. virens* and *T. longibrachaetum*) were integrated with two organic soil amendments (gliricidia leaf and neem cake).

Mathivanum *et al.* (1997) tested the efficacy of some fungicides with integration of neem oil against *R. solani* and reported that integration of carbendazim + neem oil was most effective followed by contaf + neem oil. Surulirajan *et al.* (2003) evaluated *T. viridi*, carbendazim and soil amendments (farmyard mannure and sawdust (1 per cent)), singly or in combination in pot experiment. Among the treatments, the application of *T. viridi* + carbendazim 50 WP + FYM + sawdust resulted in the lowest disease incidence and highest grain and straw yields. PF-9 (fluorescent *pseudomonas sp.*), B-44 (*Bacillus* sp.) and Chb-1 (chitinolytic bacterium) were found to be compatible with carbendazim (bavistin) at

500 and 1000ppm concentrations. Among them, PF-9 was the most effective in reducing disease severity either alone or in combination with one spray of 0.1 per cent Bavistin, followed by combination of PF-9 and B-44 (Laha and Venkataraman, 2001). Mishra and Sinha (1997) observed effective control of ShB with the application of *P. fluorescens* and Tilt.

References

Ahuja, S.C. and Payak, M.M. (1986). *In vitro* response of maize and rice isolates of *Rhizioctonia solani* to antibitics and fungitoxicants. *Int. Rice Res. Newsl.*, 11: 16.

Ahuja, S.C. and Payak, M.M. (1982). Symptoms and signs of banded leaf and sheath blight of maize. *Phytoparasitica*, 10: 41–49.

Alexopolus, C.J., Mims, C.W. and Blackwell, M. (1996). *Introductory Mycology*, 4th Edn. Wiley, New York.

Baby, U.I. and Manibhushnrao, K. (1993). Control of rice sheath blight through the integration of fungal antagonists and organic amendments. *Trop. Agric.*, 30: 240–244.

Balla, O., Vassal, S. K., Cenieeros, F.G., Sharma, R.C. and Shukuan, Z. (2000). Identifying resistance to some maize diseases in Asia. In: *Third Crop Science Congress*, 17–22, August, Hamburg, Germany.

Batsa, V.K. (2003). Integrated management of banded leaf and sheath blight of maize caused by *Rhizocotonia solani* f. sp. *sasakii*. *Ph.D. Thesis*, Indian Agricultural Research Institute, New Delhi. pp. 97.

Bertus, L.S. (1927). *Year Book. Department of Agriculture*, Ceylon, p. 44–46.

Chet, I.Y., Elad, A., Kalfon, Y. Hadar and Katan, J. (1982). Integrated control of soil-borne and bulb-borne pathogens in iris. *Phytoparasitica*, 10: 229–231.

Cramer, H.H. (1967). *Plant Protection and World Crop Production*. Bayer, Leverkusen, Germany.

Dasgupta, M.K. (1992). Plant diseases of international importance. In: *Volume I: Diseases of Cereals and Pulses*, (Eds.) Singh, U.S., Mukhopadhyay, A.N., Kumar, J. and H.S. Chaube, pp. 130–157.

Elad, Y., Katan, J. and Chet, I. (1980). Physical, biological and chemical control integrated for soil borne diseases of potato. *Phytopathology*, 70: 418–522.

Gangopadhyay, S. and Chakraborti, N.K. (1982). Sheath blight of rice. *Rev. Plant Path.*, 61: 451–460.

Gogoi, N.K. and Ali, M.S. (2005). Integrated management of sheath blight of winter rice involving *Tricoderma harzinum*, few soil amendments and captan. *Crop Res. Hissar*, 30: 423–427.

Hirrel, M.C., Lee, F.N., Dak, I.J. and Plannkett, D.E. (1988). First report of sheath blight (*Rhizoctonia solani*) on field corn in Arkansas. *Pl. Dis.*, 72: 644.

James, W.C. (1981). Estimated losses in crops from plant pathogens. In: *Handbook of Pest Management in Agriculture, Vol. 1*, (Ed.) D. Pimentel. CRC Press, Boca Raton, Florida, pp. 79–84.

Jiang, H.L, Ding, X. and Ma, H. (1991). The occurrence and chemical control of maize sheath blight (*Rhizoctonia solani*) in Rugao of Jiangsu. *Plant Prot.*, 6: 11–12.

Kannaiyan, S. and Prasad, N.N. (1981). Effect of organic amendments on seedling infection of rice caused by *R. solani*. *Plant and Soil*, 62: 131.

Kato, A. and Inoue, Y. (1995). Resistance to banded leaf and sheath blight (*Rzioctonia solani* Kühn) after fall of lower sheaths in maize (*Zea mays*). *Bull. Nat. Grassland Res. Inst., Japan*, 51: 1–5.

Khan, A.A. and. Sinha, A.P. (2006). Influence of formulations, doses and time of application of fungal bio-agents on rice sheath blight. *Annl Pl. Protec. Sci.*, 14: 157–161.

Khan, A.A. and Sinha, A.P. (2005). Comparative antagonistic potential of some biocontrol agents against sheath blight of rice. *Indian Phytopath.*, 58: 41–45.

Khan, A.A. and Sinha, A.P. (2007). Biocontrol potential of *Trichoderma* spp. against sheath blight of rice. *Indian Phytopath.*, 60:198–203.

Klassen, W. (1982). The role of biological control in integrated pest management systems. In: *Biological Control in Crop Production*, (Ed.) G.C. Papavizas. Allanheld, Osum, Granda, USA, pp. 433–466.

Laha, G.S. and Venkataraman, S. (2001). Sheath blight management in rice with biocontrol agents. *Indian Phytopath.*, 54: 461–464.

Lee, S.B., Kim, J.G., Han, M.S. and. Han, H.J. (1989). Studies on the occurrence of corn sheath blight caused by *Rhizoctonia solani* I. Influence of growth season and cultural environment on the occurrence of sheath blight disease. *J. Korean Soc. Grassland Sci.*, 9(3): 174–178.

Mathivanum, N., Srinivasan, K. and Chelliah, S. (1997). Evaluation of fungicides and their combinations with neem oil for the management of rice diseases. *Pestology*, 21: 16–17.

Meena, R.L. (2004). Evaluation of maize genotypes for resistance to banded leaf and sheath blight induced by *Rhizoctonia solani* f. sp. *sasakii* (Kuhn) Exner, *Indian Journal of Plant Protection*, 32(2): 85–88.

Meena, R.L., Rathore, R.S., Mathur, K. and Mathur, K. (2003). Efficacy of biocontrol agents against *Rhizoctonia solani* f. sp. *sasakii* causing banded leaf and sheath blight of maize. *J. Mycol. Pl. Pathol.*, 33(2): 310–312.

Meena, B. and Muthusamy, M. (1999). Effect of organic soil amendments against rice sheath blight. *Indian Phytopath.*, 52: 92–93.

Mishra, D.S. and Sinha, A.P. (1997). Biocontrol potential of some biocontrol agents against *Rhizoctonia solani* Kûhn, the cause of sheath blight of rice. In: *International Conference on Integrated Plant Disease Management for Sustainable Agriculture*, New Delhi, 10–15, November, Abstr. pp.209.

Mukhopadhyay, A.N., Shrestha, S.M. and. Mukherjee, P.K (1992). Biological seed treatments for control of soil borne plant pathogens. *FAO Plant Prot. Bull.*, 40: 1–10.

Ogoshi, A. (1987). Ecology and pathogenicity of anastomosis and interspecific groups of *Rhizoctonia solani* Kühn. *Ann. Rev. Phytopathol.*, 25: 125–143.

Payak, M.M. and Sharma, R.C. (1985). Maize diseases and approaches to their management in India. *Trop. Pest Managm.*, 31: 302–310.

Puzari, K.C., Saikia, U.N. and Bhattacharya, A. (1998). Management of banded leaf and sheath blight of maize with chemicals. *Indian Phytopathol.*, 51: 78–80.

Qing, J.L., Huang, C.W., Huang, R.J. and Li, O.C. (1994). Study on the control of maize sheath blight disease by removing infected leaves. *Plant Protc.*, 20(1): 4–6.

Sangam Lal, Barauh, P. and Butchaiah, K. (1980). Assessment of yield losses in maize cultivars due to banded leaf sclerotial disease. *Indian Phytopathol.*, 29: 129–132.

Sharma, R.C. and Hembram, D. (1990). Leaf stripping: A method to control banded leaf and sheath blight of maize. *Curr. Sci.*, 59: 745–746.

Sharma, R.C. and Rai, S.N. (1999). Chemical control of banded leaf and sheath blight of maize. *Indian Phytopathol.*, 52: 94–95.

Sharma, R.C., Carlos de Leon and Singh, N.N. (2000). Banded leaf and sheath blight of maize, its importance and current breeding efforts. In: *Proceeding of the Seventh Regional Maize Workshop*, (Eds.) Vasal *et al.* PCARRD, Los Banos, Philippines, Feb. 23–27, 1998, pp. 284.

Sharma, R.C., Carlos de Leon. and Payak, M.M. (1993). Diseases of maize in South and South East Asia: Problems and progress. *Crop Prot.*, 12: 414–422.

Sharma, R.R., Gour, H.N. and Rathore, R.S. (2004). Etiology of banded leaf and sheath blight symptoms on maize. *Journal of Mycology and Plant Pathology*, 34(1): 56–59.

Shivakumar, G. and Sharma, R.C. (2003). Induced biochemical changes due to seed bacterization by *Pseudomonas fluorescens* in maize plants. *Indian Phytopathology*, 56(2): 134–137.

Shivakumar, G., Sharma, R.C. and Rai, S.N. (2000). Bicontrol of banded leaf and sheath blight of maize by peat based *Pseudomonas fluorescens* formulation. *Indian Phytopathol.*, 53: (2): 190–192.

Singh, B. and Saxena, S.C. (2005). Evaluation of resistant and susceptible maize genotypes by laboratory and field method against banded leaf and sheath blight (*Rhizoctonia solani* f.sp. *sasakii*). *Advances in Plant Sciences*, 18(2): 923–926.

Singh, Rajbir and Sinha, A.P. (2005). Effect of *Pseudomonas fluorescens* formulations on sheath blight and yield component of rice. *Annl. Pl. Protec.*, 13: 159–162.

Sitansu, Hazra, P.S., Roy, A. and Pan, S. (2001). Efficacy of iprodione and carbendazim against rice sheath blight and their compatibility with biocontrol agents. *J. Interacademicia*, 5:11–18.

Srinivas, P. (2002). Studies on genitic and pathogenic variation in *Rhizoctonia solani* f. sp. *sasakii* incitant of banded leaf and sheath blight of maize. *Ph.D. Thesis*, Indian Agricultural Research Institute, New Delhi, pp. 110.

Sumner, D.R. and Minton, N.A. (1989). Crop losses in corn induced by *Rhizoctonia solani* AG 2–2 and nemotodes. *Phytopathology*, 79: 934–941.

Surulirajan, M. and Kandhari, J. (2003). Effects of integrated treatments against rice sheath blight severity, disease incidence and yield parameters. *Annl. Pl. Protec Sci.*, 11: 284–288.

Watanabe, R. and Matsuda, A. (1966). Studies on the grouping of *Rhizoctonia solani* Kühn pathogenic to upland crops. *Bull. Appl. Exp. (Plant Dis. Insect Pests)*, 7: 1–131.

Yang, A.G., Pan, G.T., Ye, H.G. and Tang, L. (2003). An evaluation of the resistance of inbred lines of maize to maize sheath blight and screening of resistant resources. *Plant Prot.*, 29 (1): 25–28.

Zaidi, P.H. and Singh, N.N. (2005). *Stresses on maize in tropics.* Published by Project Director, Directorate of Maize Research, Cummings Laboratory, Pusa Campus, New Delhi – 110012, pp. 500.

Zhang, P.Q., De Leon, C. and Yang, H. (1993). Investigation on the yield loss caused by *Rhizoctonia solani* on maize. Report of Gaungxi Maize Res. Inst. (Unpublished).

Plant Diseases Management for Sustainable Agriculture *Pages* **195–211**
Editor: **Dr. Shahid Ahamad**
Published by: **DAYA PUBLISHING HOUSE**

Chapter 17

Advances in Management of Guava Wilt-Complex

R.K. Prajapati[1] and R.K. Singh[2]

[1]Krishi Vigyan Kendra, Mandsaur (M.P.)
[2]Division of Crop Protection, Indian Institute of Pulses Research, Kanpur (U.P.)

Introduction

Guava (*Psidium guajava*) is a very popular fruit. It is available throughout the year except summer season. In India, it is grown almost in all the states. Bihar has the largest area *i.e.* 24.7 thousand ha. under guava cultivation followed by Uttar Pradesh (18.5 thousand ha.) and Maharashtra (14.8 thousand ha.) The per hectare productivity is the highest in Madhya Pradesh (20 tones/ha.) compared to very low in Uttar Pradesh (7.3 t/ha.). Although guava is a very hardy fruit, it suffers heavily from wilt disease, which causes about 20-100 per cent mortality under different agro-climatic and soil conditions. As the disease is soil-borne in nature, there are limitations in its control. The disease was first noticed in Babakkarpur near Allahabad (Prasad *et al.*, 1952) in India. Although guava wilt was first reported in 1935 by Jhooty *et al.*, 1984.

The guava is attacked by a large number of pathogens, mainly fungi. About 177 pathogens (167 fungi, 3 bacteria, 3 algae, 3 nematodes and one epiphyte) are reported on various parts of the crop, including fruits, causing various diseases. One of the reasons for the lower productivity is the wilt disease. Wilt of guava is one of the most important diseases of guava especially in India and loss due to this disease is substantial. Recent estimate of economic losses have been reported by different workers (Table 17.1).

Table 17.1: Economic Losses in India from Different States

State	Districts	% Losses	References
Uttar Pradesh	Allahabad, Farrukhabad and Unnao	15-30	Mathur, 1956
Uttar Pradesh	Kanpur and Jaunpur	5-15	
Uttar Pradesh	Gorakhapur, Ballia, Hardoi, Barabanki and Varanasi	5	
Uttar Pradesh	12 Districts	5-15	Singh and Lal, 1953
Uttar Pradesh	Lucknow	5-60	Misra and Shukla, 2002
West Bengal	24-Parganas, Midnapur, Kashakul, Bankura	80	Chattopadhyay and Sen Gupta, 1955; Chattopadhyay and Bhattachariya, 1968a,b

Importance and Distribution

The guava crop in India and abroad suffers from a serious malady and continues to remain an unsolved problem (Table 17.2). The disease is commonly referred to as wilt which, as well as know, is a expression of the diseased condition and not a disease in itself due to varied causes operating alone or in combination (Table 17.3). Notwithstanding this fact, the past researches have either remained tunnel visioned, concentrating around fusarioris or attempted to tackle factors not established scientifically as the cause of the trouble. The disease is known to occur for the past seven decades but the information available today has not yet proved the way to come out with any concrete solution for this problem even the actual causes are not yet confirmed.

Causal Factors

Pathological Factors

The various species of different kinds of pathogens like fungi, bacteria and nematodes have been isolated from the wilted plants by various workers in different parts of the world. Since the disease is soil-borne, a host of soil microbes most importantly fungi like -*Fusarium solani* (under heavy moisture), *Macrophomina phaseolina*(dry soils), *Collectotrichum gloeosporiodes, F. oxysporum psidii* and *Rhizoctonia solani* (moderate soil moisture), *R. bataticola, Cephalosporium* sp., *Acremonium phaseoli, Gliocladium roseum* and *Verticilium albo atrum* etc. have been responsible for guava wilt (Negi *et al.*, 2001). Many workers in guava wilt have also reported the role of bacteria and nematodes (Table 17.3).

Edaphic Factors

The soil conditions, nutrition and water are the predisposing factors for the disease. The alkaline soils with pH 7.5-9.0 are ideal for disease occurrence. Acidity or alkalinity of the soil seems to play a little role in development of wilt. In other words soil reaction at pH 6.0 and soil moisture 60-80 per cent situation are most favorable for the disease development. Besides, lower and higher nitrogen content is also favorable to it. Incorporation of organic manures and green manuring has been found effective in reducing the incidence of guava wilt.

Climatic Factors

The disease starts in the beginning of June or the starting of rainfall, but the intensity of infection and spread is accelerated with the onset of monsoon, which brings about high percentage of humidity in the atmosphere. The heaviest death of guava tree is, however, restricted to the month of September and October, beyond which it gradually diminishes. Soon after the rainy season, in September and October, The entire trees dries up within a period of 3-4 weeks (Edward, 1960 and Suhag, 1976).

Table 17.2: Report of Guava Wilt-Complex from India and Aborad

Place/Districts	*State*	*Country*	*References*
Babakkarpur (Allahabad)	Uttar Pradesh	India	Prasad *et al.*, 1952
Lucknow	Uttar Pradesh	India	Das Gupta and Rai, 1947
Allahabad, Kanpur, Lucknow	Uttar Pradesh	India	Dey, 1948
11 districts	Uttar Pradesh	India	Anonymous, 1949; 1950; Prasad *et al.*, 1952
Allahabad, Farrukhbad, Unnao, Kanpur, Jaunpur, Gorakhpur, Ballia, Hardoi, Barabanki, Varanasi	Uttar Pradesh	India	Mathur, 1956
Western parts	Uttar Pradesh	India	Edward and Srivastava, 1957
Western parts	Uttar Pradesh	India	Singh and Lal, 1953
Varanasi	Uttar Pradesh	India	Pandey and Diwevedi, 1985
Farrukhabad, Kanpur, Unnao, Allahabad, Lucknow, Agra, Aligarh	Uttar Pradesh	India	Misra and Prakash, 1986,Misra, 1987
24-Paragana, Jhargram, Midnapur	West Bengal	India	Chattopadhyay and Sen Gupta, 1955
Kashakul, Bankura	West Bengal	India	Chattopadhyay and Bhattachariya, 1968a,b, 1955
Indore	M.P.	India	Misra,2006
	Haryana	India	Suhag, 1976,Mehta, 1987
	Punjab	India	Chandra Mohan *et al.*, 1986
	Rajasthan	India	Katyal 1972,Bhargava *et al.*, 2003
	Delhi state	India	Anonymous, 1953
	Jharkhand	India	Srivastava *et al.*, 2001
	Andhra Pradesh	India	Jhooty *et al.*, 1984
	Orissa	India	Das *et al.*, 1993
	Florida	USA	Webber, 1928
		Taiwan	Hsieh *et al.*, 1976; Leu and Kao, 1979
		Cuba	Rodrigvez and Landa, 1977
		South Africa	Grech, 1985,Vos *et al.*, 2000
		Brazil	Tokeshi *et al.*, 1980; Rodriguez *et al.*, 1987; Junqueira *et al.*, 2001
		Pakistan	Ansar *et al.*, 1994
		Bangladesh	Hamiduzzaman *et al.*, 1997
	Canberra	Australia	Lim and Manicom, 2003

Epidemiology

The epidemiological factors *i.e.* soil, relative humidity, temperature are playing important role in development of guava wilt. The relationship of environment to the disease development has not yet been investigated in a precise manner. While some reports in this context such as Das Gupta and Rai(1947) reported that the wilting occurs during the rainy season. Its occurrence is maximum during and after the monsoon at the time of flowering and fruiting. The largest number of deaths occur in September and October and the disease comes almost to a stand still with the advancing winter. It appears that the cool (January to March) and hot season (April to June) prevent the spread of infection.

Table 17.3: List of Factors Involved in Guava Wilt-Complex

Causal Organism/Factors	*References*
A. Fungi	
Cephalosporium sp.	Vestal, 1941; Dey, 1948
Fusarium sp.	Das Gupta and Rai, 1947
Fusarium oxysporum (Fr.)Schl.	Prasad *et al.*, 1952
F. o.(Fr.) Schl. f.sp. *psidii*	Edward and Srivastava, 1957; Pandey and Dwivedi, 1985; Edward, 1960 b,c
Gloeosporium psidii	Tandon and Agarwal, 1954
Macrophomina phaseolina, Fusarium solani	Chattopadhyay and Bhattacharya, 1968a,b; Chattopadhyay and Sen Gupta, 1955
F. solani,F.oxysporum,M. phaseolina	Chattopadhyay and Bhattacharya, 1968a,b; Edward1960c; Chandra Mohan, 1985; Pandey and Dwivedi, 1985; Sohi, 1983a,b
Gliocladium vermoesenii Corda	Chandra Mohan, 1985
F.solani, F.o.f.sp. psidii, F.coeruleum, F.moniliforme, Rhizoctonia solani	Dwivedi, 1991a; Dwivedi, and Dwivedi, 1999
Cylindrocarpon lucidium, Gliocladium virens, Bartilinia robillardoides	Misra and Pandey, 1992
Gliocladium roseum	Misra and Pandey, 1997; Misra and Pandey, 2000a
Gliocladium virens, G. penicilloides, F.oxysporum, F.solani,Acremonium sp.	Misra and Pandey, 1999c
Botryodiplodia theobromae	Pandit and Samajpati, 2002
Verticillium albo-atrum	Gupta *et al.*, 2003
Gliocladium roseum, Sclerotium rolfsii, Rhizoctonia bataticola	Misra *et al.*, 2004a
Penicillium vermoensenii	Vos *et al.*, 2000
F.o. f.sp. *psidii*	Lim and Manicom,2003
B. Fungi + Nematodes	
F.oxysporum f.sp. *psidii* + *Heliotylenchus dihystera, H. indicus*	Hamiduzzaman *et al.*, 1997
Fusarium spp. + *Helicotylenchus dihystera*	Khan *et al.*, 2001
F.oxysporum, Collectotrichum gloeosporiodesi	Ansar *et al.*, 1994
C. Nematodes	
Meloidogyne sp.,*Helicotylenchus* sp.,*Pratylenchus* sp.	Rodriguez and Landa,1977

Contd...

Table 17.3–Contd...

Causal Organism/Factors	*References*
D. Bacteria	
Erwinia psidii	Rodrigues *et al.*,1987;Jungveria *et al.*,2001
Pseudomonas sp.	Tekeshi *et al.*,1980
E. Climatic factor	
Alkaline soils pH (7.5-9.0)	Mehta,1951
Lateritic soils pH(6.5)	Sen Verma,1954
Soil saturation 60-80 per cent moisture	Chattopadhyay and Bhattachriya,1968b
Clay loam and sandy loam	Mehta,1987
Seedling stage up to 3 years susceptible to *M. phaseolina* and *F. solani*	Chattopadhayay and Bhattachariya,1967
Five year old more susceptible	Misra and Shukla, 2002
Maximum drying in September to October	Das Gupta and Rai,1947;Edward,1960a;Suhag,1976
Maximum wilting October	Misra and Pandey,1999a,d,2000b
Temperature 31.3-33.5ºC (July-September), 23-25 ºC at 76 RH per cent	Misra and Pandey,1999a,d,2000b

High disease incidence in laterite soils at pH (6.5) with soil saturation of 60-80 per cent are found ideal for development of severe incidence of wilt. Both pH 4.0 and 8.0 reduces the disease, low incidence of the disease shows at 630 ppm N and more both at higher as well as at lower level of nitrogen. Moderate to high concentration of phosphorus (207-345 ppm) are effective in reducing the disease. Clay loam and sandy loam are more suitable for incidence of wilt in compared to heavy soil. Rainy and winter seasons are more favorable for many fungi, bacteria and nematodes, which survived better in association with root bits in adverse climatic conditions. The maximum temperature (31.3-33.5 °C), minimum (23-25°C) and 75 per cent humidity are required for the complete wilting of plants.

The plant stage is also important in wilt complex of guava, for example the seedling of guava is more susceptible to *Macrophomina phaseolina* as well as *F. solani* than the older plants of 3-years age. While *F. solani* could infect guava plants of one month old to more than 4 years old trees. Fungi inciting wilt either individually or in combination, first colonizing the surface of the roots and then enters into its epidermal cells by mechanical penetration. Therefore, intercellular mycelium establishes first in epidermal cells and then spreads into cortical cells, which are considerably damaged and filled with the mycelium. The fungus enters the xylem vessels, grows inside and blocks them. Some pathogens first invade the phloem and destroy it. The disease is commonly referred to as wilt which is an expression of the diseased condition and not a disease in it self. Basically, wilting, a generalized symptom can develop due to varied operating alone or in combination (Table 17.3).

Symptomatology

External Symptoms

Slow Wilting

It is seen on young as well as the old trees. Tree with young shoots, leaves and fruits at tender stage are more prone to attack. The typical symptoms include withering of upper most branch of the tree from top downwards and whole plant withers within a year. The growing tips turn dark brown and necrotic areas extend backwards. The branches get infected one by one till the whole tree dies. The die-back/slow wilting/decline symptoms have been detailed by Singh and Lal (1953),Tandon and Agarwal (1954) and Edward(1960). Misra and Pandey (2000b) also studied variations in the symptoms during different time of year. They noticed yellowing of the leaves with inter-veinal chlorosis during the month of August, which drop even with the slight shaking of the plants. During September, general drooping of the leaves takes place. During October, complete wilting of plants is seen with almost dried leaves and small dried black fruits hanging on the branch.

Partial Wilting

A few plants also show partial wilting, which is a very common symptom of wilt in guava. Some plants may also show variable degree of wilting, leaf yellowing, drooping of leaves, drying of terminal branches or partial wilting during different months but later escape/resist wilting. These plants start recovering from December onwards. It was recorded that out of total wilted plants, around 17 per cent plants, which initially show some symptoms of wilting ultimately escape/resist wilting (Misra and Pandey, 2000b).

Quick Wilting

The plants die immediately after the infection is noticed. Sometimes, there is a sequential death of branches, one after another, till the whole tree dries. This is usually noticed in the rainy season crop when the tree dies within a week or a month. Sometimes the decline may prolong even upto one year. Occasionally, partial wilting of branches is also observed. The entire plant becomes defoliated and eventually dries. It requires almost sixteen days for complete wilting. Some affected trees linger on even up to 252 days and then die (Misra and Pandey, 2000b).

Internal Symptoms

The internal portion of the bark turns dark and the discoloration in the form of streak extends into the cambium or even black streaks or uniformly blackens under favourable weather conditions. Pink masses of spores are produced on or beneath the bark of the wilted trees. The finer roots show black streaks, which become prominent on removing the bark (Das Gupta and Rai, 1947). The roots also show rotting at the basal region and the bark is easily detachable from the cortex. The cortical regions of the stem and root show distinct discoloration. Light brown discoloration is noticed in vascular tissues (Chattopadhyay and Bhattacharya, 1968a,b). Wilted plants later show bark splitting. The pathogen attacks young as well as old fruits bearing trees but older trees are more prone to the disease (Misra and Shukla, 2002).

Disease Management Options

Since, the guava wilt is not a disease caused by single factor, the management options included cultural, chemical, biological and biotechnological approaches. Management options against wilt-complex are essentially a management issue. There can hardly be an affordable chemical control options. While researches are continuing with some bio-control fungi but the possibility of pathogens being different in different places, no uniform bio-agent may be successful at all places. Considering that all past control attempts with emphasis on cultural management, have not given desirable results, alternative options may be searched for resistance/tolerance/root stock in wild compatible species of guava against the pathogens. There is scope to attempt for developing a management practice integrating rootstock, resistant/tolerant, biocontrol agents (against the most important/wide spread pathogens), cultural practice, organic matter, nutritional, irrigation and biotechnological approaches. Development of such strategies shall require multidisciplinary and interdisciplinary approaches.

Chemical Options

Attempt has been made to control guava wilt complex by the use of various fungicides. Nematicides, oil cakes, micro nutrients and fungicides. The minimization of the incidence with the application of fungicides/Nematicides/micronutrients have been reported by various workers with fungicides *viz.* Thiophanate methyl, Carbendazim, captafol and Tiobendazole (Bhargava *et al.,* 2003 and Leu *et al.,* 1979. Antibiotics *viz.* actiodion. (Dwivedi, 1990) and heavy metals like Hg, Ld and Cu (Dwivedi, 1991b) were found effective for control of wilt (Table 17.4)

Table 17.4: List of Effective Chemicals/Oil Cakes for Management of Guava Wilt

Name of Chemical	*Application*		
Chaubatia paste	Use in aerial part	Wilt-complex	Anonymous, 1949
8-Quinalinol sulphate (water-soluble)	Injection @ 0.1 per cent	*F.o.f.sp.psidii*	Jain, 1956
Benlate or Bavistein	Spraying @0.2% after severe pruning and drenching 4 times in a year	Wilt-complex	Suhag, 1976
Metasystox and Zinc sulphate	Spraying twice in a year	Wilt-complex	Suhag, 1976
Bavistein, thiram, topsin-M, blitox, indofil M-45	*In vitro*	Wilt-complex	Misra and Pandey, 1999b
Thiophanate methyle	*In vitro*	Wilt-complex	Bhargava *et a.,l* 2003
Captafol, carbendazim, thiobendazole	*In vitro*	Wilt-complex	Leu *et al.,* 1976
Tebuconazole, Propiconazole, Prochoraz, Triforine, Carbendazim+Flusilazole	*In vitro*	Wilt-complex	Joubert and Frean, 1993
Antibiotic, Heavy metals (Hg, Cd, Cu)	*In vitro*	Wilt-complex	Dwivedi, 1990,1991b
DBCP or Methan Sodium	@52.8 ml/10 squ. M or 252.5 ml/sq m (soil application)	Nematodes	Rodriguez and Landa, 1977
Lime or gypsum	@ 1.82 kg/tree	Soil treatment	Mathur *et al.,* 1964
Neem cake gypsum	@ (6+2)kg/plant	Soil treatment	Misra and Pandey, 1994-95
Neem, Mahua, Kusum, cakes with urea	@10 (cake) and 1 kg (urea)	Soil treatment	Das Gupta and Ghoshal, 1977
N and Zn	Judicious amendment	Nutrient	Suhag and Khere, 1986

Cultural Options

Cultural practices like orchard sanitation, pruning and up rooting of wilted plants are advocated. The proper sanitation in the orchard could be controlled wilt. Wilted trees should be uprooted, burnt and trench should be dug around the tree trunk while transplanting, roots of plants should not be severely damaged. Maintenance of proper tree vigour by timely and adequately managing inter-culture and irrigation enable them to withstand infection. The pits may be treated with formalin and kept covered for about 3 days and then transplanting should be done after two weeks. Symptoms of the disease do not appear under green manuring and the disease development is less when organic sources of nitrogen are used. Soil solarization with 30-mm transparent polythene sheets during May-June is better for the control of wilt pathogen. Intercropping with turmeric or marigold is useful to check the guava wilt.

Varietal Options

The intensity of infection may vary depending upon the variety and the prevailing environmental conditions. The feature of pathogen, along with its capacity to live in soil for considerable period of time in the absence of host presents a great obstacle in adoption of the effective, easy and cheap control measures.

Table 17.4: Varietal Reaction Against Guava Wilt-Complex

Cultivars/Variety/ Root Stock/Species	*Name of Cultivars/ Variety/Root Stock/Species*	*Type of Reaction*	*Place*	*References*
Cultivar	White Guava No.6229, Supreme, Clone 32-12, Webber and Popeno	Tolerant	Florida (USA)	Mathur and Jain (1960)
	Hart and Rolf		Florida(USA) acclimatized at Allahabad	
	Riverside and Rolf		California(USA)	
	Safeda		Sri Lanka	
	Banarasi		Andhra strain	
	Dholka, Sindh and Nasik		Bombay strain	
	Chittidar, Hfsi, Safeda, Riverside, Rolf and Stone acid	Susceptible	Allahabad	Edward (1961)
Root stock	*Psidium cattleianum* var. *lucidium*	Resistant		
	Syzigium cumini			Edward and Gaurishanker (1964)
Root Stock	*Syzigium cumini*, *Lagerstroemia indica*, *P. cattleianum* (*P. molle*), *P.quianense*, *Chinese guava* (*P. friendrichsthalianum*) and Phillippine guava			
Root stock	cv. Safeda with all compatible except *Lagerstroemia indica* incompatible			
Cultivar	Lucknow-49	Immune	U.P (India)	Singh *et al.*, 1977
	Allahabad Safeda	Resistant		
	Karela, Behat Coconut	Susceptible		
Species	*P. aracae, P.cattleianum, P.cattleianum* var. *lucidium, P.corecium, P. cujavillus, P. quineese* and *P. fridichsthalanum*	Resistant		
Variety	Variety-1	Moderately resistant	West Bengal (India)	Sen Gupta *et al.*, 1989
	Variety-4	Moderartely resistant		
	Variety-5	Moderate resistant		

Contd...

Table 17.4–Contd...

Cultivars/Variety/ Root Stock/Species	*Name of Cultivars/ Variety/Root Stock/Species*	*Type of Reaction*	*Place*	*References*
Cultivar	Allahabad Safeda, Florida seedling. Guinees. Hafsi, Karela, Mirzapuri seedling, Nasik, Pear shaped, Sindh, Superior and White Fleshed	Highly susceptible	Lucknow	Misra (1998-99)
	Behat coconut and Pourtgal	Susceptible		
	Apple colour, Chittidar, Seedless and Spear Acid, Superior Sour Lucidium, Red Flesh and Smooth Green	Tolerant		
Seedling	F1 (*P. molle* x *P. guajava*)	Resistant		Misra *et al.*, 2003c
Variety	Pei-Pa	Resistant	Taiwan	Leu and Kao(1979)
Root Stock	*P. friedrichsthhianum*		–	Normand (1994)
Root Stock	*P. cattleianum*			
Cultivar	*Fan Retief*	Highly susceptible	South Africa	Du Preez(1995)
Selection	3-selection among 30- thousand seedling	Tolerant	-	Vos *et al.*, 2000

Lack of varietals resistance also exists in commonly grown guava varieties. Therefore, easy method for combating the disease successfully would be identification and use of resistant rootstock. None of the guava cultivars/varieties is found free from wilt. Wilt of guava is complex disease and option of resistant root stock seems to be great useful. As, Edward (1961) suggested that *Psidium cattleianum* var. *lucidium* and *Syzigium cuminii* might be an effective way for the control of wilt-complex in guava. Again, Edward and Gaurishankar (1964) found guava species as rootstock for combating the wilt. Some other source of root stock *i.e.* Pei Pa (variety) *P. friendrichsthlianum*, population of *P. molle* x *P. guajava* as graft compatibility is very successful. This resistant root stock is very useful for the control of wilt.

Biological and Integrated Options

An antagonist in a microorganism that adversely affects target pathogen. Such type of microorganisms is used in biological management of plant diseases. According to Kerr (1982) an ideal biological control organism should possess the following characteristics.

It should survive for a prolonged period in soil either in an active or passive form. There must be high probability of its contacting the pathogens either directly or indirectly by diffusion of metabolites.

Multiplication in laboratory should be simple and inexpensive

It should be amenable to a simple efficient and inexpensive process of packing, distribution and application.

If possible, it should be specific to the target organism.

It should not be health hazards in its preparation, distribution and application.

It should be active under required environmental conditions and

It should be efficient and economical

Various soil inhabitant bio-control agents such as *Trichoderma* spp., *Arachniotus* spp., *Streptomyces chibaensis, Penicillium citrinum, Aspergillus niger* and plant products *i.e.* neem, tulsi, garlic have been found effective against the wilt pathogens mostly under *in vitro* conditions. Recently Dwivedi and Shukla (2002) and John, *et al.* (2006) found a native strain of *T. viride* was most effective against guava wilt in pot experiments.

Table 17.5: Effective Bioagents for Guava Wilt Management

Bioagent/Bio-tools	*Application*	*References*
Trichoderma spp	Soil	Dwivedi, 1992
Streptomyces chibaensis		
Foeniculum vulgare (seed oil)	Soil	Dwivedi, 1993a
Topsin-M +*T. harzianum*	Soil	Ansar *et al.*, 1994
Arachnioltes sp with wheat straw	Soil amendment	
Wilt Nema (seven plant extract)	Soil	Logani *et al.*, 2002
VAM-symbiont	@ 5 kg/tree	Srivastava *et al.*, 2001
Aspergillus niger Strain AN-17	Soil	Misra *et al.*, 2000, 2004
T. harzianum, Penicillum citrinum	*In vitro*	Prakash *et al.*, 2002
Aspergillus niger Strain AN-17	Most effective	Misra and Prasad 2003
T. harzianum, T. viride, Gliocladium virens	Best	Singh *et al.*, 2003
T. viride and A. niger	Field	John, D. *et al.*, 2006
Technique for bioagents multiplication	Field	Misra *et al.*, 2003b, d; Misra and Singh, 2005
A. niger (AN-17)+ resistant root stock (*P. molle+P.guajava*)	Field	Misra *et al.*, 2003c, 2004b; Misra, 2005
Intercropping with turmeric+marigold	Field	

Biotechnological Options

Modern biotechnological techniques offer many possibilities for the introduction of variability in culture system and to select out the desirable ones in controlling conditions. However, efforts are being made for screening the existing genetic variability to identify a source of resistance to guava wilt and then using in breeding program for the development of guava wilt resistant cultivars.

The techniques employed in biotechnological options are as below:

1. Mass culture of guava wilt pathogens for pathogenicity test or biological control prepared on maize grain or sorghum seed. *In vitro* propagation the most applied technique for multiplication and regeneration of selected plants can be used for colonel propagation of wilt resistant rootstock.
2. Soma clone (Larkin and Scoweroft, 1981) variation for genetic variability.
3. Cell isolation (Fowler1977, Stret, 1977)-suspension culture–large number of cell can be screened for isolation of useful variants.
4. Molecular approaches: Useful to make fungal resistant transgenic plants, which produce anti-fungal proteins and toxins. There strategies can be made for use in future when regeneration protocol is standardized.

5. Cell culture (Steward11958a, Vasil and Hilderbrandt, 1965).
6. Callus culture (Earle and Demarly, 1982).

Future Gaps

1. The efforts on etiology should be taken into consideration all available options.
2. Association of a pathogenic entity should be confirmed by the required tests.
3. The possibility of a non-living agent should also be investigated scientifically.
4. Management recommendation should be based on trials all location of major guava cultivation area.

References

Ansar, M., Saleem, A. and Iqbal, A. (1994). Cause and control of guava decline in Punjab (Pakistan). *Pakistan J. Phytopathology,* 6: 41–44.

Anonymous (1949). *Annual Administrative Report* of the Department of Agriculture, United Provinces for the Year 1947–48, pp 89.

Anonymous (1950). *Annual Administrative Report* of the Department of Agriculture, United Provinces for the Year ending 30th June, 1949, pp 125.

Anonymous (1953). *Report of the Directorate of Plant Protection,* Quarantine and Storage for the period 1946–1951. Govt. of India Press, Delhi, pp. 44.

Bhargava, A.K., Sobit, A.K. and Ghsolia, R.P. (2003). Studies on guava (*Psidium guajava* L.) drying/ wilt disease in orchards of Pushkar Valley. *J. Phytological Res.,* 16: 81–84.

Chakraborty, D.K. and Singh, R.N. (1989).Guava wilt correlation between variation in disease syndrome and edaphic factors. *Indian Phytopath.,* 42: 310.

Chandra Mohan (1985). Studies on guava decline in Punjab with special reference to wilt. *Ph.D. Thesis,* Punjab Agric. Univ., Ludhiana, India, pp. 90.

Chandra Mohan, Jhooty, J.S. and Chand, T. (1986). Prevalence of guava decline in Punjab. *Pl. Dis. Res.,* p. 177–178.

Chattopadhyay, S.B. and Bhattachariya, S.K. (1967). Investigation into the wilt disease of guava in West Bengal: Susceptibility at different ages of host. *Bull. Bot. Soc. Beng.,* 21: 107–112.

Chattopadhyay, S.B. and Bhattachariya, S.K. (1968a). Investigation on the wilt disease of guava (*Psidium guajava* L.) in West Bengal I. *Indian J. Agric. Sci.,* 38: 65–72.

Chattopadhyay, S.B. and Bhattachariya, S.K. (1968a). Investigation on the wilt disease of guava (*Psidium guajava* L.) in West Bengal II. *Indian J. Agric. Sci.,* 38: 176–183

Chattopadhyay, S.B. and Sengupta, S.K. (1955). Studies on wilt of *Psidium guajava* L. in West Bengal. *Indian J. Hort.,* 12: 76–79.

Das Gupta, M.K. and Ghoshal, B.K. (1977). Is it possible to control guava wilt through oil cake amendments? *Sci. and Cult.,* 43: 131–133.

Das Gupta, S.N. and Rai, J.N. (1947). Wilt disease of guava *Psidium guajava. Curr. Sc.,* 16: 256–258.

Dey, P.K.(1948). *Plant Pathology*, Adm.Rep., Agric.Dept., U.P., 1945–46, pp. 43–46.

Du Preez, R.J. (1995). *Guava Cultivars.*

Dwivedi, B.P. and Shukla, D.N. (2002). Biocontrol of *Fusarium* wilt of guava (*Psidium guajava L.*) using *Trichoderma* and *Gliocladium* species. *Karnataka J. Agric. Sci.*, 15: 399–400.

Dwivedi, S.K. (1990). Efficacy of some antibiotics on *Fusarium oxysporum* f.sp. *psidii* causing wilt disease of guava. *Hindustan Antibiotic Bull.*, 32: 88–90.

Dwivedi, S.K. (1991a). Population dynamics of microfungi including pathogenic forma in the beds of completely healthy, partially wilted and completely wilted guava tress grown on a line. *Intern. J. Trop. Pl. Dis.*, 9: 95–109.

Dwivedi, S.K. (1991b). Effect of some heavy metals on growth of *Fusarium oxysporum* f.sp. *psidii* causing wilt disease of guava. *Intern. J. Trop. Pl. Dis.*, 9: 127–130.

Dwivedi, S.K. (1992). Effect of culture filtrate of soil microbes of pathogens inciting wilt disease of guava (*Psidum guajava* L.) under in vitro conditions. *Nati. Acad. Sci. Letter*, 15: 33–35.

Dwivedi, S.K. (1993a). Fungioxicity of *Foeniculum vulgate* seed oil used against a guava wilt pathogen. *Nati. Acad. Sci. Letter*, 16: 7–8.

Dwivedi, S.K. (1993b). Soil solarization adversely affects some fungal pathogens causing wilt disease of guava. (*Psidium guajava* L.). *Soil Bio. Biochem.*, 25: 11.

Dwivedi, S.K. and Dwivedi, P. (1999). Wilt disease of guava: a national problem. *J. Applied Hort.*, 1: 151–154.

Dwivedi, S.K. and Dwivedi, R.S. and Tiwari, V.P. (1990). Studies on pathogenic fungi inciting guava wilt in Varanasi, *Indian Phytopath., 43: 116–117.*

Edward, J.C. (1960a). Wilt diseases of guava. *The Allahabad Farmer*, 34: 269–293.

Edwards, J.C. (1960b). Variation in the guava wilt pathogen, *Fusarium oxysporumi* f.sp. *psidii. Indian Phytopath.*, 13: 30–36.

Edwards, J.C. (1960c). Penetration and establishment of *fusarium oxysporumi* f.sp. *psidii.* In guava root. *Indian Phytopath.,* 13: 168–170.

Edward, J.C. and Gaurishanker (1964). Rootstock trial for guava (*Psidum guajava* L.). *The Allahabad Farmer*, 38: 249–250.

Edward, J.C. and Srivastava, R.N. (1957). RE 502–Studies on guava wilt. *The Allahabad Farmer,* 31: 1414–146.

Grech, N.M. (1985). First report of guava rapid death syndrome caused by *Septofusidum* sp. in South Africa. *Pl. Disease, 69: 726.*

Gupta, P., Dubey, V., A. and John, D. (2003). *Verticillium albo-atrum*: A new pathogen associated with guava wilt *Indian Phytopath.*, 56: 503–504.

Hamiduzaaman, M.M., Meat, M.B. and Ahmad, M.U. (1997). Effect of *Fusarium oxysporum* and nematode interaction on guava wilt. *Bangladesh J. Pl. Pathol.*, 13: 9–11.

Hsieh, S.P.Y., Liang, W.J., Kao, C.W. and Lau, L.S. (1976). Morphotogical and physiological characters of *Myxosporium psidi,* the causal organism of guava wilt. *Plant Prot. Bull.* (*Taiwan*), 18: 309–317.

Jain, S.S. (1956). A preliminary note on the inactivation of *Fusarium oxysporum* f.sp. *psidii* guava plants by chemotherapeutic treatment. *Indian J. Hort.*, 13: 102–104.

Jhooty, J.S., Chand, J.N. and Krishnamurthy, V. (1984). Report of committee constituted by ICAR on guava decline in Punjab and Haryana. Submitted to ICAR, New Delhi.

Joubert, M.H. and Frean, R.T. (1993). An *in vitro* evaluation of fungicides against guava wilt. In: *Lightings Bullentin Instituut Vir Tropiese en Subtropiese Gewasse*, (246): 3.

Junqueira, N.t.V. Andrade, L.R.M. de, Pereira, M., Lima, M.M. and Chaves, R. da. C. (2001). Diseases of guava (*Psidium guajava* L.) cultivated in Brazilian Cerrdos. *Circular Tecnica Embrapa Cerrados*, 15: 31.

Katyal, S.L.(1972). Twenty five years of research on fruit crops. *Indian Fmg.*, 22: 14–16.

Khan, R.M. and Misra, A.K. (2003). Influence of co-cultivation of marigold, garlic and turmeric on nemic population in guava cropping system. In: *Sourvenir and Abstract, Zonal Conference (East Zone). Indian Society of Mycology and Plant Pathology and Seminar on Plant Diseases of National Importance with Special Reference to Guava Wilt and Mango Malformation*. 4–5th April, CISH and IISR, Lucknow, pp. 24–25.

Khan, R.M. Kumar, S. and Parvath Reddy, P. (2001). Role of plant parasitic nematode(s) and fungi in guava wilt. *Pest Management in Horticulture Ecosystem*, 7: 152–161.

Leu, L.S. and Kao, C.W. (1979). Artificial inoculation of guava with *Myxosporium psidii*. *Pl. Dis. Report.*, 63: 1077–1079.

Leu, L.S., Kao, C.W., Wang, C.C., Liang, W.J. and Hsieh, S.P.Y. (1979). Myxosporium wilt of guava and its control. *Pl. Dis. Repotr.*, 63: 1075–1077.

Lim, T.K. and Manicom, B.Q. (2003). Diseases of guava. In: *Diseases of Tropical Fruit Crops*, (Ed.) R.C. Pioetz. CABI, Wallingford, U.K., pp. 275–289.

Logani, R., Pandey, R.C. Dwivedi, B.K. and Singh, S.P. (2002). An confirmedly management of wilt diseases of guava by wiltnema (a botanical fungicide). *Bioved*, 13: 47–50.

Mathur, R.S. (1956). Guava diseases in India. *Indian J. Hort.*, 13: 26–29.

Mathur, R.S. and Jain, S.S. (1960). Selecting guavas for wilt resistance. *Proc. Nat. Acad. Sci.*, B.B. 30: 33–36.

Mathur, R.S., Jain, S.S. and Swarup, J. (1964). Chemical treatment for guava wilt *Proc. Natn. Acad. Sci., India*. Sect. B., 34: 33–36.

Mehta, N. (1987). Distribution of guava wilt in relation to age, soil type, management practices and varieties grown in Haryana. *Plant Dis. Res.*, 2: 116–119.

Mehta, P.R. (1951). Observations on new and known diseases of crop plants in Uttar Pradesh. *Plant Prot. Bull.*, 3: 712.

Misra, A.K. (1987). Studies on diseases on fruits crops. *Annual Report*, CIHNP, Lucknow, pp. 124–125.

Misra, A.K. (1985). Guava wilt. In: *Advances in Diseases of Fruits in India*, (Ed.) S.J. Singh. Kalyani Publishers, Ludhiana, pp. 183–190 + 1 plate.

Misra, A.K. (1998–99). Screening against wilt. *Annual Report*, CISH, Lucknow, pp. 10.

Misra, A.K. (2003). Guava diseases: Their symptoms, cause and management. In: *Diseases of Fruits and Vegetables: Diagnosis and Management, Vol. 2,* (Ed.) S.A.M.H. Naqvi. Kluwer Academic Publishers, The Netherlands, pp. 81–119.

Misra, A.K. (2005). Important disease of guava in India with special reference to wilt. In: *Souvenir, 1st International Guava Symposium,* December 5–8, CISH, Lucknow, p. 75–90.

Misra, A.K. and Gupta, V.K. (2005). Faster and efficient multiplication of bioagents through use of polythene sheet for management of guava wilt. In: *International Conference on Plasticulture and Precision Farming,* 17–21, November, New Delhi, pp. 329.

Misra, A.K. and Pandey, B.K. (1992). Wilt of guava and associated pathogens. *Indian J. Mycol. Pl. Pathol.,* 22: 85–86.

Misra, A.K. and Pandey, B.K. (1994–95). Studies on Guava wilt. CIHNP, Lucknow. pp. 28.

Misra, A.K. and Pandey, B.K. (1996). Present status of wilt disease of guava. In: *Disease Scenario in Crop Plants, Vol. 1: Fruits and Vegetables,* (Eds.) V.P. Agnihotri, Om Prakash, Ram Kishun and A.K. Misra. International Books and Periodical Supply Service, New Delhi, pp. 61–70.

Misra, A.K. and Pandey, B.K. (1997). Pathogenicity and symptom production of wilt disease of guava by a new potent pathogen *Gliocladium roseum.* In: *Indian Phytopathological Society, Golden Jubilee International Conference on Integrated Plant Disease Management for Sustainable Agriculture,* IARI, New Delhi, November 10–15, p. 319.

Misra, A.K. and Pandey, B.K. (1999a). Natural wilting of guava plants during different months. In: *51st Annual Meeting and Nat. Symp. on Seed Health Care and Phytosanitation for Sustainable Agriculture,* 17–19 February, IISR, Lucknow, pp. 64–65.

Misra, A.K. and Pandey, B.K. (1999b). Pathogenicity and evaluation of fungicides against guava wilt pathogens. *J. Mycol. Pl. Pathol.,* 29: 274–275.

Misra, A.K. and Pandey, B.K. (1999d). Natural wilting of guava plants during different months. *Indian Phytopathol.,* 52: 312.

Misra, A.K. and Pandey, B.K. (2000a). Pathogenicity and symptom production of wilt disease of guava by a new potent pathogen *Gliocladium roseum.* In: *Proceedings, Indian Phytopathological Society,* New Delhi, pp. 749–750.

Misra, A.K. and Pandey, B.K. (2000b). Progressive natural wilting of guava plants during different months. *Indian Phytopathol.,* 53: 423–427.

Misra, A.K. and Prakash, Om (1986). *Guava Diseases: An Annotated Bibliography 1907–1990.* Bishen Singh Mahendra Pal Singh, Dehradun, pp. 132.

Misra, A.K. and Prasad, B. (2003). Relative efficacy of different bio-agents for the control of guava wilt. *J. Mycol. Pl. Pathol.,* 33: 494.

Misra, A.K. and Prasad, D. (2004). *Aspergillus niger* strain AN 17 potent bioagent to control wilt disease and its easy multiplication. In: *Symposium on Recent Advances in Fungal Bioagents and their Social Benefits,* 10th September, NBRI, Lucknow, pp. 12.

Misra, A.K. and Shukla, S.K. (2002). Assessment of loss due to Guava wilt around Lucknow. In: *Nat. Seminar on Production and Post-harvest Technology of Guava,* Dept. Hort. CSAUA&T, Kanpur, 9–10 January, pp. 34–35.

Misra, A.K. and Singh, V.K. (2005). The CISH, Lucknow, marches towards...Solving guava wilt and mango malformation. *Indian Horticulture* (April–June), 50(1): 25–27.

Misra, A.K., Prakash, Om and Sen, B. (2000). Biological control of guava wilt by *Aspergillus niger* strain AN 17 (Pusa Mrida). In: *Nat. Seminar on Hi-tech Hort*. 26–28th June, Bangalore, pp. 149.

Misra, A.K., Pandey, B.K., Babita Prasad and Shukla, S.K. (2003a). Pathogenic diversity in the cause of wilt disease of guava. *Indian Phytopath.*, 56: 314.

Misra, A.K., Pandey, B.K. and Prakash, Om (2003b). Wilt of guava, a disease of national importance and possible solutions. In: *Souvenir and Abstract, Zonal Conference (East Zone), Indian Society of Mycology and Plant Pathology and Seminar on Plant Diseases of National Importance with Special Reference to Guava Wilt and Mango Malformation*, 4–5th April, CISH and IISR, Lucknow, pp. 23.

Misra, A.K., Rajan, S., Babita Prasad, Shukla, S.K. and Prasad, D. (2003c). Resistant source of wilt disease of guava. In: *Souvenir and Abstract, Zonal Conference (East Zone), Indian Society of Mycology and Plant Pathology and Seminar on Plant Diseases of National Importance with Special Reference to Guava Wilt and Mango Malformation*, 4–5th April, CISH and IISR, Lucknow, pp. 51–52.

Misra, A.K. Prakash, Om, Pandey, B.K. Babita Prasad, Shukla, S.K. and Prasad, D. (2003d). Biological control of guava wilt. In: *Souvenir and Abstract, National Symposium on Organic Farming in Horticulture for Sustainable Production*, 29–30th August, CISH, Lucknow, pp. 65.

Misra, A.K., Shukla, S.K., Prasad, B. and Prasad, D. (2003e). An integrated approach for the management of wilt disease of guava (*Psidium guajava* L.). In: *Nat. Seminar on Emerging Trends in Plant Disease Research and Management*, December 11–12, IISR, Lucknow, pp. 22.

Misra, A.K., Prakash, Om, Pandey, B.K. Babita Prasad, Shukla, S.K. and Prasad, D. (2004). Biological control of guava wilt. In: *Organic Farming in Horticulture*, (Eds.) R.K. Pathak, Ram Kishun, R.M. Khan and R.A. Ram. CISH, Lucknow, pp. 302–305.

Misra, A.K., Babita Prasad, Mishra, D. and Lal, B. (2004a). Wilting in air layers of guava. In: *International Seminar on Recent Trends in Hi-tech Horticulture and Post-harvest Technology*, February 4–6, CSAUA&T, Kanpur, pp. 103.

Misra, A.K., Prasad, D., Babita Prasad and Shukla, S.K. (2004b). Effective management of wilt disease of guava. Souvenir and Abstract. In: *Natn. Symp. on Crop Surveillance: Disease Forecasting and Management*, Indian Phytopathological Society, February 19–21, Div. of Plant Pathology, IARI, New Delhi, pp. 92–93.

Negi, S.S., Misra, A.K. and Rajan, S. (2001). Guava wilt. In: *Proc. National Seminar on New Horizon in Production and Post Harvest Management of Tropical and Subtropical Fruits*, IARI, New Delhi. Held on December 8–9, 1998, Special Issue, Indian Journal of Horticulture, 54 (1&2)..

Normand, F. (1994). Strawberry guava, relevance for Reunion. *Fruits–Paris*, 49: 217–227.

Pandey, R.R. and Dwivedi, R.S. (1985). *Fusarium oxysporum* f. sp. *psidii* as a pathogen causing wilt of guava in Varanasi district in India. *Phytopath. Z.*, 114: 243–248.

Pandit, P.K. and Samajpati, N. (2002). Wilt disease of guava (*Psidium guajava* L.) by *Botryodiplodia theobromae* Pat. *J. Mycological Res.*, 40: 71–73.

Prasad, D., Shukla, S.K., Babita Prasad and Misra, A.K. (2003). Effect of intercrop and different doses of NPK on the incidence of guava wilt. In: *Souvenir and Abstract, Zonal Conference (East Zone), Indian Society of Mycology and Plant Pathology and Seminar on Plant Diseases of National Importance with Special Reference to Guava Wilt and Mango Malformation*, 4–5th April, CISH and IISR, Lucknow. pp. 53.

Prasad, N., Mehta, P.R. and Lal, S.B. (1952). Fusarium wilt of guava (*Psidium guajava* L.) in Uttar Pradesh, India. *Nature*, 169: 753.

Prakash, Om, Misra, A.K. and Shukla, S.K. (2002). *Penicillium citrinum*: A potent pathogen against wilt disease of guava. In: *Asian Congress of Mycology and Plant Pathology*, October 1–4, Univ. Mysore, Manasgangotri, Mysore, pp. 180.

Rodriguez, F.M.E. and Landa, J.B. (1977). Chemical soil disinfection against parasitic nematode in guava nurseries. *Centrol Agricola de la Facultod de Cieneias Agricolos*, 4: 57–77.

Rodrigues, N.J., Robbs, C.F. and Yamashino, T. (1987). A bacterial disease of guava (*Psidium guajava*) caused by *Erwinia psidiii* sp. nov. *Fitopathologia Brasileira*, 12: 345–350.

Sen, P.K. and Verma, B.S. (1954). Studies on the dieback disease of guava (*Psidium guajava* L.): A survey of the incidence of the disease in Jhargram area (West Bengal). In: *Proc. of the 41st Indian Sci. Cong.*, Sec. X Agril. Sci. pp. 258.

Sen Gupta, P.K., Bhattacharya, I. Das Gupta, A., Maity, T.K., Poi, A.K. and Dutta, S. (1989). Investigation of guava decline with special reference to wilt. Final report ICAR Research Project Dept. of Plant Pathology, Bidhan Chandra Krishi Vishwavidyalaya, Kalyani.

Shukla, S.K., Babita Prasad, Prasad, D. and Misra, A.K. (2003). A cheap substrate for the mass multiplication of bio-agents of guava wilt. In: *Souvenir and Abstract, Zonal Conference (East Zone), Indian Society of Mycology and Plant Pathology and Seminar on Plant Diseases of National Importance with Special Reference to Guava Wilt and Mango Malformation*, 4–5th April, CISH and ISSR, Lucknow, pp. 54–55.

Singh, B. and Lal, S.B. (1953). Wilt of guava. *Agril. and An. Hus.*, 3: 78–79.

Singh, S.H., Ved Ratan, Gaur, G.S. and Gangwar, D. (2003). Evaluation of antagonists against guava wilt pathogen. *Farm Science J.*, 12: 58–59.

Singh, U.R., Dhar, L. and Singh, G. (1977). Note on the performance of guava cultivars and *Psidium* species against wilt disease under natural field conditions. *Haryana J. Hort. Sci.*, 6: 149–150.

Sohi, H.S. (1983a). Diseases of tropical and subtropical fruits and their control. In: *Recent Advances in Plant Pathology*, (Ed.) A. Husain, K. Singh, B.P. Singh and V.P. Agnihotri. Print House, Lucknow, pp. 73–86.

Sohi, H.S. (1983b). Studies on wilt disease of guava. *Annual Report*, Indian Hort. Res., Hassarghatta. Bangalore, pp. 102.

Srivastava, A.K. Ahmad, R., Kumar, S. and Sukhada Mohandos (2001). Role of VA-Mycorrhiza in the management of wilt disease of guava the alfivols of Chotanagpur. *Indian Phytopath.*, 54: 78–81.

Suhaq, L.S. (1976). Observations in guava decline Haryana and its control. *Pesticides*, 10: 42–44.

Suhag, L.S. and Khera, A.P. (1986). Studies on the variation in nutrional level of wilted, regenerated and healthy trees of guava cultiva 'Banarasi Surkha'. *Indian Phytopath.*, 39: 90–92.

Tandon, R.N. and Agarwala, R.L. (1954). Pathological studies of *Gloeosporium psidii* causing die-back of guava. *Proc. Indian Acad. Sci.*, Sect. B., 40: 102–109.

Tokeshi, H., Valdebenito, R.M. and Dias, A.S. (1980). Occurrence of a bacterial disease of guava in Sao Paulo State. *Summa Phytopathologica*, 6: 85–87.

Vestal, E.F. (1941). *Textbook of Plant Pathology*. Kitabistan Allahabad, pp. 406.

Vos, J.E., Schoeman, M.H., Berjak, P., Watt, M.P. and Toerien, A.J. (2000). *Acta Horticultural*, 513: 69–79.

Webber, G.F. (1928). Plant Pathology. *Annual Rep.*, Florida Agric. Univ. Exp. Stat. June 30.

Plant Diseases Management for Sustainable Agriculture *Pages* 212–234
Editor: **Dr. Shahid Ahamad**
Published by: **DAYA PUBLISHING HOUSE**

Chapter 18

Rhizoctonia solani: The Incitant of Web Blight in Legumes: Preview

***Rakesh Kumar Singh*[1], *Manjul Pandey*[1], *Rakesh Kumar Prajapati*[2] *and R.G. Chaudhary*[1]**

[1]*Division of Crop Protection, Indian Institute of Pulses Research, Kanpur – 208 024 (U.P.)*
[2]*Krishi Vigyan Kendra, Mandsaur (M.P.)*

Pulses are an important constituent in predominantly vegetarian Indian diet. Besides rich source of protein they have unique body rejuvenation ability and traditional medicinal uses. Pulses are also important for sustainable agriculture enriching the soil through biological nitrogen fixation and improving its physical, chemical and biological properties. These crops also fit well in various cropping systems without disturbing the main cereal crops. In India pulses are cultivated over an area of 23m. ha with annual production of 14 million tones. It is estimated that the population of the country would be touching nearly 1350 million by 2020 AD and then minimum pulses requirement would be around 30.3 million tones (Anonymous, 1997). This target is more than double the present level of production. At the present trend of production the target is difficult to be achieved.

India being a major pulse growing country in the world shares 35-36 per cent and 27-28 per cent of the area and production of these crops. Almost all the pulses get infected with *Rhizoctonia solani* particularly, the *Kharif* season crops. Among these mungbean, (*Vigna radiata*), urd bean (*Vigna mungo*), horse gram (*Macrotyloma uniflorum*), moth bean (*Vigna aconitofolium*), cowpea (*vigna sinensis*), dry bean (*Phaseolus vulgaris*), rice bean (*Vigna unbellata*), faba bean (*Vicia faba*) and Dolichus bean (*Lablab purpurius*) crops produce web blight symptoms. In the years of heavy rain's pigeonpea (*Cajanus cajan*) also shows web blight infection (Naimuddin *et al.*, 1998). With the change in cultivation practices, crop husbandry and crop geometry, collateral hosts of *R. solani* have increased many fold leading to increase availability of primary inocula. Hence web blight, in the past a minor disease of rare occurrence is now considered to be major problem and emerged as new threat to the legume crops. The original concept that the *Rhizoctonia* inhabits only soil and causes rotting of roots and damping off of seedlings has been modified. Several species have been shown to occur on aerial plant producing diseases like leaf spots and blight, stem blight, stalk rot, bud rot, ear rot and fruit rot (Saksena, 1971). Aerial blights caused by *R. solani* have been described by various names like web blight; (because of spider web like mycelium on the infected parts), thread blight due to presence of silky thread like mycelial growth on the foliage or as leaf blight only.

Geographical Distribution

Web blight is a common and wide spread disease of legumes causing heavy losses in quality and quantity of produce. The disease has been reported on common bean (*Phaseolus vulgaris*) in Trinidad and Tobago, West Indies, Puerto Rico, Jamaica, North America, Florida, South America, Argentina, Brazil, Colombia, Mexico, Nicaragua, Panama and, Japan, Philippines, Myanmar, Sri Lanka, India and Pakistan. (Matz, 1921; Webber, 1939; Galvez *et. al.* 1979; Ploper, 1981; CIAT, 1985; Saksena and Dwivedi, 1973 and Alam *et al.*, 1985).

Table 18.1: Hosts Exhibiting Aerial Infection by *R. solani* in Different Countries

Crops	*Country*	*Reference*
Pigeon pea	India	Dwivedi and Saksena, 1975
Pearl millet	India	Singh and Saksena, 1980
Bean	North Western Argentina	Ploper, 1981
Beans and Lima bean	Florida	Webber, 1939
Black gram	India	Saksena and Dwivedi, 1973
Catharanthus roseus	Lousiana	Holcomb, and Carling, 2002
Celery	California	Houston and Kendrick, 1949
Coffee	El Salvador	Webber and Abrego, 1958
Cotton	Lousiana	Neal, 1942; Kotile, 1945
Legume and Grasses	South eastern USA	Allisons, 1951
Ginger	India	Sundram, 1953
Larch	Japan	Ito *et al.*, 1955
Lespendeza	Lousiana and North California	Allisons, 1951; Stroube, 1954
Maize	India	Ahuza, 1976
Morantha	India	Ramakrishnan and Ramakrishnan, 1948
Mungbean	India	Dwivedi and Saksena, 1974
Mungbean	Pakistan	Alam *et al.*, 1985
Rice	India	Saksena and Chaubey, 1973
Pak Choi	Trinidad	Rajnauth, 1987
Soybean	Puerto Rico	Hepperly *et al.*, 1982
Soybean	Lousiana	Atkins and Lewis, 1954
Sugar bean	Virginia and North central States	Kotile, 1947
Sugar bean	Philippines	Lee, 1922
Vigna angularis	India	Singh *et al.*, 1979
Yellow saga	Europe	Garibaldi *et al.*, 2003

In India, the disease is known to occur in Uttar Pradesh, Bihar, Rajasthan, Punjab, Haryana, Himachal Pradesh, Jammu and Kashmir and Uttaranchal (Anonymous, 1999, Anonymous, 2004). It occurs on crops like mungbean (Dwivedi and Saksena, 1974), black gram (Saksena,1971); cowpea (Dwivedi, 1977 and Lakshmanan *et al.*, 1979). Verma and Thapliyal (1976) have also reported its infection on many other host plants. However, the fungus appears to be less specialized for their habitat. In Uttar Pradesh this disease emerged as a main threat to legumes cultivation and causes heavy losses wherever crops is grown. Reports on aerial infection of several host plants by *R. solani* are given in Table 18.1.

Economic Importance

There is paucity of information in literature on yield losses due to *R. solani*. However, the available information's reveal heavy losses both in quantity and quality of mungbean (Ratan and Dwivedi, 1998) estimated as 30.5-37.4 per cent (Gupta *et al.*, 2002). The losses were higher if the plants get infected earlier *i.e.* 25 DAS than 35 and 45 DAS. Status of web blight severity of mungbean in eastern Uttar Pradesh was 1.0 to 69.0 per cent with an average of 12.7 per cent (Singh *et al.*, 2003). *R. solani* also caused reduction in number of pods plant^{-1}, number of seeds pod^{-1}, seed size and 1000 seed weight (Gupta, 2000). Reduction in seed weight has been observed as high as 28.6 per cent (Gupta and Singh, 2002). Information on losses caused by *R. solani* in other legumes are not available.

Symptoms

The fungus infects all above ground parts of the plant *i.e.* leaves, petioles, stem and pods. Symptoms on the leaves appear as initial small circular brown spots. These spots are surrounded by water soaked areas. Later the lesions expand and coalesce and dull white to brown fungal growth is seen on the lower surface of leaves and young branches. The mycelium on infected leaves appear as spider web thus called as 'Web blight. Lesions on stem and petioles generally appear when infected plants have lost many of their infected leaves or completely defoliated and are linear to oval and reddish brown in appearance. White brown sclerotia are produced abundantly on infected stem and petioles. The affected parts shrivel, dry and finally leaf canopy is completely destroyed and in severe cases, affected plants die prematurely before flowering and pod formation. Spots on young pods are light tan and irregular in shape but in mature pods they are dark brown and shrunken. Initially damp and wet weather followed by dry spell give rise to patchy blighted areas in the field. The low lying and depressed areas of the field show higher disease intensity than raised areas.

The Pathogen

Since De Candole (1815) first described the genus *Rhizoctonia solani,* more than one hundred species have been identified. These species are distinguished from one another primarily on the basis of morphology and dimension of sclerotia, monilioid cells, and pathogenicity. Progress in taxonomic studies on *Rhizoctonia* spp. has been reviewed by Ogoshi (1975). The general characteristics of the genus are:

1. Branching near the distal septum of cells in young vegetative hyphae;
2. Formation of a septum in the branch near the point of origin.
3. Constriction of the branch at the origin point
4. Dolipore septum
5. No clamps connection

6. No conidium excepted monilioid cells
7. Sclerotium root differentiated in to rind and medulla
8. No rhizomorph.

Variability

Definite physiological forms or pathogenic races are not yet known in *Rhizocotnia solani*. However anastomosis groups (AGs) for identification of isolates. AGs of *R. solani* correspond generally to their host specificity. There are some isolates which do not fit well to their AG and show different host range. Therefore, the mechanism of host specificity in *R. solani* seems to be vague. Understanding such mechanism at the molecular level would not only be interesting but would also be useful for developing host specific disease management strategies. Recently, DNA-DNA hybridization technique has been used to assess genetic relationship among microorganisms. To determine the genetic relationship among AGs and Intra specific groups (ISGs *i.e.* different pathological types) of *R. solani* many workers studied the relative homology of DNA- base sequence on the basis of DNA-DNA reassociation, (Kuninaga and Yokosawa,1982a and b; Kuninaga and Yokosawa,1983; Kuninaga and Yokosawa,1984a and b; Kuninaga and Yokosawa;1985a and b). *R. solani* is a multinucleate, sometimes heterokaryotic fungus. If the isolates used are heterokaryotic the experimental results shall be usually variable because of the variable genetic makeup of the isolates. The perfect state of *R. solani* has been found and the basidiospores formed by a homokaryotic isolate can be used for the study of infection process, lesion formation and disease development (Naito and Sugimoto,1978; Naito,S.,1940).

Rhizoctonia solani is a complex of related host specific groups and, as referred earlier, is divided into different anastomosis groups (AGs). Presently twelve anastomosis groups (AGs) of *R. solani* have been reported including groups 1-11 and BI (Carling, *et al.*, 1986; Carling, *et. al*, 1986; Ogoshi, 1987 and Rovira, *et al.*, 1986). Ogoshi (1987)recognized ISG in the *R. solani* complex which have steadily increased with 13- ISGs in 1987 to as many as 25-ISGs in 1993. These ISGs differ in their anastomosis behaviour, cultural appearance and pathogenicity. Hyphal fusion occurs only between isolates of the same anastomosis groups and each anastomosis group represents a genetically isolated population (Anderson, 1982).Parameter *et al.* (1967) pointed out that most of the bean web blight isolates from Costa Rica fell in AG-2.Similarly, Watanable and Matsumoto (1966) indicated that web blight isolate fell in AG1 IB group and Ogoshi (1987) reported AG1 IB, AG4 and AG5 from leguminous hosts. AG-11 is associated with soybean. These studies clearly indicated the existence of specificity and wide variability in *R. solani* infecting different hosts particularly the legumes. However, a thorough systematic work is required combining the blend of cultural, morphological, pathogenic and molecular tools. For deciding anastomosis groups (AGs), isolates of *R. solani* are paired 2-3 cm. apart on a water agar plate. After the mycelia grow and overlap, observation for hyphal fusion is made. If hyphal fusion occurs, the isolates belong to the same AG, and death of fused cells is observed (McRoy *et al.*, 1984; and Yokoyama and Ogoshi, 1984). If the attraction, fusion and hyphal death do not occur, these isolates belong to different AGs.

Host Range

Rhizoctonia solani has wide host range (Duggar and Stewart, 1901; Peltier, 1916; Braun, 1930; Walker, 1957;and Chupp and Sherf, 1960). Compilation of the hosts of some specialized strains is available in the literature (Matsumoto, 1921; Gratz, 1925; Storey, 1941; Durbin, 1957). Several workers have reported that the fungus *C. sasakii* (Syn. of *R.. solani*), isolated from rice, is able to infects a number of crop plants, and weeds belonging to family gramineae and similarly, the fungus isolated from other crops can infect rice (Yokogi, 1927; Tervet, 1937and Sato and Shoji, 1957).Among 30 weeds inoculated with sclerotia, 19 became infected (Kannaiyan and Prasad, 1980).Another list of 27 plants infected by *T. cucumeris* has been documented by Kim *et al.* (1981). Baruah and Lal (1981) reported the aerial strain of *R. solani* causing banded sclerotial diseases of maize, produced clear banded blight symptoms on susceptible hosts of Gramineae (*Cynodon dactylon, Oryza sativa, Saccharum officinarum* and *Sorghum bicolor*) within four and half days of inoculation while hosts of leguminaceae (*Arachis hypogea, Glycine max, Pisum sativum and Vigna radiata*) and solanaceae (*Lycopersican esculentum* and *Solanum tuberosum*) showed initial symptoms after 5 days. Other collateral hosts of rice sheath blight (*R. Solani*) strain are *Sesamum indicum, Arachis hypogea, Sesbania aculeata, Cyperus iria, Eimbristyles miliacae, Apluda aristata* and *Monochoria vaginalis* (Gokulpalan and Nair, 1983).Similarly, Dwivedi and Dubey (1986) reported that by artificial inoculation. The aerial strain of *T. cucumeris* causing web blight of groundnut in nature infected twenty one species belonging to different families. Twenty plants species from 7 families could be artificially inoculated with winged bean isolate. The symptom of infection was observed on urdbean, cowpea, clusterbean, dolichos bean, soybean, groundnut, pigeon pea, french bean, mungbean, rice, sorghum, castor, bottle gourd, bitter gourd, tomato, brinjal, chilies and okra but were not observed on maize and radish (Singh and Malhotra, 1994). Three geographically distinct isolates of *R. solani* isolated from mungbean were equally virulent on maize, rice, pigeon pea, mungbean and urdbean and produced blight symptoms on leaves (Singh, 2007).

Above extensive studies on aerial strains of *R. solani* clearly pointed that the pathogen has extremely wide host range, cosmopolitan in nature and can cause severe disease in many crop plant species if the favourable environment is available.

Infection Process

There are specific infection sites of *R. solani* on the plant surface such as epidermis, stomata and wounds by which pathogen causes infection through different infection organs such as:

Infection Cushion

Dome shaped infection cushion is one of the typical characters of *R. solani* to infect host plants. The process of formation of infection cushion is reported for rice, cotton (Kolattukudy *et al.*, 1987), barley (Edwards and Allen, 1970; Kolattukudy, 1974) and bean (Vanden and Linskens, 1974). Hyphae of *R. solani* growing on the surface of host plants attaches to the cuticle with mucilage substances and align along the junction of epidermal cell walls. These hyphae form short, swollen side branches which aggregate to form the infection cushion. Penetration through epidermis occurs from the base of the cushion with infection pegs or hyphal tips.

Lobate Appressorium

The lobate appressorium is formed from short, swellon branches of hyphae which elongates along the junction of epidermal cells. Appressoria adheres on to the plant surface with mucilage substances and the infection occur through the formation of infection pegs from the lobes of appressoria.

Single Hyphae Infection

There are very few reports on hyphal mediated infection by *R. solani.* Infection with single hyphae was observed only in repeat secondary infections. This kind of infection was not observed in primary infection on the host plants.

Basidiospores Infection

Basidiospores of *R. solani* germinate and form aspersoria, penetrate the host epidermis and form primary lesions. Hyphae developed from the first lesion reinfect host tissue and induce secondary lesions (Naito and Sugimoto, 1978). Infection behaviour of hyphae and basidiospore from the same isolate was different. The basidiospore is important in the ecology of *R. solani* because it confers air borne characteristics to the pathogen.

Epidemiology

Most of the host plants were found susceptible particularly to penetration and tissue invasion mainly during seedling stages of growth and the plants built up resistance with the advancement in the age. (Leuttrell and Garren, 1952; Fulton and Hanson, 1960; Christon, 1962; Holly and Baker, 1963; Shephard and Wood, 1963 and Pitt, 1964). Townsend (1934), however in some cases, found that relatively matured tissue may be infected. He observed that in button rot disease, the infection generally occurred when plants attained heading and maturity stage. Neal (1942) too came across with a situation where the early planted cotton seeding were infected with *R. solani* and the infection persisted and caused considerable damage to the grown up plants. These studies gather enough evidence to show that no general yard stick is possible on the suitable crop stage for the infection. It possibly varies from host to host and is influenced greatly by the prevailing weather condition in a geographical area. Higher population of *R. solani* in the field has been reported during crop growing season because of the presence of host crops and suitable temperature (Ui *et al.*, 1965).

Environmental factors play a vital role in the development of this disease. If favourable conditions prevailed for a longer period, the disease widely spreads and creates epiphytotic in a particular area. Tiwari and Tiwari (1998) reported that high rainfall, relative humidity and soil temperature (30°±2°C) favoured the development of disease in mungbean. Web blight development and its intensity was maximum at a temperature range of 25-30°C and 80-95 per cent relative humidity (Ratan and Dwivedi, 1998). Gupta *et al.* (2002) reported that high relative humidity (82.6 per cent) and rainfall (105 mm) development with temperature of 28.5°C and sunshine of 5.8 hrs/day favoured the web blight. Incidence of web blight was more in rainy season (30-55 per cent) than spring (7-15 per cent) in mungbean varieties K 851, T 44 and Pusa Baisakhi (Gupta *et al.*, 2003). Kumar (2005) also reported high rainfall, maximum humidity and frequent rainy days being positively correlated with maximum disease development in mungbean. In another study, maximum infection rate per day was observed in 37th standard week (10-16th September) with mean temperature of 28.9°C, relative humidity of 83.85 per cent rainfall 34.3 mm, 3.5 rainy days, and soil temperature of 30.5°C during both the year (Singh, 2007). Epidemiological work on *Rhizoctonia* blight in other crops has revealed that some *T. cucumeris* was most severe under high temperature and high relative humidity at active growth stage of the crop (Baker and Martinson, 1970; Stan and Rai, 1973; Galvez *et al.*, 1979 and Tewari, 1993).

The development of disease on aerial parts of plants depends largely on free moisture or nearly 100 per cent relative humidity (Baker and Martinson, 1970; Mathew and Gupta, 1996). The impact of rainfall during the flowering stage of crops encouraged the outbreak of web blight in soybean (Verma and Thapliyal, 1976; Joge, 1986 and Patel and Bhargava, 1998). Yang *et al.* (1990) studied the epidemics of *Rhizoctonia* aerial blight of soybean and found highly significant correlation between rainfall and number of primary infection. Establishment of disease at early plant growth stage was associated with heavy and frequent rainfall. Tiwari (1993) recorded 59-63 per cent web blight intensity at temperature ranging between 24-26°C and relative humidity of 65-85 per cent under field conditions in rice bean. Higher aerial temperature (28-30°C), relative humidity (>80 per cent) and soil temperature (30-33°C) favoured development of web blight severity. Rainfall (91-97mm) has a significant role in severe development of disease during the early stage of the crop. Significant and positive correlation of web blight severity with relative humidity, aerial temperature, rainfall and soil temperature was observed by Sharma and Tripathi (2001) on urdbean. In winged bean, web blight appeared during second week of September and increased gradually till maturity. Apparent infection rate was maximum during September 10 to 17. Significant negative contribution of minimum temperature was observed in the prediction of disease through multiple regression analysis and accounted for more than 80 per cent variation in disease (Kumar and Dubey, 2002). Maximum progress of *Rhizoctonia* blight was observed in 37th standard week (September 16-22), at 28.9°C mean temperature, 83.5 per cent relative humidity, 33.6 mm rainfall, 4.3 hours sunshine/day and three rainy days (Kumar, 2005).

Disease Cycle

The quantum of surviving inocula of *R. solani* in soil directly influences the disease severity in field. Sclerotia are the most important source of inocula (Allison, 1951). These sclerotia remained viable for more than 18 months to several months in soil (Muller, 1924) (Palo, 1926). Itaka and Hitomi (1930) observed that sclerotia of *Hypochnus sasakii* (*R. solani*) from rice plants were exclusively found in the field where they fell before harvest. The pathogen hibernates in the field as sclerotia and infect rice crop in the subsequent seasons. At room temperature on dry or moist soil, the sclerotia survived for at least 130 days and for 224 days when submerged under 3" of water (Park and Bertus, 1932). Sclerotia of the coffee strain placed on the soil surface of coffee nursery beds survived upto 225 days but those buried in the soil survived up to 375 days (Venkatasubbaiah and Safeeulla, 1983). Dubey and Dwivedi (1992) found that sclerotia of groundnut strain survived for 21 months in the lab and 18 months in soil kept in pots. All above findings concluded that providing anaerobic or near anaerobic conditions to the sclerotia either through burial in soils or through submergence in water increased the longevity of survival of the sclerotia of *R. solani.*

Hyakumachi and Ui (1982) observed high correlation in disease index, sclerotial number, viable sclerotial number and the presence of *R. solani* AG-2 type 2 from plant debris in soil around diseased sugarbeet after oversetting. There was no correlation between the recovery of the pathogen from plant debris and damping off in sugarbeet seedlings but there was a high correlation between the number of viable sclerotia and damping off. It was thus concluded that the sclerotia are the main source of inoculum. Besides, infection is also caused by basidiospores and hyphae.

Foliage blight of sugar beet is mainly caused by basidiospores and hypahe of *T. cucumeris* AG2-2 (root rots type,) while AG-1 (Web type) isolate attacked with hyphae alone (Naito, 1940). Seed borne nature of *R. solani* was also observed by Hedgcock (1904) in bean seeds. *Rhizoctonia* sp. has been found to be associated with diseased and discolored kernels of groundnut (Kang and Ghohan, 1966).Evans and Poole (1938) isolated *R. solani* from the lesions on the seed coat and shells along with other fungi. *Rhizoctonia* sp. was isolated from pegs, shells and seeds along with other soil borne fungi and molds (Wilson, 1947).Crosier *et. al.*(1969) showed that in artificial inoculation of wheat ears with *R. solani,* the fungus established itself inside the seed and can be isolated after surface sterilization. This showed the internal establishment of the pathogen and possibility of the fungus to become internally seed borne. Saksena and Chaubey (1973) reported that *R. solani* causing banded blight disease of paddy has been found to be carried on or inside the seed surface.

Dubey and Dwivedi (1992) reported that web blight pathogen of groundnut was isolated from 28 per cent of seeds from affected pods and were found present both internally and externally. They concluded that infected seeds, plant debris and sclerotia contributed as primary source of inoculum during the subsequent season. Weeds or rotation crops may be infected with *R. solani* and may function as source of inoculum. Boosalis and Scharen (1960) presented convincing evidence that pigweed is susceptible to *R. solani* strains pathogenic on economic crops such as been, sugar beet, alfa alfa and potato Herzog (1961).

Intensive studies (Carpenter's, 1949&1951; Kotile, 1945; Lorenz, 1948) confirmed that basidospores are a major source of inoculum dissemination of *R. solani* strains causing target leaf spot of leave rubber. The rapid spread of web blight of common bean in warm and humid conditions can also be explained by efficient production and dissemination of basidiospores (Echandi, 1965).

The early infection in the field is caused by imperfect state of *Rhizoctonia solani* and the secondary spread of the disease took place through basidiospores (Saksena and Dwivedi, 1973; Dwivedi and Saksena, 1974 and Dubey and Dwivedi, 1992). It has also been observed that the fungus grew from leaf to leaf contact (Leuttrell, 1962; Saksena and Dwivedi, 1973 and Naito 1940) and from cotyledon to stem (Webber and Abrego, 1958).Galindo *et al.* (1983) undertook studies under field with a history of severe incidence of the disease in bean (*Phaseolus vulgaries*). Sclerotia and mycelia of the pathogen *T. cucumeris*, either free in the soil or colonizing debris, were the main source of inoculum and the basidiosproes caused secondary infection. Reddy and Rao (1981) studies the disease cycle involved in damping off of groundnut, caused by *Rhizoctonia solani* and presented schematic steps of pathogenesis. It is proved that the pathogen spreads by air borne basidiosproes, mycelial bridge between plants, rain splash, sclerotia, infects debris and infected seeds. (Galvez *et. al.*,1979; Godoy–Lutz *et. al.*,1996).

Management

Prolonged survival of *R. solani* in the soil in the form of sclerotia, its seed borne nature and secondary spread through basidiospores, rain splashes and foliar contacts makes the management of this pathogen quite difficult and therefore, to deal with these situation a multifaceted approaches are required for the effective management of this pathogen which are discussed below.

Host Plant Resistance

Screening Technique

Field Screening

Takegami *et al.* (2004) used following methodology for field screening of bean for web blight.

1. A pure culture of the mycelium of *R. solani* grown on PDA medium to be used to prepare the inoculum.
2. Eighteen 4 mm diameter disc from the margin of the fungal colonies were placed in a liquid medium containing 10g of peptone, 15g of dextrose, 0.5g of KH_2PO_4 and 0.25 of $MgSO_4$ per liter of water. The solution was agitated in a shaker incubator at 27°C for a period of 14 days.
3. The mycelial growth was blended for 30 seconds at low speed, filtered and washed in distilled water. The filtrate was dried 24 h before water was added.
4. The fine concentration of the inoculum was obtained using a spectrophotometer. Water was added until 250 per cent transmittance was obtained at a wave length of 640nm. Tween 80 was added at a concentration of 1cc/l to increase the adhesion of the inoculum to the surface of the leaves.
5. At 35 days after planting, the leaves of the bean plants were inoculated in the field with a backpack sprayer as spray. The inoculum was applied late in the afternoon and the trial received 15 minutes of irrigation each morning during the first week after inoculation.

6. Leaf damage was assessed at 8th and 15th day after inoculation using the CIAT (1-9) scale. After the harvest, seed was separated into commercially acceptable and blemished seed categories to calculate percentage of damaged seeds.

Detached Leaf Technique

A detached leaf technique has been devised for screening against web blight (Galindo *et al.*, 1982; Bautista, *et al.*, 2000). Protocol used for the preparation of inoculum and the evaluation of bean plants for web blight infection using the detached leaf are recently published (Takegami *et al.*, 2004).

1. A pure culture of the mycelial state of *Rhizoctonia solani* (AG-1-1E) on potato dextrose agar medium was the source of the infected discs used as inoculum.
2. Fully expanded trifoliate leaves were detached from the bean plants and immediately placed in orchid tubes filled with water.
3. The leaf lets were positioned in 42x30x6m trays as the base of the trays contained moisture paper towels. The leaf lets were placed on top of Petri plates to avoid immersion in water. One of the leaflets in each tray should be from a line susceptible to web blight to serve as a check.
4. One 4mm diameter disc of agar colonized with *R. solani* was placed on the adaxial side and centered on each leaf let's to serve as a control. The control and infected treatments should be applied at random.
5. In addition, each tray contained a leaf let of a susceptible bean line as a means to confirm that conditions were favourable for the development of infection.
6. The trays were placed inside plastic bags after inoculation to create a high humidity favourable for the development of the disease. The trays were incubated in laboratory at 27±1°C.
7. Mean lesion size was measured at 24, 48 and 72h after inoculation. The measurement of the lesion size can be facilitated by the use of a digital image system.

Attached Leaf Technique

1. Mycelial inoculum of *R. solani* is prepared by cutting a disc from the edges of a developing colony on potato dextrose agar and transferring to deep Petri plates with 25cc of V 8 liquid medium (100 cc V8 juice, 15g $CaCO_3$ and 400 cc of demineralized sterile water). Plates are kept without shaking until the colony reaches the edges of the plate at 48 hrs.
2. After a mycelia mat covers the surface in 48 hr (before sclerotia initials develop), the mat is filtrated, blended in sterile demineralized water and diluted 500 times based on calculation of colonies that developed on water agar by the serial dilution plating method.
3. Mycelial suspension is sprayed to the plants in trays in inoculation boxes. The suspension is supplemented with 0.05 per cent Tween 20 and sprayed under light air pressure until run off. Each tray containing the pots with inoculated plants is moved into a humidity chamber (2x1m) divided in two compartments (top and bottom) designed to accommodate six trays apart from each other to prevent cross contamination. The chamber is sealed with plastic covers and the humidity kept at 100 per cent with humidifiers set at low. The humidifiers should not run for more than 28 hrs to prevent fungus overgrowth. Record room temperature inside the chamber and keep the cycle of 14 hr. fluorescent light/dark.

4. If numerous lines are to be tested against many isolates set up the experiment as a randomized complete block design with split plot and repeat over time as separate trails. Isolates serve as the main plot and lines (cultivars) the sub plots.
5. Lesions develop within 24-30 hrs. Monitor disease development and measure the area under lesion over the total trifoliate.

Breeding of varieties for resistant to web blight has been a difficult work, as donors with desirable degree of resistance against this disease are lacking. Singh (2007) screened 200 varieties/lines of mungbean against web blight but none was found resistant. However, 12 varieties/lines *viz.*, V 2306, PM-3, DND-1077, LM-249, V-4526, K-429, NDM-1, PM-4, NDM-94-2, ML-3, Diakara and 2KM3-43 were moderately resistant. Several studies have been made in the past for screening of mungbean varieties/line against web blight in India. Singh *et al.* (1989) reported four lines LS-598, LM-220, LM-412 and V3681 resistant to web blight under field conditions. Similarly, twelve varieties/lines *viz.*, NP 52, V3681, ML 267, Sam rat, ML5, MH-96-1, PDM-54, Pant nagar- 4, NDM 97-1, LM-598, LM-220 and LM-412 were moderately resistant (Kumar,2005). Few genotypes of common bean also have been found to possess moderate levels of resistance to web blight (Arnand, *et.al.*,1998; Montoya *et. al.*,1997; Takegami *et al.*, 2004), Although no specific genes for resistance have been identified, inter specific crosses are being attempted to transfer web blight resistance from scarlet runner bean (*Phaseolus coccinus* L.) to common bean. Some resistant sources are listed below in the Table 18.2.

Table 18.2: Source of Resistance to Web Blight in *Phaseolus vulgaris*

Genotype	*Characters*	*Reference*
Arroyo Loso Negro	9/ Black	Arnand *et al.* (1998)
Talmanca, HT 7719	1/ Navy	
	2m/ Pinto	
	1/Great Northern	
Vax 6	7/ Purple	Singh and Munoz (1999)
Bribri	6/ small red	
	5/ pink	
	2R/ Cranberry	
A-193	6M/ Red mottled	Arya and Arya (2000)
	6K/ Dark red Kidney	
	1/ Snap	

Cultural Management

Tillage practices, seed bed preparation, time of sowing, plant population, weeding, mulching crop rotation and intercropping are the useful practices that influence the disease development in legumes caused by *R. solani.* (Leach and Garber, 1970). Mulching with rice husks (2.5 cm thick) greatly reduced the inoculum of *R. solani* and lowered the web blight severity in bean.

Biological Management

Rudakov (1970) reported mycophilic properties of *Gliocladium* as hyper parasitic and source of antifungal antibiotic. *G. virens* parasitized *R. solani* mycelium which is strongly inhibited by antibiosis

(Howell, 1982).Of 39 isolates of *Trichoderma viride* used, 17 were highly antagonistic to *R. solani* on sugar beat, bean and cotton seedlings (Dumitras and Sesan, 1980).

Table 18.3: Crop Wise Chronology of Antagonists Evaluated Against *Rhizoctonia solani*

Name of the Crop	*Antagonists (Employed)*	*Reference*
Citrus	*T. lingorum*	Weindling, 1932
Mustard	*T. harzianum*	Akhtar, 1969
Lupine, Tomato and groundnut	*T. harzianum*	Wells *et al.* (1972)
Cotton, Linseed	*T. harzianum* and *B. subtilis*	Agarwal *et al.*, 1977
Bean, Eggplant and Tomato	*T. harzianum*	Hadar *et al.*, 1979
Radish, Lucerne, Sugar beat	*T. harzianum*	Chet and Baker, 1980
Bean, Cotton, Tomato	*T. harzianum*	Elad *et al.* (1980)
Potato	*T. harzianum*	Elad *et al.* (1981)
Bean, Radish and Pea	*T. harzianum*	Chet and Baker, 1981
Strawberry	*T. harzianum*	Elad *et al.* (1981)
Radish	*T. harzianum*	Hanis *et al.* (1981)
Snap bean	*T. harzianum*	Marshall, 1982
Soybean, Maize, Pea, Snap bean and Squash	*T. harzianum*	Ruppel *et al.* (1983)
Potato	*T. viride* and *G. virens*	Beagle and Papavizas, 1985
Tomato	*T. harzianum*	Strashnov *et al.*, 1985
Apple and Capsicum	*T. harzianum* and *Laetisaria arvalis*	Conway, 1986
Snap bean	*T. hamatum* and *T. harzianum*	Cardoso and Echandi, 1987
Pine	*T. harzianum*	Smith and Wehner, 1987
Potato	*T. koningii* and *T. viride*	Bari, 1989
Chickpea	*T. harzianum*	Kaur, 1989

Tu and Vaartaja (1981) reported *G. virens* to be a hyperparasites of *R. solani* using light and electron microscopy. When hypae of *G. virens* touched those of *R. solani,* the following occurred: contact with host cells, formation of appresoria and penetration of host cells. Three modes of antagonism which include:

1. Competition for nutrition, space and biological oxygen,
2. Antibiosis through production of certain antibiotics, enzymes or other harmful chemicals,
3. Direct exploitation of host fungi in the form of predation, hyper parasitism and mycoparasitism.

Three folds antagonistic weapons of *Trichoderma* spp. *were* non-volatile antibiotics (Trichodermin, peptide), volatile antibiotics (acetaldehyde) and hyphal interaction (coil round the host hyphae) (Dennis and Webster, 1971 a, b, c). Gliovirin a new diketopiperazine antibiotic was isolated from *G. virens* (Stipanovic and Howell, 1982). Gokulpalan and Nair (1984) reported that *Trichoderma viride* was the most antagonists inhibiting linear growth of *R. solani in vitro.* In tests with 28 strains of *Trichoderma* spp. and 3 strains of *R. solani* it was found that mycoparasites were the most effective mechanism for destroying hyphae (Camporota, 1985). Lewis and Papavizas (1987) reported mycelial preparations of 8 of 14 isolates of *Trichoderma* spp. and G. *virens that* reduced survival of *R. solani* at least 50 per cent in pathogen infested seed and in soil. Water extract of *G. virens* affected growth of *R. solani* adversely (Lewis and Papavizas, 1987). In *in vitro* studies *R. solani* was strongly inhibited by antagonistic activity of *Trichoderma viride and T. harzianum* both when the pathogen and antagonists were grown in pairs on the same medium and when grown on separate media in a confined environment (Marchetti *et al.* (1992). Elad *et al.* (1987) reported that hyphae of antagonist *T. harzianum* when approached to that of *R. solani,* they formed branches which grew directly towards the host and fibrillar material was deposited between the hyphae during the first interaction. Similar results have been found by Benhamou and Chet (1993) who reported that coiling of the antagonists (*T. harzianum*) around its host (*R. solani*) was an early event proceeding hyphal damage. Ultra structural observations of hyphal interactions showed that contact between the two fungi was mediated by a fine, extra cellular matrix originating from cells of *R. solani*. Hypahe of *T. harzianum* isolates coiled around the *R. solani* hyphae or grew parallel to end along them, sometimes forming bulbous as hook like structures, penetrating and growing within the host (Zhong *et al.,* 1990). *Gliocladium virens* was found to colonize, penetrate and sporulate inside the sclerotia of the *R. solani* (Mukherjee *et al.,* 1995). Manzali *et al.* (1993) reported *in vitro* antagonism of 41 strains of *Trichoderma* isolated from soil against 5 strains of *R. solani*. After 4 days contact between the two fungi, the antagonism were included lysis, formation of barriers and growth of *Trichoderma* hyphae on *R. solani.*

Chemical Management

No variety is known to have desirable degree of resistance in different legumes. If favourable conditions prevail there is strong possibility of breaking down the resistance and development of disease epiphytotic any time during the crop season. Under such circumstances use of chemical fungicides is a practical approach to manage the disease and reduce losses below economic threshold level.

In vitro

Several attempts have been made in past to evaluate numerous fungicides *in vitro* to assess their effectiveness against *R. solani*. Copper carbonate, Copper sulphate, Mercuric chloride, Agrosan GN, Quintozene, Kasugamycin, Carboxin, Pyracarbolid, Carbendazim, Chloronebm, benomyl, and tetrazole were most inhibitory to mycelial growth and had EC 50 value of < 1mu g a.i./m. (Gupta *et al.,* 2002).Similarly, Carbendazim, Benomyl, Furmecyelox, Oxyzasboxin, Carboxin, Phenyl mercury acetate and Zineb completely inhibited growth of the pathogen (Singh *et al.,* 1993). Alone and a few combinations were also found inhibitory against pathogen among them Carbendazim, Kitazin, Topsin-M Carbendazim + Indofil M-45 and Carbendazim + Captaf completely inhibited the growth of the pathogen (Dubey and Prasad, 1997). Kumar, 2005 and Singh, 2007 reported Propiconazol, Thiophanatemethyl and Carbendazim completely inhibiting the mycelium growth even at very low concentration (100ppm). Singh (2007) reported seed treatment with *T. viride* (10^7cfu) carbendazim 2g/kg seed alone or in combination along with two foliar sprays of Tilt. Topsin-M was best for management of disease as well as promotion of growth in mungbean. Similar trend was also found by Jhamaria and Sharma, (2002). Sharma and Tripathi (2001) and Shailbala and Tripathi (2004) on others leguminous hosts Web blight severity in French bean was greatly reduced with foliar application of Carbendazim (0.1 per cent) and Tebuconazole (0.5 per cent) (Upamanyu, 2003). *In vitro* assayed Jkstin (0.1 per cent), Ridomil (0.1), Carbendazim(0.1 per cent) Topsin-M (0.1 per cent) Indofil M-45(0.2 per cent), Indofil Z-78(0.2 per cent) Caresan wet (0.25 per cent) Captafol (0.2 per cent) and Blitox-50 (0.2

per cent) against *Thanatephorus cucumeris.* All fungicide significantly inhibited fungal growth. Jkstin, Ridomil, Carbendazim, Topsin–M and Indofil M-45 resulted in 100 per cent fungal growth inhibition. Dubey and Dwivedi (1988) found 6 sprays of Carbendazim (0.2 per cent) as most effective in reducing the web blight incidence followed by Dithane Z-78(0.25 per cent) and Benlate (0.2 per cent). Among 19 fungicides evaluated against web blight pathogen *in vitro* Carbendazim, Benomyl, Furmecyclox, Oxycarboxin, Carboxin, Quintozine, Phenyl mercuric acetate and Tineb completely inhibited the growth of pathogen.

In vivo

Ratan and Dwivedi (1998) reported seed treatment with Benlate or Vitavax @ 2g Kg seed eliminates the seed borne infection completely and enhanced the germination. In mungbean, disease severity can be greatly reduced by seed treatment with Carbendazim (0.5 per cent) and two foliar spray of Carbendazim 0.1 per cent or three spray of Dithane M-45(0.2 per cent) at 10 days intervals. (Gupta *et al.*, 2002; Dubey and Dwivedi, 1988; Tiwari and Tiwari, 1998). Ratan and Dwivedi (1998) reported that seed treatment with Benlate or Vitavax @ 2 g/kg seed eliminated the seed borne infection completely and enhanced the germination. Two sprays of Jkstin (0.2 per cent) at 14 days interval after the first appearance of the disease was most effective in minimizing the disease. Gupta *et al.* (2002) found that Carbendazim, Vitavax and Mancozeb gave 95-100 per cent inhibition in colony growth of *R. solani.* Out of 6 fungicides seed treatment with carbendazim (0.05 per cent) and two foliar application of carbendazim (0.1 per cent) or three spray of Mancozeb (0.2 per cent) at 10 days interval was most effective in reducing web blight in mungbean (Gupta and Singh, 2002). Jhamaria and Sharma (2002) found that Carbendazim was the best in minimizing the disease severity in mungbean followed by Topas, Tilt and Kavach. The highest yield was obtained in plots sprayed with Carbendazim followed by Topas. Sharma and Tripathi (2001) and Shailbala and Tripathi (2004) worked on urdbean and found lowest disease severity and highest grain yield in Tilt (0.1 per cent) followed by Carbendazim (0.1 per cent) and Dithane M-45 per cent (0.25) sprayed plots. Carbendazim was found highly remunerative giving maximum returns in this crop. In pot experiment Benomyl, Quintozine, Zineb, Oxycarboxin and Carboxin gave effective control of the disease (Singh *et al.*, 1993). Tiwari and Tiwari (1998) reported two sprays of Carbendazim as significantly superior in reducing the web blight disease followed by one spray of Carbendazim. However, the plots with two sprays of Carbendazim displayed more green foliage and delayed maturity by 10 days than rest of the treatments. Seed treatment either Carbendazim or with Vitavax (2g/kg seed) and a single spray of Carbendazim (0.05 per cent) at 35-40 day crop stage reduced web blight incidence and significantly improved yield in mungbean. The cost benefit ratio was also high. Upamanyu (2002) reported foliar sprays of Carbendazim (0.1 per cent) and Tebuconazole (0.05 per cent) as most effective in reducing web blight severity, while seed treatment with Carbendazim (0.2 per cent) combined with foliar spray was less effective. Use of fungicides not only reduce the pathogen but also induced growth and changed the polyphenolic constituents of the hypocotyls of infected plants.

Integrated Management

As the web blight pathogen *R. solani* has very wide host range, fast developing nature in a short time and low level of resistance of available varieties in legume there is always a risk of disease epiphytotic to develop. A single approach to manage this disease does not help much and therefore, an integrated approach could be a better option to manage it in an efficient manner.

Web blight severity in French bean could be effectively controlled by integration of Mustard cake + Carbendazim (ST+FS), followed by Tebuconzole + *T. virens* (ST) + Carbendazim (FS).Sprays of

Carbendazim (0.1 per cent) on mulch and foliage were highly effective (Upamanyu *et al.*, 2002). Mungbean and urdbean web blight were managed through integration of soil application of Karanj cake (2q ha^{-1}), seed treatment with *G. virens* (10^6 spore ml^{-1} 10 g $seed^{-1}$) + Carboxin (1g kg^{-1}) + *Rhizobium* sp.(25 g kg^{-1} seed) and foliar spray of Karanj leaf extract (10 per cent) which enhanced seed germination, plant vigour and reduced disease severity. However, soil application of Karanj cake or *G. virens* (multiplied on pulse bran- saw dust- tap water) (3:1:4: w/w/v), seed treatment with *G. virens* + Carboxin, + *Rhizobium* sp. and foliar spray of Karanj leaf extract separately also proved effective for the management of disease (Dubey, 2003). Singh, 2007, integrated Carbendazim (0.1 per cent) and *T. viride* (10^7 cfu/ml water) as seed treatment followed by two foliar spray of Propiconazole (0.1 per cent) or Topsin-M (0.1 per cent) as best combination for management of web blight disease of mungbean. Cotton cake + *T. virens* and Carboxin a seed treatment, Mustard cake +*T. virens* and Carbendazim (ST), seed treatment with combination of *T. viride* and Carboxin, *T. harzianum* + Tebuconazole, *T. virens* + Tebuconazole and Mustard cake + Carbendazim (ST) were effective in controlling root rot caused by *R. solani*. Web blight severity was best controlled by Mustard cake + Carbendazim (ST+FS), followed by Tebuconozole + *T. virens* (ST) + Carbendazim (FS). Sprays of Carbendazim (0.1 per cent) on mulch and foliage were highly effective. The integration of soil application of Karanj cake seed treatment with *G. virens* (10^6 spores ml^{-1}/10 g seed) + Carboxin (1 kg^{-1}) + *Rhizobium* (25g kg^{-1} seed) and foliar spray of Karanj leaf extract (10 per cent) increased 35.7-96 per cent seed germination, reduced 92.4-93 per cent disease intensity.

References

Agrawal, S.C., Khare, M.N. and Agrawal, P.S. (1977). Biological control of *Sclerotium rolfsii* causing rot of lentil. *Indian Phytopath.*, 30: 176–179.

Ahuza, S.C. (1976). Investigation on banded leaf and sheath blight of maize Ph.D. Thesis, Division of Plant Pathology, IARI, New Delhi, 92 pp.

Akhtar, C.M. (1969). Biological control of *Rhizoctonia solani* causing seed decay and damping off of mustard. *W. Pakistan. J. Agric. Res.*, 7: 29–52.

Alam, S.S., Qureshi, S.H. and Bashir, M. (1985). A report on web blight of mungbean in Pakistan. *Pakistan J. Bot.*, 17(1): 165.

Allison, J.L. (1951). Rhizoctonia blight of forage legumes and grasses. *Plant Dis. Reptr.*, 35: 372–373.

Anderson, N.A. (1982). The genetics and pathology of *R. solani*. *Ann. Phytopath.*, 20: 329–347.

Anonymous (2004). *Annual Report* (*Kharif*). All India Coordinated Research Project on MULLaRP (ICAR), IIPR, Kanpur. 112 pp.

Anonymous (1997). *Vision 2020: IIPR Perspective Plant*, IIPR, Kanpur. 10 pp

Anonymous (1999). *Technology for Increasing of Pulses Production in India*, 2004, IIPR, Kanpur. 108 pp.

Arnand-Santana, E., Coyne, D.P. Steadman, J.R. (1998). Inheritance and heritability of the reaction to web blight. *Ann. Rep. Bean Improv. Coop.* 41: 29–30.

Arya, C.M. and Arya, Y.R. (2000). Advances enla selection de fuentes de resistance a las principles enfermedades de frijol common (*Phaseolus vulgaris* L.) en Costa Rica. *Agronomia Mesomaericana*, 11(2): 25–29.

Atkins, J.G. and Lewis, W.D. (1954). *Rhizoctonia* aerial blight of soybean on Lonisiana. *Phytopath.*, 44: 215–218.

Baker, K.F. (1947). Seed transmission of *Rhizoctonia solani* in relation to control of seedling damping off. *Phytopath.*, 37: 912–924.

Baker, R. and Martinson, M. (1970). Epidemiology of diseases caused by *Rhizoctonia solani*. In: *Rhizoctonia solani, Biology and Pathology*, (Ed.) J.R. Parameter Jr. Univ. of California Press, Berkeley, Los Angeles and London, 172–188 pp.

Bari, A.M. (1989). Integration of biological and chemical methods for the control of black scurf of potato. *Ph.D. Thesis*, G.B. Pant University of Agriculture and Technololgy, Pantnagar, 168 pp.

Baruah, P. and Sangam Lal (1981). Host range of *Rhizoctonia solani* f.sp. *sasakii*, the incitant of banded sclerotial disease of maize. *Indian Phytopath.*, 34: 494–496.

Bautista, Perez-M-de-J. Echvez, Badel, R., and Macthiavelli, R.E. (2000). Virulence of *R. solani* isolates associated with web blight of common bean. *J. Agril. Univ. of Puerto Rico*, 84(1–2): 95–97.

Beagle, Ristanio, J.E. and Papavizas, G.C. (1985). Biological control of *Rhizoctonia solani* stem canker and black scurf of potato. *Phytopath.*, 75: 560–564.

Benhamou, N. and Chet, I. (1993). Hyphal interactions between *Trichoderma harzianum* and *Rhizoctonia solani*: Ultra structure and gold cyto-chemistry of the myco-parasitic process. *Phytopath.*, 83: 1062–1071.

Boosalis, M.G. and Scharen, A.L. (1960). Methods for microscopic detection of *Aphanomyces euteiches* and *Rhizoctonia solani* associated with plants debris. *Phytopath.*, 49: 192–198.

Braun, H. (1930). Der Wurzeltoter der kartoffel. Monograph. *Pflanzebschutz.*, 5: 1–136.

Camporota, P. (1985). Antagonism *in vitro Trichoderma* spp. to *Rhizoctonia solani* Kühn. *Agronomic.*, 5: 613–620.

Cardoso, J.E. and Echandi, E. (1987). Biological control of *Rhizoctonia* root rot of snapbean with binucleate *Rhizoctonia* like fungi. *Plant Disease.*, 71: 167–170.

Carling, D.E., Leiner, R.H. and Kebler, K.M. (1986). Characterization of an undescribed anastomosis group of *Rhizoctonia solani*. *Phytopath.*, 76(1): 64.

Carpenter, J.B. (1949). Production and discharge of basidospores by *Pellicularia filamentosa* (Pat.) Rogers on *Hevea* rubber. *Phytopath.*, 39: 980–985.

Carpetner, J.B. (1951). Target leaf spots of the *Hevea* rubbler tree in relation to host development, infection, defoliation and control. *U.S. Dept. Agr. Tech. Bull.*, p. 1028–1034.

Chet, I. And Bakar, R. (1980). Introduction of suppressiveness to *Rhizoctonia solani* in soils. *Phytopath.*, 70: 994–998.

Chet, I. and Baker R. (1981). Isolation and bio-control potential of *Trichoderma hamatum* from soil naturally suppressive to *Rhizoctonia solani*. *Phytopath.*, 71: 286.290.

Christon, T. (1962). Penetration and host parasite relationships of *Rhizoctonia solani* in the bean plant. *Phytopath.*, 52: 381–389.

Chupp, C. and Sherf, A.P. (1960). *Vegetable Diseases and their Control*. The Ronald Press Co., New York, 693 pp.

CIAT (1985). *Annual Report*, 1984. In: Bean programme Centro International de Agricultue Tropical, Cali Columbia. 263–267 pp.

Conway, K.E. (1986). Use of fluid drilling gels to deliver biological control agents to soil. *Plant Disease,* 70: 835–839.

Crosier, W.F., Natte, J.J. and Crosier, D.C. (1969). Occurrence *Rhizoctonia* species in seeds of small grains and other Plants. *Phytopath.,* 59: 808.

De Condole, A.P. (1815). Memoire sur les rhizoctonis, nouveau genre de champignons qui attaque les raciness, des plants at in particulier celle de la Luzerne cultivee. *Men Mus C. Hist Nat.,* 2:209–216.

Dennis, C. and Webster, J. (1971a). Antagonistic properties of species group of *Trichoderma* I. Production of non–volatile antibiotics. *Trans. Brit. Mycol. Soc.,* 57: 25–39.

Dennis, C. and Webster, J. (1971b). Antagonistic properties of species group of *Trichoderma* II. Production of volatile antibiotics. *Trans. Brit. Mycol. Soc.,* 57: 41–48.

Dennis, C. and Webster, J. (1971c). Antagonistic properties of species group of *Trichoderma* III. Hyphal interaction. *Trans. Brit. Mycol. Soc.,* 57: 363–369

Dubey, S.C. and Dwivedi, R.P. (1988). Chemical control of web blight of blackgram. *Farm Sci. J.,* 3(2): 182–183.

Dubey, S.C. and Prasad, R.B. (1997). Evaluation of different fungitoxicants against web blight of horsegram. *I.J. Pt. Patho.* 31(3): 161–166.

Dubey, S.C. (2003). Integrated management of web blight of urd and mungbean. *Indian Phytopath.* 56(4): 413–417.

Dubey, S.C.and Dwivedi, R.P. (1992). Survival of *Thanatephorus cucumeris* causing web blight of ground nut. *Indian J. Mycol. Pl. Pathol.,* 22: 49–53.

Duggar, B.M. and Stewart, F.C (1901). The sterile fungus *Rhizoctonia* as a cause of plant diseases in America New York (Cornell). *Agr. Expt. Sta. Bull.,* 186: 51–76.

Dumitras, L. and Sesan, T. (1980). Features of antagonism of *Trichoderma viride* Pers. ex fr. to *Rhizoctonia solani* Kühn. *Biologic Vegetala,* 32: 87–92.

Durbin, R.D. (1957). Importance of variation and quality of pathogens. *California Agr. Expt. Sta. Manual,* 23: 255–262.

Dwivedi, R.P. and Dubey, S.C (1986). Web blight of groundnut caused by *Thanatephorus cucumeris. Farm. Sci. J.,* 1: 33–57.

Dwivedi, R.P. and Saksena, H.K. (1974). Occurrence of web blight caused by *Thanatephorus cucumeris* on mungbean. *Indian J. Farm Sci.,* 2: 100.

Dwivedi, R.P. (1977). Studies on *Thanatephosus* species attacking cowpea plants. *Ph.D. Thesis,* C.S. Azad Univ. of Agri. Tech., Kanpur. 106 pp.

Echandi, E. (1965). Basidiospores infection by *Pellicularia filamentosa* (= *Corticium microsclerotia*), the incitant of web blight of common bean. *Phytopath.,* 55: 698–699.

Edwards, H.H. and Allen, P.J. (1970). A fine structure study of the primary infection process during infection of barley by *Erisiphae graminis* f. sp. *horde. Phytopath.,* 60: 1504.

Elad, Y., Chet, I. and Heris, Y. (1981). Biological control of *Rhizoctonia solani* in strawberry field by *Trichoderma harzianum. Plant and Soil,* 60: 245–254.

Elad, Y., Sadowsky, Z. and Chet, I. (1987). Scanning electron microscopical observations of early stages of interaction of *Trichoderma harzianum* and *Rhizoctonia solani. Trnas. Br. Mycol. Soc.*, 88: 259–263.

Elad, Y, Chet, I. and Katan, J. (1980). *Trichoderma harzianum* a biological agent effective against *Sclerotium rolfsii* and *Rhizoctonia solani. Phytopath.*, 70: 119–121.

Evans, M.M. and Poole, R.F. (1938). Some parasitic fungi hard board by peanut seed stocks (Abstr.). *J. Elisha Mitchell Sci. Soc.*, 54: 190–191.

Fulton, N.D. and Hanson, E.W. (1960). Studies on root rots of red colour in Wisconsin. *Phytopath.*, 50: 541–550.

Galindo, J.J. (1982). Variability among isolates of *Rhizoctonia solani* associated with soybean hypocotyls and soils in New York. *Plant Dis.* 66(1): 390–394.

Galindo, J.J., Abaui, G.S., Thurston, H. D. and Galrez, G. (1982). Source of inoculum and development of been web blight in Costa Rice. *Pl. Dis.*, 67: 1076–1021.

Galindo, J.J., Abawi, G.S., Thurston, H.D. and Galveg, G. (1983). Effect of mulching on web blight of bear in Costa Rice. *Phytopath.*, 73: 610–615.

Galvez, G.F., Guzman, P. and Castano, M. (1979). Web blight in bean production problems In: *Disease, Insect, Soil and Climatic Constraints of Phaseolus vulgaris* (Eds.) H.F. Schwariz and G.E. Galver. Cali, Columbia, Centro International de Agriculture Tropical, 101–110 pp.

Galvez, G.E., Mora, B. and Pastor Corraleo, M.A. (1979). Web blight. In: *Bean Production Problems in the Tropics*, (Eds.) H.F. Schwart and M.A. Pastor Corrales. CIAT, Calis. Colmbia, 195–259 pp.

Garibaldi, A., Minuto, A., Bertetti, D., Nicoletti, R. and Gullino, M.L. (2003). First report of web blight on yellow saga (*Lantana camara*) caused by *Rhizoctonia solani* in Europe. *Plant Dis.*, 87(7): 275.

Gokulpalan, C. and Nair, M.C. (1984). Antagonism of a new fungi and bacteria against *Rhizoctonia solani* Kühn. *Indian J. Microbiol.*, 24: 57–58.

Gratz, L.O. (1925). Wirestem of Cabbage. New York (Cornell). *Agro Expt. Sta. Mem.*, 85: 1–60.

Godoy-Lutz, G., Arias, J., Standman, J.R. and. Eskridge, K.M (1996). Role of natural seed infection by the web blight pathogen in common bean seed damage, seedling emergence, and easily disease development. *Plant Dis.*, 80: 887–890.

Gupta, A.B. (2000). Studies on foliar diseases in mungbean [*Vigna radiata* (L. Wilczek] under different climatic conditions. *M. Sc. (Ag.) Thesis*, N.D.U.A. &T., Kumarganj, Faizabad (U.P.).

Gupta, A.B., Gupta, R.P. and Singh, R.V. (2003). Epidemiology of yellow mosaic virus and assessment of yield losses in mungbean. *New Botanist*, 30: 169–178.

Gupta, A.B., Gupta, R.P. and Singh, R.V. (2002). Effect of season on web blight development and its impact on yield in mungbean. IPS (MEZ) Annual Meet and National Symposium on "*Integrated Management of Plant Diseases of Mid-Eastern India*", 5–7, Dec. held at NDUA&T, Kumarganj, Faizabad, Souvenir and Abstract, p. 30.

Gupta, R.P. and Singh, R.V. (2002). Assessment of yield losses in mungbean due to *Rhizoctonia solani. J. Mycol. Pl. Pathol.*, 32(1):142

Hadar, Y., Elad, I. Chet and Y. Henis (1979). Biological control of *Rhizoctonia solani* damping off with wheat bran culture of *Trichoderma harzianum. Phytopath.*, 69: 64–68.

Hanis, Y., Gaffer, A. and Baker, R (1981). Factors affecting *Trichoderma hamatum* applied to seeds as a bio-control agent. *Phytopath.*, 71: 569–572.

Hedgcock, G.C. (1904). A note on *Rhizoctonia. Science*, 19: 268.

Hepperly, P.R., Mignucci, J.S., Sinclair, J.B., Smith, R.S. and Judy, W.H. (1982). Rhizoctonia web blight of soybean in Puerto Rice. *Plant Dis.*, 66(3): 256–257.

Herzog, W. (1961). Das uberdauern und der saporphytismus des wureltoters, *Rhizoctonia solani* Kühn in Boden. *Phytopath. Z.*, 40: 379–415.

Holcomb, G.E. and Carling, D.E. (2002). First report of web blight caused by *Rhizoctonia solani* on *Catharanthus rosens* in Lousioma. *Plant Dis.*, 86: 111–272.

Holly, W.D. and Baker, R. (1963). *Carnation Production including the History, Breeding, Culture and Marketing of Carnations*. WmC. Brown Company, Dubuque, Iowa, 142 pp.

Houston, B.R. and Kendrick J.B. (1949). A crater spot of celery petioles caused by *Rhizoctonia solani. Phytopathol.*, 39: 470–474.

Howell, C.R. (1982). Effect of *Gliocladium virens* on *Pythium ultimum*, *Rhizoctonia solani* and damping off of cotton seedlings. *Phytopath.*, 72: 456–498.

Hyakumachi, M. and Ui, T. (1982). The role of the over wintered plant debris and sclerotia as inoculum in the field occurred with sugarbeet root rot. *Ann. Phytopath. Soc. Japan*, 48: 628–633.

Itaka, S. and Hitomi, T (1930). On the mode of primary infection through sclerotia and field observations on the basidiospores formation in the earliest stage of monko disease (*Hypochnus sasakii Shirai*) of rice plants. *J. Pl. Prot. Tokyo*, 17: 17–28.

Ito, K., Kontani, S. and Kondo, H. (1955). Web blight fungus of Japanese larch seedlings. *Bull. Forestry Expt. Sta. Meguro*, 79: 43–70.

Jhamaria, S.L. and Sharma, D.P. (2002). Management of web blight of mungbean through chemicals and plant product. *Indian Phytopathl.* 55(4): 526.

Joge, G.F. (1986). Management of Rhizoctonia aerial blight of soybean and biology of sclerotia of *Rhizoctonia solani* Kûhn. *Ph.D. Thesis*, Louisiana State Anio, Boton Rough, 81 pp.

Kang, M.S. and Ghohan, J.S. (1966). Some pathogenic fungi from groundnut seeds in Punjab. *J. Res. Ludhiana*, 3: 158–161.

Kannaiyan, S. and Prasad, N.N. (1980). Effect of fungicides seed treatment on rice seedling growth. *Int. Rice. Res. Newsletter*, 5: 21.

Kaur, Narinder, P. (1989). Integration of biological and chemical methods for the control of chickpea wilt complex. *Ph.D. Thesis*, G.B. Pant Univ. of Agric. and Tech., Pantnagar, 159 pp.

Kim, W.G., Kim, C.K. and Yu, S.H. (1981). Anastomosis grouping and cultural characteristics of isolates of *Rhizoctonia solani* Kuhn from sclerotia over wintered in paddy fields. *Korean J. Pl. Pathl.*, 4: 136–142.

Kolattukudy, P.E. (1974). Enzymatic penetration of the plant cuticle by fungal pahogens. *Ann. Rev. Phytopathol.*, 12: 247.

Kolattukudy, P.E., Ettinger, W.F. and Sebastian, J. (1987). Cuticular lipids in plant microbe interactions. In: *The Metabolism, Structures and Function of Plant Lipids*, (Eds.) Stuff, P.K., Mudd, B.D. and W.E. Ness. Plenum Publishing Corporation, New York, pp. 473.

Kotile, J.E. (1945). Cotton leaf spot *Rhizoctonia* and its perfect stages on sugarbeet. *Phytopath.*, 35: 741–743.

Kotile, J.E. (1947). Rhizoctonia blight of sugarbeet. *J. Agric. Res.*, 74: 289–314.

Kumar, M. and Dubey, S.C. (2002). Relationship of disease intensity with weather and management of Web blight of winzed bean. *Indian Phytopath.*, 55: 152–157.

Kumar, S. (2005). Succession and management of major diseases on mungbean [*Vigna radiata* (L.) Wilczek]. *M.Sc. (Ag.) Thesis*, N.D.U.A.&T., Kumarganj, Faizabad (U.P.), 98 pp.

Kuninaga, S. and Yokasawa, R. (1982a). DNA base sequence homology in *Rhizoctonia solani* Kühn I. Genetic relatedness with anastomosis group 1. *Ann. Phytopathol. Soc. Jpn.*, 48: 659–67.

Kuninaga, S. and Yokasawa, R. (1982b). DNA base sequence homology in *Rhizoctonia solani* Kühn II. Genetic relatedness with anastomosis group 2. *Ann. Phytopathol. Soc. Jpn.*, 48: 668–73.

Kuninaga, S. and Yokasawa, R. (1983). DNA base sequence homology in *Rhizoctonia solani* Kühn III. Genetic relatedness was within AG–3, AG–5, AG–7 and AG–BI, *Ann. Phytopathol. Soc. Jpn.*, 49: 648–652.

Kuninaga, S. and Yokosawa, R. (1985a). DNA base sequence homology in *Rhizoctonia solani* VI. Genetic relatedness seven anastomosis groups. *Ann. Phytopathol. Soc. Jpn.*, 51: 127–32.

Kuninaga, S. and Yokosawa, R. (1984). DNA base sequence homology in *Rhizoctonia solani* Kühn V. Genetic relatedness with in AG–6. *Ann. Phytopathol. Soc. Jpn.*, 50: 346–52.

Kuninaga, S.and Yokosawa, R. (1985b). DNA base sequence homology in *Rhizoctonia solani* VII. Genetic relatedness between AG–BI and other anastomosis groups. *Ann. Phytopathol. Soc. Jpn.*, 51: 133–38.

Lakshmanan, P., Nair, M.C. and Menon, M.R. (1979). Collar rot and web blight of cowpea caused by *Rhizoctonia solani* in Kerala, India. *Plant Dis. Repor.*, 63: 410–413.

Leach, L.D. and Garber, R.H. (1970). Control of *Rhizoctonia*. In: *Rhizoctonia salani: Biology and Pathology*, (Ed.) J.R. Parameter, Jr. Univ. of California Press, Berkeley, 24 pp.

Lee, H.A. (1922). Observations on previously unreported or noteworthy plant diseases in Philippines. *Philipp. Agri. Spec. Public. Coll. Agric.*, Taiwan University, 10: 161–177.

Leuttrell, E.S. (1962). Rhizoctonia blight of lall rescrue grass. *Plant Sci. Rept.*, 46: 661–664.

Lewis, J.A. and Papavizas, G. C. (1987). Effect of mycelia preparation of *Trichoderma* and *Gliocladium* on populations of *Rhizoctonia solani* and the incidence of damping off. *Phytopath.*, 75: 812–817.

Lorenz, R.C. (1948). A new leaf disease of *Hevea* in Peru. *J. Forestry*, 46: 27–30.

Luttrell, E.S. and Garren, K.H. (1952). Blights of snap bean in Georgia. *Phytopath.*, 42: 607–613.

Maiti, I.B., and Kolattukudy, P.E. (1979). Prevention of fungal infectious of plants by specific inhibitions of cuticular. *Sciences*, 205: 507.

Manazali, D., Nipoti, P., Pisi, A., Filippine, G. and Ercole, N.D. (1993). Scanning electron microscopy study of *in vitro* antagonisms of *Trichoderma* spp. Strains *Rhizoctonia solani* Kühn. *Phytopathologia Mediterranea*, 32: 1–6.

Marchetti, R., Nipoti, P, Ercole, N. D. and Guerzoni, M. E. (1992). Competition at atmospheric level as bio-control mechanism in *Trichoderma* spp. *Petria*, 2: 137–147.

Marshall, D. S. (1982). Effect of *Trichoderma harzianum* seed treatment and *Rhizoctonia solani* concentration on damping off of snap bean in acidic soil. *Plant Disease, 66*: 788–789.

Mathew, A. and Gupta, S.K. (1996). Studies on web blight of French bean caused by *Rhizoctonia solani* and its management. *Indian J. Mycol. Pl. Pathol.*, 26(2): 171–177.

Matsumoto, T. (1921). Studies in the physiology of the fungi. XII. Physiological specialization in *Rhizoctonia solani* Kuhn. *Ann. Missouri Botan. Garden*, 8: 1–62.

Matz, J. (1921). Una engermeded danina da habichenla, Est. Exp. Hrs. Rio. Pedras, P.R. Cire. No. 57. San Juan, Puenta Rice, 8 p.

McRoy, R.J. and Kraft, J.M. (1984). Resistance in *Pisum sativum* to epicotyl caused by *Rhizoctonia solani. Plant Dis.*, 68: 491.

Milden hall, J. P. and Williams P.H. (1973). Effect of soil temperature and host maturity on infection of carrot by *Rhizoctonia solani. Phytopathology*, 63:198–262.

Montoya, C.A., Beaver, J.S., Rodriguez, R., Miklas, P.N. and Godaylutz, G. (1997). Heritability of resistance to web blight in fire common bean populations. *Crop Sci.*, 37: 780–783.

Mukherjee, P.K., Mukhopadhyaya, A.N. Sarmah, D.K. and Shresha, S.M. (1995). Comparative antagonistic properties of *Gliocladium virens* and *Trichoderma harzianum* on *Sclerotium rolfsii* and *Rhizoctonia solani* its relevance to understanding the mechanism of bio-control. *J. Phytopath*, 143: 275–279.

Muller, K.O. (1924). Untersuchungen ZUR Entwickelungs geschichte and Biologie Von *Hypochnus salani* P.U.D. *(Rhizoctonia salani* K.). *Arbeit Biol. Reichsanst Land–U. Forstw.* 13:198–262.

Murray, D.I.L. (1982). Penetration of barley root and coleoptile surface by *Rhizoctonia solani. Trans. Bri. Mycol. Soc.*, 79: 354.

Naimuddin, Chaudhary, R. G. and Vishwa Dhar (1998). Role of mixed cropping in the development of web blight of pigeonpea. In: *National Symposium on Management of Biotic and Abiotic Stresses in Pulse Crops*, June 26–28, ISPRD and IIPR, Kanpur, Abstract No.I–B 7, 59 pp.

Naito, S. (1940). Ecological role of basidiospores of *Thanatephorus cucumeris* (Frank) Donk in the incidence of foliage blight of sugarbets in Japan. J.A.R.Q., *Japan Agric. Res. Quarterly*, 23: 268–225.

Naito, S. and Sugimato, T. (1978). Basidiospores infection and lesion development on sugarbeet leaves by *Thanatephorus cucumeris* (Frank) Donk. *Ann. Phytopathol. Soc., Jpn.*, 44: 426.

Neal, D. C. (1942). *Rhizoctonia infection* of cotton and symptoms accompanying the disease in plants beyond the seedling stage. *Phytopath.*, 32: 641–642.

Neergaard, P. (1958). Infection of Danish seed by *Rhizoctonia solani* Kühn. *Plant Dis. Report.*, 42: 1276–1278.

Ogoshi, A. (1975). A Grouping of *Rhizoctonia solani* Kühn and their perfect stages. *Rev. Plant. Prot. Res.*, 8: 93.

Ogoshi, A. (1987). Ecology and Pathogenicity of anastomosis and interaspecific groups of *Rhizoctonia solani. Kûhn. Ann. Rev. Phytopathol.*, 25: 125–143.

Palo, M.A. (1926). Rhizoctonia disease of rice. I.A. study of the disease and of the influence of certain conditions upon the viability of the sclerotial bodies of the causal fungus. *Philippine Agr.*, 15: 361–376.

Parameter, J.R., Jr. Whitney, H.S. and Platt, W. D. (1967). Affinities of some *Rhizoctonia* species that resemble mycelium *Thanatephorus cucumeris. Phytopath.*, 57: 218–223.

Park, D. and Betetus, L.S. (1932). Sclerotial diseases of rice in Celon. I. *Rhizoctonia solani* Kühn. *Ann. Rov. Botan. Garden* (Paradeniya), 11: 319–331.

Patel, B.L. and Bhargava, P.K. (1998). Aerial blight of soybean caused by *Rhizoctonia solani. Indian J. Agric. Sci.*, 68(50): 277–278.

Pitt, D. (1964). Studies on sharp eye spot disease of cereals I. Disease symptoms and pathogen city of isolates of *Rhizoctonia solani* Kühn and the influence of soil factors and temperature on disease development. *Ann. Apple. Biol.*, 54: 77–89.

Ploper, L.D. (1981). La Maustia Hilachosa, Nueva enfermeded en los cultivos de porota (*Phaseolus vulgaris* L.) del noroerste Argentina Revista. *Industrialy Agricola de Tucuman*. 58: 101–111

Rajnauth, G. (1987). Web blight: An important disease of bean and pak-choi in Trinidad. *Trop. Agric.*, 64(4): 356–358.

Ramakrishnan, K. and Ramakrishnan, T.S. (1948). Banded leaf blight of arrow-root (*Moranta arundinacia*). *Indian Phytopathol.*, 129–136.

Ratan, V. and Dwivedi, R.P. (1998). Development of web blight disease of mungbean. In: *National Symposium on Management of Biotic and Abiotic Stresses in Pulses Crops*, Kanpur, June, 26–28. 121–122 pp.

Reddy, M.N. and Rao, A.S. (1981). Disease cycle involved in damping off of groundnut caused by *Rhizoctonia salani. Plant and Soil.*, 62: 465–467.

Rovira, A.D., Ogoshi, A. and Mc Donald, H.J. (1986). Characterization of isolates of *Rhizoctonia solani* from cereal in south Australia and New South Works. *Phytopathol.*, 76: 1245–1248.

Rudakov, O. L. (1970). Mycophyllic properties of the genus *Gliochadium. Mikol i. fitopatol.*, 4: 293–296.

Ruppel, *E.G.*, Baker R., Harmen, G.E., Hubbard, J. P., Hecker, R. J. and Chet, J. (1983). Field test of *Trichoderma harzianum* Rifai aggr. as a biocontrol agent of seedling disease in several crops and Rhizoctonia root rot of sugar beet. *Crop Protection*, 2: 339–408.

Saksena, H.K. (1971). Progress in the study of parasitism by *Rhizoctonia. Indian Phytopath.* 19: 1–15.

Saksena, H.K.and Chaubey, R.D. (1973) Banded blight disease of Paddy in North India. *Int. Rice Res. Conf. Manila*, Phillipines, April, 1973.

Saksena, H.K.and Dwivedi, R.P. (1973). Web blight of blackgram caused by *Thanatephorus cucumris. Indian J. Farm. Sci.*, 1: 58–61.

Sato, K. and Shoji, T.S. (1957). Pathogencity of *Rhizoctonia solani* Kühn isolated from Graminceous weeds in forestry nurseries. *Bull. Forestry. Exp. Sta. Megaro.*, 96: 89–104.

Shailbala and Tripathi, H.S. (2004). Fungicidal management of web blight of urdbean. *Indian Phytopath.* 57(1): 99–100.

Shephard, M.C., and Wood, R.K.S. (1963). The effect of environment and nutrition of pathogen and host in the damping off of seedlings by *Rhizoctona salani. Ann. Apple. Biol.*, 52: 389–402.

Singh, A., Malhotra, S.K.and Singh, A. (1993). Chemical control of web blight of winged bean. *Crop Research, Hisar*, 6(3): 563–565.

Singh, A., Malhotra, S.K. (1994). Host range studies of *Rhizoctonia solani* causing web blight in winged bean. *Bhartiya Krishi Anusandhan Patrika,* 9: 113–116.

Singh, B.M., Shyam, K.R. and Sud, V.K. (1979). Web blight of *Vigna angularis* (wight) Wilczek caused by *R. solani* Kûhn. *Tropical Grain Legume Bulletin,* 16: 14–15.

Singh, D.P., Singh, B.B. and Gupta, R.P. (1989). Evaluation of mungbean germplasm of diseases. *Indian J. Pulses Res.,* 2(10): 10–14.

Singh, Jai, Singh, R.G., Singh, S.K. and Singh, J. (2003). Status of web blight of mungbean in eastern Uttar Pradesh. *Ann. Pl. Protec. Sci.,* 11(1): 173–175.

Singh, S.B. and Saksena, H.K. (1980). A new sheath and leaf blight of bajra. *Indian Phytopath.,* 33: 127–129.

Singh, S.P. and Munoz, C.G. (1999). Resistance to common bacterial blight among *Phaseolus* species and common bean improvement. *Crop Sci.,* 39: 80–89.

Singh, R.K. (2007). Studies on Rhizoctonia blight of mungbean [*Vigna radiata* (*L.*) *Wilczek*]. *Ph.D. Thesis,* N.D.U.A.&T., Kumarganj, Faizabad (U.P.), 89 pp.

Smith, E.M. and Wehner, F.C. (1987). Biological and chemical integrated with deep soil cultivation against crater disease of wheat. *Phytophlactica,* 19: 87–90.

Stan, G.and Rai, C. (1973). *Rhizoctonia solani* Kûhn Ia fasol *Revist ade Horticultura si viticulture,* 22: 80–86.

Stipanovic, R.D. and Howell, C.R. (1982). The structure of Gliovirin, a new antibiotic from *Gliocladium virens. Jr. Antibiotics,* 35: 1326–1330.

Storey, I.F. (1941). A comparative study of strains of *Rhizoctonia solani* Kühn with special reference to their parasitism. *Ann. Apple. Biol., 28*: 219–228.

Strashnov, Y., Elad, Y., Sivan, A., Rudich, Y. and Chet, T. (1985). Control of *Rhizoctonia solani* fruit rot of tomato by *Trichoderma harzianum* Rifai. *Crop Protection,* 4: 356–364.

Stroube, W.H. (1954). *Rhizoctonia* aerial blight, a threat to the Lespedeza production in Lonisiana (Abstr.). *Phytopathol.,* 44: 507.

Sundram, N.V. (1953). Threat blight of ginger. *Indian Phytopathol.,* 6: 80–85.

Takegani, J.C., Beaver, J.S., Godoy-Lutz., G., Badel, Echavez and Steedman, J.R. (2004). Inheritance of web blight resistance in common bean. *J. Agric. Univ.* of *Puerto Rica* 88: 45–54.

Tervet, I.W. (1937). An experimental study of some fungi injurious to seedling flax. *Phytopath.,* 27: 531–546.

Tiwari, Anamika and Tiwari, A. (1998). Incidence of web blight disease on *Kharif* legume crops in Madhya Pradesh a study. *Bhartiya Krishi Anusandhan Patrika,* 13(3–4): 117–121.

Tiwari, T.N., (1993). Studies on web blight of rice bean caused by *Thanatephorus cucumeris* (Fank) Donk. *Ph.D. Thesis,* CSAUA&T, Kanpur, 103 pp.

Townsend, G.R. (1934). Bottom rot of lettuce, New York, (Carnell). *Agr. Expt. Sta. Mem.,* 158: 46.

Tu, J.C. and Vaartija, O. (1981). The effect of hyperparasite (*Gliocladium virens*) on *Rhizoctonia solani* and Rhizoctonia root rot in white beans. *Canadian J. Botany,* 59: 22–27.

Ui, T., Mitsui, Y., Suzui, T., Ogoshi, A. and Ro, T. (1965). Variation in activity of *Rhizoctonia solani* Kühn in sugar beat field soils. *Bull. Sugar Beet Res.*, 4: 101–12.

Upamanyu, S., Gupta, S.K., Shyam, R.R. and Upamanyu, S. (2002). Innovative approaches for (*Rhizoctonia solani*) of French bean. *J. Mycol. Pl. Pathol.*, 32(3): 317–331.

Vanden Ende, G. and Linskens, H.F. (1974). Cutinolytic enzymes in relations to pathogenesis *Ann. Rev. Phytopath.*, 12: 247.

Venkatasubbaiah, P. and Safeeulla, K.M. (1983). Studies on the viability of Sclerotia of *Rhizoctonia solani* in the soil. *J. Coffee Research*, 13: 30–32.

Verma, H.S. and Thapliyal, P.N. (1976). Rhizoctonia aerial blight of soybean. *Indian Phytopath.*, 29: 389–391.

Walker, J.C. (1957). *Plant Pathology, 2nd edn*. McGraw Hill Book Company, New York, 437–447 pp.

Watanable, B. and Matsumoto, A. (1966). Studies on the grouping of *Rhizoctonia salani* Kühn pathogenic to upland crops. *Appointed Experiment. (Plant Disease and Insect Pests)*. 7:10–131.

Webber, G.F. and Absergo, L. (1958). Damping off and thread blight of coffee in El-salvadon. *Plant. Dis. Reptr.*, 43: 1378–1381.

Webber, G.F. (1939). Web blight a disease of beans caused by *Corticium microslcerotia*. *Phytopatho.*, 29: 559–575.

Weindling, R. (1932). *Trichoderma lignorum* as a parasite of other soils fungi. *Phytopath.*, 22: 837–845.

Wells, H.D., Bell, D.K. and Jawski, C.A. (1972). Efficiency of *Trichoderma harzianum* as a biocontrol agent for *Sclerotium rolfsii*, 62: 442–447.

Wilson, C. (1947). A survey of the fungi associated with pegs and seed roots of peanut in Southern Alabama (Aster.). *Phytopath.*, 37: 24.

Yang, Y.B., Snow, J.P. and Berggren, G.T. (1990). Analysis of epidemics of Rhizoctonia aerial blight of soybean in Louisiana. *Phytopathol.*, 80(54): 386–392.

Yokogi, K. (1927). On the Hypochnus disease of soybean and its comparison with that of rice plant. *J. Pl. Prot. Tokyo.*, 14: 12.

Yokoyama, K. and Ogoshi, A. (1984). Studies on hyphal anastomosis of *Rhizoctonia solani*. IV. Observation of imperfect fusion. *Ann. Phytopathol. Soc. Jap.*, 50: 348.

Zhong, J.P., Chen, J.H., Xu, T. and Li, D.B. (1990). Hyperparasitism of *Trichoderma* spp. of *Rhizoctonia solani. Acta Agriculturea University Zhejiangensi*, 16: 73–77

Sharma, J and Tripathi, H.S. (2001). Influence of environmental factors on web blight development in urdbean [*Vigna mungo* (L.) Hepper]. *J. Mycol. Pl. Pathol.*, 31(1): 54–58.

Plant Diseases Management for Sustainable Agriculture *Pages* **235–242**
Editor: **Dr. Shahid Ahamad**
Published by: **DAYA PUBLISHING HOUSE**

Chapter 19

Diseases of Minor Millets and their Control Measures

***S. Kumar*[1]*, P.K. Sharma*[2] *and U.R. Sangle*[2]**

[1]*Department of Plant Pathology, CSK Himachal Pradesh Krishi Vishvavidyalaya, Palampur – 1760*

[2]*ICAR Research Complex for North Eastern Hill Region, Manipur Centre, Imphal*

The term "Millets' refers to any of the small-seeded cereal and forage grasses used for food, feed or forage. In earlier times, sorghum and even maize were included in this category. Synonyms like "miscellaneous cereals" or "coarse grains" are also used to describe the millets. However, with increased use and advances in improvement, first maize and later sorghums received separate status; although in many regions of the world, crop production statistics are still reported as a total for the millets and sorghum. It is likely that other millet species will gain separate status as they begin receiving increased attention in improvement and are more extensively used.

Generalizing on the common characteristics and utilization of the fourteen species of millets and miscellaneous cereals can be somewhat misleading. However, it might seem desirable to justify this categorization. The following statements may be considered applicable to one or more millets at this time:

1. Millets are small-seeded, annual cereal grasses used for food, feed, forage, industrial or other products in tropical or temperate regions. Miscellaneous cereals include species of *Eragrostis, Digitaria* and *Coix* not ordinarily considered millets, but which are somewhat restricted in use and otherwise fit this description.
2. They are frequently short-term, warm-season (summer) crops sometimes used in emergencies following crop failure from some unfavorable weather conditions or other calamities.
3. Individual millet species or varieties frequently possess some unusual characters for adaptation or use, like tolerance or resistance to drought, high temperatures, low soil fertility and diseases or pests; or for making special foods or beverages.

4. The millets have received very little attention as regards their improvement, notwithstanding their relative importance and future potential for increasing food production in both the developing and developed regions of the world.

The term "millets and minor cereals" are considered to be the members of the grass family Gramineae. These are listed below together with the most widely used common and scientific names (Mansfield, 1952).

Table 19.1

Sl.No.	*Common Name*	*Scientific Name*
1.	Ragi or finger millet	*Eleusine coracana* Gaertn
2.	Italian or foxtail millet	*Setaria italica* Beauv
3.	Little millet	*Panicum miliare* Lam
4.	Kodo or ditch millet	*Paspalum scribiculatum*
5.	Jungle rice or shama millet	*Echinochloa colona* (L)
6.	Fonio or hungry rice	*Digitaria iburua* Stapf
7.	Fonio or hungry rice	*Digitaria exilis* Stapf
8.	Adlay or Job's tears	*Coix lachrymal*-jobi Linn
9	Proso or common millet	*Panicum miliaceum*

The tribe Paniceae with seven genera and eleven species includes five of the most important millets–Pearl, Italian, Proso, Japanese and Koda millets in addition to sorghum the "great millet".

Finger Millet

Common Name

Ragi, Nagli, Mandua

Scientific Name

Eleusine coracana

Blast

Symptoms

The disease affects the crop in all stages of growth from the seedling to the time of grain formation. In the nursery, lesions are formed on the leaves. These are spindle-shaped and of varying sizes. The spots are at first grayish green in the centre, with yellowish margins. Later, the central portion turns whitish and gradually disintegrates. Under humid conditions an olive-grey overgrowth of the fungus is seen over the central portions of the spots. This is evident more on the upper surface. The lesions are isolated to begin with, but may soon coalesce, covering extensive areas. The upper nodes of the infected shoots often become blackened as in rice blast. The loss in yield depends on the time of infection, being more in those affected early.

Causal Organism

Piricularia eleusinis

Control Measures

Seed disinfection with organomercurials like ceresin is likely to minimize the incidence of the disease in the nurseries. The latter could also be protected satisfactorily by spraying fungicides. However, use of mercury fungicides on crops is banned. Bordeaux mixture has been successfully used for this purpose. At the time of transplanting, the infected leaves are clipped off and the seedlings dipped in Bordeaux mixture. This treatment is effective in preventing the immediate spread of infection to the transplanted crop. However, it does not prevent the secondary infection caused by air-borne infection at a later stage of the crop. Dusting with ceresan-lime mixture (Hashioka and Ando, 1955) prove useful against this disease.

Blight and Rots

Symptoms

All the parts of the plant–roots, base of the plant, culm, leaf sheath, leaf blade, neck of the panicle and the fingers (spikes) are liable to be infected. The disease occurs in all stages of growth from the seedling to the mature plant. Germinating grains may be killed before the seedling emerges above the soil. After emergence, minute lesions may develop in large numbers on the leaves. These are at first oval and light brown in colour. They may coalesce, forming patches or stripes and turn dark brown in colour. The infected leaves wither prematurely. Seedlings may readily succumb within 15 days of infection. Lesions are common on the leaf blade and leaf sheath of older plants also. Grayish- brown woolly masses of fungal growth develop in the centre of the lesions. The smaller spots coalesce into bigger patches, and heavily infected leaves wither prematurely. The spots on the leaf sheaths are not well defined and are usually located at the junction of the blade and the sheath. The lesions are oblong on the stem. The tissues collapse and the portion above the lesion or point of infection may die. According to Mitra and Mehta (1934 B), the fungus affects the roots and base of the shoot, resulting in the development of root rot and foot rot, and quick death of the infected plants. But Venkatakrishnayya (1939) could not observe any infection of the roots by *H. nodulosum* in Mysore.

Pathogen

Helminthosporium nodulosum

Helminthosporium leucostylum

Control Measures

Use of copper fungicides on this crop can reduce the incidence of the disease.

Wilt or Foot Rot

Symptoms

The affected plants appear pale green and stunted. The infection occurs at the base of the plants, involving the leaf sheath and culm. These soften and turn brown at the place of infection; eventually, the plants wilt, lodge and dry up. White, fan-like mycelia growth is evident between the sheaths or on the stem at the basal region. Later minute, mustard seed like, tan-coloured sclerotia are formed on the surface.

Pathogen

Sclerotium rolfsii

Downy Mildew

Symptoms

The affected plants are stunted with shortened internodes. The leaves arise close together and the plants assume a bushy and bunched appearance. The leaves are often pale green, but are not shredded as in sorghum. The disease is easily recognized after the formation of the ears. Instead of the normal spikes bearing the spikelets, varying degrees of proliferation are exhibited by the spikelets. The lemma and palea, and sometimes the glumes are converted in to leafy structures. The proliferation is first noticed in the basal spikelets, and gradually others also become involved, so that finally the whole ear presents a brush-like appearance (Thirumalachar and Narasimhan, 1949).

Pathogen

Sclerospora macrospora

Smuts

Symptoms

The smut is avaricious and only a few grains in an ear become affected. The infected grains are scattered in the spikes. These are converted into enlarged globose sacs, greenish at first, but later turning dirty black and 5-15 mm in diameter. Lysigenous cavities are formed in the tissues of the hypertrophied structures. Spores are formed in these cavities. At first the matrix is gelatinous, but later the mass becomes pulverulent (Thirumalachar and Mundkur, 1947).

Pathogen

Ustilago eleusinis

Control Measures

Since the infection is air-borne, the disease is difficult to control. Different varieties of *ragi* exhibit differences in their reaction to the smut. It has been observed in Mysore that the varieties, H.2, H.22 and H. 40- the local Hullubele- are susceptible (Narasinmhan, 1934).

Foxtail Millet

Common Name

Kangni, Korra, Rala

Scientific Name

Setaria italica

Smuts

Symptoms

The sori are confined to the flowers and the basal parts of the palea. Every shoot produced by the infected plant and every grain in the ears become smutted, though rarely the terminal portion of the spike may escape. The sori are slightly larger than the grains, of a pale grayish colour, and 2-4 mm in diameter. In later stages, the sori rupture exposing the black spore mass.

Pathogen

Ustilago crameri

Control Measures

In the earlier years, steeping in copper sulphate solution (2 per cent) for 10 to 30 minutes or in formalin solution (0.5 per cent) were being recommended. In Europe the latter method was found to be more satisfactory. But Melchers (1927) observed that formalin had an undesirable effect on the germination of the grain. Hot- water treatment of the seeds at 58°C for six minutes was also reported to be effective.

Downy Mildew or Green Ear

Symptoms

Two types of infection are prevalent. The primary infection which starts with the seedling is systemic, while on older plants secondary local infection occurs. The former causes large-scale chlorosis of the affected plants and the leaves may even turn whitish. The terminal spindle does not unroll but remains yellow or whitish and later turns brown and becomes shredded. The affected plants rarely produce normal ears. Grains are not formed by the affected plants. Secondary infection causes formation of restricted chlorotic lesions on the younger leaves. On these the downy fungal growth may be seen under humid conditions.

Pathogen

Sclerospora graminicola.

Control Measures

Fresh oospores (in dry condition) exposed to a temperature of 77°C for one hour gave 52 per cent infection, while wet spores lost their viability at 50°C for the same duration. Takasugi (1935) reported that hot–water treatment of seeds of 55° C for one hour and steeping in 0.1 per cent mercuric chloride solution proved effective in controlling the disease. Crop rotation is also recommended. But the viability of the oospores persists for several years, so this method may not be helpful. The cultivation of resistant varieties will help to reduce losses significantly. Eradication and destruction of infected plants can be adopted when the incidence is low, as this will lead to the reduction of the potential inoculum.

Rust

Symptoms

Minute, brown urediosori are formed on both sides of the leaves as oblong cinnamon–brown pustules, often arranged in rows. They are produced on the leaf sheaths and culms also. Earlier records indicated that the telia are not common in Japan and North India. But in South India, they are abundant, but more so in the colder months as black sori, long covered by the epidermis. The rust affects the crop in all stages of growth. The greatest extent of damage occurs when the infection commences before the flowering stage. When the infection takes place after the grains are set, no appreciable reduction in yield is noticed. Premature drying of the foliage is caused by early and heavy infection, and the plants may dry up before heading or even the ears that are formed may develop only few grains.

Pathogen

Uromyces setariae-italicae

Control Measures

Some varieties have been found to be immune or highly resistant. The cultivation of such varieties, if suitable to the locality, will help in avoiding the disease. Further their use as parents in hybridization with local varieties may lead to the evolution of resistant varieties.

Proso or Common Millet

Common Name

Chena, Hog- millet, Samai

Scientific Name

Panicum miliaceum.

Head Smut

Symptoms

The entire inflorescence is converted into a spore sac containing the spores and blackened fibres representing the remnants of the vascular bundles. The sorus is long concealed inside the leaf sheath till about harvest time. It is covered by a shining grayish–white membrane which ruptures early and exposes the spore mass. Abnormal development of hair is evident on the leaf sheaths of the diseased plants. According to Buchheim (1935) the smut stimulates the growth the stem and leafy shoots, but suppresses the panicle. The affected plants remain green for a much longer period. In some instances, the panicle is not completely destroyed and a few aborted flowers on tortuous stalks are present.

Pathogen

Sphacelotheca destruens.

Control Measures

Seed treatment with various fungicides has controlled the incidence of the smut. In Russia 5 per cent magnesium arsenite dust is reported to have reduced the incidence of smut from 61.8 per cent in the untreated to 5.4 per cent in the treated plots. According to Richter and Muller, seed dressing with abavit, ceresan or germisan or by steeping seeds in or sprinkling them with the liquid formulations of these fungicides were very effective. Cersan, fusariol, germisan, agrosan G and lunasan have all been found to be effective. Hot water treatment at 55°C for 7-12 minutes has also been recommended. Formaldehyde 1: 300 as a steep for 10-15 minutes or 400 litres of 1 in 100 solution sprinkled over 100 kg of seed as a partly dry treatment prevented the smut infection (Todovora, 1953).

Downy Mildew

Symptoms

Long, yellowish stripes appear on the leaves. Very soon these become conspicuously white on account of the development of conidiophores and conidia. Later the leaves dry up. Shredding of the leaf is prevalent but rare. In many instances, the entire plants dry up before flowering. The ears, when formed, are chaffy. Sometimes infection is partial and confined to some of the shoots of a plant. In such cases normal ears are formed from the healthy shoots. It is apparent that primary and secondary infections occur. The primarily infected plants are completely damaged and do not produce ears. In cases of secondary infection, the lesions are confined to the leaves or, in some, may extend to one or two lateral shoots (Ramakrishanan and Srinivasan, 1952).

Pathogen

Sclerospora graminicola.

Blast

Symptoms

The disease is characterized by formation of typical, spindle shaped brown lesions on the leaves. Blasting of the panicle is evident in varying degrees, depending on the intensity of infection. Sometimes only portions of the panicle fail to develop the grains.

Pathogen

Pyricularia grisea.

Little Millet

Common Name

Kutki, Small millet

Scientific Name

Panicum miliare.

Rust

Symptoms

Numerous minute, narrow, brown or black pustules are arranged in linear rows on the upper surface of the leaf blades.

Pathogen

Uromyces linearis.

Kodo Millet

Common Name

Kodu, Kodra

Scientific Name

Paspalum scrobiculatum.

Head Smut

Symptoms

The entire inflorescence is transformed into a long sorus, 50-75 mm long and 6 mm broad. When young, a cream-coloured membraneous covering envelops the sorus. Some times, it is enclosed in the boot leaf and may not emerge fully. In later stages, the membrane brusts and exposes the black masses of spores.

Pathogen

Sorosporium paspali.

Control Measures

Seed treatment has been found to control the disease effectively. Steeping the seeds in 1.5 per cent solutions of copper sulphate or dusting with copper carbonate were equally effective (Sattar 1940). There was no reduction in the germination of the seeds even if stored for two months after the treatment.

Rust

Symptoms

The uredia are formed on the upper surface of the leaf blade and on the leaf sheath as oval, erumpent, brown sori. Under sheltered conditions, these can be seen throughout the year on the grass-form of the host; and from this it can readily pass on to the cultivated millet. The telia are formed on the under surface of the leaf blade and on the leaf sheath during December-February. These are brown, and remain covered by the epidermis for a long time.

Pathogen

Puccinia substriata

Ergot or Sugary

Symptoms

Individual spikelets are affected. Some times the majority of the spikelets in an ear may be involved. Pearly drops of honey-dew ooze out of the infected spikelets. These may overflow over the lemma and palea, and finally harden into reddish-brown crusts. A mass of fungal growth replaces the ovary. In later stages, as the ears mature, dark- grey sclerotia are present in some of the spikelets.

Pathogen

Claviceps paspali

Control Measures

Destruction of infected ears minimizes the inoculum

References

Buchheim, A. (1935). The influence of smut infection on growth and habit of the host. *Phytopath.*, 8: 615–621.

Hashioka, Y. and Ando, K. (1955). Phytopharmocology of rice diseases. *Hofchen-Briefe Bayer Pflanzenschutz-Nachrichten* (Eng. Ed.), 4: 177–190.

Mansfield, R. (1952). "Zur Systematik and Nomenklatur der Hirsen (Systematics and Nomenclature of the Millets)". *Zuchter*, 22:304–315.

Melchers, L.E. (1927). Studies on the control of the millet smut. *Ibid*, 17: 739–741.

Mitra, M. and Mehta, P.R. (1943B). Diseases of *Eleusine coracana* Gaertn. and *E. aegyptica* Desf. caused by species of *Helminthosporium. Ibid*, 4: 943–975.

Ramakrishnan, T.S and Srinivasan, K.V. (1952). Additions to the fungi of Madras. XIII. *Proc. Indian Acad. Sci.* B., 36: 85.

Sattar, A. (1940). *Sorosporium paspali* McAlp. On *Paspalum scrobiculatun* (Kodra). *Bult. Imp. Agric. Res. Inst. Pusa*, 201: 16.

Takasugi, H. (1935). The relation of environmental factors and the treatment of oospores to the infection by oospores of *Sclerospora graminicola* (Studies in Nipponese Peronosporales. IV). *Ibid*, 2: 459–480.

Thirumalachar, M.J. and Mundkur, B.B. (1947). Morphology and the mode of transmission of the ragi blight. *Phytopathology*, 37: 481–486.

Thirumalachar, M.J. and Narasimhan, M.J.(1949). Downy mildew of *Elusine coracana* and *Iseilema laxum* in Mysore. *Indian Phytopath.*, 2: 47–51.

Todovora, V. (1953). Seed disinfection experiments against smut on millet and Italian millet. *Rev. Inst. Rech. Sci. Minist. Agric. Sofia.* 1: 159–168 (cf. *Rev. Appl. Mycol.* 34: 365).

Venkatakrishnayya, N.S. (1939). Leaf blight and foot rot of ragi. *J. Mysore Agric. Exp. Union*, 15: 125–129.

Plant Diseases Management for Sustainable Agriculture
Editor: **Dr. Shahid Ahamad**
Published by: **DAYA PUBLISHING HOUSE** *Pages* **243–250**

Chapter 20

Bean Anthracnose: A Dreaded Disease of French Bean and its Management

Sachin Gupta, C.S. Kalha, V.K. Razdan and Vishal Gupta
Division of Plant Pathology, FOA-Udheywalla, SKUAST-J, Jammu (J&K)

Anthracnose is one of the most destructive disease of french bean. *Colletotrichum lindemuthianum*, the incitant of anthracnose of french bean was discovered by Lindemuth (1875) and first described by Saccardo (1878). Bean anthracnose and *C. lindemuthianum* have been extensively studied in South America and Europe.

Status of the Disease

Anthracnose is one of the most widespread and economically important fungal disease of beans in all the bean growing regions of the world (Schwartz *et al.*, 1982). In India, the disease has been reported from Himachal Pradesh, Haryana, North-Eastern states and Jammu and Kashmir (Anonymous, 2000). The incidence of disease has been reported varied between 24.59 to 51.72 per cent (Sindhan and Bose, 1981; Sindhan, 1984; Sharma and Sugha, 1995) in different parts of Uttaranchal and Himachal Pradesh. In Tanzania, highly susceptible bean variety T 8 recorded yield loss of 86 per cent while moderately susceptible variety Mexican 142 registered 27 per cent yield loss (Shao and Teri, 1985). Araya *et al.* (1986) studied the incidence of anthracnose on bean under different inoculum pressures and found higher seed transmission and incidence in the rainy season and noticed the rotting of pod directly proportional to the number of infection foci. A significant positive correlation was observed between the percentage of infected pods and seed transmission, but none between seed transmission and incidence on leaves. Schwartz and Pastor-Corrales (1989) stated that bean anthracnose is a worldwide disease in susceptible varieties grown in locations with cool to moderate temperatures and high relative humidity causing cent per cent loss when severely infected seed is planted.

Sharma *et al.* (1994) studied the status and distribution of bean anthracnose in Himachal Pradesh and recorded the disease incidence range from 5 to 65 per cent and observed that disease severity varied according to location, cultivar and cropping pattern. In Kenya, 38-95 per cent crop losses in bean due to anthracnose (*C. lindemuthianum*) have been observed (Makini, 1995). Bashir (1997) reported from Ethiopia that bean anthracnose was widely spread and economically most important disease of Haricot Bean and the severity varied between 17.2 to 76.6 per cent, thus causing 67.2 per cent loss. He also observed in 2003 that anthracnose caused complete yield loss on susceptible cultivars when contaminated seeds are planted. Issa and Ahmed (2003) found that anthracnose was wide spread in different bean growing areas of Jordan. Investigations on yield loss assessment in kidney bean due to anthracnose were made and it was noticed that the disease drastically affected the growth parameters and yield components of kidney bean causing significant reduction in yield with more losses in dwarf type of cultivars than the pole type (Sharma *et al.*, 2005)

Integrated Disease Management

Biological Control

Although research into biological control has advanced considerably in many major plant diseases in recent years, yet it is not true for bean anthracnose. May be, the disease has been effectively brought under control by breeding for disease resistance (Tu, 1992). However, in perusal of literature a few reports on this aspect have been referred.

Lacicowa and Pieta (1985) found that the fungus *Trichoderma konigii* caused the death of *C. lindemuthianum* after three days and it was observed that *Gliocladium catenulatum* and *G. roseum* were mycoparasites of *C. lindemuthianum*. Morshed (1985) demonstrated antagonism between *Trichoderma* spp. and *C. lindemuthianum* and noted that growth of *T. viride* was vigorous in dual culture and it was an effective hyper parasite penetrating and coiling its hyphae around the *C. lindemuthianum* hyphae. Barros *et al.* (1995) tested the technique of paired cultures and cellophane paper to select *Trichoderma* spp. with better antagonistic performance against *C. lindemuthianum* races *in vitro* and concluded that in the pairing test, five *Trichoderma* spp. caused morphological changes in the hyphae of two phytopathogenic races indicating an antagonistic reaction. The cellophane paper test showed that *T. harzianum* and *T. viride* produced a noticeable decrease in colony diameter compared to other species. Wickeremsinghe *et al.* (1999) isolated and identified fungi from mushroom composts and evaluated their biological activity and found that crude aqueous extracts of *Aspergillus fumigatus* and *T. harzianum* showed antifungal activity against *C. lindemuthianum*. Ravi *et al.* (1999) observed that *T. viride* caused maximum inhibition of mycelial growth followed by *Pseudomonas fluorescens* and *T. harzianum* in dual culture technique. The culture filtrate of *T. viride* exerted the maximum inhibition of pathogen spore germination and mycelial growth. Infected *P. vulgaris* seeds soaked in 10 per cent culture filtrate, treated with 0.4 per cent talc formulation of *T. viride* recorded minimum seed infection and maximum seed germination. Adebanjo and Bankole (2004) evaluated the efficacy of some fungal and bacterial isolates in inhibiting the *in vitro* and *in vivo* growth of *C. lindemuthianum* and reported that *A. niger* inhibited the growth of pathogen on contact while *T. viride*–TH 14 and *T. viride*–TH 31 hyper parasited the pathogen. The two isolates of *T. viride* and all tested bacteria significantly reduced seedling infection from anthracnose infected seeds in pot experiments while spray application of *T. viride*–TH 31 suppressed incidence as well as severity of disease. The antagonist was found more effective when applied twice weekly than once a week.

Plant Extracts

Plant extracts have assumed tactical significance in the present day strategy of developing ecologically safe method of plant disease management. Ahmed and Grainge (1982) analyzed 5280 plant species and observed that many of these plants possessed insecticidal, fungicidal, antibacterial and antiviral properties.

Zambonelli *et al.* (1996) tested the essential oils of *Thymus vulgaris, Lavandula* R.C. hybrid and *Mentha piperita in vitro* against *C. lindemuthianum* and found that all the oils inhibited fungal growth with thyme oil most effective due to thymol. Scanning Electron Microscopy observations revealed that the oils caused degeneration of fungal hyphae. Aqueous extracts of *Melia azadichta* and *Trichilia emetica* were found good inhibitor of *C. lindemuthianum* (Lovang and Wildt, 1998). Amadiohe and Obi (1999) reported that hot and cold water leaf extracts of *Cymbopogan citratus* and *Ocimum gratissimum* were effective in reducing the spore germination and radial growth of *C. lindemuthianum in vitro* and the growth of pathogen *in vivo*. Onifade (2000) studied the antifungal effect of *Azadirachta indica* extracts on *C. lindemuthianum* and found that extract was directly inhibitory to the pathogen *in vitro*. The percentage inhibition was distributed in a decreasing order of potency among the seed, fruit, leaf, root and bark of *A. indica* and water extract was found superior to organic solvent extracts in reducing fungal spore germination, mycelial growth and dry weight while among organic solvents, methanol extract was rated best. Amadiohe (2003) evaluated water and alcohol extracts of *Piper nigrum, Ocimum sanctum* and *Citrus limon* against *C. lindemuthianum* and concluded that these plants could act as source of pesticides (plant origin) to control this pathogen in the field.

Chemical Management

Several attempts have been made to manage bean anthracnose by using fungicides. Due to seed borne nature of the fungus, seed treatment with systemic fungicides has been found to be quite effective.

Halsted (1895) and Whetzel (1908) employed bordeaux mixture for bean anthracnose control and found that this mixture had little effect on mycelium within the seeds. Bedford (1900) and Sevey (1918) used different concentrations of formalin and reported reduction in amount of disease without any appreciable effect on germination. Muncie (1917) concluded that any chemical that would destroy the organism would also reduce the germination considerably while Kreuzpointer (1922) found that steeping in uspulum controlled anthracnose causing the yield to increase five folds (Scheinpflug, 1924., Bondar, 1924). Effective control of disease, stimulation of seed germination and good seedling vigour was found while treating the seeds for half an hour in 0.1 per cent solution of ceresan (Steinberg, 1934). Hadorn (1944) found that none of the disinfectants tested by him gave a complete control of anthracnose. Ceresan dust (0.5g/100g seed) eliminated the pathogen only when the seed coat was involved, but not when hyphae penetrated the cotyledons or much deeper. Sweep (1945) obtained good control of bean anthracnose by immersing the bean seeds in 0.25 per cent ceresan at 20°C for 10 minutes. Voelcker (1953) and Johnston (1953) practiced bordeaux mixture as foliar spray (3-4 sprays) to prevent leaf and pod infection.

Bruinsma and Labruyere (1953) recommended four to six applications of dithane Z-78 per season for the effective control of disease. Bremer (1955) reported promising results by treating seeds with zineb and captan and recommended cerenox together with ceresan to control the anthracnose of beans. Application of zineb (0.5 per cent), copper oxychloride (0.5 per cent) shortly before flowering as foliar spray provided better protection against *C. lindemuthianum* infection (Anderson and Olofsson, 1959). Gabryl (1961) found that fungitox-T as a seed treatment and in combination with four sprays of zineb (0.25 per cent) yielded best results and application of brestan (2.5 Kg ha^{-1}) provided better protection against anthracnose of bean than orthophalton (3 Kg ha^{-1}). Acimovic (1964) reported that thiram and ceretam as seed dressing provided significant control of anthracnose. Bernat and Raicu (1974) managed bean anthracnose with thiram and found cryptodin as effective fungicide for controlling the disease and benlate and thiram were found highly toxic to mycelial growth, sporulation and conidial germination of *C. lindemuthianum*. Navarro and Puserta (1977) reported that benlate and difoltan 80 significantly reduced bean anthracnose on pods and maximum yields were obtained with benlate at 0.5 g l^{-1} sprayed every eight or fifteen days. Benomyl and thiobendazole have been found to be highly effective in inhibiting spore germination of *C. lindemuthianum* (Tu and Jarvis, 1979). Navarro *et al.* (1981) recorded control of bean anthracnose by applying four sprays of benomyl and captafol in rotation or with mancozeb at 15 day interval after 30 days of sowing. Rolim *et al.* (1981) found that chlorothalonil and captafol were effective against bean anthracnose followed by thiobendazole and chlorothalonil + copper oxychloride. Benlate followed by bavistin, ziram, vitavax, ferbam and lime sulphur as foliar spray were found effective in reducing bean anthracnose disease incidence and increased seed yield (Sindhan and Bose, 1981). They further reported that benlate followed by bavistin, vitavax, ziram and agrosan GN applied as seed dressers were found effective in increasing seed germination and yield, and decreasing disease incidence.

Issa (1983) recorded the best control of bean anthracnose with four sprays of maneb in the rainy season at 10 day interval. Use of thiram as seed dressing fungicide resulted in significant control of bean anthracnose followed by benomyl, bitertanol, captafol and carboxin (Robert *et al.*, 1985). Sharma and Sugha (1992) found that seed treatment with bavistin followed by two or three sprays of dithane M-45 reduced the bean anthracnose at different growth stages of crop and significantly increased the yield of beans.

Hidalgo and Araya (1993) observed that application of fungicides at three and six week intervals gave the best disease control and yield and stages R_5 and R_7 (before flowering and during pod formation, respectively) were the most suitable for treatment. Prakasam (1994) obtained significant control of bean anthracnose with three sprays of carbendazim (0.1 per cent) starting after 45 days of sowing at 15 days interval. Efficacy of various fungicides for the control of bean anthracnose was tested and found that thiophanate-methyl, chlorothalonil, fentin triphenyl hydroxide, maneb, benomyl, prochloraz and propiconazole controlled disease and correspondingly increased yield (Balardin and Balardin, 1994). Morshed (1995) observed that carbendazim, thiram and thiophanate-methyl were effective against *C. lindemuthianum* and also these fungicides were compatible with *Rhizobium phaseoli*. Sharma and Sugha (1995) recommended seed treatment with carbendazim and carbendazim + thiram and found that seed treatment followed by two or three foliar sprays of mancozeb (0.25 per cent) after 45, 60 and 75 days of sowing effectively manage the disease resulting in manifold increased in yield of dry beans.

Ghimire *et al.* (1995) developed a strategic spray schedule with mancozeb (Indofil M-45) for anthracnose and found that maximum marginal return was achieved when chemical spray was started from the expansion of true leaf until 4th harvest at an interval of 20 days. Benomyl (0.1 per cent) managed 90 per cent of anthracnose when sprayed two days after inoculation with the test fungi (Canteri *et al.*, 1998). Rava *et al.* (1998) tested *in vitro* efficiency of fungicides against *C. lindemuthianum* and reported that fluazinam gave the greatest mycelial growth reduction followed by benomyl while chlorothalonil was most efficient for inhibiting spore germination followed by fluzinam. Barros *et al.* (1999) compared the efficiency of fungicides for control of anthracnose under field conditions and reported that disease was controlled by tebuconazole + fentin hydroxide, azoxystrobin and fentin hydroxide. Mahanta *et al.* (2000) concluded that seed treatment with bavistin was most effective in decreasing per cent disease incidence and increasing the yield. Gupta *et al.* (2000) in their field experiments reported that a combination of seed treatment with tebuconazole (0.1 per cent) and foliar spray of tricyclazole (0.03 per cent) was most effective in reducing the severity on leaves as well as pods followed by phenyl pyrrole (0.1 per cent) and hexaconazole (0.05 per cent).

Integrated Management

Integration of all the available practices of disease management can be helpful in devising an Integrated Management Strategy for the control of disease. Ravi *et al.* (2000) evaluated various leaf extracts, bioagents and fungicides under pot culture and field conditions against french bean anthracnose (*C. lindemuthianum*) and showed that seed treatment followed by foliar sprays of different leaf extracts, biocontrol agents and fungicides were more effective than the seed treatment alone. The botanicals tested proved to be better than *T. viride*, *P. fluorescens*, carbendazim and mancozeb.

References

Acimovic, M. (1964). Causes of anthracnose of bean in Yugoslavia and susceptibility of domestic bean varieties to this disease. *Zast. Billia*, 15: 229–261.

Adebanjo, A. and Bankole, S. (2004). Evaluation of some fungi and bacteria for biocontrol of anthracnose disease of cowpea. *Journal of General Microbiology*, 44: 3–9.

Ahmed, N. and Grainge, M. (1982). Some promising plants for pest control under small scale farming operation in developing countries. *Working Paper*. Resource Systems Institute, East West Centre, Honolulu, U.S.A. pp 2–6.

Amadiohe, A.C. (2003). Evaluation of some plant leaf extracts against *Colletotrichum lindemuthianum* in cowpea. *Acta-Phytopathologica-et-Entomologica-Hungarica*, 38: 259–265.

Amadiohe, A.C. and Obi, V. I. (1999). Control of anthracnose disease of cowpea by *Cymbopogan citrates* and *Ocimum gratissimum*. *Acta-Phytopathologica-et-Entomologica-Hungarica*, 34: 85–89.

Andersen, F. and Olofsson, B. (1959). Control experiments against bean anthracnose. *Vointskyddsnotiser Stockly*, 23: 3–4.

Anonymous. (2000). *Handbook of Horticulture*. I.C.A.R. Publication, New Delhi, pp. 399–402.

Araya, C.M., Dhingra, O.D. and Kushalappa, A.C. (1986). Incidence of bean anthracnose on bean under different inoculum pressures. *Fitopathologia-Brasileira*, 11: 795–801.

Balardin, R.S. and Balardin, C.R.R. (1994). Fungicide efficiency for *Colletotrichum lindemuthianum* control in common bean crop. *Ciencia-Rural*, 24: 19–23.

Barros, B.C., Castro, J.L. and Patrico, F.R.A. (1999). Comparison of fungicides on control of bean diseases. *Arquivos do Insttituto Biologico Sao Paulo*, 66: 53–58.

Barros, S.T., Oliveira, N.T. and Bastos, S.T.G. (1995). *Trichoderma* spp. in the biological control of *Colletotrichum lindemuthianum*, agent of the bean anthracnose. *Boletin Micologico*, 10: 5–11.

Bashir, T. (1997). Loss assessment study on haricot bean due to anthracnose. *Pest Management Journal of Ethiopia*, 1: 69–72.

Bedford, S.A. (1900). The vegetable garden: Formalin as a germicide for bean anthracnose. *Minnesota Agriculture Experimental Farms Report*, 1899: 323–325.

Bernat, N. and Raicu, C. (1974). The control of bacteriosis and anthracnose in bean crops. *Horticultura*, 23: 36–39.

Bondar, J. (1924). Disinfection experiments with uspulum (Novenyve delem). *Fortschr der landri*, 17: 550.

Bremer, H. (1955). Fungicidal control of bean anthracnose. Nachral dtsch Pfisch Disnt. (Braunschw) *Stuttgart* 7: 129.

Bruinsma, F. and Labruyere, R.B. (1953). Control of anthracnose in seed beans (*Colletotrichum lindemuthianum*). *Meded Dir. Tuinb*, 16: 243–252.

Canteri, M.G., Dalla, P.M., Amorim, L. and Bergamin, F.A. (1998). Curative effect of fungicides to control anthracnose and angular leaf spot on bean in growth chamber. *Fitopatologia Brasileira*, 23: 471–473.

Gabryl, J. (1961). Effect of seed treatment and plant spraying on the control of bean anthracnose. *Biul. Warzywn. Inst. Uprawy. Nawoz. Glebozn*, 5: 335–342.

Ghimire, S.R., Pradhanang, P.M. and Lohar, D.P. (1995). Use of fungicides and identification of resistant varieties for the management of bean rust and anthracnose diseases in common bean. *Working Paper Lumle Regional Agricultural Research Centre*, 95: 42.

Gupta, S.K., Mathew, K.A., Shyam, K.R. and Gupta, D.K. (2000). Role of fungicides in the management of anthracnose of french bean. In: *Proceedings of Indian Phytopathological Society, Golden Jubilee. International Conference on Integrated Plant Disease Management for Sustainable Agriculture*, 11–15 November 1997, pp. 705–707.

Hadorn, C. (1944). Bean diseases and control experiments with seed disinfectants. *Forsch Ergebn Gartenb*, 4: 3–33.

Halsted, B.D. (1895). Field experiment with fungicides on beans. *New Zealand Agricultural Experimental Station Bulletin*, 108: 24–30.

Hidalgo, R. and Araya, C.M. (1993). Optimum growth stage of common bean for chemical control of anthracnose (*Colletotrichum lindemuthianum*) and angular leaf spot (*Isariopsis griseola*) in San Carlos, Costa Rica. *Agranomia Costarricense*, 17: 75–80.

Issa, E. (1983). Effect of time of fungicide application on the control of bean (*Phaseolus vulgaris*) diseases. *Biologico*, 49: 69–73.

Issa, Q.S. and Ahmed, M.A. (2003). The occurrence and seed transmission of bean anthracnose caused by *Colletotrichum lindemuthianum* in Jordan. *Dirasat Agriculture Science*, 30: 205–210.

Johnston, A. (1953). The control by spraying of some diseases of high land vegetables. *Malay Agriculture Journal*, 36: 28–35.

Kreuzpointer, J. (1922). Notes on seed steeping. *Zbl. Agrikchem*, 8: 208.

Lacicowa, B. and Pieta, D. (1985). Fungi of the genera *Trichoderma* and *Gliocladium* as mycoparasites of some bean pathogens. *Ochrona-Roslin*, 29: 16–17.

Lindemuth, J. (1875). Danish fungi as represented in the herbarium of E Rorturp. p. 648.

Lovang, U. and Wildt, P.T. (1998). The effect of aqueous extracts of *Melia azadirachta* and *Trichilia emetica* on selected pathogens of tomato, bean and maize. In: *Botanical Pesticides*. International Office, Swedish University of Agricultural Science, p. 23.

Mahanta, I.C., Dhal, A., Mohanty, A.K. and Mohapatra, S.S. (2000). Integrated management of diseases of kidney bean or rajmash (*Phaseolus vulgaris* L.) in North Central plateau zone of Orissa. *Orissa Journal of Horticulture*, 28: 25–28.

Makini, F.W. (1995). Bean production and constraints in Kenya with emphasis on diseases. In: *Proceedings of a Regional Workshop for Eastern, Central and Southern Africa*. Njoro, Kenya, October 2–6, pp. 104–109.

Morshed, M.S. (1985). *In vitro* antagonism of different species of *Trichoderma* on some seed borne fungi of bean. *Bangladesh Journal of Botany*, 14: 119–126.

Morshed, M.S. (1995). Effect of fungicides on seed borne fungi and nodule formation of bean (*Phaseolus vulgaris*). *Bangladesh Journal of Plant Pathology*, 11: 39–40.

Muncie, J.H. (1917). Experiments on the control of bean anthracnose and bean blight. *Michigan Agriculture Experiment Station Technical Bulletin*, 38: 50.

Navarro, A.R., Puerta, E.O.D. and Isaza, L. (1981). Alternate use of fungicides to control bean anthracnose (*Colletotrichum lindemuthianum*) on bean (*Phaseolus vulgaris*). *Fitopathologia Colombiana*, 10: 3–6.

Navvaro, A.R. and Puserta, E.O.D. (1977). Effect of fungicides in controlling anthracnose of bean under conditions in Eastern Autioquia. *Fitopatologia-Colombiana*, 6: 87–92.

Onifade, A.K. (2000). Antifungal effect of *Azadirachta indica* extracts on *C. lindemuthianum*. *Global Journal of Pure and Applied Science*, 6: 425–428.

Prakasam, V. (1994). Management of anthracnose disease (*Colletotrichum lindemuthianum*) in french bean (*Phaseolus vulgaris* L.). In: *National Symposium on Ultrastructure, Cytology, Sexuality and Variability in Plant Pathogens*, T.N.A.U., Coimbatore, 27–29 January, p. 47.

Rava, C.A., Sartorato, A. and Botelho, S.A. (1998, . *In vitro* and *in vivo* efficiency of fungicides for the control of *Colletotrichum lindemuthianum*. *Summa Phytopathologica*, 24: 45–48.

Ravi, S., Sabitha, D., Valluvaparidasan, V., Jayalakshmi, C. and Doraiswamy, S. (1999, . Effect of biocontrol agents on seed borne *Colletotrichum lindemuthianum* in french bean. *Plant Disease Research*, 14: 146–151.

Ravi, S., Valluvaparidasan, V., Sabitha, D. and Jayalakshmi, C. (2000, . Production of *Colletotrichum lindemuthianum* free french bean seeds. *Legume Research*, 23: 170–173.

Robert, R., Flori, P. and Giordani, G. (1985, . Bean seed treatment against *Colletotrichum lindemuthianum* (Sacc. and Magn.) *Difesa delle Piante*, 8: 3–13.

Rolim, P.R.R., Briganani, N.F., Roston, D.J. and Oliveira, D.A. (1981). Chemical control of bean (*Phaseolus vulgaris*) anthracnose on rainy season bean. *Biologico*, 47: 245–248.

Saccardo, P.A. (1878, . Fungi veneti novi vel, critici, vel mycolgise venetae addandi, *Isariopsis griseola* Sacc. *Michelia*, 1: 273.

Scheinpflug, E. (1924). Results and experiments with the seed steep uspulum. *Dt. Obst. Gemuseb Ztg.*, 16: 177–178.

Schwartz, H.F. and Pastor-Corrales, M.A. (1989). *Bean Production Problems in the Tropics, 2nd edn*. CIAT, Columbia, pp. 77–104.

Schwartz, H.F., Pastor-Corrales, M.A. and Singh, S.P. (1982). New sources of resistance to anthracnose and angular leaf spot of beans (*Phaseolus vulgaris* L.). *Euphytica*, 31: 741–754.

Sevey, G.C. (1918). Insects and pests with remedies. *His Bean Culture*, pp. 55–70.

Shao, F.M. and Teri, J.M. (1985). Yield losses in *Phaseolus* beans induced by anthracnose in Tanzania. *Tropical Pest Management*, 31: 60–62.

Sharma, P.N. and Sugha, S.K. (1992). Management of bean anthracnose (*Colletotrichum lindemuthianum*) through chemicals. In: *Symposium on Disease Management and Plant Health*, University of Horticulture and Forestry, Solan, 25–26 September, p. 15.

Sharma, P.N. and Sugha, S.K. (1995). Management of bean anthracnose through chemicals. *Indian Phytopathology*, 48: 304–307.

Sharma, P.N., Patial, S., Sharma, O.P. and Kapil, R. (2005). Yield loss assessment in kidney bean due to bean anthracnose (*Colletotrichum lindemuthianum*) under sub-temperate conditions of Himachal Pradesh. *Indian Journal of Mycology and Plant Pathology*, 58: 373.

Sharma, P.N., Sharma, O.P. and Tyagi, P.D. (1994). Status and distribution of bean anthracnose in Himachal Pradesh. *Himachal Journal of Agriculture Research*, 20: 91–96.

Sindhan, G.S. (1984). Evaluation of fungicides against occurrence of angular leaf spot of french bean caused by *Phaseriopsis griseola*. *Indian Journal of Plant Pathology*, 2: 39–42.

Sindhan, G.S. and Bose, S.K. (1981). Evaluation of fungicides against anthracnose of french bean caused by *Colletotrichum lindemuthianum*. *Indian Phytopathology*, 34: 325–329.

Steinberg, J. (1934). Fruhere und hohere bohnenerntess durch beizung. *Uber Schadlings Bekampf*, 9: 82–85.

Sweep, J. (1945). Ascochyta spot disease of bean (*Phaseolus*). *Tijdschr. Pl. Zlekt*, 1: 1–16.

Tu, J.C. (1992). Bean anthracnose. In: *Plant Diseases of International Importance*, Vol. 2, (Eds.) Chaube, H.S., Kumar, J., Singh, U.S. and A.N. Mukhopadhyay. Prentice Hall International, Eaglewood Cliffs, New Jersey, pp. 1–17.

Tu, J.C. and Jarvis, W.R. (1979). Response of *Colletotrichum lindemuthianum* to benomyl. *Canadian Journal of Plant Pathology*, 1: 12–16.

Voelcker, C.J. (1953). *Report of the Department of Agriculture*, Malaya, pp. 71.

Whetzel, H.H. (1908). Bean anthracnose. *Agricultural Experimental Station Bulletin*, New York (Cornell) 255: 428.

Wickeremsinghe, R., Abeywickrama, K. and Abeytunga, D.T.U. (1999). Isolation and identification of fungi from mushroom composts and evaluation of their biological activity. *Journal of National Science Foundation of Sri Lanka*, 27: 29–40.

Zambonelli, A., Zechini, D., Bianchi, A. and Albasini, A. (1996). Effect of essential oils on phytopathogenic fungi *in vitro*. *Journal of Phytopathology*, 144: 491–494.

Plant Diseases Management for Sustainable Agriculture *Pages* **251–254**
Editor: **Dr. Shahid Ahamad**
Published by: **DAYA PUBLISHING HOUSE**

Chapter 21

Epidemiology and Management of Alternaria blight of Rapeseed-Mustard

Brajesh Kumar[1] ***and Shahid Ahamad***[2]
[1]*S.D.J. Post Graduate College, Chandeshwar, Azamgarh, U.P.*
[2]*RARS, SKUAST–J, Rajouri – 185 131*

Rapeseed (*Brassica campestris*) and mustard (*Brassica juncea*) are the major components of *Brassica* group. They contribute about 38 per cent towards national production. Presently, the experimental average potential yield of most improved varieties ranges from 20 to 30 q ha^{-1}. But the national average yield is reported to be less than 10 q ha^{-1}. Obviously, the low yield of these crops is due to several biotic and abiotic stresses. Among biotic stresses, Alternaria blight disease occurs as endemic in all rapeseed–mustard growing states of the country and is ranked first in adversely affecting production of these crops.

It has been reported that this disease causes an average yield loss of 35 to 40 per cent in Indian mustard and up to 70 per cent in yellow *sarson* (Kolte,1985b; Kolte *et al.,.,*1987). The disease is also reported to be of major economic importance in Australia,France and Poland and of moderately importance in Canada and United Kingdom (Sharma and Kolte,1984; Verma and Saharan,1994).In addition to direct yield loss, the disease also adversely affects the seed quality by reducing seed size, oil content and seed discolouration. It has been found that losses in seed oil content of infected plants ranges from 14.5 to 35.97 per cent (Ansari *et al.*,1988).

Biology, Epidemiology and Disease Cycle

Alternaria blight disease of rapeseed–mustard is incited by the fungus, *Alternaria brassicae* (Berk.) Sacc.There are three other species *viz. A. brassicicola* (Schw) Wilts, *A. alternata* (Fr.) Keissler and *A. raphani* (Groves and skolko) which are also reported to be associated with affected plants under field conditions in India (Kolte,1985a).

The incitant *Alternaria brassicae* is an imperfect fungus that belongs to Division–Eumycota, Sub-division–Deuteromycotina, Class–Hyphomycetes, Order–Moniliales, Family–Dematiaceae.

Table 21.1: Brief Biology of *A. brassicae* and *A. brassicicola*

A. brassicae	*A. brassicicola*
Mycelium- Brown to white	Brown to white
Conidiophores–Arise in form of fascicles	Branched
Conidia–Brown, more septation than *A. brassicicola,* long beak, muriform, single or in chain of 2-3 conidia	Dark brown, few septation, muriform, without beak, long chain of 8-10 conidia
Chlamydospores–formed less frequently	Not known

The existence of variability in *A. brassicae* was established by Kolte *et al.* (2001). They identified three distinct isolates *viz.* A, C and D. Later, Vishwanath *et al.* (1997a) reported the isolate 'A' as virulent, 'C' moderately virulent and 'D' avirulent based on true leaf inoculation. Biochemical variation among three isolates was reported by Vishwanath (1997b).

The onset of disease appears on lower most leaves of the plant in the form of numerous black brown spots of pin head size after 30 to 40 days after sowing. These spots gradually enlarge with growth of plants. The spots produced by *A. brassicae* are small, round and brown in colour with concentric rings as compared with large, round and shooty velvety with concentric rings produced by *A. brassicicola.* The elongated lesions without concentric rings also developed on stems and pods of affected plants. Seeds from severely affected plants become discoloured, shrivelled and may get low germination ability.

The inoculum of *Alternaria* spp. overwinters as mycelium, chlamydospores/microsclerotia in infected plant parts in the field the seeds. If the pathogen (*A.brassicae*) is carried with seeds, it gets eliminated during storage from April to September in Northern India because of prevailing high temperature (35-40°C) as reported and confirmed by several workers (Chahal, 1981; Kolte, 1985b). Thus, the seed has no effective role in initiation of primary infection of *A. brassicae* unless it is stored in cold storage or in region of temperate climate. The primary infection arises from either chlamydospores/ microsclerotia present in diseased stubbles or from neighbouring volunteer crucifer plants. The formation and survival of chlamydospores/microsclerotia in plant debris has been reported by Kolte (1985a).

Secondary infection takes place through secondary inoculum (wind–borne conidia) and infection cycle is repeated several times during favourable weather conditions. As conidia of *A. brassicae* come in contact with leaves of plant, they germinate and the germtube penetrates susceptible tissues through stomata. Contrary to this, *A. brassicicola* enters the plants directly or through stomata. Conidial germination of *A. brassicae* is optimal at 17-19°C in both light and dark conditions. The disease develops best during rosette to flowering stages of the crop. Increase in age beyond 30 days is positively correlated with severity of the disease (Awasthi and Kolte, 1994). The maximum disease development was recorded during 62 to 72 DAS on leaves and from 100 to 110 DAS on pods (Kumar and Kolte, 2006). Relative humidity 67-73 per cent, rainfall 770mm, sunshine 5-7h per day and minimum temperature of 7–10 °C concomitant with maximum temperature of 20-30 °C have been found positively correlated with severe occurrence of the disease.

Disease Management

An unified programme of disease management practices can minimize crop losses by the Alternaria blight disease.

Field Sanitation

Field sanitation is an integral part of cultural practices. Since the pathogen is soil borne, field sanitation by removing the plant debris and eradicating the collateral hosts (*e.g.*, hedge mustard) are essential for effective check of the disease.

Integrated Use of Plant Nutrients

It has been found that plant nutrients have tremendous effects on Alternaria blight development. Experiments have demonstrated that potassium imparts very high level of resistance in rapeseed–mustard against Alternaria blight disease and therefore, potash should be essentially included in fertilizing the crop at recommended doses (Sharma and Kolte,1994 a).

Use of Disease Resistant/Tolerant Varieties

Growing of disease resistant/tolerant varieties is an ecofriendly and economical approach for management of plant diseases. However, presently most commercial varieties are susceptible to this disease, though,there is certain degree of resistance in some varieties. For this purpose, varieties,such as Kranti, PR 8988 and PR 9024 could be included in effective management strategy.

Early Sowing of Crop

Early sowing of rapeseed-mustard has been found very useful and it enables the crop to escape from Alternaria blight disease. The sowing of the crop must be done at the first week of October to prevent from the disease occurrence.

Fungicidal Application

At present time, disease management strategy is mainly dependent upon use of fungicidal application. Just one or two sprayings of Iprodion @ 0.2 per cent at pod formation and at development stages of Alternaria blight have greater potential as compared to commonly used Dithiocarbamate in disease control. However, in terms of cost/benefit ratio and availability of fungicides in the market, Dithiocarbamates such as mancozeb, zineb and copper oxychloride fungicides presently offer the best possibility in controlling Alternaria blight.

Induction of Resistance

Recent research reports indicate that there is a strong possibility of Alternaria blight control by 'systemic acquired resistance' (SAR) through some selective chemicals such as salicylic acid or jasmonic acid and 2,6-dichloro-isonicotinic acid. Resistance in mustard against Alternaria black spot disease can also be induced by using an avirulent *A. brassicae* isolate-D, as demonstrated by Vishwanath *et al.* (1999).

References

Ansari, N.A., Khan, M.W. and Muheet, A. (1988). Effect of Alternaria blight on oil content of rapeseed and mustard. *Curr. Sci.*, 57: 1023–1024.

Awasthi, R.P. and Kolte, S.J. (1994). Epidemiological factors in relation to development and prediction of Alternaria blight of rapeseed-mustard. *Indian Phytopathol.*, 47: 395–399.

Chahal, A.S. (1981). Seed borne infection of *Alternaria brassicae* in Indian mustard and its elimination during storage. *Curr. Sci.*, 50: 621–623.

Kolte, S.J. (1985a). *Diseases of Annual Edible Oilseed Crops, Vol 2: Rapeseed-mustard and Sesame Diseases.* CRC Press, Boca Raton, Florida, pp. 135.

Kolte, S.J. (1985b). Disease management strategies for rapeseed-mustard crops in India. *Agric. Rev.*, 6: 81–88.

Kolte, S.J., Bardoloi, D.K. and Awasthi, R.P. (1991). The search for resistance to major diseases of rapeseed-mustard in India. In: *Proc. GCIRC 8th. Int. Rapeseed Congr.*, July 9–11, Saskatoon, Canada.p. 60 (Abstr.), Proc. Vol. 1: 219–225.

Kumar, Brajesh and Kolte, S.J. (2006). Development of Alternaria blight in genotypes of Indian mustard [*Brassica juncea* (L.) Czern.& Coss.] under field conditions. *Indian Phytopath.*, 59:314–317.

Sharma, A.K., Gupta, J.S. and Maheshwari, R.K. (1984). The relationship of *Streptomyces arabicus* to *Alternaria brassicae* (Berk.) Sacc. and *Alternaria brassicicola* (Schew.) Wiltshire on the leaf surface of yellow *sarson* and Taramira. *Geobios*, 3: 83–84.

Kolte, S.J. (1987). Important diseases of rapeseed and mustard in India. Present research progress and future research needs. In: *Proc. IDRC, Canada, 3rd Oil Crops Network Workshop*, held in Addis Abada, Ethiopia, Oct. 6–10, 1986, p. 91–106.

Sharma, S.R. and Kolte, S.J. (1994a). Effect of soil applied NPK fertilizers on severity of black spots disease caused by *Alternaria brassicae* (Berk.) Sacc. and yield of oilseed rape. *Plant and Soil*, 167: 313–320.

Verma, P.R. and Saharan, G.S.(1994). *Monograph of Alternaria Diseases of Crucifers.* Research Branch, Agriculture and Agriculture Food, Canada, Saskatoon Research Centre, p. 162.

Vishwanath and Kolte, S.J. (1997a).Variability in *Alternaria brassicae*. Response to host genotypes, toxin production and fungicides. *Indian Phytopath.*, 50:373–381.

Vishwanath and Kolte, S.J.(1997b). Biochemical variation among three isolates of *A. brassicae. Indian Phytopath.*, 50: 437–438.

Vishwanath, Kolte, S.J., Singh, M.P. and Awasthi, R.P. (1999). Induction of resistance in mustard (*Brassica juncea*) against Alternaria black spot with in avirulent *Alternaria brassicae* isolate-D. *European J. Pl. Path.*, 105: 217–220.

Plant Diseases Management for Sustainable Agriculture *Pages* 255–279
Editor: **Dr. Shahid Ahamad**
Published by: **DAYA PUBLISHING HOUSE**

Chapter 22
Fungi as Potential Biocontrol Agents of Phytopathogens

C.R. Raghavender and B.N. Reddy*

Mycology and Plant Pathology Laboratory, Department of Botany, Osmania University, Hyderabad – 500 007

Introduction

The success of affluent countries in meeting the increasing food demands of their populace has largely been achieved by increased agricultural production and the control of losses during storage and distribution. New chemicals and new technologies have played an indispensable role in achieving this agronomic 'miracle' (Walker, 1987) and this is reflected in the global figures for the agrochemical market which amounted to some $ 17.4 billon. Of this figure, 44 per cent of the market share was for herbicides, 31 per cent for insecticides, 19 per cent fungicides and 6 per cent for plant growth regulators (Wood MacKenzie, 1987). Over the last few decades, as the demand for food has risen, the global market for pesticides has also steadily increased. This market expansion shows a sign of slowing as the developed world's population has stabilized and the agrochemical products discovered have satisfied most major niches in the market (Powell and Faull, 1989).

Agricultural practice is changing as a result of demands to reduce the use of chemical pesticides, including fungicides, and to provide abundant feed, food and ûbre using environmental friendly, sustainable systems. Much of this change has been initiated and mandated or encouraged by national and international legislatures. The potential roles of biological methods in these evolving practices of pest control *i.e.* plant pathogens, weeds and insects (pests) are detrimental to agricultural production. (Whipps and Lumsden, 2001).

* *Corresponding author*: E-mail: reddybn1@yahoo.com.

Threat of Chemical Control

Currently the only reliable and relatively cheap solution to problems of pests and diseases in large-scale agriculture is chemical treatment. Environmental and organic foods groups have lobbied for change but there is little alternative to chemical treatment if large-scale agriculture is to continue to provide stable food supplies in the future. Agriculture, therefore, has a problem. High yields depend upon effective use of agrochemicals but the costs of registration of new products are becoming increasingly expensive (Powell and Faull, 1989). Moreover, the number of pests, diseases and weeds that can be targeted by chemicals is limited by their potential market size, and a list of those which the agrochemical industry considers as justifiable targets can be found (Jutsum, 1988).

Different approaches may be used to prevent, mitigate or control plant diseases. Beyond good agronomic and horticultural practices, growers often rely heavily on chemical fertilizers and pesticides. Such inputs to agriculture have contributed significantly to the spectacular improvements in crop productivity and quality over the past 100 years. However, the environmental pollution caused by excessive use and misuse of agrochemicals, as well as fear-mongering by some opponents of pesticides, has led to considerable changes in peoples attitudes towards the use of pesticides in agriculture. Today, there are strict regulations on chemical pesticide use, and there is political pressure to remove the most hazardous chemicals from the market. Additionally, the spread of plant diseases in natural ecosystems may prevent successful application of chemicals, because of the scale to which such applications might have to be applied (Pal and Gardener, 2006). Although restrictions are being imposed to protect food quality and the environment, chemicals are still our only recourse at present to prevent diseases of food and fibre crops. In recent years, the need to develop disease control measures as alternative to chemicals has become a priority of scientists worldwide.

The market niches that can also be defined for biocontrol agents targeting include pests, diseases and weeds which have no current effective chemical control agents, have chemical controls that are too expensive or toxic to be considered, or where targets have become resistant to the existing chemical control. Another potential market niche for biocontrol agents are the so-called protected environments where environmental parameters are much more controlled than in the field, and a certain degree of manipulation of growing conditions can give better results for these biocontrol agents than that achievable in the field (Powell and Faull, 1989). Consequently, some pest management researchers have focused their efforts on developing alternative inputs to synthetic chemicals for controlling pests and diseases. Among these alternatives are those referred to as biocontrol agents.

Need for Biological Control

Though there is an increasing interest in biological control of plant diseases in recent years but little commercial success in the field. This failure may reside in the lack of understanding of the mode of action of antagonists, their ecology as well as inadequate knowledge of methods of production and formulation for commercial use (Whipps, 1986; Whipps, Lewis and Coke, 1988; Lumsden and Lewis, 1989). Integrated studies of applied eco-physiology are required rather than any single empirical approach. Nevertheless, the use of model systems and critical *in vivo* studies are an essential prerequisite to successful biological control. Such systems can provide a detailed knowledge of mode of action and factors that affect a control system. They also highlight future possibilities of genetic manipulation and the *de novo* selection of desirable traits between isolates (Lewis, Whipps and Cooke, 1989).

Biocontrol agents may utilize several modes of action, therefore it is important to know the proportion and timing of each mode of action that may occur. Information of this type can be obtained from *in vitro* studies or using plants grown under *in vivo* conditions where the potential activity of biocontrol agents can be assessed. However, such studies do not provide information on their modes of action *in vivo*, particularly with in plants where separation of plant response or antagonist activity is not always possible or in soil where direct observation and chemical analyses are difficult. These limitations must be borne in mind when extrapolating results obtained in the laboratory to the natural environment (Lewis, Whipps and Cooke, 1989).

Looking to the future, biological control agents (BCAs) offer a potential solution to the dilemma from both the environmental and the industrial viewpoint. The environmental impact of BCAs is assumed to be less than that of agrochemicals. Their usage is presumed to reduce the global consumption of chemical pesticides, and lead to fewer resistance problems in target organisms. BCAs can also be used in integrated pest management (IPM) schemes, where in combination with good crop husbandry and hygiene, a high quality product can be produced with low chemical input. IPM requires the acceptance by the public of a degree of damage on a product rather than the perfection normally expected (Soper and Ward, 1981). On the industrial side, attraction includes cheaper discovery and registration of BCAs relative to chemical pesticides, with a consequently shorter lead time. The effect of these factors on the financial viability of a market can be quite dramatic (Powell and Faull, 1989). It may be instructive to consider some of the features of the successful BCAs that are currently in the market place. Like agrochemicals, BCAs fall into the categories of fungicides, insecticides, nematicides and herbicides. Different aspects of each class will be examined to illustrate the various important features of successful BCAs.

Most plant diseases are caused by fungi belonging to about 50 genera. Potential pathogens have gained considerable attention and appearing to be promising as a biological control (Table 22.1). Especially, using fungal antagonists against fungal plant viable supplement or alternative to chemical control (Papavizas, 1981, 1985). In the broad sense of biological control (Cook and Baker, 1983), crop resistance to disease is the ideal means of controlling diseases. However, many crops have little or no resistance to certain plant pathogens, so that biological control using microorganisms against plant pathogens is an attractive method for disease control. Considerable research has gone into the study of biological control of plant diseases over the past 60 years.

In order to establish biocontrol efficacy there are many inherent obstacles to overcome. These include basic aspects of strain selection, efficient production of biomass, formulation, storage ability, and methods of application. Also, a better understanding of mechanisms of action, nutrition and ecology of biological control agents is needed. Further, the potential for genetic manipulation of fungi to create genetically superior strains or hybrids that can perform better than wild types empirically selected from the environment needs to be examined (Elzein and Kroschel, 2003).

Mycorrhizas as Biocontrol Agents

Over the last 30 years there has been an increasing interest in the potential role that both arbuscular mycorrhizal and ectomycorrhizal fungi can play in the control of plant diseases and various aspects of this concept have been reviewed (Borowicz, 2001; Graham, 2001; Sylvia and Chellemi, 2001; Barea *et al.*, 2002). More than 80 disease biocontrol agents are on the market worldwide (Whipps, 2004), but none of these contain mycorrhizal fungi. This is despite ample evidence that both these fungi can control a number of plant diseases. The commercial use of mycorrhizal fungi as biocontrol agents is still in its infancy. There are several reports of mycorrhizal symbioses providing disease control in a great number of plant pathogen-plant species combinations and few lists of these have been published earlier (Perrin, 1990; Singh *et al.*, 2000). More recent examples along with some key earlier references are given in Tables 22.2 and 22.3 to illustrate both AM fungi and ectomycorrhizal fungi respectively.

Table 22.1: Fungi Developed or Being Developed for the Biological Control of Diseases (Adapted from Burges, 1998; Butt and Copping, 2000; Butt *et al.*, 2001)

Product	*Fungus*	*Target*	*Producer*
		Mycoparasites	
Rotstop	Phlebiopsis (=*Peniophora gigantea*)	*Heterobasidium annosus*	Kemira Agro Oy, Finland
Primastop	*Gliocladium catenulatum*	Several plant diseases	Kemira, Agro Oy, Finland
SoilGard (= GlioGard)	*Gliocladium virens*	Several plant diseases Damping off and root pathogens	ThermoTrilogy, USA
Cotans WG	*Coniothyrium minitans*	*Sclerotinia* spp.	Prophyta, Germany; KONI, Germany
AQ10 Biofungicide	*Ampelomyces quisqualis*	Powdery mildews	Ecogen Inc., USA
YieldPlus	*Cryptococcus albidus*	*Botrytis* spp., *Penicillium* spp.	Anchor Yeast, South Africa
Aspire	*Candida oleophila*	*Botrytis* spp., *Penicillium* spp.	Ecogen Inc., USA
Endothia parasitica	*Endothia parasitica* (non-pathogenic strain)	*Endothia parasitica* (Chestnut blight)	CNICM
Fusaclean	*Fusarium oxysporium*	*Fusarium oxysporium*	Natural Plant Protection, France
Biofox C	*F. oxysporium*	*F. oxysporium, Fusarium moniliforme*	SIAPA, Italy
Polygandron polyversum	*Pythium oligandrum*	*Pythium ultimum*	Plant Protection Institute, Slovak Republic
Trichoderma 2000	*Trichoderma harzianum*	*Rhizoctonia solani, Sclerotium rolfsii, Pythium*	Mycontrol (EfA1) Ltd, Israel
Trichopel	*T. harzianum*	Wide range of fungal diseases	Agrimm Technologies Ltd, New Zealand
T-22 and T-22HB Bio-Trek, Root Shield	*T. harzianum*	*Pythium, Rhizoctonia, Fusarium, Sclerotinia*	BioWorks (= TGT Inc.), Geneva, USA
Supresivit	*T. harzianum*	Various fungi	Borregaard and Reitzel, Denmark; Fytovita, Czech Republic
Trichodowels, Trichoject, Trichoseal and others	*T. harzianum* and *T. viride*	*Chondrostereum purpureum* and other soil and foliar pathogens	Agrimms Biologicals, New Zealand and others
Binab T	*T. harzianum, T. polysporum*	Fungi causing wilt, wood decay and take-all	Bio-Innovation, Sweden
Trichodex	*T. harzianum*	Fungal diseases, *e.g.Botrytis cinerea*	Makhteshim-Agan, Israel, several European companies, *e.g.* DeCeuster, Belgium

Table 22.2: Recent Examples of Arbuscular Mycorrhizas Exhibiting Control of Plant Diseases (Adapted from Whipps, 2004)

Mycorrhizal Fungus	*Pathogen*	*Disease*	*Host Plant*	*Effect on Plant*	*System*	*Reference*
			Soilborne pathogens			
Gigaspora margarita	*Fusarium oxysporum* f. sp. *asparagi* Cohen	Root rot and wilt	*Asparagus officinalis* L. (asparagus)	Decrease disease incidence	Sterilized soil-vermiculate mix; glasshouse	Mastsubara *et al.*, 2001
	Helicobasidium mompa Tanaka	Violet root rot	*Asparagus officinalis* L. (asparagus)	Decrease disease index	Sterilized soil mix; glasshouse	Mastsubara *et al.*, 2000
	Verticillium dahliae Kleb.	Wilt	*Solanum melongena* L. (egg plant)	Decrease disease Symptoms	Inoculated plants transferred to field soil	Mastsubara *et al.*, 1995
Glomus sp.	*Cylindrocladium spathiphylli*	Root rot	*Musa acuminata* Colla (banana)	Decrease disease severity	Pasteurized soil	Declerck *et al.*, 2002
	Schoulties *et al.* *Fusarium oxysporum* f. sp. *asparagi* Cohen	Root rot and wilt	*Asparagus officinalis* L. (asparagus)	Decrease disease incidence and index	Sterilized soil-vermiculate mix; glasshouse	Mastsubara *et al.*, 2001, 2002
	Helicobasidium mompa Tanaka	Violet root rot	*Asparagus officinalis* L. (asparagus)	Decrease disease index	Sterilized soil mix; glasshouse	Matsubara *et al.*, 2000
	Phytophthora cinnamomi Rands	Root rot	*Ananas comosus* (L.) Merr (pineapple)	Decrease disease and increase growth	Sterilized soil; controlled conditions	Guillemin *et al.*, 1994
	Sclerotium cepivorum Berk.	White rot	*Allium cepa L.* (onion)	Decrease disease and increase yield	Fumigated field soil	Torres-Barragan *et al.*, 1996
Glomus aggregatum Schenck and Smith emend. Koske	*Cylindrocarpon destructans* (Zinssm.) Scholten	Root rot	*Prunus persicaria* (L.) Batsch. (peach)	Decrease disease incidence	Sterilized soil or potting mix; controlled environment	Traquair, 1995
		Yellowing disease	*Cymbopogon winterianus* Jowett (*Java citronella*)	Decrease disease levels and increase biomass	Autoclaved soil; glasshouse	Ratti *et al.*, 1998

Contd...

Table 22.2–Contd...

Mycorrhizal Fungus	*Pathogen*	*Disease*	*Host Plant*	*Effect on Plant*	*System*	*Reference*
Glomus clarum Nicolson & Schenck	*Rhizoctonia solani* Kühn	Root rot	*Vigna unguiculata* (L.) Walp. (cowpea)	Decrease root necrosis and sclerotia production	Sterilized soil; glasshouse	Abdel-Fattah and Shabana 2002
Glomus coronatum Giovanetti	*Rhizoctonia solani* Kühn	Various Root and stem rots	*Vigna radiata* (L.) Wilzeck (mungbean)	Decrease disease severity	Sterilized soil-sand mix; glasshouse	Kasiamdari *et al.*, 2002
	Rhizoctonia solani (binucleate)	Weak root rot	*Vigna radiata* (L.) Wilzeck (mungbean)	Decrease disease severity	Sterilized soil-sand mix; glasshouse	Kasiamdari *et al.*, 2002
Glomus etunicatum Becker & Gerd.	*Verticillium dahliae* Kleb	Wilt	*Solanum melongena* L. (egg plant)	Decrease disease symptoms	Inoculated plants transferred to field soil	Mastsubara *et al.*, 1995
	Phytophthora fragariae var. *fragariae* Hickman	Root rot and red core	*Fragaria ananassa* Duch. (strawberry)	Decrease root necrosis	Autoclaved soil-sand mix; glasshouse	Norman *et al.*, 1996
Glomus fasciculatum (Thaxter sensu Gerd.) Gerd. & Trappe emend. Walker & Koske	*Fusarium oxysporum* f. sp. *asparagi* Cohen	Root rot and wilt	*Asparagus officinalis* L. (asparagus)	Decrease disease incidence	Sterilized soil-vermiculite mix; glasshouse	Matsubara et al. 2001
	Helicobasidium mompa Tanaka	Violet root rot	*Asparagus officinalis* L. (asparagus)	Decrease disease index	Sterilized soil mix; glasshouse	Matsubara *et al.*, 2000
	Phytophthora fragariae var. *fragariae* Hickman	Root rot and red core	*Fragaria ananassa* Duch. (strawberry)	Decrease root necrosis	Autoclaved soil-sand mix glasshouse	Norman *et al.*, 1996
	Fusarium moniliforme Sheldon	Damping-off	*Elettaria cardomomum* (L.) Maton (cardomom)	Decrease disease severity	Sterilized sand-soil-manure mix; shade house	Thomas *et al.*, 1994
Glomus fistulosum Skou & Jakobson	*Phytophthora fragariae* var. *fragariae* Hickman	Root rot and red core	*Fragaria vesca L.* (Wild strawberry)	Increase resistance to pathogen	Sterilized peat potting mix; glasshouse	Mark and Cassells, 1996

Contd...

Table 22.2–Contd...

Mycorrhizal Fungus	*Pathogen*	*Disease*	*Host Plant*	*Effect on Plant*	*System*	*Reference*
Glomus intraradices Schenck & Smith	*Aphanomyces euteiches* Drechsler	Root rot	*Pisum sativum* L. (pea)	Decrease disesase symptoms	Autoclaved soil; controlled environment and glasshouse	Kjoller and Rosendahl 1996
	Fusarium oxysporum f. sp. *dianthi* (Prill.& Delacr.) Snyd. & Hans.	Wilt	*Dianthus caryophyllus* L. (carnation)	Decrease disesase	Clay; glasshouse	St-Arnaud *et al.*, 1997
	Fusarium oxysporum f. sp. *lini* (Bolley) Snyd. & Hans.	Wilt	*Linum usitatissimum* L. (linseed)	Increase disease resistance	Sand-clay-compost mix; glasshouse	Dugassa *et al.*, 1996
	Fusarium oxysporum f. Sp. *radicis. Lycopsersici* Jarvis & Shoemaker	Crown and root rot	*Lycopersicon esculentum* Mill. (tomato)	Decrease disease incidence	Fumigated field soil	Datnoff *et al.*, 1995; Nemec *et al.*, 1996
	Fusarium sambucinum Fuckel	Tuber dry rot	*Solanum tuberosum* L. (potato)	Decrease tuber rot	AM fungi inoculated microtubers in sterile potting mix; controlled conditions	Niemira *et al.*, 1996
	Fusarium solani f. sp. *phaseoli* (Burk.) Snyd. & Hans.	Root rot	*Phaseolus vulgaris* L. (common bean)	Decrease disease symptoms and decrease pathogen biomass	Autoclaved soil; glasshouse	Filion *et al.*, 2003
	Phytophthora parasitica Breda de Haan	Root rot	*Citrus sinensis* L. Osb. (sweet orange)	Decrease disease	Transplanted to nursery field	Nemec *et al.*, 1996
	Pythium ultimum Trow.	Root rot	*Tagites patula* L. (French marigold)	Decrease in pathogen infection and propagules	Autoclaved perlite-vermiculate mix; screen house	St-Arnaud *et al.*, 1994

Contd...

Table 22.2–Contd...

Mycorrhizal Fungus	*Pathogen*	*Disease*	*Host Plant*	*Effect on Plant*	*System*	*Reference*
Glomus mosseae (Nicolson & Gerd.) Gerd. & Trappe	*Aphanomyces euteiches* Drechsler	Root rot	*Pisum sativum* L. (pea)	Decrease pathogen propagules and energy production	Irradiated sand-soil mix; glasshouse	Larsen and Bodker, 2001
	Helicobasidium mompa Tanaka	Violet root rot	*Asparagus officinalis* L. (asparagus)	Decrease disease index	Sterilized soil mix; glasshouse	Matsubara *et al.*, 2000
	Fusarium solani Mart.	Root and pod rot	*Arachis hypogea* L. (peanut)	Decrease inoculum of pathogen	Autoclaved soil-sand mix; glasshouse	Abdalla and Abdel-Fattah, 2000
	Phytophthora nicotianae f.sp. *parasitica* Breda de Haan	Root rot	*Lycopersicon esculentum* Mill. (tomato)	Decrease disease symptoms	Sterilized sand-soil mix; controlled conditions	Pozo *et al.*, 2002a
				Decrease in root necrosis	Sterilized sand; controlled conditions	Vigo *et al.*, 2000
	Rhizoctonia solani Kühn	Root and pod rot	*Arachis hypogea* L. (peanut)	Decrease inoculum of pathogen	Autoclaved soil-sand mix; glasshouse	Abdalla and Abdel-Fattah, 2000
	Rhizoctonia solani (Binucleate)	Weak root rot	*Pisum sativum* L. (pea)	Decrease pathogen development on root	Irradiated soil; controlled conditions	Morandi *et al.*, 2002
	Verticillium dahliae Kleb.	Wilt	*Gossypium hirsutum* L. and *G. barbadense* L. (cotton)	Decrease disease incidence	Sterilized soil; pots outside	Liu, 1995
			Lycopersicon esculentum Mill. (tomato)	Increase fresh and dry mass and plant height	Sterilized soil; glasshouse	Karagianndis *et al.*, 2002
			Solanum melongena L. (egg plant)	Increase fresh and dry mass and plant height	Sterilized soil; glasshouse	Karagianndis *et al.*, 2002
Glomus proliferum Dalpe & Declerck	*Cylindrocladium spathyphylli* Schoulties et al.	Root rot	*Musa acuminata* Colla. (banana)	Decrease disease severity	Pasteurized soil	Declerck *et al.*, 2002
Glomus versiforme (Karsten) Berch	*Verticillium dahliae* Kleb.	Wilt	*Gossypium hirsutum* L. and *G. barbadense* L. (cotton)	Decrease disease incidence	Sterilized soil; pots outside	Liu, 1995

Contd...

Table 22.2–Contd...

Mycorrhizal Fungus	*Pathogen*	*Disease*	*Host Plant*	*Effect on Plant*	*System*	*Reference*
Sclerocystis sinuosa (Gerd., & Bakshi) Almeida & Schenck	*Verticillium dahliae* Kleb.	Wilt	*Gossypium hirsutum* L. and *G. barbadense* L. (cotton)	Decrease disease incidence	Sterilized soil; pots outside	Liu, 1995
Natural arbuscular mycorrhizas	*Aphanomyces euteiches* Drechsler	Root rot	*Psium sativum* L. (pea)	Decrease pathogen propagules; no correlation with disease severity	Field soil with carbendazim added to inhibit arbuscular mycorrhizas	Bodker *et al.*, 2002
			Foliar pathogens			
Glomus constrictum Trappe	*Uromyces* (*phaseoli*) *appendiculatus* (Pers.) Unger	Rust	*Phaseolus vulgaris* L. (common bean)	Increase rust infection and plant mass	Sand-soil mix; glasshouse	Meyer and Dehne, 1986
Glomus etunicatum Becker & Gerd	*Blumeria graminis* f. sp. *hordei* (DC.) Speer	Powdery mildew	*Nicotiana tabacum* L. (tobacco)	Increased conidial production but not infected leaf areas	Sterilized compost-clay mix; glasshouse and netted field site	Gernns *et al.*, 2001
	Bremia lactucae Regel	Downy mildew	*Lettuce sativa* L. (lettuce)	Increased infected area and fresh mass	Sand-soil mix; glasshouse	Meyer and Dehne, 1986
Glomus fasciculatum (Thaxter sensu Gerd.) Gerd & Trappe emend. Walker & Koske	*Uromyces* (*phaseoli*) *appendiculatus* (Pers.) Unger	Rust	*Phaseolus vulgaris L.* (common bean)	Increase rust infection and plant mass	Sand-soil mix; glasshouse	Meyer and Dehne, 1986
Glomus intraradices Schenck & Smith	*Botrytis cinerea* Pers.: Fr.	Grey mould	*Nicotiana tabacum* L. (tobacco)	Increase necrotic lesions	Sand; controlled conditions	Shaul *et al.*, 1999
	Oidium lini Skoriae	Powdery mildew	*Linum usitatissimum* L. (linseed)	Increase tolerance to pathogen and increased susceptibility	Sand-compost-clay mix; glasshouse	Dugassa *et al.*, 1996
	Tobacco mosaic virus	Virus	*Nicotiana tabacum* L. (tobacco)	Increase necrotic lesions	Sand; controlled conditions	Shaul *et al.*, 1999
Glomus mosseae (Nicolson & Gerd.) Gerd & Trappe	Phytoplasma	Yellows disease	*Lycopersicon esculentum* Mill. (tomato)	Decrease disease symptoms and increased tolerance	Sand; controlled conditions	Lingua *et al.*, 2002

Table 22.3: Recent Examples of Ectomycorrhizas Exhibiting Control of Plant Diseases (Adapted from Whipps, 2004)

Mycorrhizal Fungus	*Pathogen*	*Disease*	*Host Plant*	*Effect on Plant*	*System*	*Reference*
Clitocybe claviceps (Fr.) Kummer	*Fusarium moniliforme* Sheldon	Damping-off and root rot	*Picea glauca* (Moench) Voss (white spruce)	Increase seedling survival and decrease root-rot severity	Sterile peat-vermiculate; glasshouse	Chakravarty *et al.*, 1999
			Pinus contorta Dougl. var. *latifolia* Engelm. (lodgepole pine)	Increase seedling survival and decrease root-rot severity	Sterile peat-vermiculate; glasshouse	Chakravarty *et al.*, 1999
	Fusarium oxysporum Schlecht.: Fr.	Damping-off and root rot	*Picea glauca* (Moench) Voss (white spruce)	Increase seedling survival and decrease root-rot severity	Sterile peat-vermiculate; glasshouse	Chakravarty *et al.*, 1999
			Pinus contorta Dougl. var. *latifolia* Engelm. (lodgepole pine)	Increase seedling survival and decrease root-rot severity	Sterile peat-vermiculate; glasshouse	Chakravarty *et al.*, 1999
Hebeloma crustiliniforme (Bull. ex Fr.) Quel	*Fusarium moniliforme* Sheldon	Damping-off and root rot	*Pinus sylvestris* L. (Scots pine)	Increase seedling survival and decrease root-rot severity	Brick pellets-peat mix; glasshouse	Chakravarty and Unestam, 1987b
	Phytopthora cambivora (Petri) Buism.	Ink disease	*Castanea sativa* Mill. (chestnut)	Decrease disease symptoms and increase shoot and root development	Sterilized peat-sand mix; controlled conditions	Brazanti *et al.*, 1999
	Phytopthora cinnamomi Rands	Ink disease	*Castanea sativa* Mill. (chestnut)	Decrease disease symptoms and increase shoot and root development	Sterilized peat-sand mix; controlled conditions	Brazanti *et al.*, 1999
	Rhizoctonia solani Kühn	Root rot	*Pinus sylvestris* L. (Scots pine)	Decreased mortality	Brick pellets-peat mix; glasshouse	Chakravarty and Unestam, 1987b

Contd...

Table 22.3–Contd...

Mycorrhizal Fungus	*Pathogen*	*Disease*	*Host Plant*	*Effect on Plant*	*System*	*Reference*
Hebeloma sinapizans (Paulet & Fr.)	*Phytopthora cambivora* (Petri) Buism.	Ink disease	*Castanea sativa* Mill. (chestnut)	Decrease disease symptoms and increase shoot and root development	Sterilized peat-sand mix; controlled conditions	Brazanti *et al.*, 1999
	Phytopthora cinnamomi Rands	Ink disease	*Castanea sativa* Mill. (chestnut)	Decrease disease symptoms and increase shoot and root development	Sterilized peat-sand mix; controlled conditions	Brazanti *et al.*, 1999
Laccaria bicolor (R. Maire) Orton	*Fusarium moniliforme* Sheldon	Damping-off and root rot	*Picea glauca* (Moench) Voss (white spruce)	Increase seedling survival and decrease root-rot severity	Sterile peat-vermiculate; glasshouse	Chakravarty *et al.*, 1999
			Pinus contorta Dougl. var. *latifolia* Engelm. (lodgepole pine)	Increase seedling survival and decrease root-rot severity	Sterile peat-vermiculate; glasshouse	Chakravarty *et al.*, 1999
	Fusarium oxysporum Schlecht.: Fr.	Root rot	*Picea abies* (L.) Karst. (Norway spruce)	Decrease disease	Fumigated soil; field	Sampagni *et al.*, 1986
			Picea glauca (Moench) Voss (white spruce)	Increase seedling survival and decrease root-rot severity	Sterile peat-vermiculate; glasshouse	Chakravarty *et al.*, 1999
			Pinus contorta Dougl. var. *latifolia* Engelm. (lodgepole pine)	Increase seedling survival and decrease root-rot severity	Sterile peat-vermiculate; glasshouse	Chakravarty *et al.*, 1999
			Pseudotsuga menziesii (Mirb.) Franco (Douglas-fir)	Decrease disease	Fumigated soil; field	Sampagni *et al.*, 1986
				Temperature dependant decrease in lethal infections with nonagressive strain	Petri dish agar synthesis; *in vitro* controlled conditions	Strobel and Sinclair, 1991b

Contd...

Table 22.3–Contd...

Mycorrhizal Fungus	*Pathogen*	*Disease*	*Host Plant*	*Effect on Plant*	*System*	*Reference*
Laccaria laccata (Scop. ex Fr.) Berk & Br.	*Cylindrocarpon destructans* (Zinssm.) Scholten	Root rot	*Pinus sylvestris* L. (Scots pine)	Decreased root death rate	Brick pellets-peat mix; glasshouse	Chakravarty and Unestam 1987a
	Fusarium moniliforme Sheldon	Damping-off and root rot	*Pinus sylvestris* L. (Scots pine)	Increase seedling survival and decrease root-rot severity	Brick pellets-peat mix; glasshouse	Chakravarty and Unestam 1987b
	Fusarium oxysporum Schlecht.: Fr.	Root rot	*Picea abies* (L.) Karst. (Norway spruce)	Decrease disease	Fumigated soil; field	Sampagni *et al.*, 1986
			Pseudotsuga menziesii (Mirb.) Franco (Douglas-fir)	Decrease disease	Fumigated soil; field	Sampagni *et al.*, 1986
	Phytopthora cambivora (Petri) Buism.	Ink disease	*Castanea sativa* Mill. (chestnut)	Decrease disease symptoms and increase shoot and root development	Sterilized peat-sand mix; controlled conditions	Brazanti *et al.*, 1999
	Phytopthora cinnamomi Rands	Ink disease	*Castanea sativa* Mill. (chestnut)	Decrease disease symptoms and increase shoot and root development	Sterilized peat-sand mix; controlled conditions	Brazanti *et al.*, 1999
	Rhizoctonia solani Kühn	Root rot	*Pinus sylvestris* L. (Scots pine)	Decreased mortality	Brick pellets-peat mix; glasshouse	Chakravarty and Unestam 1987b
Paxillus involutus (Batsch) Fr.	*Cylindrocladium floridanum* Sobers & C.P. Seymour	Root rot	*Picea mariana* (Mill.) BSP (black spruce)	Decrease infected number seedlings	Sterile peat-vermiculate; controlled conditions	Morin *et al.*, 1999
	Fusarium moniliforme Sheldon	Damping-off and root rot	*Picea glauca* (Moench) Voss (white spruce)	Increase seedling survival and decrease root-rot severity	Sterile peat-vermiculate; glasshouse	Chakravarty *et al.*, 1999
			Pinus banksiana Lamb. (Jack pine)	Increase seedling survival	Sterile peat-vermiculate; controlled conditions	Hwang *et al.*, 1995

Contd...

Table 22.3–Contd...

Mycorrhizal Fungus	*Pathogen*	*Disease*	*Host Plant*	*Effect on Plant*	*System*	*Reference*
			Pinus contorta Dougi. var. *latifolia* Engelm. (lodgepole pine)	Increase seedling survival and decrease root-rot severity	Sterile peat-vermiculate; glasshouse	Chakravarty *et al.*, 1999
				Increase seedling survival and decrease root-rot severity	Sterile peat-vermiculate; controlled conditions	Pedersen *et al.*, 1999
			Pinus resinosa Ait. (red pine)	Increase seedling survival	Sterile peat-vermiculate; controlled conditions	Chakravarty *et al.*, 1991
	Fusarium oxysporum Schlecht.: Fr.	Damping-off and root rot	*Picea glauca* (Moench) Voss (white spruce)	Increase seedling survival and decrease root-rot severity	Sterile peat-vermiculate; glasshouse	Chakravarty *et al.*, 1999
			Pinus contorta Dougi. var. *latifolia* Engelm. (lodgepole pine)	Increase seedling survival and decrease root-rot severity	Sterile peat-vermiculate; glasshouse	Chakravarty *et al.*, 1999
				Increase seedling survival and decrease root-rot severity	Sterile peat-vermiculate; controlled conditions	Pedersen *et al.*, 1999
			Pinus resinosa Ait. (red pine)	Supressed disease and increase survival	Autoclaved potting mix; controlled conditions	Farquahar and Peterson, 1991
				Increase seedling survival	Sterile peat-vermiculate; controlled conditions	Chakravarty *et al.*, 1991
	Phytopthora cambivora (Petri) Buism.	Ink disease	*Castanea sativa* Mill. (chestnut)	Decrease disease symptoms and increase shoot and root development	Sterilized peat-sand mix; controlled conditions	Brazanti *et al.*, 1999

Contd...

Table 22.3–Contd...

Mycorrhizal Fungus	*Pathogen*	*Disease*	*Host Plant*	*Effect on Plant*	*System*	*Reference*
	Phytopthora cinnamomi Rands	Ink disease	*Castanea sativa* Mill. (chestnut)	Decrease disease symptoms and increase shoot and root development	Sterilized peat-sand mix; controlled conditions	Brazanti *et al.*, 1999
Pisolithus tinctorius (Pers.) Coker & Couch	*Cylindrocarpon destructans* (Zinssm.) Scholten	Root rot	*Pinus sylvestris* L. (Scots pine)	Decreased root death rate	Brick pellets-peat mix; glasshouse	Chakravarty and Unestam, 1987a
	Fusarium moniliforme Sheldon	Damping-off and root rot	*Pinus sylvestris* L. (Scots pine)	Decreased mortality	Brick pellets-peat mix; glasshouse	Chakravarty and Unestam, 1987b
	Rhizoctonia solani Kühn	Root rot	*Pinus sylvestris* L. (Scots pine)	Decreased mortality	Brick pellets-peat mix; glasshouse	Chakravarty and Unestam, 1987b

Many studies have been published but the underlying mechanisms are poorly understood. Some basic mechanisms have been suggested *i.e.* improvement of plant nutrition and competition for photosynthates (Azcon-Aguilar and Barea, 1996), however, AM induced suppression of root pathogens and stimulation of saprotrophs and plant growth promoting microorganisms may also be important (Kapoor and Mukerji, 1998). Some authors have discussed other mechanisms, which include anatomical or morphological changes in the root system induced by the AM fungus; and local elicitation of plant defense mechanisms by AM fungi (Gianinazzi-Pearson *et al.*, 1994). Due to difficulties in producing huge quantities of pure culture of AM inoculum, relatively few studies have explored the practical use of AM fungi as inoculants to increase plant resistance to root-rotting pathogens. Combined inoculation of AM fungi with growth stimulating bacteria may favour the inoculum production (Reddy *et al.*, 2006). Several studies have demonstrated that some AM fungi exhibit biocontrol properties (Niemira, Hammerschmidt and Safir, 1996; Newsham, Fitter and Watkinson, 1995; Caron, 1989) against root pathogens. Whether AM fungi could be used as biocontrol agents practically, or possibly function as vectors for associated bacteria with biocontrol properties remains to be explored (Johansson, Paul and Finlay, 2004).

Prospects of Entomogenous Fungi

The importance of entomogenous fungi as natural regulators of insect populations has been recognized for many years (Gillespie and Moorhouse, 1989). Over 2000 years ago, the Chinese were aware of the infection by species of *Cordyceps* and *Isaria* that caused mortality in populations of silk worms and cicadas (Roberts and Humber, 1981). The possibilities of using entomogenous fungi as biocontrol agents were first explored by Metchnikoff (1879) and Krassilstschik (1888) in the Ukraine. These two pioneers mass produced *Metarhizium anisopliae* and tested it against wheat cockchafer and sugarbeet curculionid. Over the past 10 years there has been an unprecedented increase in activities related to the commercial development of entomopathogenic fungi. More than 30 trade-named mycoinsecticide products now registered or under development world- wide (Wraight *et al.*, 2001). Entomopathogenic hyphomycetes have been employed against a variety of different insect pests in a variety of different agroecosystems and they have demonstrated varying degrees of success (Inglis *et al.*, 2001). The prospects of entomogenous fungi are tabulated in Table 22.4.

Fungi will not be cure-alls for pest problems on all crops and in all agricultural settings, and it is unlikely that they will ever totally supplant the management of insect pests with chemical insecticides. Nevertheless, they represent a valuable management resource to be utilized within an IPM framework, and will contribute significantly to reductions in chemical pesticide use (Lacey and Goettel, 1995). Research to identify constraints on efficacy and the utilization of biologically based strategies to overcome these constraints will continue to increase the consistency of insect suppression with entomopathogenic hyphomycetes. Furthermore, the integration of this group of fungi with other management practices will require detailed compatibility studies and the development of effective guidelines for their use. For example, in a typical agroecosystem, a range of pest species, including phytopathogens and insects, are encountered and have to be combated simultaneously. Thus the role of mycoinsecticides in pest control must be viewed within the context of a total crop management programme. Such integration will be more complex than the use of traditional chemical control tactics, and will rely upon efficient monitoring to optimize the levels of control obtained from each application. Recent developments portend well for the future use of entomopathogenic hyphomycetes in pest management programmes.

Table 22.4: Fungi Developed or Being Developed for the Biological Control of Diseases (Adapted from Burges, 1998; Butt and Copping, 2000; Butt *et al.*, 2001)

Product	*Fungus*	*Target*	*Producer*
		Entomogenous fungi	
Mycotal	*Verticillium lecanii*	Whitefly and thrips	Koppert, the Netherlands
Vertalec	*V. lecanii*	Aphids	Koppert, the Netherlands
Biogreen	*Metarhizium anisopliae*	Scarab larvae on pasture	Bio-care Technology, Australia
Metaquino	*M. anisopliae*	Spittlebugs	Brazil
Bio-Path	*M. anisopliae*	Cockroaches	EcoScience, USA
Bio-Blast	*M. anisopliae*	Termites	Ecoscience, USA
Cobican	*M. anisopliae*	Sugar-cane spittlebug	Probioagro, Venezuela
Conidia	*Beauveria bassiana*	Coffee-berry borer	Live Systems Technology, Colombia
Ostrinil	*B. bassiana*	Corn-borer	Natural Plant Protection (NPP), France
CornGuard	*B. bassiana*	European corn-borer	Mycotech, USA
Mycotrol GH	*B. bassiana*	Grasshoppers, locusts	Mycotech, USA
Mycotrol WP and BotaniGard	*B. bassiana*	Whitefly, aphids, thrips	Mycotech, USA
Naturalis-L	*B. bassiana*	Cotton pests including boll-worms	Troy Biosciences, USA
Proecol	*B. bassiana*	Army worm	Probioagro, Venezuela
Boverin	*B. bassiana*	Colorado beetle	Former USSR
Boverol	*B. bassiana*	Colorado beetle	Czechoslovakia
Beverosil	*B. bassiana*	Colorado beetle	Czechoslovakia
Engerlingspliz	*Beauveria brongniartii*	Cockchafers	Andermatt, Switzerland
Schweizer	*B. brongniartii*	Cockchafers	Eric Schweizer, Switzerland
	Beauveria spp.		
Melocont	*B. brongniartii*	Cockchafers	Kwizda, Austria
Green Muscle	*Metarhizium Flavoviride*	Locusts, grasshoppers	CABI BioScience, UK
PFR-97	*Paecilomyces fumosoroseus*	Whitefly	ECO-tek, USA
Pae-Sin	*P. fumosoroseus*	White fly	Agrobionsa, Mexico
Laginex	*Lagenidium giganteum*	Mosquito larvae	AgraQuest, USA

Potentials of Nematophagous Fungi

The effective natural control of specific nematode pests in intensive agricultural systems has been well documented and the causal microbial parasites and pathogens have often been identified (Kerry, 1987; Stirling, 1991; Dickson *et al.*, 1994). The recognition of suppressive soils in which biotic factors prevent nematodes multiplying on susceptible crops has demonstrated that the biological control of nematodes has potential as a management strategy. Research has concentrated on the use of nematode-trapping fungi for the control of cyst and root-knot nematodes, as these are the most economically important nematode pest species. However, these fungi proved difficult to manipulate and it was often not possible to ensure that they produced traps when the infective nematode juveniles were migrating towards roots (Kerry, 2001). The fungi developed for the biological control of nematode pests are provided in Table 22.5.

Table 22.5: Fungi Developed or Being Developed for the Biological Control of Diseases (Adapted from Burges, 1998; Butt and Copping, 2000; Butt *et al.*, 2001)

Product	*Fungus*	*Target*	*Producer*
		Nematophagous fungi	
Ditera	*Myrothecium verrucaria*	Plant-parasitic nematodes	Valent (Sumitomo), USA, Japan
Product under development	*Duddingtonia flagrans*	Animal-parasitic Nematodes	Christian Hansen, Denmark
Product under development	*Verticillium chlamydosporium*	Plant-parasitic nematodes	DeCeuster, Belgium

There is a need to make surveys for nematophagous fungi, preferably in soils where the pest nematode is indigenous and in soils where it causes less of a problem than expected. Such surveys should result in the isolation of new potential biological control agents (Cook and Baker, 1983) and in increasing knowledge of the factors affecting natural control. Biological agents are not a replacement for nematicides and are unlikely to give such rapid and effective control. However, combined with resistant and tolerant cultivars, they could make an important contribution to integrated pest management systems for several important plant nematode pests and greatly decrease the current dependence on chemical nematicides.

Progress in Mycoherbicide Research

During the last 30 years, extensive research has been conducted to develop a 'bioherbicide approach' for controlling weedy plants. This approach uses the annual application of indigenous plant pathogens to control weed species (Templeton, 1982). Weed scientists and plant pathologists have identified more than 100 microorganisms which are candidates for development as commercial bioherbicidal agents (Templeton, 1982; Charudattan, 1991). These organisms have been selected because they exhibit specificity toward their host weed and are usually highly aggressive in inciting disease. The regulatory issues and market demand have hindered the development of some potential mycoherbicides, the overall lack of commercial success in using living microbial BCAs stems from difficulties in producing and stabilizing these agents and from the lack of consistently effective weed control in field situations (Zorner *et al.*, 1993; Auld and Morin, 1995).

Selecting weed targets that cause significant economic losses in higher-value crops is imperative if additional commercial successes in mycoherbicide development are to be realized. The ongoing development of herbicide-resistant crops, such as soybean and cotton, will only increase the importance of selecting promising weed-crop systems for mycoherbicide development. Some of the important mycoherbicides used for the successful control of weeds are given in Table 22.6. The amenability of potential mycoherbicides to low-cost production methods must be evaluated early in the selection and development process. This factor will weigh heavily in determining commercial potential. If commercial development is the ultimate goal, extensive studies on mycoherbicides that lack potential for low-cost production should be discouraged.

Table 22.6: Some of the Important Mycoherbicides Used for the Successful Control of Weeds (Data from Templeton and Heiny, 1989)

Mycoherbicide	*Fungus*	*Weed*	*Field/Crop*
Collego	*Colletotrichum gloeosporioides* f. sp. *aeschynomene*	Northern jointvetch (*Aeschynomene virginica*)	Rice
DeVine	*Phytophthora palmivora*	Milk weed vine (*Morrenia odorata*)	Citrus
Velgo	*Colletotrichum coccodes*	Velvet leaf (*Abutilon theophrasti*)	Corn, Soybean
Luboa 2	*Colletotrichum gloeosporioides* f. sp. *cuscutae*	Dodder(*Cuscuata chinensis* & *C. australis*)	Soybean
Biomal	*Colletotrichum gloeosporioides* f. sp. *malvae*	Round leaved mallow (*Malva pusilla*)	Wheat, Lentils
Casst	*Alternaria cassiae*	Sickle pod and Coffee senna (*Cassia obtusifolia* & *C. occidentalis*)	Soybean,Peanuts
–	*Acremonium diospyri*	Persimmon trees (*Diospyros virginiana*)	–
–	*Colletotrichum orbiculare*	Bathurst burr (*Xanthium spinosum*)	–
–	*Cercospora rodmanii*	Water hyacinth (*Eichhornia crassipes*)	–
–	*Chondrostereum purpureum*	American Black Cherry (*Prunus serotina*)	Forests

Modest progress in development of natural strains of fungi as mycoherbicides can be anticipated; there are many to explore and develop. However, all are likely to have innate biological limitations that affect utility. There has been a great number of naturally occurring strains researched for possible use as mycoherbicides but only a small proportion has been developed to commercial products (Templeton and Dana, 1989). The future of mycoherbicides will be dependent on technological advance and market opportunities. The continuing development of herbicide-resistant crops will certainly limit the market demand for these products. Potential markets for the use of fungal biopesticides will include organic farms, rangelands and low-value public lands with serious invasive weed problems. As Zorner *et al.* (1993) point out, research efforts must now be shifted from the discovery of mycoherbicides to solving the production, storage and efficacy problems that plague all mycoherbicides. Commercial interest and user acceptance of mycoherbicides as weed management tools are dependent on the development of low-cost, stable products that provide consistent weed control under field conditions. These biological constraints are not unique to mycoherbicides but, rather, general problems that have long impeded the development of microbial BCAs. Solving these problems will require research strategies that involve weed scientists, plant pathologists, microbiologists, fermentation specialists, biochemists and formulation scientists. Advances in mycoherbicide stabilization and field efficacy will require additional participation by formulation and application scientists.

Conclusions

Although biological control of fungal plant pathogens with fungal antagonists has not dramatically solved agricultural problems so far, we are at a turning point in technology to make significant advances. It may be several years before microbial biocontrol products are accepted for agricultural use but it seems likely that the time is coming. Several important considerations must be implemented in order to expedite future use of microbial agents for disease control. New approaches must be introduced to develop improved production, formulation and shelf-life of biocontrol agents. Increased understanding is essential concerning persistent problems related to dependability and reliability of microbial agents. Models need to be developed for analysis of epidemiological, economic and biological data to determine maximum benefits and how they might be introduced into integrated pest management systems. We need to determine mechanisms of action of biological control agents, including factors responsible for naturally occurring suppressive soils, to understand better the ways of improving microbial action against pathogens both through manipulation of their biology and their genetic constitution. Investigation of new technologies is needed, especially genetic manipulation and hybridization; to induce new biotypes for improved performance, adaptation of microbial agents to new disease control requirements and improved compatibility with chemical fungicides (Lumsden and Lewis, 1989).

Most pathogens will be susceptible to one or more biocontrol strategies, but practical implementation on a commercial scale has been constrained by a number of factors. Cost, convenience, efficacy, and reliability of biological controls are important considerations, but only in relation to the alternative disease control strategies. Cultural practices (*e.g.* good sanitation, soil preparation, and water management) and host resistance can go a long way towards controlling many diseases, so biocontrol should be applied only when such agronomic practices are insufficient for effective disease control (Pal and Garderner, 2006). It is imperative that successful cases of biological control be introduced in agriculture to set precedents for industry and crop producers to show that this approach is worthy of efforts to register and commercialize these products. These successes should also demonstrate that microbial biocontrol agents are economically feasible, safe and above all, effective.

References

Abdalla, M.E. and Abdel-Fattah, G.M. (2000). Influence of the endomycorrhizal fungus *Glomus mosseae* on the development of peanut pod rot disease in Egypt. *Mycorrhiza*, 10: 29–35.

Abdel-Fattah, G.M. and Shabana, Y.M. (2002). Efficacy of the arbuscular mycorrhizal fungus *Glomus clarum* in protection of cowpea plants against root rot pathogen *Rhizoctonia solani. J. Plant Dis. Prot.*, 109: 207–215.

Auld, B.A. and Morin, L. (1995). Constraints in the development of bioherbicides. *Weed Technol.*, 9: 638–652.

Azcon-Aguilar, C. and Barea, J.M. (1996). Arbuscular mycorrhizas and biological control of soil-borne plant pathogens: An overview of the mechanisms involved. *Mycorrhiza*, 6: 457–464.

Barea, J.M., Azcon, R. and Azcon-Aguilar, C. (2002). Mycorrhizosphere interactions to improve plant fitness and soil quality. *Antonie Leeuwenhoek*, 81: 343–351.

Bodker, L., Kjoller, R. and Rosendahl, S. (1998). Effect of phosphate and the arbuscular mycorrhizal fungus *Glomus intra-radices* on disease severity of root rot of peas (*Pisum sativum*) caused by *Aphanomyces euteiches. Mycorrhiza*, 8: 169–174.

Bodker, L., Kjoller, R., Kristensen, K. and Rosendahl, S. (2002). Interactions between indigenous arbuscular mycorrhizal fungi and *Aphanomyces euteiches* in field-grown pea. *Mycorrhiza*, 12: 7–12.

Borowicz, V.A. (2001). Do arbuscular mycorrhizal fungi alter plant-pathogen relations? *Ecology*, 82: 3057–3068.

Brazanti, M.B., Rocca, E. and Pisi, A. (1999). Effect of ectomycorrhizal fungi on chestnut ink disease. *Mycorrhiza*, 9: 103–109.

Burges, H.D. (1998). *Formulation of Microbial Pesticides*. Kluwer Academic Publishers, Dordrecht, The Netherlands.

Butt, T.M. and Copping, L. (2000). Fungal biological control agents. *Pesticide Outlook*, 11: 186–191.

Butt, T.M., Jackson, C. and Magan, N. (2001). Introduction–Fungal biological control agents: Progress, problems and potential. In: *Fungi as Biocontrol Agents,* (Eds.) Butt, T.M., Jackson, C. and N. Magan. CAB International, USA, pp. 1–8.

Caron, M. (1989). Potential use of mycorrhizae in control of soil-borne diseases. *Can. J. Plant Pathol.*, 11: 177–179.

Chakravarty, C., Peterson, R.L. and Ellis, B.E. (1991). Interaction between the ectomycorrhizal fungus *Paxillus involutus,* damping-off fungi and *Pinus resinosa* seedlings. *J. Phytopathol.*, 132: 207–218.

Chakravarty, P. and Unestam, T. (1987a). Mycorrhizal fungi prevent disease in stressed pine seedlings. *J. Phytopathol.*, 118: 335–340.

Chakravarty, P. and Unestam, T. (1987b). Differential influence of ectomycorrhizae on plant growth and disease resistance in *Pinus sylvestris* seedlings. *J. Phytopathol.*, 120: 104–120.

Chakravarty, P., Khasa, D., Dancik, B., Sigler, L., Wichlacz, M., Trifonov, L.S. and Ayer, W.A. (1999). Integrated control of Fusarium damping-off in conifer seedlings. *J. Plant Dis. Prot.*, 106: 342–352.

Charudattan, R. (1991). The mycoherbicide approach with plant pathogens. In: *Microbiological Control of Weeds*, (Ed.) D.O. Te Beest. Chapman and Hall, New York, pp. 24–57.

Cook, R.J. and Baker, K.F. (1983). *The Nature and Practice of Biological Control in Crop Production*. American Phytopathological Society, St. Paul, Minnesota.

Datnoff, L.E., Nemec, S. and Pernezny, K. (1995). Biological control of Fusarium crown and root rot of tomato in Florida using *Trichoderma harzianum* and *Glomus intraradices. Biological Control*, 5: 427–431.

Declerck, S., Risede, J.M., Rufyikiri, G. and Delvaux, B. (2002). Effects of arbuscular mycorrhizal fungi on severity of root rot of bananas caused by *Cylindrocladium spathiphylli. Plant Pathol.*, 51: 109–115.

Dickson, D.W., Oostendorp, M., Giblin-Davis, R.M. and Mitchell, D.J. (1994). Control of plant–parasitic nematodes by biological antagonists. In: *Pest Management in the Subtropics Biological Control: A Florida Perspective,* (Eds.) Rosen, D., Bennett, F.D. and J.L. Capinera. Intercept, Andover, UK, pp. 575–601.

Dugassa, G.D., Vonalten, H. and Schonbeck, F. (1996). Effects of arbuscular mycorrhiza (AM) on health of *Linum usitatissimum* L. infected by fungal pathogens. *Plant Soil*, 185: 173–182.

Elzein, A. and Kroschel, J. (2003). Progress on management of parasitic weeds. In: *Weed Management for Developing Countries,* (Ed.) R. Labrada. FAO Plant Production and Protection Paper, 120, Addendum 1, Food and Agriculture Organization of the United Nations, Rome, pp. 290.

Farquhar, M.L. and Peterson, R.L. (1991). Later events in suppression of *Fusarium* root rot of red pine seedlings by the ectomycorrhizal fungus *Paxillus involutus*. *Can. J. Bot.*, 69: 1372–1383.

Filion, M., St-Arnaud, M. and Jabaji-Hare, S.H. (2003). Quantification of *Fusarium solani* f. sp. *phaseoli* in mycorrhizal bean plants and surrounding mycorrhizosphere soil using real-time polymerase chain reaction and direct isolations on selective media. *Phytopathology*, 93: 229–235.

Gernns, H., Von Alten, H. and Poehling, H.M. (2001). Arbuscular mycorrhiza increased the activity of a biotrophic leaf pathogen is a compensation possible? *Mycorrhiza*, 11: 237–243.

Gianinazzi-Pearson, V., Gollotte, A., Dumas-Gaudot, E., Franken, P. and Gianinazzi, S. (1994). Gene expression and molecular modifications associated with plant responses to infection by arbuscular mycorrhizal fungi. In: *Advances in Molecular Genetics of Plant-Microbe Interactions*, (Eds.) Daniels, M., Downic, J.A. and A.E. Osbourn. Kluwer, Dordrecht, pp. 179–186.

Gillespie, A.T. and Moorhouse, E.R. (1989). The use of fungi to control pests of agricultural and horticultural importance. In: *Biotechnology of Fungi for Improving Plant Growth*, (Eds.) Whipps, J.M. and R.D. Lumsden. Cambridge University Press, Great Britain, pp.55–84.

Graham, J.H. (2001). What do root pathogens see in mycorrhizas? *New Phytol.*, 149: 357–359.

Guillemin, J.P., Gianinazzi, S., Gianinazzi-Pearson, V. and Marchal, J. (1994). Contribution of arbuscular mycorrhizas to biological protection of micropropagated pineapple *Ananas comosus* (L.) Merr against *Phytophthora cinnamomi* Rands. *Agric. Sci. Finl.*, 3: 241-251.

Hwang, S.P., Chakravarty, P. and Chang, K.F. (1995). The effect of two ectomycorrhizal fungi, *Paxillus involutus* and *Suillus tomentosus*, and of *Bacillus subtilis* on Fusarium damping-off in jack pine seedlings. *Phytoprotection*, 76: 57–66.

Inglis, G.D., Goettel, M.S., Butt, T.M. and Strasser, H. (2001). Use of hypomycetous fungi for managing insect pests. In: *Fungi as Biocontrol Agents*, (Eds.) Butt, T.M., Jackson, C. and N. Magan. CAB International, USA, pp. 23–69.

Johansson, J.F., Paul, L.R. and Finlay, R.D. (2004). Microbial interactions in the mycorrhizosphere and their significance for sustainable agriculture. *FEMS Microbiology Ecology*, 48: 1–13.

Justum, A.R. (1988). Commercial application of biological control: status and prospects. *Philosophical Trans. Royal Soc. London, Series B*, 318: 357–373.

Kapoor, R. and Mukerji, K.G. (1998). Microbial interactions in mycorrhizosphere of *Anethum graveolens* L. *Phytomorphology*, 48: 383–389.

Karagiannidis, N., Bletsos, F. and Stavropoulos, N. (2002). Effect of Verticillium wilt (*Verticillium dahliae* Kleb.) and mycorrhiza (*Glomus mosseae*) on root colonization, growth and nutrient uptake in tomato and eggplant seedlings. *Sci. Hortic.*, 94: 145–156.

Kasiamdari, R.S., Smith, S.E., Smith, P.A. and Scott, E.S. (2002). Influence of the mycorrhizal fungus, *Glomus coronatum*, and soil phosphorus on infection and disease caused by binucleate *Rhizoctonia* and *Rhizoctonia solani* on mung bean (*Vigna radiata*). *Plant Soil*, 238: 235–244.

Kerry, B.R. (1987). Biological control. In: *Principles and Practice of Nematode Control in Crops*, (Eds.) Brown, R.H. and B.R. Kerry. Academic Press, New York, pp. 233–263.

Kerry, B.R. (2001). Exploitation of the nematophaghous fungal *Verticillium chlamydosporium* Goddard for the biological control of root-knot nematodes (*Meloidogyne* spp.). In: *Fungi as Biocontrol Agents*, (Eds.) Butt, T.M., Jackson, C. and N. Magan. CAB International, USA, pp. 155–167.

Kjoller, R. and Rosendahl, S. (1996). The presence of the arbuscular mycorrhizal fungus *Glomus intraradices* influences enzymatic activities of the root pathogen *Aphanomyces euteiches* in pea roots. *Mycorrhiza*, 6: 487–491.

Krassilstschik, J. (1888). La production industrielle des parasites vegeteux pour la destruction des insectes nuisible. *Bulletin Science France Belgique*, 19: 461–472.

Lacey, L.A. and Goettel, M.S. (1995). Current developments in microbial control of insect pests and prospects for the early 21st century. *Entomophaga*, 40: 1–25.

Larsen, J. and Bodker, L. (2001). Interactions between pea root-inhabiting fungi examined using signature fatty acids. *New Phytol.*, 149: 487–493.

Lewis, K., Whipps, J.M. and Cooke, R.C. (1989). Mechanisms of biological disease control with special reference to the case study of *Pythium oligandrum* as an antagonist. In: *Biotechnology of Fungi for Improving Plant Growth*, (Eds.) Whipps, J.M. and R.D. Lumsden. Cambridge University Press, Great Britain, pp.191–217.

Lingua, G., D'Agostino, G., Massa, N., Antosiano, M. and Berta, G. (2002). Mycorrhiza-induced differential response to a yellows disease in tomato. *Mycorrhiza*, 12: 191–198.

Liu, R.J. (1995). Effect of vesicular-arbuscular mycorrhizal fungi on verticillium wilt of cotton. *Mycorrhiza*, 5: 293–297.

Lumsden, R.D. and Lewis, J.A. (1989). Selection, production, formulation and commercial use of plant disease biocontrol fungi: problems and progress. In: *Biotechnology of Fungi for Improving Plant Growth*, (Eds.) Whipps, J.M. and R.D. Lumsden. Cambridge University Press, Great Britain, pp.171–190.

Mark, G.L. and Cassells, A.C. (1996). Genotype-dependence in the interaction between *Glomus fistulosum*, *Phytophthora fragariae* and the wild strawberry (*Fragaria vesca*). *Plant Soil*, 185: 233–239.

Matsubara, Y., Kayukawa, Y., Yano, M. and Fukui, H. (2000). Tolerance of asparagus seedlings infected with arbuscular mycor-rhizal fungus to violet root rot caused by *Helicobasidium mompa*. *J. Jpn. Soc. Hortic. Sci.*, 69: 552–556.

Matsubara, Y., Ohba, N. and Fukui, H. (2001). Effect of arbuscular mycorrhizal fungus infection on the incidence of Fusarium root *rot* in asparagus seedlings. *J. Jpn. Soc. Hortic. Sci.*, 70: 202–206.

Matsubara, Y., Tamura, H. and Harada, T. (1995). Growth enhance-ment and verticillium wilt control by vesicular-arbuscular mycorrhizal fungus inoculation in eggplant. *J. Jpn. Soc. Hortic. Sci.*, 64: 555–561.

Metchnikoff, E. (1879). Maladies des hannetons duble. *Zapiski imperatorskogo obshcestaa sel'skago Khlzyaistra yuzhnoi rossii*, p. 17–50.

Meyer, J.M. and Dehne, H.W. (1986). The influence of VA mycorrhizae on biotrophic leaf pathogens. In: *Physiological and Genetical Aspects of Mycorrhizae* (Eds.) Gianinazzi-Pearson, V. and Gianinazzi, S. Institut Na-tional de la Recherche Agronomique, Paris, pp. 781–786.

Morandi, D., Gollotte, A. and Camporota, P. (2002). Influence of an arbuscular mycorrhizal fungus on the interaction of a binucleate *Rhizoctonia* species with Myc^+ and Myc^- pea roots. *Mycorrhiza*, 12: 97–102.

Morin, C., Samson, J. and Dessureault, M. (1999). Protection of black spruce seedlings against cylindrocladium root rot with ectomycorrhizal fungi. *Can. J. Bot.*, 77: 169–174.

Nemec, S., Datnoff, L.E. and Strandberg, J. (1996). Efficacy of biocontrol agents in planting mixes to colonize plant roots and control root diseases of vegetables and citrus. *Crop Prot.*, 15: 735–742.

Newsham, K.K., Fitter, A.H. and Watkinson, A.R. (1995). Arbuscular mycorrhiza protect an annual grass from root pathogenic fungi in the field. *J. Ecol.*, 83: 991–1000.

Niemira, B.A., Hammerschmidt, R. and Safir, G.R. (1996). Postharvest suppression of potato dry rot (*Fusarium sambucinum*) in prenuclear minitubers by arbuscular mycorrhizal fungal inoculum. *Amer. Potato J.*, 73: 509–515.

Norman, J.R., Atkinson, D. and Hooker, J.E. (1996). Arbuscular mycorrhizal fungal–induced alteration to root architecture in strawberry and induced resistance to the root pathogen *Phytophthora fragariae. Plant Soil*, 185: 191–198.

Pal, K.K. and Gardener, B.S. (2006). Biological control of plant pathogens. *The Plant Health Instructor* DOI: 10.1094/PHI-A-2006-1117-02.

Papavizas, G.C. (1981). *Biological Control in Crop Production*. Allanheld Osmun and Co., Totowa, New Jersey.

Papavizas, G.C. (1985). *Trichoderma* and *Gliocladium:* Biology, ecology and potential for biocontrol. *Ann. Rev.Phytopathol.*, 23: 23–54.

Pedersen, E.A., Reddy, M.S. and Chakravarty, P. (1999). Effect of three species of bacteria on damping–off, root rot development, and ectomycorrhizal colonization of lodgepole pine and white spruce seedlings. *Eur. J. Forest Pathol.*, 29: 123–134.

Perrin, R. (1990). Interactions between mycorrhizae and diseases caused by soil–borne fungi. *Soil Use Manag.* 6: 189–195.

Powell, K.A. and Faull, J.L. (1989). Commercial approaches to the use of biological control agents. In: *Biotechnology of Fungi for Improving Plant Growth,* (Eds.) Whipps, J.M. and R.D. Lumsden. Cambridge University Press, Great Britain, pp.259–275.

Pozo, M.J., Cordier, C., Dumas-Gaudot, E., Gianinazzi, S., Barea, J.M. and Azcon-Aguilar, C. (2002). Localized versus systemic effect of arbuscular mycorrhizal fungi on defence responses to *Phytophthora* infection in tomato plants. *J. Exp. Bot.*, 53: 525-534.

Pozo, M.J., Dumas-Gaudot, E., Slezack, S., Cordier, C., Asselin, A., Gianinazzi, S., Gianinazzi-Pearson, V., Azcon-Aguiiar, C. and Barea, J.M. (1996). Induction of new chitinase isofonns in tomato roots during interactions with *Glomus mosseae* and (or) *Phytophthora nicotianae* var. *parasitica. Agronomie,* 16: 689–-697.

Ratti, N., Alam, M., Shanna, S. and Janardhanan, K.K. (1998). Effects of *Glomus aggregatum* on lethal yellowing disease of *Java citronella* caused by *Pythium aphanidermatum. Symbiosis*, 24: 115–126.

Reddy, B.N., Raghavender, C.R. and Hindumathi, A. (2006). Mass multiplication of mycorrhizal inoculum: Use of sorghum roots for rapid culturing of *Glomus fasciculatum. Natl. Acad. Sci. Lett.*, 29(9 and 10): 355–359.

Roberts, D.W. and Humber, R.A. (1981). Entomogenous fungi. In: *Biology of Conidial Fungi,* (Eds.) Cole, G.T. and B. Kendrick. Academic Press, London, pp.201–236.

Sampangi, R., Perrin, R. and Le Tacon, F. (1986). Disease suppression and growth promotion of Norway spruce and Douglas-fir seedlings by the ectomycorrhizal fungus *Laccaria laccata* in forest nurseries. In: *Physiological and Genetical Aspects of Mycorrhizae*, (Eds.) Gianinazzi-Pearson, V. and S. Gianinazzi. Institut National de la Recherche Agronomique, Paris, pp. 799–806.

Shaul, O., Galili, S., Volpin, H., Ginzberg, I., Elad, Y., Chet, I. and Kapulnik, Y. (1999). Mycorrhiza-induced changes in disease severity and PR protein expression in tobacco leaves. *Mol. Plant-Microbe Interact.*, 12: 1000–1007.

Singh, C.S. (1992). Mass inoculum production of vesicular-arbuscular (VA) mycorrhizae. 2. Impact of N_2-fixing and P-solubilizing bacterial inoculation on VA-mycorrhiza. *Zentralblatt für Mikrobiologie*, 147: 503–508.

Singh, R., Adholeya, A. and Mukerji, K.G. (2000). Mycorrhiza in control of soil-borne pathogens. In: *Mycorrhizal Biology*, (Eds.) Mukerji, K.G., Chamalo, B.P. and J. Singh. Kluwer Academic–Plenum Publishers, New York, pp. 173–196.

Slezack, S., Dumas-Gaudot, E., Rosendahl, S., Kjoller, R., Paynot, M., Negrel, J. and Gianinazzi, S. (1999). Endoproteolytic activities in pea roots inoculated with the arbuscular mycorrhizal fungus *Glomus mosseae* and (or) *Aphanomyces euteiches* in relation to bioprotection. *New Phytol.*, 142: 517–529.

Soper, R.S. and Ward, M.G. (1981). Production, formulation and application of fungi for insect control. In: *Biological Control in Crop Production*, (Ed.) G. Papavizas. Allanheld Osmun and Co., Totowa, New Jersey, pp. 161–180.

St-Arnaud, M., Hamel, C., Vimard, B., Caron, M. and Fortin, J.A. (1997). Inhibition of *Fusarium oxysporum* f. sp. *dianthi* in the non-VAM species *Dianthus caryophyllus* by co-culture with *Tagetes patula* companion plants colonized by *Glomus intraradices. Can. J. Bot.*, 75: 998–1005.

St-Arnaud, M., Hamel, C., Caron, M. and Fortin, J.A. (1994). Inhibition of *Pythium ultimum* in roots and growth substrate of mycorrhizal *Tagetes patula* colonized with *Glomus intraradices. Can. J. Plant Pathol.*, 16: 187–194.

Stirling, G.R., Licastro, K.A., West, L.M. and Smith, L.J. (1998). Development of commercially acceptable formulations of the nematophagous fungus *Verticillium chlamydosporium. Biological Control*, 11: 215–221.

Strobel, N.E. and Sinclair, W.A. (1991). Influence of temperature and pathogen aggressiveness on biological control of Fusarium root rot by *Laccaria bicolor* in Douglas-fir. *Phytopathology*, 81: 415–420.

Sylvia, D.M. and Chellemi, D.O. (2001). Interactions among root-inhabiting fungi and their implications for biological control of root pathogens. *Adv. Agron.*, 73: 1–33.

Templeton, G.E. (1982). Status of weed control with plant pathogens. In: *Biological Control of Weeds with Plant Pathogens*, (Eds.) Charudattan, R. and H.L. Walker. John Wiley and Sons, New York, pp. 29–44.

Templeton, G.E. and Heiny, D.K. (1989). Improvement of fungi to enhance mycoherbicide potential. In: *Biotechnology of Fungi for Improving Plant Growth*, (Eds.) Whipps, J.M. and R.D. Lumsden. Cambridge University Press, Great Britain, pp.127–151.

Thomas, L., Mallesha, B.C. and Bagyaraj, D.J. (1994). Biological control of damping-off of cardamom by the VA mycorrhizal fungus, *Glomusfasciculatum. Microbiol. Res.*, 149: 413–417.

Torres-Barragan, A., Zavaleta-Mejia, E., Gonzalez-Chavez, C. and Ferrera-Cerrato, R. (1996). The use of arbuscular mycorrhizae to control onion white rot (*Sclerotium cepivorum* Berk.) under field conditions. *Mycorrhiza*, 6: 253–257.

Traquair, J.A. (1995). Fungal biocontrol of root diseases: Endomy-corrhizal suppression of cylindrocarpon root rot. *Can. J. Bot.*, 73: S89–S95.

Trotta, A., Varese, G.E., Gnavi, E., Fusconi, A., Sampo, S. and Berta, G. (1996). Interactions between the soilborne root pathogen *Phytophthora nicotianae* var. *parasitica* and the arbuscular mycorrhizal fungus *Glomus mosseae* in tomato plants. *Plant Soil*, 185: 199–209.

Vigo, C., Norman, J.R. and Hooker, J.E. (2000). Biocontrol of the pathogen *Phytophthora parasitica* by arbuscular mycorrhizal fungi is a consequence of effects on infection loci. *Plant Pathol.*, 49: 509–514.

Walker, R. (1987). Foreword. In: *Toxicological Aspects of Food*, (Ed.) K. Miller. Elsevier Applied Science, London, pp. v–ix.

Whipps, J.M. (1986). Use of microorganisms for biological control of vegetable diseases. *Aspects Appl. Biol.*, 12: 75–94.

Whipps, J.M. (2004). Prospects and limitation for mycorrhizas in biocontrol of root pathogens. *Can. J. Bot.*, 82: 1198–1227.

Whipps, J.M. and Lumsden, R.D. (2001). Commercial use of fungi as plant disease biological control agents: Status and prospects. In: *Fungi as Biocontrol Agents*, (Eds.) Butt, T.M., Jackson, C. and N. Magan. CAB International, USA, pp. 9–22.

Whipps, J.M., Lewis, K. and Cooke, R.C. (1988). Mycoparasitism and plant disease control. In: *Fungi in Biological Control Systems*, (Ed.) M.N. Burge. Manchester University Press, Great Britain, pp. 161–187.

Wood Mackenzie (1987). *Agrochemical Service*, March 1986. Wood Mackenzie, Edinburgh.

Wraight, S.P., Jackson, M.A. and de Kock, S.L. (2001). Production, stabilization and formulation of fungal biocontrol agents. In: *Fungi as Biocontrol Agents*, (Eds.) Butt, T.M., Jackson, C. and N. Magan. CAB International, USA, pp. 253–287.

Zorner, P.S., Evans, S.L. and Savage, S.D. (1993) Perspectives on providing a realistic technical foundation for the commercialization of bioherbicides. In: *Pest Control with Enhanced Environmental Safety*, (Eds.) Duke, S.O., Menn, J.J. and J.R. Plimmer. American Chemical Society, Washington, DC, pp. 79–86.

Plant Diseases Management for Sustainable Agriculture *Pages* **280–291**
Editor: **Dr. Shahid Ahamad**
Published by: **DAYA PUBLISHING HOUSE**

Chapter 23

Potential Entomopathogenic and Antagonistic Fungi

Nasim Ahmad[1], A.A. Jafri[1], O.M. Bambawale[1] and Shahid Ahamad[2]
[1]National Centre for Integrated Pest Management,
IARI Campus, New Delhi – 110 012
[2]Regional Agricultural Research Station, SKUAST-J, Rajouri – 185 131

Fungi can be facultative or obligate parasite of different living organisms such as plants, vertebrates, invertebrates, and other microorganisms. It is well to note that these organisms can be plant or animal "pathogens" and the disease symptoms or pathologies covered by these fungi are generally termed "mycoses". Mycoses covered by parasitic fungi of insects and related arthropods sometimes are severe enough to eliminate a host population in a given habitat. Such outbreaks, termed epizootic, are generally attributable to a combination of circumstances that favor the fungal disease. Paramount among these are physical factors of the environment, particularly moisture and a dense host population. Biological control of pathogens, that is, the total or partial destruction of pathogen populations by other organisms, occurs routinely in nature. There are, for example, several diseases in which the pathogen cannot develop in certain areas either because the soil, called suppressive soil, contains microorganisms antagonistic to these pathogens or because the plant that is attacked by a pathogen has also been naturally inoculated with antagonistic microorganisms before or after the pathogen attack. Sometimes, the antagonistic microorganisms may consist of avirulent strains of the same pathogen that destroy or inhibit the development of the pathogen, as happens in *hypovirulence* and cross protection. In some cases, even higher plants reduce the amount of inoculum either by trapping available pathogens (trap plants) or by releasing into the soil substances toxic to the pathogens. Agriculturists have increased their efforts to take advantage of such natural biological control, which can be used effectively against several plant diseases. Biological antagonisms, although subject to numerous ecological limitations, are expected to become an important part of the control measures employed against many more diseases.

The main problems encountered by the mass producers are with the taxonomic identification of various fungi, correct execution of microbial techniques, quality control and registration. The mass production of these group of Bioagents is comparatively not much complicated and a standard procedure with minor variations is applicable to them. Entrepreneurs can get a lot of information on different aspects of these Bioagents through several websites on the internet.

Entomopathogenic Fungi

Fungi which have a parasitic or pathogenic association with arthropods, particularly the true insects (Hexapoda), and mites (Acrania) are termed as "entomogenous" or "entomopathogenic" fungi. Here we will discuss only those fungi which are associated with true insects. With renewed interest in entomological biological control and integrated pest management, more research has been directed to the role of naturally occurring entomogenous fungi in the regulation or suppression of population of insect pests. In addition, entomogenous fungi are receiving more attention as artificially introduced microbial insecticides and also are being identified as key natural mortality factors in the ecology of many important invertebrate pests. It should be pointed out, however, that entomogenous fungi hold a relatively modest position as compared to bacteria, as microbial control agents today. There are approximately 750 species of fungi found through out the five major taxa comprising the division Eumycota. Eumycota have been identified as being entomogenous or having confirmed or suspected entomogenous spp. The major taxa are the Mastigomycotina (zygosporic fungi), the Ascomycotina (sac fungi), the Basidiomycotina (club fungi) and the Deuteromycotina, better known as Fungi Imperfecti. The potential entomogenous fungi are given in Table 23.1.

Table 23.1: Entomogenous Fungi Being Considered as Potential Microbial Insecticides

Fungus	*Infective stage*	*Insect host*	*Habitat*
Beauveria	Conidia	Beetles Caterpillars	Foliage
Metarhizium	Conidia	Froghoppers Beetles	Foliage
Nomuraea	Conidia	Caterpillars	Foliage
Verticillium	Conidia	Aphids Whitefly Scales	Glasshouse Foliage

Beauveria

During the last 15 years, special attention has been given to the use of *Beauveria bassiana* as a component of an integrated pest management strategy against several agriculturally important pests including whiteflies and the Lepidopteran pests including *Helicoverpa armigera*. Generally, two treatments made at 15-day intervals with 1.5 kg/ha concentrated product of *B. bassiana* ($3.\ 0 \times 10^9$ conidia/g) are effective.

Metarhizium

The green muscardine, *Metarhizium anisopliae* var. *anisopliae* and var. major like the white muscardine fungi have been studied for 100 years. However, promising field results with this fungus have occurred only within the last 15 years. The rhinoceros beetle, *Orycetes rhinoceros*, a serious pest in coconut growing regions of the South Pacific, has been successfully controlled by *Metarhizium anisopliae* var. *major*. *M. anisopliae* was successfully used to control spittel bug on sugarcane and in pastures for nearly eight years. As much as 50,000 ha in a single province was aerially treated at 50 l/ha with 6×10^{11} to 1.2×10^{12} conidia/l. Although *M. anisopliae* occurs in natural epizootics, introducing the fungus early in the season appears to keep spittel bugs bellow damaging levels. *M. anisopliae* has also been effective in fields trials against mosquitoes. It has reduced pasture cockchafer populations when applied as spore bait. In China, *M. anisopliae* was pathogenic to *Colasposoma* when applied to the soil at a rate of 75 kg/ha. From 50 to 60 per cent mortality of pecan weevil also has been obtained in the U.S. with applications of *M. anisopliae*. At this time, there is considerable interest in developing *Metarhizium anisopliae* into a commercial product in our country for the management of white grubs in sugarcane and for the control of jassids and termites and the program is expanding to include control of other insects.

Nomuraea

Epizootics of *Nomuraea rileyi* on lepidopteran pests occur regularly in agro-ecosystems in India and other countries. Extensive research has been conducted on the survival of the fungus in soybean fields and a model has been proposed for predicting the level of infection in velvetbean caterpillar population in Florida. Early close planting of soybeans and cotton, that influences microclimatic field conditions, also enhanced the development of *N. rileyi* generally occur after caterpillars have caused damage and too late to prevent losses in soybean yield, the prophylactic use of *N. rileyi* has been tested using high conidial doses with encouraging results. A dust formulation of *N. rileyi* with 12.4×10^{10} conidia/g applied early in the season at 1.1×10^{13}/acre to soybeans caused a premature epizootic. These studies suggest that a prophylactic treatment 2 to 4 weeks before economic levels of injury are anticipated may provide control of defoliating caterpillars.

Verticillium

Verticillium (=*Cephalosporium*) *lecanii*, reported from scale insects for some time, has become important as a microbial insecticide for use in glasshouses. A spore preparation containing 41×10^7 active spores/g either undiluted or as a 10 per cent concentration (diluted with talc or water) was applied to citrus leaves infested with *Coccus hesperidum* L. and in the greenhouse, 85 to 100 per cent mortality was achieved within 3 weeks. Although *V. lecanii* killed all stages of the brown soft scale, it was unable to complete its development because of lack of nutrients. Hence the fungus, unable to survive and spread within the scale population, must be repeatedly introduced. Hall and Burges 1979 in a well-conceived research programme, achieved excellent control of *Myzus persicae*, a major pest; of *Chrysnathemums* in glasshouses with *Verticillium lecanii*. Conidia (10^5 to 10^7/ml) and blastospores (10^7 to 10^8/ml), suspended in phosphate buffer containing 0.02 per cent Triton X-100 as a wetting agent, sprayed to run-off gave consistent results when sprayed in the evening and when plants were covered with polyethylene balckout sheets to boost humity. In similar trials on *Chrysanthemums*, *Macrosiphoniella sanborni* and *Brachycaudus helychrysi* were also controlled a slow to high population densities during all seasons. Although differences in species' susceptibility and effect of spore concentration were detected in Laboratory assay, no differences appeared in glasshouse trials. In trials on cucumbers, *Verticillium lecanii* was also effective against *Aphis gossypii* (within 14 days of treatment), and the whitefly, *Trialeurodes vaporariourum* in England.

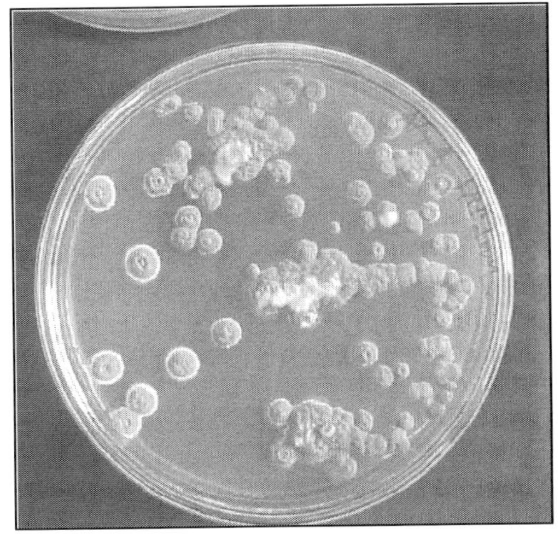

Nomuraea rileyi

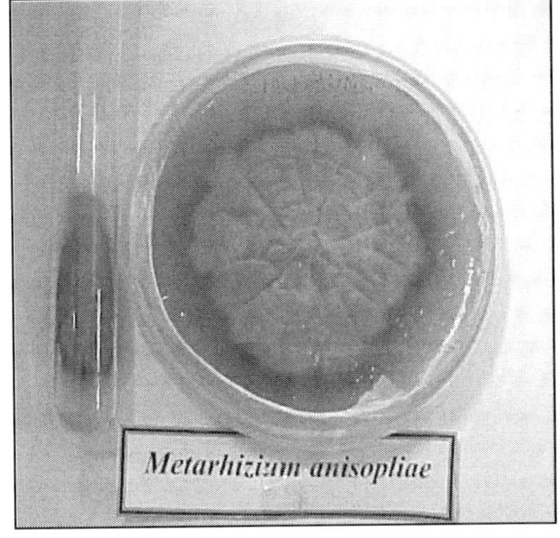

Metarhizium anisopliae

Beauveria bassiana

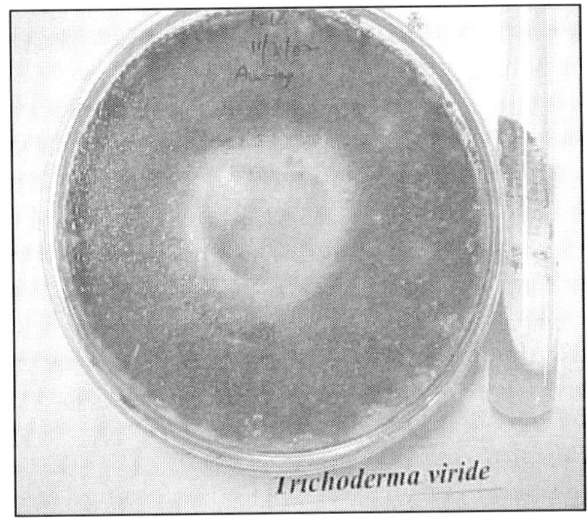

Trichoderma viride

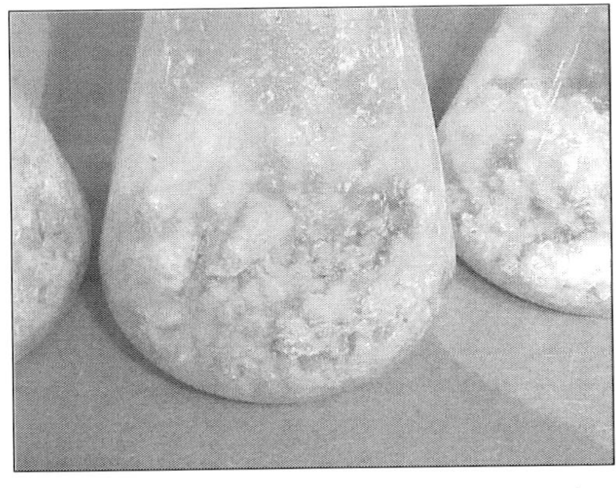

Nomuraea rileyi multiplied on jowar seeds

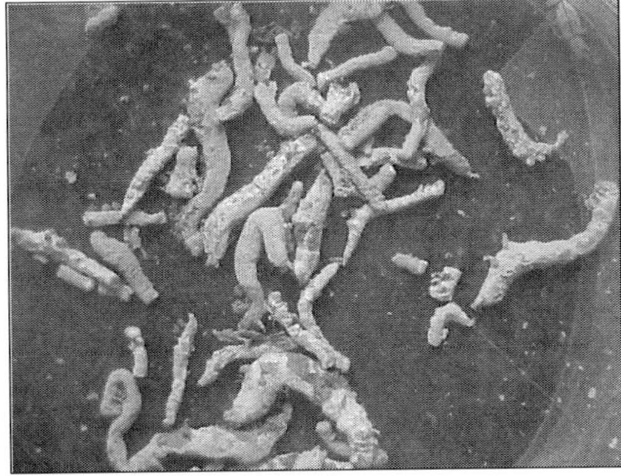

Nomuraea rileyi infected larvae

Figure 23.1

Systematics of Genera *Nomuraea, Beauveria, Metarhizium* and *Verticillium* Species Used as Biocontrol Agent for Insects

The genera *Nomuraea, Beauveria, Metarhizium* and *Verticillium* belong to taxa Deuteromycotina–Hyphomycetes- Moniliales (Ainsworth *et al.*, 1973).

The genus *Nomuraea* has wide host range particularly among *Lepidopterous* insect pests. The following medium is used for the growth of *Nomuraea* species:

Sabouraud Maltose Agar with Yeast (SMAY) Medium

Agar	20 g
Maltose	40 g
Peptone	10 g
Yeast extract	5 g
Water	1 ltr

I. Nomuraea Maublanc–*Bull Soc. Mycol.* Fr. 19: 296, 1903

The genus colour and phialides shape *N. rileyi* indicate a close resemblance with *Penicillium* but differs in numerous conidial structures in compact cluster along the conidiophores.

Key to species of *Nomuraea*

A. Culture green, conidiophores mononematous, conidia broadly elliptical to cylindrical 3.5–4.5 x 2–3.5 mm size–*N. rileyi.*

AA. Culture in purple shades, conidiophore mono or synnematous, conidia cylindrical, slightly curved, 4.0–6.0 x 1.2–1.5 mm size–*N. atypicola*.

***Nomuraea rileyi* (Farlow) Samson, *Studies in Mycology* 6: 81-85, 1974**

= *Botrytis rileyi* Farlow–*Rep. U.S. Comm. Agric.* 1883: 121.

Colonies

Slow growing, at first-white, changing to olive.

Conidiophores

Erect, septate. 160 mm long, 2-2.5 μm in diam, forming dense cluster of branches each bearing 2-3 compact phialides, around the stalk.

Conidiogenous Cell

Structures smooth, hyaline to slight green phialides 4.5–6.5 x 2-3 μm, cylindrical, occasionally with swollen base, neck short or absent.

Conidia

Dry in divergent chains, broadly ellipsoidal, cylindrical, smooth, pale green to green in mass, 3-5 x 2-3 μm size.

II. Beauveria Vuill. Bull. Soc. Bot. Fr. 59: 40,1912.

Type: *B. bassiana* (Bals–Criv.) Vuill (= *Botrytis bassiana* Bals–Criv);

B. effusa (Beauv) Vuill (=*Botrytis effusa* Beauv)

The fungus grows well on PDA, Sabourad's agar at 24°C and can be isolated from infected insect, plating them in humid chamber and isolating the spore mass under stereo binocular or dissecting microscope.

Beauveria bassiana (Bals–Criv.) Yuill

Mycelium

Hyaline, septate, 3.5 µm wide.

Conidiophores

Single or branched, abundant, arising from vegetative hyphae bearing groups of clustered conidiogenous cells.

Conidiogenous Cells

Polyblastic, globose to flask shaped with well developed rachis, upto 20 mm long 1-1.5 mm wide, 3-5 x 3-7 mm in diam.

Conidia

Develop basifugal series, blastic, smooth, hyaline globose to broadly ellipsoidal 1-4 x 1.5-5.5 mm in size.

III. Metarhizium Sorokin (1878)

The fungus is found infecting numerous insect species and causes diseases which Metschinikoff called "Green mascardine" due to dark green colour of the conidial masses, who initially recorded on *Cockchufer* of wheat, *Anisopliae austrica* from Russia as a *Entomophthora anisopliae* which later Sorokin (1878) made the type species *Metarhizium anisopliae* (Metsch.) Sorokin, of the genus *Metarhizium.*

This fungus can be grown on Potato Dextrose Agar and Rabbit food Agar (Rabbit food, Commercial pellets 25.0 g, Agar 15.0 g, Distilled water 1.0 liter) by boiling the rabbit food, steep for one hour, then filter through cheesecloth and add agar to filtrate.

***Metarhizium anisopliae* var. *anisopliae*:**

(*a*) Conidiogenous cell: Cylindrical to narrow ellipsoid

(*b*) Conidia

- (*i*) Shape: Cylindrical (often narrowed centrally)
- (*ii*) Conidial length: 7-9 mm
- (*iii*) Colour of conidial mass: Grass dark or olive green, Grey-green
- (*iv*) Colony growth and characteristics: Subhymenium with inflated hyphal bodies absent
- (*v*) Conidial chain and texture of conidial mass: Chain adhere laterally in columns, spore mass smooth surfaced

Fungal Antagonists

Biological control practices for direct protection of plants from pathogens involve the deployment of antagonistic microorganisms at the infection court before or after infection takes place. Although hundreds, possibly thousands, of microorganisms have been shown to interfere with the growth of plant pathogens in the laboratory, greenhouse, or field, and to provide some protection from the diseases caused by them, so far only few microorganism have been registered and are commercially available for use. These microorganisms include fungi: *Gliocladium virens,* for control of seedling

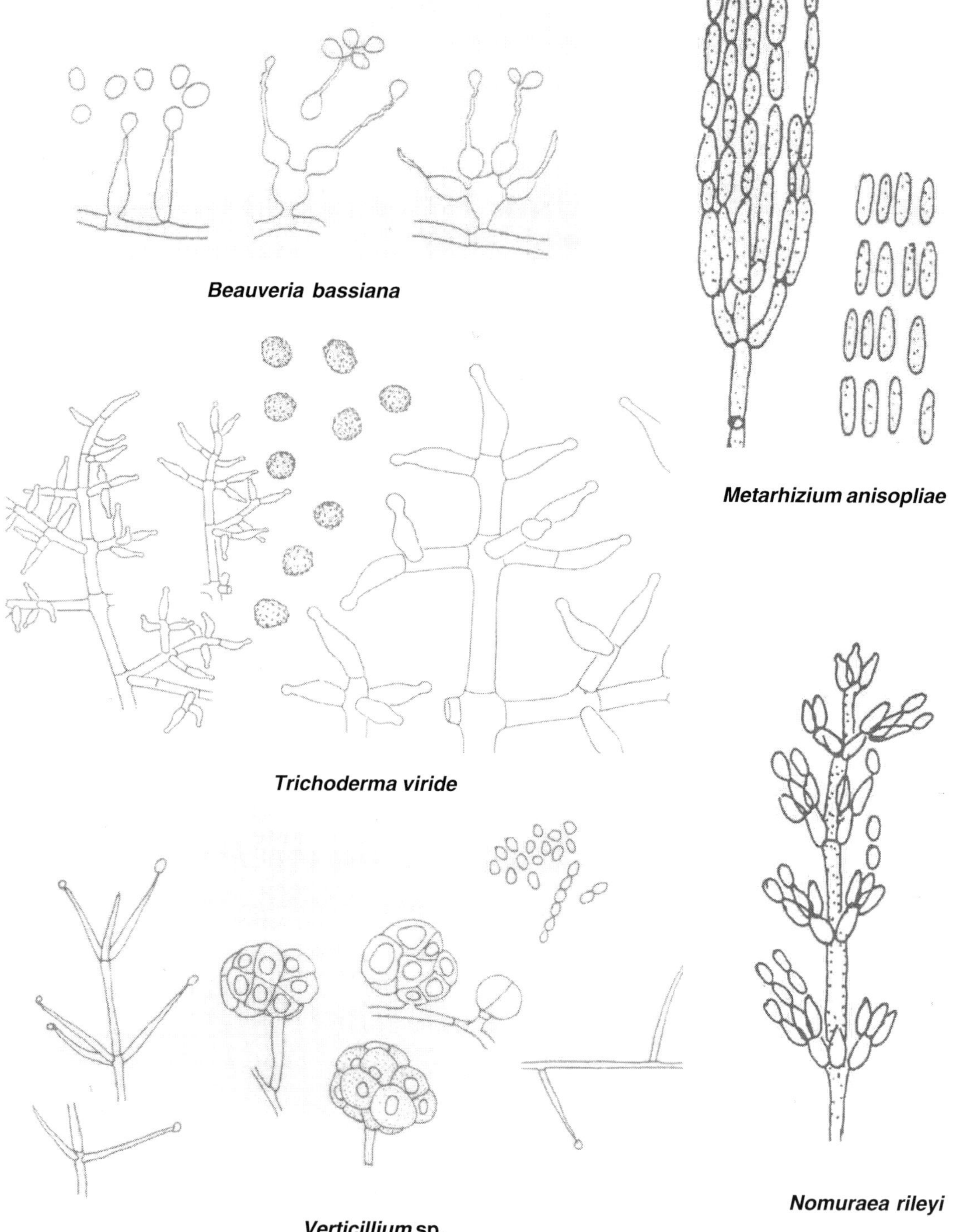

Figure 23.2: Microscopic Features of Important Fungal Bioagents

diseases of ornamental and bedding plants; *Trichoderma harzianum/T. viride*, for control of several soilborne plant pathogenic fungi and *Trichoderma harzianum/T. polysporum*, for control of wood decays. The other commercially available microorganisms are bacteria: *Agrobactrium radiobacter* for use against crown gall; *Pseudomonas fluorescsens*, for use against *Rhizoctonia* and *Pythium* damping-off of cotton; and *Bacillus subtilis*, used as a seed treatment. Although the actual use of these biological control products is rather limited, it is expected that these and other such products will find wide acceptance and will fill a real need in the not too distant future.

Soilborne Pathogens

The mycelium and resting spores (oospores) or sclerotia of several phytopathogenic soil fungi such as *Pythium, Phytophthora, Rhizoctonia, Sclerotinia,* and *Sclerotium* are invaded and parasitized (mycoparasitism) or are lysed (mycolysis) by several fungi, which as a rule are not pathogenic to plants. Several non-plant pathogenic fungi, including some *oomycetes, chytridiomycetes,* and *hyphomycetes*, and some pseudomonas and actinomycetous bacteria infect the resting spores of several plant pathogenic fungi. Among the most common mycoparasitic fungi are *Trichoderma* sp., mainly *T. barzianum* and *T. viride*. The latter fungus has been shown to parasitize mycelia of *Rhizoctonia* and *Sclerotium*, to inhibit the growth of many other fungi, for example, *Pythium, Phytophthora, Fusarium,* and *Heterobasidion* (Fomes), and to reduce the diseases caused by most of these pathogens. Other common mycoparasitic fungi *are Laetisaria arvalis* (*Corticium sp.*), a mycoparasite and antagonist of *Rhizoctonia and Pythium* also, *Sporidesmium sclerotiorum, Gliocladum virens*, and *Coniothyrium minitants*, all destructive parasites and antagonists of *Sclerotinia* sclerotiorum and all effectively controlling several of the *Sclerotinia* diseases; and the *Talaromyces flavus*, which parasitizes *Verticillium* and controls *Verticillium* wilt of eggplant. Also, some *Pythium* species parasitize species of *Phytophthora* and other species of *Pythium*. Several yeast, for example, *Pichia gulliermondii*, also parasitize and inhibit the growth of plant pathogenic fungi like *Botrytis and Penicillium.*

Aerial Pathogens

Many fungi have been shown to antagonise and inhibit numerous fungal pathogens of aerial plant parts. For example, *Chaetomium* sp. and *Athelia bombacina* suppress *Venturia inaequalis* ascospre and conidia production in fallen and growing leaves, respectively. *Tuberculina maxima* parasitizes the white pine blister rust fungus *Cronartium ribicola; Darluca filum* and *Verticillium lecanii* parasitize several rusts; *Ampelomyces quisqualis* parasitizes several powdery mildews; *Tilletiopsis* sp. parasitizes the cucumber powdery mildew fungus *Spaerotheca fuliginea*; and *Nectria inventa* and *Gonatobotrys simplex* parasitize two pathogenic species of *Alternaria.*

Mechanisms of Action

The Mechanisms by which antagonistic microorganisms affect pathogen populations are not always clear, but they are generally attributed to one of four effects:

1. Direct parasitism or lysis and death of the pathogen.
2. Competition with the pathogen for food.
3. Direct toxic effects on the pathogen by antibiotic substances released by the antagonist.
4. Indirect toxic effects on the pathogen by volatile substances, such as ethylene, released by the metabolic activities of the antagonist.

Many of the antagonistic microorganisms are naturally present in crop soils and exert a certain degree of biological control over one or many plant pathogens regardless of human activities. Humans, however, have been attempting; increase the effectiveness of antagonists either by introducing new and larger populations of antagonists, for example, *Trichoderma harzianum* in fields where they are lacking, and/or by adding soil amendments that serve as nutrients for, or otherwise stimulate growth of, the antagonistic microorganisms and increase their inhibitory activity against the pathogen. Unfortunately, although both approaches are effective in the laboratory and in the greenhouse, neither has been particularly successful in the field. New microorganisms added to the soil of the field cannot compete with the existing microflora and cannot maintain themselves for very long. Also, soil amendments, so far, have not been selective enough to support and build up only the populations of the introduced or existing antagonists. Thus, their potential for eventual disease control is quite limited. There are several cases of successful biological control of plant pathogens when the antagonistic microorganism is used for direct protection of the plant from infection by the pathogen.

Biological Control of Soil Borne Diseases

The principal fungi used as biological control agents against soilborne diseases are mainely, *Gliocladium virens* and *Trichoderma harzianum*. They are used in potting mixes, are mixed with soil, or are used and solid matrix in seed treatments. They are effective against damping-off diseases of ornamentals and vegetables caused by *Pythium*, *Phytophthora*, *Sclerotium*, and some other fungi. In addition, *Sporidesmium sclerotiorum*, *Conithyrium minitants*, *Talaromyces flavus*, and others give experimental control of some diseases caused by *Sclerotinia*, *Rhizoctonia*, and *Verticillium*. Some species of *Pythium*, such as *Pythium nunn* and *P. oligandrum*, protect potted ornamentals and vegetable from plant pathogenic species of *Pythium*, and *Talaromyces flavus* and binucleate *Rhizoctonias* protect plants from the pathogenic *Rhizoctonia solani*. Experimental control of several other soilborne diseases has been obtained with many other fungi. Finally, *Fusarium* wilts of several crops, such as celery, cucumber, and sweet potato, caused by the respective formae specialis for *Fusarium oxysporum* have been reported to be successfully controlled by inoculating transplants or cutting with nonpathogenic strains of the same fungus. Some of these strains have been isolated from the vascular tissue of host plants that remained healthy while nearby plants had been killed by the wilt-inducing strains of the fungus. It is believed that the nonpathogenic strains not only compete with the pathogenic ones in the rhizosphere and for infection sites, but they also enhance the resistance of the host toward the pathogenic strains.

Biological Control of Diseases of Aerial Plant Parts

Many filamentous fungi and yeasts have been shown to be effective antagonists of fungi infecting the aerial parts of plants, for example, inoculation of postbloom, dead tomato flowers with conidia of *Cladosporium herbarum* or *Pencillium* sp. almost completely suppressed subsequent infection of developing fruits by *Botrytis cinerea*. Similarly, spraying spores of common bark saprophytes, such as *Cladosporium* sp. and *Epicoccum* sp., and of the soil fungus *Trichoderma* on pruning cuts of fruits trees has prevented infection by canker causing pathogens such as *Nectria galligena* and *Leucostoma* (*cytospora*) sp. sprays with *Trichoderma* in the fields also reduced *Botrytis* rot of strawberries and of grapes at the time of harvest and in storage. Several foliar diseases have also been reduced significantly (by more than 50 per cent) when the leaves were sprayed with spores of common phylloplane fungi, for example, *Alternaria*, *Cochliobolus*, *Septoria*, *Colletotrichum*, and *Phoma*; or with spores of hyperparasites, for example, the cucumber powdery mildew fungus *Sphaerotheca fuligiena* with spores of *Ampelomyces quisqualis*, or *Tilletiopsis*, the wheat leaf rust fungus *Puccinia recondita* with spores of *Darluca filum*, and the carnation rust fungus with *Verticillium lecanii*. None of the above, nor any of the numerous other know cases of fungal antagonism by fungi, is used as yet for practical control of any disease of aerial plant parts.

Biological Control of Post-harvest Diseases

Postharvest rots of several fruits could be reduced considerably by spraying with spores of antagonistic fungi and saprophytic yeasts at different stages of fruit development, or by dipping the harvested fruit in the inoculum. For example, yeast treatments reduced postharvest rotting of peach and apple. Also, significant reduction of citrus green mold (caused by *Penicillium digitatum*) was obtained by treating the fruit with the antagonist *Trichoderma viride*, whereas pre-harvest and post-harvest Botrytis rot of strawberries was reduced by seaveral sprays of *Trichoderma* spores on strawberry blossoms and young fruit. *Penicillium* rot of pineapple was reduced considerably by spraying the fruit with non-pathogenic strains of the pathogen. Similarly, several antagonistic yeasts protected grapes and tomatoes from *Botrytis, Penicillium*, and *Rhizoctonia* rots. In addition, the yeast *Candida oleophila* was approved for post harvest decay control in citrus and apples under the trade name Aspire.

Systematics of *Trichoderma* Species for Biocontrol of Plant Pathogens

The anamorphic stage belongs to Hypocreaceae while teleomorphs to genus Hypocrea that has nine species especially in soil and found wide spread.

Trichoderma species can readily be isolated from soil either through soil plate or soil dilution plate methods. It can be grown on number of media. However, *Trichoderma* Selective agar Medium (TSM) by Elad and Chet (1983) is more effective and well suited to laboratory ($MgSO_4.7H_20$ 0.2 g, K_2HPO_4 0.9 g, KCl 0.15 g, Ammonium nitrate 1.0 g, Glucose 3.0 g, Chloremphenicol 0.25 g, Dexon 60 WP (or) Aprol (Metalaxyl) 0.3 g, Penta chloro nitro benzene 0.20 g, Rose bengal 0.I5 g, Agar 18.0 g and distilled water 1 litre. Malt Extract Agar (MEA) (25.0g malt extract, agar 20.0 g oxide malt extract (powdered) in 1 litre water and PDA are some of the media for cultivation and maintenance of cultures.

Trichoderma harzianum Rifai. *Mycological Papers* 116,1969

Colonies

Growing rapidly, white green to bright green to dull green.

Mycelium

Septate, colourless, smooth. 1.5–2.5 mm.

Chlamydospores

Mostly globose, smooth, 6-12 mm in diam.

Conidiophores

Loose tuft, main branch produced numerous side branches specially lower portion.

Phialides

Aerial false verticellate upto five, short, skittle shaped, narrow at the base, attenuate abruptly, sharp pointed neck,–75 x 3-4 mm.

Phialospores

Accumilated at the tip of phialides, subglobose, short obovoid, often broad truncate base, smooth, pale green, much darker in mass, 2.8–3.2 x 2.5- 2.8 mm.

***Trichoderma viride* Per.ex.S. F.Gray *Nat. Arrang. br;* 1:560,1821.**

Colonies

Fast growing, surface smooth, become hairy, dark green typical coconut odour is emitted in old culture.

Mycelium

Hyaline, smooth, septate much branched.

Chlamydospores

Intercalary, globose, rarely ellipsoidal, 10–15 mm in diam.

Conidiophores

Arise in compact or loose tuft, main branches produced several side branches in groups of 2-3, all branches stand at wide angles.

Phialides

In false whorls beneath each terminal phialides, usually more than 2-3 phialides, 8-15 x 2-3 mm in size, curved, pin shaped, narrower at the base, widening above the middle, attenuated into long neck.

Phialospores

Globose or short obovoid broadly ellipsoidal, at distant apiculus like base, minute roughing of their wall, 3.5-4.5 mm in size, accumulated at the tip of each phialides pale green, smooth.

References

Fargues, J., Smits, N., Vidal, C., Vey, A., Vega, F., Mercadier, G. and Quimby, P. (2002). Effect of liquid culture media on morphology, growth, propagule production and pathogenic activity of the Hyphomycete, *Metarhizium flavoviride*. *Mycopathologia,* 154(3): 127–138.

Lacey, L.A., Erutos, R., Kaya, H..K. and Valis, P.(2001). Insect pathogens as biocontrol agents. *Biological Control,* 21: 230–248.

Lingappa, S. and Patil, R.K. (2002). *Nomuraea rileyi*: A potential mycoinsecticide. Department of Agricultural Entomology, University of Agricultural Sciences, Dharwad, Karnataka, India, pp. 30.

Mazumdar, D., Puzari, K.C. and Hazarika, L.K. (1995). Mass production of *Beauveria bassiana* and its potentiality on rice hispa. *Indian Psychopathology,* 48(3): 275–278.

McCoy, C.W., Sampson, R.A. and Boucias, D.G. (1988). Entomogenous fungi. In: *Handbook of Natural Pesticides, Vol. 5: Microbial Pesticides, Part A: Entomogenous Protozoa and Fungi,* (Eds.) C.M. Ignoffo and N.B. Mandava. CRC Press, Boca Raton, FL, pp. 151–236.

Patel, K.C., Yadav, D. N., Dube, H.C. and Patel, R.J. (1990). Laboratory and mass production studies with *Metarhizium anisopliae*. *Annale of Biology,* 6(2): 135–140.

Puzari, K.C., Sarmah, D.K. and Hazarika, L.K. (1997). Medium for mass production of *Beauveria bassiana* (Balsamo) Vuillemin. *Journal of Biological Control,* 11(1–2): 97–100.

Vimla Devi, P.S. and Prasad, Y.G. (1997). The entomofungal pathogen *Nomuraea rileyi*. In: *Information Bulletin,* Directorate of Oilseeds Research, Rajendranagar, Hyderabad, pp. 5.

Vimla Devi, P.S. and Prasad, Y.G. (1998). Fungus to control bollworm in cotton. In: *Science and Technology–Agriculture, The Hindu*, Dated 14th May.

Vimla Devi, P.S., Chowdary, A. and Prasad, Y.G. (2001). Cost effective multiplication of the entomopathogenic fungus *Nomuraea rileyi* (F) Samson. *Mycopathologia,* 151(1): 35–39.

Wraight, S. P. and Carruthers, R.I. (1998). Production, delivery and use of micoinsecticides for control of insect pests in field crops. In: *Biopesticides: Use and Delivery,* (Eds.) F.R. Hall and J.J. Menn. Humana Press, Totowa, pp. 233–269.

Wraight, S.P., Jackson, M.A., Kock, S.L. (2001). Production, stabilization and formulation of fungal biocontrol agents. In: *Fungi as Biocontrol Agents,* (Eds.) Butt, T.M., Jackson, C.W. and N. Magan. CABI Publishing, Wallingford, Oxon, U.K., pp. 385.

Plant Diseases Management for Sustainable Agriculture *Pages* **292–298**
Editor: **Dr. Shahid Ahamad**
Published by: **DAYA PUBLISHING HOUSE**

Chapter 24

Biopesticides: An Ecofriendly, Sustainable and Cost Effective Approach for Integrated Disease and Insect Pest Management of Agricultural Crops

B.K. Goswami[1], Rajesh Kumar Pandey[1], Sheoraj Singh[2] and Kabindra Singh Rathour[3]

[1]Amity Centre for Biocontrol and Plant Disease Management, Amity University, U.P.
[2]Division of Plant Pathology, [3]Division of Nematology,
Indian Agricultural Research Institute, New Delhi – 110 012

Bio-pesticides are a class of environmentally friendly pesticides that are used to control pests (fungi, insects, weeds and diseases) using living organisms in this case. In fact, fungi have long been known to infect and kill fungal, insects and mites diseases. The Chinese recorded infections of *Cordyceps* in the 7th century. During the 19th century in Italy, Agostino Bassi recorded the infection of silk worms by the pathogenic fungus *Beauveria bassiana.* Elie Metchnikoff won the Nobel Prize for his work in developing a biocontrol agent, the fungus *Metarhizium anisopliae*, as a method for the control of the cereal cockchafer. However, as stated earlier, fungi can be used to control insects, other fungi, and weeds.

It is hearting to observe the growing awareness among the farmers and policy makers about ecologically sustainable methods of pest management. More and more farmers are coming to realize the short-term benefits and long-term positive effects of the use of bioagents and other ecologically safe methods to tackle pests. The present article 'Biopesticides' is of much relevance in this context. 'Biopesticides' are certain types of pesticides derived from such natural materials as animals, plants, bacteria, and certain minerals. These include for example; fungi such as *Beauveria* sp., bacteria such as *Bacillus* sp., neem extract and pheromones. Similarly Canola oil and baking soda have pesticide applications and are considered as biopesticides. The use of these materials is widespread with applications to goliage, turf, soil, or other environments of the target insect pests. In a much simpler way we can say that these are pest management tools that are based on beneficial microorganisms (bacteria, viruses, fungi and protozoa), beneficial nematodes or other safe, biologically based active ingredients. Benefits of biopesticides include effective control of insects, plant diseases and weeds, as well as human and environmental safety. Biopesticides also play an important role in providing pest management tools in areas where pesticide resistance, niche markets and environmental concerns limit the use of chemical pesticide products.

Antagonistic Biocontrol Agents and their Available Bioformulations

Fungi to Control other Pathogenic Fungi

Several fungi are used to control the plant fungal pathogens. One example is the damping-off disease that affects seedlings. Other plant diseases form galls in and rots in plants while others infect the leaves (Table 24.1). There are fungal organisms that actually attack these different plant pathogens and have been used to control them. *Trichoderma* sp. have been used to control *Fusarium oxysporum, Rhizoctonia solani* and *Pythium* on a variety of plants.

Table 24.1: Name of Registered Microbial Products

Acremonium spp.	*Metarhizium anisopliae*
Aspergillus niger	*Metarhizium flavoviride*
Aspergillus terreus	Nuclear Polyhedrosis Viruses
Fusarium nivale	*Bacillus thuringienis*
Trichoderma harzianum	*Bacillus sphaericus*
Trichoderma viride	*Pseudomonas fluorescens*
Trichoderma virens	*Chladosporium oxysporum*
Paecilomyces lilacinus	*Verticillium lecanii*
Beauveria bassiana	VAM (*Glomus fasciculatum*)

Fungi to Control Harmful Insects

Many fungi are capable of infecting and killing insects. Mass inoculation using insect fungal pests was initiated by Krassilstchik in 1888. However, the advance in fungal insect biopesticides did not progress as quickly as initially thought. This was mainly due to the development of chemical pesticides. However, due to the environmental concerns of applying chemical pesticides, fungal insecticides are being revisited and explored. Some of the most widespread fungi used to control insects are *Beauveria bassiana, Metarhizium* spp., *Dermolepida albohirtum*, and *Lagenidium giganteum*. The application of fungal insecticides is usually done using fungal spores. However, fungal mycelia are also used as inoculant to control insect pathogens. Table 24.2 provides a few examples of fungal pesticides, the insects they target and some of the companies which supply the biopesticide.

Nematodes to Control Pathogenic Fungi

These are also called as Fungivores or myceliophagous nematodes. This group of nematodes has a spear similar to plant parasitic nematodes, but it is used to feed on fungal hyphae and spores. These nematodes by feeding upon harmful fungi accelerate the decomposition process and help in recycling of minerals and other nutrients from fungi.

Examples

Aphelenchus avenae

Aphelenchoides composticola

Ditylenchus myceliophagus

Ditylenchus triformis

Table 24.2: Name of Registered Microbial Products

Acremonium spp.	*Trichogramma japonicum*
Beauveria bassiana	*Trichogramma brasilliensis*
Metarhizium spp.	*Trichogramma achaeae*
Lagenidium giganteum	*Trichogramma pretiosum*
Verticillim lecanii	*Trichogramma exiguum*
Nomuraea rileyi	*Bracon* species
Trichogramma chilonis	

Nematodes to Control Harmful Insects

These are also called as entomophilic or entomopathogenic nematodes. Entomopathogenic nematodes belonging to the families Steinernematidae and Heterorhabditidae, are soil dwelling lethal parasites of insects that are used for inundative, augementative or inoculative biological control. They are environment friendly.

The members of family Steinernematidae and Heterirhabditidae are unique in their ability to carry and release specific bacteria of the genus *Xenorhabdus* and *Photorhabdus* respectively) inside the insect haemocoel, where they cause septicaemia and kill the insect host, and in turn, provide nourishment for the developing nematodes in the insect cadaver. Till now about 56 species of the genus *Steinernema* and 13 of *Heterorhabditis* have been reported throughout the world. Some of these nematodes are being mass multiplied on artificial media and have been commercialized for the biological control of insects. These nematodes have been found to be the highly effective against the insect pests of various crops.

The following species of these nematodes are being used as bio control agents for controlling the insect pests of various crops and other important plants in India and throughout the world.

Steinernema carpocapsae (Weiser,1955) Wouts *et al*, 1982

Steinernema glaseri (Steiner,1929) Wouts *et al*, 1982

Steinernema feltiae (Filipjev, 1934) Wouts *et al*, 1982

**Steinernema thermophilum* Ganguly and Singh, 2000

**Heterorhabditis indica* Poiner *et al*, 1992

Heterorhabditis bacteriophora Poiner,1976

(*Described from India)

Besides the members of the genus *Steinernema* and *Heterorhabditis* there are a few other nematodes are known to control the insect pests. These are:

1. *Mermis nigrescens*–a parasite of grasshopper
2. *Romanomermis culicivorax*–a parasite of mosquitoes

Relevance of Bio-pesticides

25 million cases of acute occupational pesticide poisoning in developing countries are being reported each year (WHO, 1990). 14 per cent of all known occupational injuries and 10 per cent of all fatal injuries are caused by pesticides (ILO,1996). Obsolete pesticides are being stored in developing countries-20,000 tonnes in Africa alone. Pesticide residues in agricultural commodities are being the issue of major concern besides their harmful effect upon human life, wild life and other flora and fauna. Equally worrying thing is about development of resistance in pest to pesticides. The only solution of all these is use of 'Biopesticide' that can reduce pesticide risks, as-

1. Biopesticide would sustain the soil health and improve the diversity of beneficial microorganisms without leaving any toxic residues, therefore, the environmental quality and plant vigour improves and soil degradation is arrested.
2. Biopesticides are best alternatives to conventional pesticides and usually inherently less toxic than conventional pesticides.
3. Biopesticides generally affect only the target pest and closely related organisms, in contract to broad spectrum, conventional pesticides that may affect organisms as rent as birds, insects, and mammals.
4. Biopesticides often are effective in very small quantities and often decompose quickly, thereby resulting in lower exposures and largely avoiding the pollution problems caused by conventional pesticides.
5. When used as a fundamental component of Integrated Pest Management (IPM) programs, biopesticides can greatly decrease the use of conventional pesticides, while crop yields remain high.
6. Amenable to small-scale, local production in developing countries and products available in small, niche markets that are typically unaddressed by large agrochemical companies.

Concept of Bio-pesticides Production Technology

The Plant Protectionists have been advocating for an eco-friendly, non-chemical and economic approach for plant disease control since the time most of the "chemical killer pesticides" *viz.*, DBCP, MBr, BHC, DDT, etc., were banned. Prior to this, high chemical fertilizer inputs like application of chemical pesticides in agriculture caused:

Economic Problems

Affected by availability and affordability of chemical fertilizers through imports and subsidies causing unstable burden on the fiscal system. Moreover, the resource poor farmers of rural India could hardly afford to purchase costly chemical fertilizers.

Crop Yield

Over the years, crop production under high chemical input has resulted in soil degradation and loss of fertility leading to a decline in seed germination and crop yield.

Health Hazards

Extensive input of chemical fertilizers in soil has resulted in accumulation of pollutants such as heavy metals, pesticides etc. in the soil posing as potential hazards of chemical contamination of our food and ground water.

It is, therefore, necessary to devise strategic methods for:

1. Enhancement of agricultural sustainability
2. Environmental safety
3. Reduction of disease incidence

To attain the above mentioned objectives of getting better and safer product of biopesticide-cum-bio-fertilizer, as substitutes of chemical fertilizers and chemical pesticides respectively, organic agriculture was initiated which apply methods for improving soil quality in terms of both physico-chemical and biological properties. These systems minimize the use of purchased inputs by substituting them through use of farm-generated inputs, or by exploiting biological systems such as management by biopesticide-cum-bio-fertilizer of A-Mycorrhizal fungi (*Glomus* spp.) and oilseed cakes (Karanj-*Pongamia pinnata*) with antagonistic fungal bioagents (*Trichoderma* spp. *Paecilomyces lilacinus* etc.) to enhance nutrient availability, plant health and vigour, reduction of agriculture inputs and also for reduction of diseases incidence.

Nature of Bio-pesticide

Biopesticides fall into three major classes:

1. Microbial pesticides consist of a naturally occurring or genetically controlled microorganism (*e.g.*, a bacterium, fungus, virus or protozoan) as the active ingredient. These pesticides can control many different kinds of pests, although each separate active ingredient is relatively specific for its target pest(s). For example, there are fungi that control certain weeds, and other fungi that kill specific insects.

   They suppress pest by:

   (*a*) Producing a toxin specific to the pest.

   (*b*) Causing a disease.

   (*c*) Preventing establishment of other microorganisms through competition or

   (*d*) Other modes of action.

   An example of a most widely used microbial pesticide is subspecies and strains of *Bacillus thuringiensis*, or "Bt". It is a naturally occurring soil bacterium that is toxic to the larvae of several species of insects but not toxic to untargeted organisms. Bt can be applied to plant foliage or incorporated into the genetic material of crops and as discovered, it is toxic to the caterpillars (larvae) of moths and butterflies. These also can be used in controlling mosquitoes and black flies. Several strains of Bt have been developed and now strains are available that control fly larvae. While some Bt's control moth larvae found on plants, other Bt's are specific for larvae of flies and mosquitoes. The target insect species are determined by whether the particular Bt produces a protein that can bind to a larval gut receptor, thereby causing the insect larvae to starve.

2. Plant-Incorporated-Protectant (PIPs) is pesticide substances that plants produce from genetic material that has been added to the plant. For example, scientists can take the gene for the Bt pesticide protein, and introduce the gene into the plant's own genetic material. Then the plant, instead of the Bt bacterium, manufactures the substance that destroys the pest.

3. Biochemical pesticides are naturally occurring substances that control pests by nontoxic mechanisms. Conventional pesticides, by contrast, are generally synthetic materials that directly kill or inactivate the pest. Biochemical pesticides include substances, such as insect sex pheromones, that interfere with mating as well as various scented plant extracts that attract insect pests to traps. Man-made pheromones are used to disrupt insect mating by creating confusion during the search for mates, or can be used to attract male insects to traps. Pheromones are often used to detect or monitor insect populations, or in some cases, to control them. Farmers in their traditional wisdom have identified and used a variety of plant products and extracts for pest control, especially in storage. As many as 2121 plant species are reported to possess pest management properties, 1005 species of plants exhibiting insecticide properties, 384 with antifeedant properties, 297 with repellant properties, 27 with attractant properties and 31 with growth inhabiting properties have been identified. The most commonly used plants are neem (*Azadirachta indica*), pongamia (*Pongamia glabra*) and mahua (*madhuca indica*). 2-5 per cent neem or mahua seed kernel extract has been found effective against rice cutworm, tobacco caterpillar, rice green leafhopper, and several species of aphids and mites. The efficacy of vegetable oils in preventing infestation of stored product pests such as bruchids, rice and maize weevils has been well documented. Root extracts of Tagetes or Asparagus as nematicide and Chenopodium and Bougainvillea as antivirus have also been reported.

Criteria of Developing Potential Bio-pesticide

The efficacy of many of the bio-pesticide can equal that of conventional chemical pesticides. However, the mode of action will be different. With many of the bio-pesticides, the time from exposure to morbidity and death of the target insect may be 2 to 10 days. With the mission to develop a successful and appropriate bioformulations to ease its application under farmers' field conditions in the present investigation attempt is made. The powder-based formulation was desired to possess the essential characters like long retention of viability ease of store and apply with an excellent dispersible nature. Biological control of diseases and insect pests has been achieved by seed treatment with the *Trichoderma* and *Paecilomyces* in recent years successfully by several workers. However, the success of a biocontrol agent depend on its ability to produce inoculum in excess and to survive, grow and proliferate well on seed coat, spermosphere and rhizo-sphere of the growing plants. Such types of antagonistic establishment need a substrate during seed treatment which can provide a better coating on seed surface and food base to antagonist for better proliferation. The use of microbial biomass as alginate priUs and granules (Lumsden *et al.,* 1995, 1996;) starch granules (Lewis *et al.,* 1995), wheat bran alginate pellet (Smith, 1996) pyrax (Lewis and Fravel, 1996), Talc powder (Tewari and Mukhopadhyay) extruded granules (Lewis and Larkin, 1997) and cellulose granules (Lewis *et al.,* 1998) has been successfully used against soil-borne plant pathogens as seed/soil application in various crops.

Biopesticides in general:

1. Have a narrow target range and a very specific mode of action.
2. Are slow acting.
3. Have relatively critical application times.
4. Suppress, rather than eliminate, a pest population.
5. Have limited field persistence and a short shelf life.
6. Are safer to humans and the environment than conventional pesticide.
7. Present no residue problems.

Understanding the fundamental differences in the mode of action of bio-pesticide vs. traditional pesticides is important since the use patterns of a bio-pesticide may be different from traditional pesticides to control a particular pest species. It is important to be careful when using any pesticide, even organic or natural or bio-pesticide. Even if this product is considered to be organic in origin, it is still a pesticide. Just because a product is though to be organic, or natural, does not mean that it is not toxic. Some organic pesticides are as toxic, or even more toxic, than many synthetic chemical pesticides. Organic pesticides have specific modes of action; just as do synthetic pesticides have specific modes of action, just as do synthetic pesticides. While some organic pesticides may be nontoxic or are only slightly toxic to people, they may be very toxic to other animals. For instance, the organic pesticide ryania is very toxic to fish. Also, some organic pesticides may be toxic to beneficial insects, such as honeybees, if they are combined with other materials, such as combining pyrethrins with rotenone. The use of an Integrated Pest Management Program (IPM) is important to insure success.

Conclusion

It is fact that any transition from the existing norms takes some time and if the proper testing etc. has been done for the product only then the community accepts with cost effective and eco-friendly view. Dissemination of the information should be properly done and the community should also develop faith in the cost effective, sustainable and eco-friendly multifarious potential products once a faith is developed, it is obvious that the community will continue to accelerate the growth of the bio-pesticides-cum-bio-fertilizers because protection of crop yield is equally important of production.

References

Cook, R.J. and Baker, K.F. (1983). *The Nature and Practice of Biological Control of Plant Pathogens.* APS, St. Paul. M.W., 539 pp.

Fravel, D.R. (2005). Commercialization and implementation of Biocontrol. *Annu. Rev. Phytopathology,* 43: 337–359.

Jones and Lisansky, S. (1947). Microbial biopesticides. In: *Microbial insecticides: Novelty or Necessity?"* British Crop Protection Council Proceedings/Monograph Series, 68: 3–10.

Jumanah, F. (1994). Pesticide policies in developing countries: Do they encourage excessive use? World Bank Discussion Papers 238, The World Bank, Washington DC.

Lewis, J.A., Larkin, R.P. and Rogers, D.L. (1998). A formulation of *Trichoderma* and *Gliocladium* reduce damping off caused by *Rhizoctonia solani* and saprophytic growth of the pathogen in the soilless mix. *Plant Dis.,* 82: 501–506.

Lewis, J.A. and Larkin, R.P. (1997). Extruded granular formulation with biomass of bio-control *Gliocladium virens* and *Trichoderma* spp. to reduce damping off of eggplant caused by *Rhizoctonia solani* and saprophytic growth of the pathogen in soilless mix. *Biocontrol. Sci. Tech.,* 7: 49–60.

Lewis, J.A. and Papavizas, G.C. (1987). Application of *Trichoderma* and *Gliocladium* in alginate pellets for control of Rhizoctonia damping off. *Plant Pathol.,* 36: 397–404.

Tewari, A.K. and Mukhopadhyay, A.N. (2001). Testing of different formulation of *Gliocladium virens* against chickpea wilt complex. *Indian Phytopath.,* 54: 67–71.

Sen, Bineeta (2000). Biological control: A success story. *Indian Phytopath.,* 53: 243–249.

Plant Diseases Management for Sustainable Agriculture *Pages* **299–327**
Editor: **Dr. Shahid Ahamad**
Published by: **DAYA PUBLISHING HOUSE**

Chapter 25

Plant Growth Promoting Rhizobacteria in Crop Plants Against Pests and Diseases

***Amit Kumar Jain*[1], *Sudhir Kumar*[2] *and J.D.S. Panwar*[3]**

[1]*Department of Microbiology, Janta Vedic College, Baraut (Baghpat), U.P.*
[2]*Indian Institute of Pulses Research, Kalyanpur, Kanpur, U.P.*
[3]*Principal Scientist, Division of Plant Physiology, IARI, New Delhi*

ABSTRACT

To keep pace with the ever-increasing demand for crop yield by the exploding population, use of chemical fertilizers and pesticides in agricultural production has become indispensable. Production cost of agriculture has skyrocked owing to the abandoning of green, organic farmyard manure and use of chemicals. However, the problems associated with chemical pesticides are abundant. The excessive application of chemical fertilizers causes soil sealing, fertility diminishing and residual problems. The leftover chemical fertilizers and pesticides have seriously affected the quality of agricultural products, people's health and caused environmental pollution. The overuse of biofertilizers has also damaged the soil's original micro-ecological balance and deteriorated the diseases spread by soil. While some pesticides must be abandoned because of their unacceptable non-target effect, there will always be a need in agriculture for safe and selective chemicals to limit the effects of pests. More significantly, it is becoming increasingly more difficult and expensive to find new kinds of synthetic chemical pesticides. Implications of chemical pesticides in ecological, environmental and human health problems have increased public awareness and concern regarding the continued use of agrichemicals that are damaging to human or environmental health thereby obviating the search for effective alternative approaches which have minimal deleterious effects, more environmentally friendly, and will contribute to the goal of sustainability in agriculture. In this line Plant growth-promoting rhizobacteria (PGPR) present immense potential and promise as effective substitutes for chemicals.

Plant growth promoting rhizobacteria (PGPR) belonging to *Pseudomonas* spp. are being exploited commercially for plant protection to induce systemic resistance against various pests and diseases to phytopathogens and is used as inoculants for biofertilization, phytostimulation and biocontrol. Mixtures of different PGPR strains have resulted in increased efficacy by inducing systemic resistance against several pathogens attacking the same crop. Seed-treatment with PGPR causes cell wall structural modifications and biochemical/physiological changes leading to the synthesis of proteins and chemicals involved in plant defense mechanisms. Lipopolysaccharides, siderophores and salicylic acid are the major determinants of PGPR mediated ISR. The performance of PGPR has been successful against certain pathogens, insect and nematode pests under field conditions.

Keywords: *PGPR, ISR, Biotic and Abiotic Stress, Genetics.*

Introduction

The new millennium brings with it, renewed hope for a better livelihood for the populations of this earth. Hence the themes often discussed at international fora of human welfare and agriculture range from sustainability, food security and safety to the provision of a productive and health environment to humankind and its future generations. Current scientific strategies to maintain and improve yields in support of high input agriculture place great emphasis on "far safe" techniques for each component of the production sequence with little consideration of the integration of these comments in a holistic systems approach. The fact remains that stable and lower human population is an integral component of sustainable agriculture and more so in the Indian context. Thanks due to the high yielding varieties we are still self-sufficient in rice and wheat, but for how long? High input agriculture is increasingly recognized as an environment degrading and not profitable. We now recognize that technical progress may have social and environmental costs we cannot pay. In any case considerable crop damage by insects and pathogens still prevails. In an Indian context, use of bioinoculants assumes special significance because the size of individual farm holdings is going down gradually due to growth in population. Farmers are struggling to earn enough income from farming to meet the minimum essential requirements of their families. The cost of input up is going on while the price of out put is going down. Indeed, this is the very reason for much of the research into biological methods of using bioinoculants as possible alternative to intensive chemical fertilizers, fungicides and pesticide use. Under the present scenario, microbes will probably assume an increasingly prominent role as we took forward to new responsible technologies, if agriculture is to evolve towards both economic and environmental sustainability. The interaction between microbes and plants developed with the process of evolution in plants, and hence the use of in bioinoculants is pre-adapted to fit into long-term sustainable agricultural systems. Thus, biological approaches through the use of bioinoculants are pre-adapted to fit into long term sustainable agricultural systems.

The term rhizosphere was introduced by Hiltner in 1904, and is defined as a volume of soil surrounding plant roots in which bacterial growth is stimulated (Sorensen, 1997). The rhizosphere has attracted much interest since it is a habitat where several biologically important processes and interactions take place. The rhizosphere is populated by a diverse range of microorganisms, and the bacteria colonizing this habitat are called rhizobacteria (Schroth and Hancock, 1982). The colonizing microorganism form symbiotic, associative, neutralistic or parasitic relations with the plant, which in turn is dependent on the type of microorganism, available nutrients, plant defense responses and environment. The symbiotic association between legumes and *Rhizobium* is a well known phenomenon which contributes to biological nitrogen fixation (BNF) and soil fertility. Some of the free living rhizobacteria utilize nutrients secreted by the plant roots and influence the plant in a direct or indirect way resulting in growth stimulation. This set of beneficial term as plant growth promoting rhizobacteria (PGPR), was coined by Kloepper *et al.* (1980) that comprise of a broad spectrum of bacterial genera. Additionally, PGPR also protect the plant from infection by phytopathogenic organisms within the vicinity of the root system.

Root exudates are believed to determine which microorganisms colonize roots in the rhizosphere (Kunc and Macura, 1988). It is now known that plant roots also generate signals, it has been shown that zoospores of oomycetic pathogens take advantage signals to guide their movements towards the root surface ((Low *et al.*, 1999). It has been estimated that about 30 per cent of plant photosynthate production is released via root exudation (Smith *et al.*, 1993) which attracts microorganisms (Lemanceau *et al.*, 1995).

Plant growth colonizing bacteria can function as harmful, deleterious rhizobacteria (DRB) or beneficial, plant growth promoting rhizobacteria. Rhizobacteria that inhibit plant growth have been described as deleterious rhizobacteria (Schroth and Hancock, 1982). However, there are no published experiments supporting the existence of DRB, plant growth promotion by rhizobacteria can occur directly and indirectly (Glick, 1995; Persello-Cartieaux *et al.*, 2003). There are several ways by which plant growth promoting bacteria affect plant growth directly. These include fixation of atmospheric nitrogen, solubilization phosphorus, production of siderophores that solubilize and sequester iron, or production of plant growth regulators (hormones) that enhance plant growth at various stages development (Enebak and Carey, 2000). Indirect growth promotion occurs PGPR promote plant growth by improving growth restricting conditions (Glick *et al.*, 1999). It can happen directly by producing antagonistic substances, or indirectly by inducing resistance to pathogens (Glick, 1995). A bacterium can affect plant growth by one or more of these mechanisms, and also use different abilities for growth promotion at various times during the life cycle of the plant (Glick *et al.*, 1999).

Effect of PGPR in Plant Growth Promotions

A. Directly affected as

- ☆ Nitrogen Fixation
- ☆ Phosphorus solubilization
- ☆ Enhanced iron availability
- ☆ Production of plant hormones

B. Indirectly affected as

- ☆ Antibiosis
- ☆ Parasitism and Predation
- ☆ Induced resistance
- ☆ Iron scavenging
- ☆ Competition for nutrients and niche

Mechanism of PGPR in Direct Plant Growth Promotions

The ways by which PGPR can influence plant growth directly may differ from species to species as well as strain to strain. Symbiotic plant colonizers such as rhizobia mostly contribute to plant growth by nitrogen fixation. Free living rhizobacteria usually do not rely on single mechanisms of promoting plant growth (Glick *et al.*, 1999). In addition to nitrogen fixation, several PGPR are also able to provide the plant with sufficient iron in iron-limited soils or other important minerals, *e.g.* phosphate (Singh and Singh, 1993).

Organic substances capable of regulating plant growth produced either endogenously or applied exogenously are called plant growth regulators. They regulate growth by affecting physiological and morphological processes at very low concentrations (Arshad and Frankenberger, 1998). Several microorganisms are capable of producing auxins, cytokinins, gibberellins, ethylene (ET), or abscisic acid (ABA). Auxins are produced by several rhizobacterial genera, *e.g.*, *Azospirillum Agrobacterium*, *Pseudomonas* and *Erwinia* (Costacurta and Vanderleyden, 1995).

Production of IAA by *Azospirillum* appears to be the most likely explanation for the observed increase in seedling's root length, root hairs, root branching and root surface areas. However, it is highly unlikely that IAA alone explains yield increases that have been reported on a large number of crops (Bashan and de-Hashan, 2004). Other plant growth regulators and N-fixation are other attributes of *Azospirillum* that likely contribute to yield increases and enhanced nutrient uptake of inoculated plants. Panwar and Sirohi (1989) suggested additive hypothesis to explain growth and yield promotion with *Azospirillum* (Panwar *et al.*, 2001). According to this hypothesis, growth promotion is the result of multiple mechanisms working together. Hence, when a single mechanism is tested, the full growth promotion response that occurs with inoculation of *Azospirillum* is not seen. Rather, the observed plant growth promotion results from the sum effect of the multiple mechanisms *i.e.* the levels of N-fixation are not considered adequate to account for the observed growth promotion of plants. In addition, *Azospirillum* strains produce IAA and other plant growth regulators in culture and association with the plant which may contribute growth promotion,

Glick *et al.* (1998) have defined the role of PGPR-produced IAA and the response of the host plant. With this system, bacterial production of IAA may stimulate plant cell proliferation or elongation and results in plant production of ACC (1-aminocyclo propane-1-carboxylate). ACC is an immediate precursor of ethylene, and ethylene inhibits root elongation of seedlings. IAA produced by *Pseudomonas putida* GR12-2, together with endogenously produced plant IAA, activates ACC synthase, leading to production ACC. ACC is convened to ethylene by ACC oxidase. However, the plant produced ACC is taken up by the PGPR strain and is hydrolyzed by ACC deaminase that converts ACC to ammonia and alpha-ketobutyrate. This conversion lowers the pool of ACC, thereby, reducing the level of ethylene and ethylene's inhibition of root elongation, resulting in increased root length.

Cytokinins and gibberellins are produced in the rhizosphere by several bacteria. eg. *Azospirillum*, *Agrobacteriurn* and *Pseudomonas* genera (Gaudin *et al.*, 1994). Cytokinins promote root formation, but a minor over production of cytokinins instead leads to inhibition of root development, and severely deficient cytokinin mutant plants do not survive (Binns, 1994). Cytokinins are believed to be the signals involved in mediating of environmental stresses from roots to shoots (Jackson, 1993). Nevertheless, more studies need conducted before cytokinins signaling can be fully understood.

In addition to that, co-immobilization of *Chlorella vulgaris* and *Azospirillum brasilense* in alginate beads under synthetic wastewater culture conditions significantly increased the removal of ammonium and soluble phosphorus compare over mobilization of the microalgae alone (De-Bashan *et al.*, 2002). Co-immobilization of the two microorganism reaching removal of up to 100 per cent ammonium, and 92 per cent phosphorus within 6 days (De-Bashan *et al.*, 2003).

Mechanism of PGPR in Indirect Plant Growth Promotion

PGPR are known to control a wide range of phytopathogens like fungi, bacteria, viruses, nematodes etc., by biocontrol mechanism which may be by producing antibiotics, parasitism, by competition or by induction of systemic resistance against pathogens throughout the entire plant system.

Antibiotics

According to Pierson and Pierson (1996); Wood and Pierson (1996) an auto-inducer signal required for induction of antibiotic synthesis. Also, other environmental sensors such as the regulating proteins Gac A and Apd A can influence the production of secondary metabolites involved in pseudomonas biocontrol (Corbel and Loper, 1995, Haas *et al.*, 2002). In addition, house-keeping factor sigma and the stress related sigma have critical role in production of antibiotic metabolites in disease suppression (Schnider *et al.*, 1995).

Various antibiotics can be produced by rhizobacteria like *Bacillus* and *Pseudomonas*, like bacilysin, iturin like lipopeptides, diacetyl phioroglucinol and phyrrolnitrin, HCN, phenazine-1–carboxylate (Thornshow *et al.*, 1990). The most widely studied group of rhizospheric bacteria with respect to the production of antibiotics is that of the fluorescent pseudomonas. The first antibiotics described as being implicated in biocontrol were phenazine derivatives produced by fluorescent pseudomonads. Their role has been elucidated by transposon insertion mutations which results in a defect in production of phenazine-1-carboxylate, thus reducing disease suppressive activity (Pierson and Pierson, 1996).

Parasitism and Predation

An additional mechanism by which biocontrol agents can reduce plant disease is parasitism on pathogens, mostly fungi. Digestion of the parasite cell wall is accomplished by several excreted enzymes including proteases, chitinases and glucanases. Individually, all these enzymes display antifungal activity. But they often act synergistically with antibiosis (Lorito *et al.*, 1994). De Meyer *et al.* (1999) reported that rhizosphere colonization by *Pseudomonas aeruginosa* 7NSK 2 activated phenylalanine ammonia lyase (PAL) in bean roots and increased the salicylic acid levels in leaves. The *Pseudomonas fluorescens* strain was found to produce antibiotics (Phenazine and 2, 4 diacetyl phloroglucinol), HCN and lytic enzymes *viz.* chitinase and glucanase. These bacreria metabolites were associated with direct inhibitory action against several *Fusarium, Pythium, Phytophthora* and *Colletoirichum* pathogens (Samiyappan, 2003).

Competition for Nutrients and Niches

In addition to the above described and commonly reported antibiosis mechanism there are other ways by which rhizobacteria can inhibit pathogens. One example concerns competition for nutrients and suitable colonization niches on the root surface. Such mechanisms are often overlooked, in part because they are difficult to study in systems. Competition for nutrients supplied by root exudates is probably a significant factor in most interactions between PGPR and pathogens. Populations of bacteria established on a plant root could act as a sink for the nutrients in the rhizosphere, hence reducing the nutritional element availability for pathogen stimulation or subsequent colonization of the root. This mechanism is most probably often used by fluorescent pseudomonads due to their nutritional versatility, and because of their high growth rates in the rhizosphere (Sorensen, 1997).

Inducing Systemic Resistance

Induced protection of plants against various pathogens by biotic or abiotic agents has been reported since 1930s, when Chester (1933) proposed the term "acquired physiological immunity". Since then several terms have been used to describe the phenomenon of induced resistance such as "systemic acquired resistance" (Ross, 1961), "translocated resistance" (Hurbert and Helton, 1967) and "plant immunization" (Tuzun and Kuc, 1991). Systemic resistance refers to an increased level of resistance at sites within that plant system to those at which induction had occurred. How bacteria trigger systemic resistance is still largely unknown. Two general plant defense mechanisms are induced systemic resistance (ISR) and systemic acquired resistance (SAR). ISR is rhizobacterially mediated systemic resistance that does not involve any damage to plant. By contrast, SAR is induced by foliar pathogens and results in activation of resistance mechanisms in uninfected plants. Thus, in systemic acquired resistance a first infection predisposes the plant to resist further attacks. Thus the induction of systemic resistance by rhizobacteria is referred as ISR, whereas that by other agencies is called SAR (Van Loon *et al.*, 1998). SAR is expressed to a maximum level when the inducing organism causes necrosis (Cameron *et al.*, 1994), whereas ISR by PGPR typically do not cause any necrotic symptoms on the host plants (Van Loon *et al.*, 1998).

The biotic inducers of SAR include virulent pathogens, non-pathogens and elicitors of fungal cell wall metabolites. Abiotic agents are salicylic acid (SA), ethylene, dichloro-isonicotinic acid and benzothiadiazole (Sticher *et al.*, 1997). The utilization of pathogenic organisms as inducing agents has not been successful under field conditions. Generally, the duration of protection is less following induction of a pathogen than that with PGPR-mediated ISR and prior inoculation of pathogen may provide a source of secondary inoculum (Wei *et al.*, 1991). For successful disease management, it is important to find more effective, practical and economical ways to protect plants from various pests or diseases. PGPR are root colonizing bacteria with beneficial effects including plant growth promotion and biological disease control. In recent years, the use of PGPR as an inducer of systemic resistance in crop plants against different pathogens has been demonstrated under field conditions (Wei *et al.*, 1991). The utilization of natural PGPR strains as inducers of plant defence responses may increase the chance of their applicability and offer a practical way to deliver immunization.

Induction of Systemic Resistance by PGPR against Diseases, Insect and Nematode Pests

PGPR induce resistance in plants against fungal, bacterial and viral diseases (Maurhofer *et al.*, 1998), and insect (Zehnder *et al.*, 1997) and nematode pests (Sikora, 1988). Induction of systemic resistance by selected strains of PGPR against plant diseases and insect pests has been proved by spatially separating the pathogen and PGPR in the plants.

Diseases PGPR Provide Different Mechanisms for Suppressing Plant Pathogens

They include competition for nutrients and space (Elad and Baker, 1985), antibiosis by producing antibiotics *viz.*, pyrrolnitrin, pyocyanine, 2,4-diacetyl phloroglucinol and production of siderophores (fluorescent yellow green pigment), *viz.*, pseudobactin, which limits the availability of iron necessary for the growth of pathogens (Kloepper *et al.*, 1980). Other important mechanisms include production of lytic enzymes such as chitinases and β-1, 3-glucanases which degrade chitin and glucan present in the cell wall of fungi (Frindlender *et al.*, 1993, Velazhahan *et al.*, 1999) and degradation of toxin produced by pathogen (Duffy and Defago, 1997).

Several studies have been carried out to elicit ISR by PGPR in plants. In carnation, application of *Pseudomonas* sp. strain WCS 417r protected plants systemically against *Fusarium* wilt caused by *Fusarium oxysporum* f. sp. *dianthi* (Van Peer *et al.*, 1991). PGPR strains applied as a seed-treatment resulted in a significant reduction in anthracnose disease caused by *Colletotrichum orbiculare* in cucumber (Wei *et al.*, 1991). Similarly, induction of systemic resistance by *Pseudomonas putida* strain 89B-27 and *Serratia marcescens* strain 90-166 reduced *Fusarium* wilt of cucumber incited by *Fusarium oxysporum* f.sp.

cucumerinum (Liu *et al.*, 1995a). PGPR as a seed-treatment alone or as seed-treatment plus soildrenching has protected cucumber plants against anthracnose disease (Wei *et al.*, 1996).

Induction of systemic resistance by PGPR against viral diseases has been reported in cucumber and tobacco plants. Seed-treatment with *P. fluorescens* strain 89B-27 and *S. marcescens* strain 90-166 has consistently reduced the number of cucumber mosaic virus-infected plants (CMV) and delayed the development of symptoms in cucumber and tomato (Raupach *et al.*, 1996). Soil-application of *P. fluorescens* strain CHAO has induced systemic protection against inoculation with tobacco necrosis virus (TNV) in tobacco (Maurhofer *et al.*, 1994). These experiments show that PGPR strains initiate ISR against a wide array of plant pathogens causing fungal, bacterial and viral diseases.

Insect Pests

Reports on PGPR-mediated ISR against insects are restricted to very few crops. Generally, fluorescent pseudomonads influence the growth and development of insects at all stages of their growth. *Pseudomonas maltophila* affects the growth of larval stage of *Helicoverpa zea*, the corn earworm, leading to more than 60 per cent reduction in adult emergence while pupae and adults that emerged from bacteria-infected larva were smaller (Bong and Sikorowski, 1991). Similarly, the relative growth rate, consumption rate and digestibility of feed by *Helicoverpa armigera* have been affected when larvae fed on cotton plants treated with *Pseudomonas gladioli* due to an increase in their polyphenol and terpenoid content (Qingwen *et al.*, 1998). The populations of the striped cucumber beetle have significantly reduced by Induction of systemic resistance by PGPR strains, *viz.*, *P. putida* strain 89B-27, *S. marcescens* strain 90-166, *Flavomonas oryzihabitans* strain INR-5 and *Bacillus pumilus* strain INR-7. As certain fluorescent pseudomonads are effective rhizosphere colonizers and are endophytic in nature in the plant system, attempts have been made to transfer the insecticidal crystal protein from *Bacillus thuringiensis* to *P. fluorescens*. *P.fluorescens*, thus genetically engineered was effective against a lepidopteran insect pest. Transgenic *P. cepacia* strain 526 with the crystal protein gene has consistently shown insecticidal activity against tobacco hornworm (Stock *et al.*, 1990). Thus, PGPR treatment of crops can be effective for insect pest management and has a great potential for future use.

Nematodes

PGPR also induce systemic resistance against nematode pests (Oostendorp and Sikora, 1990, Sikora and Hoffmann-Hergarten, 1992). *P. fluorescens* has induced systemic resistance and inhibited early root penetration of *Heterodera schachtii*, the cyst nematode in sugar beet (Oostendorp and Sikora, 1990). Similarly, *B. subtilis* has induced protection against *Meloidogyne incognita* and *M. arenaria* in cotton (Sikora, 1988). Though attempts to use PGPR for nematode control are limited, the use of PGPR as biological control agents of plant parasitic nematodes especially for sugar beet and potato cyst nematode has been reported as a successful strategy in management of these nematodes (Sikora, 1992). Treatment of rice seed with PGPR alone or in combination with chitin and neem cake has reduced the root and soil population of the rice root nematode, *Hirschmanniella oryzae* (Swarnakumari *et al.*, 1999). The level of infestation of root-knot nematode *M. incognita* on tomato was reduced with fewer galls and egg masses in the soil following root dipping with *P. fluorescens* strain Pf1 (Santhi and Sivakumar, 1995). Similarly, application of *P. chitinolytica* reduced the root knot nematode infection in tomato crop (Spiegel *et al.*, 1991). These experiments show that PGPR-mediated ISR is effective in both dicotyledonous plants, *viz.*, *Arabidopsis*, bean, carnation, cucumber, radish, tobacco and tomato and certain monocotyledonous plants, *viz.*, rice, maize and sugarcane etc.

Synergistic Effect of PGPR Strain Mixtures

It is likely that most of naturally occurring biological control results from mixtures of antagonists rather than from high populations of a single antagonist. Similarly, application of a mixture of introduced biocontrol agents would more closely mimic the natural situation and might broaden the spectrum of biocontrol activity and enhance the efficacy and reliability of control (Dufy and Weller, 1995). Application of a mixture of three PGPR strains *viz.*, *Bacillus pumilus* strain INR 7, *B. subtilis* strain GB03 and *Curtobacterium yaccumfaciens* strain ME1 (as a seed-treatment) has resulted in much more intensive plant growth promotion and disease reduction, when compared to strains tested singly. This might be due to different mechanisms of action for each PGPR strain (Raupach and Kloepper, 1998). Use of chitinase-producing *Streptomyces* spp. and *Bacillus cereus* isolates used in combination with antibiotic producing *P.fluorescens* and *Burkholderia* (*Pseudomonas*) *cepacia* isolates have shown a synergistic effect on the suppression of rice sheath blight incited by *Rhizoctonia solani* (Sung and Chung, 1997). Mixtures of *Pseudomonas* spp have suppressed take-all disease of wheat (*Gaeumannomyces graminis* var *tritici*) than treatment using a single *Pseudomonas* sp. (Pierson and Weller, 1994). Combination of *P. fluorescens* strains, *viz.* Pf1 and FP7 gave effective control of rice sheath blight disease when compared to each strain applied singly (Nandakumar, 1998). From the economic point of view, a biocontrol product composed of a mixture of strains may be more expensive than a product containing a single strain. In developing mixtures of strains, greater emphasis must be given on their efficacy against disease suppression by ISR.

Spectrum of Protection by PGPR

Recently, work on the spectrum of PGPR-mediated ISR against different pathogens in different crop plants has been gaining importance (Hoffland *et al.*, 1996, Kloepper *et al.* (1997). *P. fluorescens* strain WCS 417 (seed treatment) has protected radish through induction of systemic resistance not only against the fungal root pathogen *F. oxysporum* f. sp. *raphani*, but also against the avirulent bacterial leaf pathogen *P. syringae* pv. *tomato* and fungal leaf pathogens *Alternaria brassicola* and *F. oxysporum* (Hoffland *et al.*, 1996). It implies that the same PGPR strain can induce resistance against multiple pathogens in the same crop. Later studies showed that the same PGPR strains induced systemic protection against angular leaf spot caused by *P. syringae* pv. *lachrymans* (Liu *et al.*, 1993a), *Fusarium* wilt incited by *F. oxysporum* f.sp. *cucumerinum* (Liu *et al.*, 1993b) and cucurbit wilt caused by *Erwinia tracheiphila* (Kloepper *et al.*, 1993). *S. marcescens* strain 90-166 (seed treatment) has shown ISR in cucumber against anthracnose, cucumber mosaic virus, bacterial angular leaf spot and cucurbit wilt diseases (Kloepper *et al.*, 1993). The broad spectrum of PGPR-mediated ISR is more rewarding than narrow spectrum of disease protection. Hence selecting a suitable strain having potential to induce systemic resistance against multiple pathogens and pests is the most important task in the delivery of microbial agents to the field condition.

PGPR as Endophytes

Endophytic bacteria are defined as bacteria which reside within the living plant tissues without doing substantive harm or gaining benefit other than residency (Kado, 1992). In addition to rhizosphere and rhizoplane colonization, certain PGPR are reported to be endophytes localized in the intercellular spaces of the root epidermal cells and vascular tissue. The internal tissues of plants provide a relatively uniform and protected environment when compared with the rhizosphere and rhizoplane (Chen *et al.*, 1995), where ectophytic bacteria must compete for nutrients with other microbes and endure fluctuations of temperature and moisture, as well as exposure to ultraviolet radiation on above ground surfaces. In spite of these advantages, the potential of bacterial endophytes has only been explored to a limited extent. Application of endophytic bacteria by stem injection in cotton plants reduced root rot caused by *Rhizoctonia solani* and vascular wilt caused *by F. oxysporum* f. sp. *vasinfectum* (Chen *et al.*, 1995). These bacteria move upward and downward from the point of application and by colonizing the internal tissues, can exclude the entry of a pathogen in the vascular stele. Endophytic bacteria have brought about significant control against *F. solani* in cotton and *Sclerotium rolfsii* in beans (Pleban *et al.*, 1995). Seed treatment of tomato with endophytic bacterium *Bacillus pumilus* strain SE 34 prevented the entry of vascular wilt fungus *F. oxysporum* f. sp. *Radicislycopersici* into the vascular stele and the mycelial growth was restricted to the epidermis and outer root cortex (Benhamou *et al.*, 1998). The use of an endophytic strain for inducing systemic resistance is more beneficial in vegetatively propagated crops, like banana, sugarcane, and tapioca, for example Viswanathan (1999) and Viswanathan and Samiyappan (1999a) revealed the utility of endophytic *P. fluorescens* strain EP1 isolated from stalk tissues of sugarcane in inducing systemic resistance against red rot (*Colletotrichum falcatum*). Further, when the endophytic bacteria are introduced into the vegetatively propagated seed, the bacteria survives and moves in the vegetative part and subsequently the propagative seed will also have the introduced bacteria, thus minimizing the need for frequent application of bacterial strains.

Mechanism of ISR-Mediated by PGPR

PGPR bring about ISR through fortifying the physical and mechanical strength of the cell wall as well as changing the physiological and biochemical reaction of the host leading to the synthesis of defence chemicals against the challenge pathogen.

Structural and Ultra-structural Cell Wall Modifications in the Host Plants

The success of a plant in warding off invading pathogens relies primarily on its ability to build a line of defence rapidly for protecting cell walls against the spread of a pathogen (Benhamou *et al.*, 1996a). It is well known that PGPR induces cell wall structural modification in response to pathogenic attack (Benhamou *et al.*, 1996b). Treatment of pea seeds with *P. fluorescens* strain 63-28 has resulted in formation of structural barriers, *viz.*, cell wall apposition (papillae) and deposition of newly formed callose and accumulation of phenolic compounds at the site of penetration of invading hyphae of *Pythium ultimum* and *F. oxysporum* f. sp. *pisi* (Benhamou *et al.*, 1996a). In tomato, cell wall thickening, deposition of phenolic compounds and formation of callose restricted growth of *F. oxysporum* f. sp. *radicis-lycopersici* to the epidermal cell and outer cortex in the root system in the treated plants (M'Piga *et al.*, 1997). Similarly, seed-treatment using *Bacillus pumilus* strain SE 34 has also induced strengthening of cell walls in tomato against *F. oxysporum* f. sp. *Radicislycopersici* (Benhamou *et al.*, 1998). Such a rapid defence reaction at sites of fungal entry delays the infection process and allows sufficient time for the host to build up other defence reactions to restrict pathogen growth to the outer most layers of root tissue.

Biochemical/Physiological Changes in the Host Plants

Application of PGPR results in biochemical or physiological changes in the plants. Normally ISR by PGPR is associated with the accumulation of pathogenesis-related proteins (Maurhofer *et al.*, 1994), synthesis of phytoalexin and other secondary metabolites (Van Peer *et al.*, 1991). Colonization of bean root by fluorescent bacteria is correlated with induction of PR proteins and systemic resistance against *Botrytis cinerea* (Zdor and Anderson, 1992). Maurhofer *et al.* (1994) reported that ISR by *P. yuorescens* strain CHAO against tobacco necrosis virus (TNV) in tobacco was associated with accumulation of PR proteins namely β-1, 3 glucanases and endochitinases. Seed treatment with *P. fluorescens* strain 63-28 in pea has produced hydrolytic enzymes such as chitinases and β-1, 3-glucanases. These host lytic enzymes accumulate at the site of penetration of the fungus, *F. oxysporum* f. sp. *pisi* resulting in the degradation of fungal cell wall (Benhamou *et al.*, 1996b). Inoculation of tomato plants with the same strain has similarly induced the production of plant chitinases when challenged with the wilt pathogen, *F. oxysporum* f. sp. *radicislycopersic* (M'Piga *et al.*, 1997). ISR has been correlated with a two-fold increase in activity of pathogenesis-related peroxidase and chitinase proteins. Similarly, in sugarcane, PGPR-mediated ISR against *C. falcatum*, enhanced levels of chitinase and peroxidase were noticed and specific induction of two new chitinase isoforms were found when inoculated with *C. falcatum* (Viswanathan and Samiyappan, 1999a). Accumulation of phytoalexin in response to *Pseudomonas* sp. strain WCS 417r treatment in carnation results in protection of carnation from wilt disease incidence (Van Peer *et al.*, 1991). Increased peroxidase activity as well as an increase in the level of mRNAs encoding for phenylalanine ammonia lyase (PAL) and chalcone synthase (which lead to synthesis of phytoalexin) are recorded in the early stages of interaction between bean roots and various bacterial endophytes (Zdor and Anderson, 1992).

In summary, induction of systemic resistance by PGPR especially against soil-borne pathogens is associated with ultra-structural cell wall modification that prevents the ingress of mycelium of the pathogen in the vascular stele followed by the biochemical changes, *viz.* accumulation of PR proteins and/or phytoalexins. Defence mechanisms induced against insect pests in cucumber plants are different from that against pathogens. PGPR do not kill insects, but application of PGPR brings about some physiological changes in the host plant that prevents the insects from feeding as has been demonstrated in cucumber against cucumber beetles. Due to PGPR treatment, there is a shift in the metabolic pathway in cucumber plants away from the cucurbitacin synthesis and towards that of other plant defence compounds, resulting in fewer beetles being attracted (Zehnder *et al.*, 1997). In nematode control, PGPR induce resistance by altering root exudates or inducing the host to produce repellents that affect nematode attraction or recognition of the host (Oostendorp and Sikora, 1990) and altering the syncytial development or sex ratio in the root tissue (Wyss, 1989). In rice, seed treatment with PGPR strains increases the chitinase enzyme activity and phenolic content and it is correlated with the reduced nematode infestation (Swarnakumari, 1996).

PGPR Determinants in ISR

There are several bacterial determinants involved in the induction of systemic resistance by PGPR the most important being lipopolysaccharides present in the outer membrane of bacterial cells, siderophore and salicylic acid production.

Lipopolysaccharides (LPS)

Lipopolysaccharides present in the outer membrane of PGPR are the major determinants of ISR in certain PGPR strains. For induction of ISR in radish, l0 CFU of the inducing *Pseudomonas* sp strain per gram of root tissue was reported as a minimum threshold (Raazijmakers *et al.*, 1995). It has also been shown that treatment of radish plant roots even with nanogram quantities of salicylic acid was sufficient to induce systemic resistance against *Fusarium oxysporum* (Leeman *et al.*, 1996).

LPS of *P. fluorescens* strain WCS 417 have induced systemic resistance in carnation against *Fusarium* wilt caused by *F. oxysporum* f. sp. *dianthi* (Van Peer and Schippers, 1992). Similarly, LPS of *P. fluorescens* strains WCS 374 and WCS 417 have induced systemic resistance in radish against *F. oxysporum* f. sp. *raphani* (Leeman *et al.*, 1995). They also established that mutant of *P. fluorescens* strain WCS 417, lacking the o-antigen side chain of the LPS, has not induced resistance in radish indicating the o-antigen side chain of he LPS might have served as a signal or trigger in the induction of defence mechanism in plants. In contrast, LPS of *P. putida* strain WCS 358 having o-antigen side chain do not induce systemic resistance in radish. In another study, LPS of WCS 417r and mutant of WCS 417r lacking O-antigen side chain of LPS elicit defence mechanism in *Arabidopsis* (Van Wees *et al.*, 1997). It has been shown for *Pseudomonas* strains that an outer membrane lipopolysacharide with strain specific carbohydrate side chains is the elicitor involved in induced resistance (Leeman *et al.*, 1996). This indicates that ISR by LPS of PGPR varies with different host plant and lipopolysaccharide is not the only trait in determining the ISR. Other traits of PGPR are also involved in ISR.

Siderophores

Iron is abundant in the Earth's crust but most of it is in the high insoluble form of ferric hydroxide, and thus unavailable to organisms in soil solution. Some bacteria have developed iron uptake systems (Neilands and Nakamura, 1991), which involve a siderophore an iron binding ligand and an uptake protein, needed to transport iron into the cell. It has been suggested that the ability to produce specific siderophores and/or to utilize a broad spectrum of siderophores, may contribute to the root colonizing ability of Pseudomonas strains. The production of siderophores that chelate and thereby scavenge, the ferric iron in the rhizosphere, may result in growth inhibition of other microorganism whose affinity for iron is lower (Kloepper *et al.*, 1980).

Thus, siderophores are produced by PGPR under iron-limited conditions. Leeman *et al.* (1996) reported that LPS of *P. fluorescens* strains WCS 374 and WCS 417 are the major determinants of ISR under iron-replete conditions but under iron-limited conditions, LPS of these bacteria were not involved in ISR in radish against *Fusarium* wilt. They also found that pyoverdin-type pseudobactin, siderophore, produced by these bacteria was responsible for ISR. Application of purified pseudobactin alone, isolated from strain WCS 374, to the roots of radish induced resistance. Thus, different bacterial determinants in inducing systemic resistance in radish vary depending upon iron availability. Induction of ISR by LPS and siderophores seems to be complementary rather than additive and full induction of resistance by one determinant masks contributions by other(s).

Salicylic Acid, Jasmonate and Ethylene

Signalling molecules like salicylic acid (SA), jasmonate (JA) and ethylene (ET) when accumulating coordinate the defense responses and, when applied exogenously, are even sufficient to induce resistance. It has been shown that these signalling molecules activate specific sets of defense-related genes: SA induces genes encoding pathogenesis-relate the proteins (PRs). These proteins have anti-microbial activity (Kombrik and Somssich, 1995). Certain PGPR strains are capable of producing salicylic acid and are responsible for the induction of ISR in plants (Maurhofer *et al.*, 1994). Salicylic acid production by *P. aeruginosa* strain 7NSK2 is essential for induction of resistance to *Botrytis cinerea*. Mutants of the same strain lacking salicylic acid production have lost their ability to induce systemic resistance in bean (DeMeyer and Hofte, 1997). Introduction of pch A and pch B gene which encode for the synthesis of salicylic acid in *P. fluorescens* strain P3, renders this strain capable of salicylic acid production and significantly improved its ability to induce systemic resistance in tobacco against TNV. *P. fluorescens* strain CHAO naturally produces salicylic acid under conditions of iron limitation and also induces ISR in tobacco against TNV (Maurhofer *et al.*, 1998). In contrast to these examples, mutants of *S. marcescens* strain 90-166 lacking salicylic acid production induce the same level of resistance in cucumber as the wild strain of *S. marcescens* (salicylic acid producing strain) does in cucumber and tobacco. In another study, salicylic acid producing strain of *S. marcescens* 90-166 induce resistance both in wild type tobacco and NahG-tobacco (tobacco plant transgened with NahG-gene encoding salicylic acid hydroxylase which converts salicylic acid to catechol) (Press *et al.*, 1997). Similarly, *P. fluorescens* strains WCS 417r and WCS 358r, which produce salicylic acid, induce resistance in both wild type *Arabidopsis* and NahG-gene transformed *Arabidopsis* plants (Van Wees *et al.*, 1997). This suggests that ISR induced by *P. fluorescens* strains WCS 417r and WCS 358r is independent of salicylic acid production in *Arabidopsis*. Expression of ISR by *P. fluorescens* strain WCS 417r requires ethylene dependent signaling pathway but not salicylic acid-dependent signaling at the site of application (Knoester *et al.*, 1999). It has also been shown that treatment of radish plant roots even with nanogram quantities of salicylic acid was sufficient to induce systemic resistance against *Fusarium oxysporum* (Leeman *et al.*, 1996).

ET is involved in the expression of the genes encoding *Hel* (a heveime like protein) (Potter *et al.*, 1993), *Chi B* (basic chitinase) (Samac and Shah, 1994), and *pdfl.2* (a plant defensive) (Penninckx *et al.*, 1996). Also JA has been shown to activate the genes encoding these three proteins, all of which possess antifungal activity (Penninckx *et al.*, 1996). SA induces SAR and controls activity of regulatory (activator) protein NPRI. Besides that, JA and ET act synergistically in inducing genes for seven PR proteins (Norman Setterbiad *et al.*, 2000).

Various *Pseudomonas* spp., strains induce broad-spectrum disease resistance in plants in a bacterial strain/plant species-specific manner. In *Arabidopsis thaliana,* specific strains of non-pathogenic fluorescent *Pseudomonas* bacteria develop an induced systemic resistance (ISR) that unlike pathogen-induced systemic acquired resistance (SAR), is independent of salicylic acid (SA) but requires sensitivity to jasmonic acid (JA) and ethylene (ET). Different bacterial determinants appear to be involved in triggering *e.g.* lipopolysaccharides, siderophores, and flagella. Combining ISR and SAR increased resistance against pathogens that are resisted by both JA/ET and SA-dependent defenses.

All these experiments show that different determinants of PGPR are involved in the induction of systemic resistance and ISR by these bacterial determinants varies with iron-limiting conditions, bacterial strains, host plants, and their cultivars.

Volatile Organic Compounds

Recently, Ryu *et al.* (2003) reported that PGPR strains *Bacillus subtilis* and *Bacillus amyloliquefaciens* released a blend of volatile organic compounds (VOCs) that promote growth in *Arabidopsis* seedlings and induce resistance against *Erwinia carotovora* pv. *carotovora*. The volatile components 2, 3 butanediol and 3-hydroxy-2-butanone (also referred to as acetoin) were released exclusively from PGPR strain trigger the greatest level of growth promotion and induced diseased resistance.

Harpins

Several phytopathogenic bacteria produce harpins, a class of acidic, glycine rich, and heat stable proteinaceous elicitors of hypersensitive response (HR) in various plant pathogen interactions (Wei *et al.*, 1992). Harpins have also been shown to trigger SAR. Their role in pathogenesis and defense signaling has been difficult to access since it may be masked by massive-production of pectinolytic enzymes in pathogens such as *Erwinia carotovora.* Recently the role of harpins as elicitors of plant defense responses has gained strong support. Apparently, the elicitor harpin HrPN produced by *Erwinia* acts in concert, and cooperatively, with oligogalacturonides (OGAs) in plant defense signaling in *Arabidopsis* (Kariola *et al.*, 2003).

PGPR Formulation and Methods of Application

An important area of microbiological research with regard to biocontrol is the development of formulations that would preserve microbial activity for a period long enough to enable delivery of an effective product for field application. *P.fluorescens* can be applied in the form of a bacterial suspension (Thompson, 1996) and as a powder formulation (Kloepper and Schroth, 1981). It is desirable from the consumer's perspective to formulate and package PGPR in ways similar to chemical pesticides. Mass multiplication of PGPR in a suitable medium and development of a powder formulation was first used as seed-treatment and soil-application. The survival of PGPR in a dried formulation and the effectiveness of methylcellulose in a powder formulation for coating sugar beet seed has been well documented (Suslow, 1980). When the stability and efficacy of the product was tested under field conditions root colonization was better with the powder formulation than aqueous preparations resulted by Kloepper and Schroth (1981). Seed-treatment followed by soil application of talc-based powder formulation effectively checked chickpea wilt and pigeonpea wilt under field conditions and has increased the yield (Vidhyasekaran *et al.*, 1997b). A peat-based formulation has also effectively controlled the rice sheath blight disease and increased the yield (Rabindran and Vidhyasekaran, 1996). Thus PGPR can be formulated and delivered effectively to the field for systemic protection of crop. The methods of application of formulated product include seed-treatment (Rosales and Mew, 1997), root dip (Maurhofer *et al.*, 1994), seed-treatment in sugarcane (Viswanathan and Samiyappan, 1999a), sucker-treatment in banana (Raguchander *et al.*, 1997), soil-application (Vidhyasekaran *et al.*, 1997a, b) and foliar application (Chatterjee *et al.*, 1996). Combinations of different methods of application could be more effective in disease management than a single method of application (Vidhyasekaran and Muthamilan, 1999).

Durability of ISR

Resistance mechanisms attain their maximum effectiveness at four to 4-5 days after the application of an inducing agent, but the level of persistence of resistance generally decreases over time. These criteria determine the number of applications of PGPR needed to maintain the resistance level in the crop plants (Dalisay and Kuc, 1995). In rice, seed-treatment with *P. fluorescens* strain Pf1 induces resistance which is observed upto 45 days after sowing. Foliar sprays of *P. fluorescens* formulations should be given at every 15 days intervals for managing rice foliar diseases. A parallel experiment conducted by Nayar (1996) indicated that induction of defence mechanisms using *P. fluorescens* persisted upto 60 days by seed-treatment, 30 days by root-dipping and 15 days by foliar spray. In cucumber, PGPR-mediated ISR resulted in protection of cucumber from anthracnose disease for 5 weeks from the time of sowing. But the consistency of ISR over time varies with PGPR strains (Liu *et al.*, 1995c). In sugarcane, induction of resistance by PGPR persists for 90 days of crop growth (Viswanathan, 1999). Generally, the durability of resistance by PGPR differs from crop to crop and also due to different bacterial strains.

PGPR-mediated ISR Under Field Conditions

In most of the cases, ISR has been studied mainly in the laboratory and greenhouse. However, some reports indicate that ISR by PGPR can protect the crop plants under field conditions. In a field trial, PGPR strains, *viz.*, *P. putida* strain 89B-27, *S. marcescens* strain 90-166 and *Flavomonas oryzihabitans* strain INR-5 have caused systemic protection against angular leaf spot and bacterial wilt besides increasing the plant growth (Kloepper *et al.*, 1993). Similarly, application of PGPR strains either as seed-treatment alone or as seed-treatment plus soil-drenching at the time of transplanting have protected cucumber plants inoculated with the *P. syringae* pv. *lachrymans*, the angular leaf spot pathogen and reduced the levels of anthracnose disease in the field (Wei *et al.*, 1996). In rice, seed-treatment and root-dipping of rice seedling with PGPR strain mixtures, *viz.*, *P. fluorescens* strains Pf1 and PB2 reduced rice sheath blight disease incidence and improved the grain yield under field conditions (Nandakumar, 1998). In sugarcane, application of PGPR as seed treatment induced systemic resistance against *C. falcatum* in addition to enhanced seed germination, tillering and growth of the cane both under controlled conditions as well as field conditions. These studies clearly show that PGPR-mediated ISR and plant growth promotion can operate under field conditions.

Role of PGPR in Crop Plants

PGPR strains *Bacillus, Azospirillum, Azotobactors, Pseudomonas* spp. etc. are found to improve seed germination, root/shoot length, nodulation, biomass, yield and nutrient mobilization in various crops (Table 25.1). The beneficial bacteria also protect the crops from the attack of soil borne pathogen among which *Fusarium spp, Rhizoctonia solani, Rhizoctonia bataticola* and *Sclerotium rolfsii* are highly destructive (Table 25.2). Pyrrolnitrin, Phenazine-1-carboxylic acid and 2, 4-diacetlyphloroglucinol produced by *Pseudomonas* spp were the major antibiotics involved.

Table 25.1: Plant Growth Promoting Rhizobacteria (PGPR) in Enhancing the Growth of Crop Plants

Crop	*Bacterial Isolate*	*Effects on Growth and Yield*
Wheat	*Bacillus subtilis*	Seed germination and grain yield
Pigeonpea	*Bacillus substilis AFI*	Increase in shoot and root length and biomass
Mungbean, pea	*Azotobacter*	Increase in yield
Chickpea	*Pseudomonas* sp.	Enhanced seed germination, root and shoot dry weight, yield
	Azospirillum brasilense	Nodulation
	Azotobacter	Number of shoots/plant, yield, seed protein
Soybean	*Bacillus cereus UW8S*	Increase in root growth, nodulation and yield
Fababean	*Azospirillum brasilense*	Increase in growth of root and shoot and nodulation
Groundnut	*Bacillus subtilis, Bacillus amyloliquefaciens, Pseudomomas aeruginosa, Pseudomonas fluorescens*	Increase in root and shoot length, biomass and yield
Sunflower	*Azospirillum*	Increase in yield
	Azotobacter	
	Pseudomonas	Higher seed yield
Black pepper ginger, turmeric, spice Crop	*Pseudomonas fluorescens Phosphobacteria IISR 4000*	Nutrient mobilization

PGPR: Usefulness Under Biotic Stress

Combined inoculation of other PGPR with rhizobia was observed to exert positive effects on the growth of legumes. Co-inoculation of PGPR along with *Rhizobium* increased the root nodulation. The beneficial effects of co-inoculation are strain dependent. The basic mechanisms involved in this synergistic activity were not completely known and remains a challenge. One possibility is that PGPR, by altering the host's secondary metabolism and/or creating antibiosis in the rhizosphere, out compete the pathogens and eliminate competition of *Rhizobium* with deleterious micro or for colonize of the plant parts. Alteration of the plant flavinoid metabolism was proposed as another mechanism of synergistic activity of PGPR and rhizobia. The applied rhizobacteria stimulate formation of additional infection sites that may after be occupied by rhizobia, thus increasing the nodulation. Inoculation of *Azospirillum* sp. produces large quantities of auxins and stimulates the formation of epidermal cells that become root hair cells or additional infection sites for rhizobial colonization. Unlike chemical fertilizers, PGPR exert a beneficial effect on rhizobial nodulation of legumes and should be exploited for the economic subsistence farming systems. A complete understanding of the mechanisms of the synergistic activity between PGI'R facilitates the better utilization and improvement of bioinoculants in legume based cropping systems.

Table 25.2: Plant Growth Promoting Rhizobacteria (PGPR) in Control of Root Diseases of Crop Plants

Crop	Used PGPR strain	Disease
Rice	*Pseudomonas fluorescens* *Pseudomonasputida* *Bacillus polymyxa* *Bacillus pumulus* *Bacillus coagulans* *Enterobacter agglornerans* *Bacillus cereus*	Suppression of plant pathogens blast, sheath blight, sheath rot stem rot, bacterial blight tungro disease, downey mildew
Sorghum	*Pseudomonas fluorescens*	Head mould
Pigeonpea	*Bacillus subtilis,* *Pseudomonas aeruginosa,* *Pseudomonas fluorescens*	Wilt
Chickpea	*Azotobacter chroococcum,* *Azospirillum lipoferum,* *Pseudomonas fluorescens*	Wilt
Pea	*Bacillus inegaterium*	Root rot
Soybean Cowpea	*Bravibacterium linens,* *Pseudomonas fluorescens*	Damping-off
Bean	*Pseudomonas capacia*	Stem and root rot
Bean	*Bacillus subtilis AF 1*	Root rot
Groundnut	*Pseudomonas fluorescens*	Crown rot
Lentil	*Pseudomonas aeruginosa*	Wilt
Urdbean	*Pseudomonas fluorescens*	Root rot
Mustard	*Trichoderma viride* *Bacillus amyloliquefaciens*	White rust Bud rot
Coconut	*Pseudomonas fluorescens*	Leaf rot

PGPR's and their effectiveness has been established on foot rot and slow decline of black pepper and rhizome rot of ginger and dump rot of cardamom. Strains of *Pseudomonas fluorescens* were found to increase the growth and vigor of these crops apart from suppressing the soil borne diseases (Sarma *et al.*, 2000).

In coconut and cocoa mixed crop increased activity of *Beijerinckia* sp., *Aspergillus niger* and *Pseudomonas sp.*, which also include growth promoting substances like IAA and GA (Nair and Subba Rao, 1977) which reflected on the yield increase in the mixture. Similar increase with nitrogen fixers and phosphorus solubilizers were reported in coconut tree spices (Cinnamon, clove, nutmeg) mixed crop system (Iyer, 1983). The role of *Pseudomonas fluorescens* mediated nutrient flux in the soil microcosm in plant growth promotion was studied with the higher uptake of nutrients by the bacterizoid plants.

PGPR: Usefulness Under Abiotic Stress

Microbial colonization of plant roots is affected by biotic and abiotic factors such as root exudates, competition, inorganic nutrients, soil pH and temperature etc. Among the role of agriculturally important abiotic stresses, drought, heat, soil salinity, and alkalinity rank as growth the most detrimental and cause loss of crop productivity all over the world.

Kulkarni and Nautiyal (2000) conducted a study to examine the growth response of a *Rhizobium* sp. NBR 1330 isolated from root nodules of *Prosopis juliflora* growing in alkaline soil. The individual stress survival limit of the strain in a medium containing 32 per cent NaCl was 8 hrs and at 55°C up to *M* also conducted an ecological survey (Kulkarni *et al.*, 2000) to characterize 5000 *Rhizobium* sp., *Sesbania* strains of diverse geographical origin, isolated from the root nodules of *Sesbania aculeata* growing in neutral (pH 7) and alkaline (pH 8.5 and above) soils. The rhizobia from the alkaline soil showed significantly higher salt tolerance than those isolated from neutral soil. High temperature was tolerated efficiently by *Rhizobium* sp. NBRl 2505 *Sesbania*, in the presence of salt at higher pH. Studies were conducted by Rehman and Nautiyal (2002) to elucidate the nature of drought tolerance in *Rhizobium* sp. NBR 2505 *Sesbania* and its transposan Tn5 induced mutant *Rhizobium* sp. NBRI 2505 *Sesbania* T 112 containing single Tn5 insertion, selected after screening about 1000 clones. Our results demonstrate positive effect of calcium on the survival of *Rhizobium* sp., *Sesbania* under acidic stress conditions. Mutant strain TI 12 produced ineffective symbiosis with the plant host in the presence of 2.5 and 5 per cent PEG indicating that drought tolerance is required for effective symbiosis.

Several soil bacteria produce osmolytes to protect themselves against frequent fluctuations in osmotic conditions *e.g. Paenibacillus polyinyxa* (previously *Bacillus polymyxa*) and *Bacillus subtilis* produces glycine betaine (Lucht and Bremer, 1994). This compound lowers the water potential outside the cell wall and inhibits intercellular spaces. The increased concentration of these compounds could be sensed by the plant as a local dehydration. Consequently, dehydration genes such as ERD 15 and RAB 18 would be activated in plants in order to tolerate drought.

Tripathi *et al.* (1998) have studied in detail salinily stress responses in the plant growth promoting rhizobacteria *Azospirillum* sp. They have shown that in order to adapt to the fluctuations in soil salinity/osmolarity, the bacteria accumulate compatible solutes such as glutamate, proline, glycine, betain, trehalose etc. Proline seems to play a major role in osmoadoptation. With increase in osmotic stress the dominant osmolyte in *Azospirillum brasilense* shifts from glutamate to proline. Accumulation of proline in *Azospirillum brasilense* takes place both uptake and synthesis.

Rangarajan *et al.* (2001) analysed populations of *Pseudomonas* for their biochemical characters and genetic diversity using molecular tools including RAPD and PCR-RFLP. It was observed that increasing salinity caused a predominant selection of salt tolerant species, in particular *Pseudomonas pseudoalcaligenes* and *Pseudomonas alcaligenes,* irrespective of the host rhizosphere. Rangarajan *et al.* (2003) have recently isolated indigenous *Pseudomonas* strain from rhizosphere of rice cultivated in the coastal agri-ecosystem and screened for *in vitro* antibiosis against *Xanthomonas ozyzae* pv. *oryzae* and *Rhizoctina solani.*

Phosphorus (P) is one of major essential macronutrients for biological growth and development. P in soil is immobilized or becomes less soluble either by absorption, chemical precipitation or both. Seed or soil inoculation with phosphate-solubilizing bacteria (PSB) known to improve solubilization of fixed soil phosphorus and applied phosphates, resulting in higher crop yields. However, the establishment and performance of these microbes are affected severely under stress such as high salt, p and temperature prevalent in degraded ecosystems such as alkaline soils with tendency to fix phosphorus. Therefore, it would be desirable to have PSB capable of solubilizing P under stress conditions. Gaind and Gaur (1991) have described thermo tolerant phosphate solubilizers microorganism and their interaction with mung bean. Nautiyal *et al.* (2000) characterized NBRI 2601 isolated from the rhizosphere of chickpea and alkaline soils which has the capability to solubilize phosphorus in the presence of 10 per cent salt, pH 12 or 45°C.

PSB that could tolerate high salinity have also been isolated (Kumar, 2003). P-starvation inducible quinoprotein glucose dehydrogenase (Gcd) has been shown to be responsible for the high gluconic acid secretion resulting efficient P solubilization. Gluconic acid producing PSB show P-solubilizing ability on a mixture of low concentrations of several sugars as a consequence of the broad substrate specificity of the Gcd enzyme. Since, Gcd in certain bacteria is under catabolite repression by TCA cycle intermediates, the P-solubilizing phenotype of these organism is adversely affected by the presence of organic acids such as succinate, malate etc. Also, expression of PQQ synthase gene has been demonstrated to be sufficient to confer P-solubilization ability in transgenic bacteria. Alternatively, mineral phosphate solubilizing genes have been cloned from *Synechocyslis* PCC 6803 in *Escheria coli* from genomic DNA library (Kumar, 2003). Attempts are also being made to develop P-solubilization ability by the over expression of phosphoenol pyruvate carboxylase gene. Conditional repression of TCA cycle genes by antisense RNA could also help in developing PGPRs with P-solubilization ability.

Genetics of PGPR Strains

Rhizospheric Competence

Efficient PGPR should survive in the rhizosphere, make use of nutrients exuded by the plant root, proliferate, be able to colonize the entire root system and compete with indigenous microbes (Bolemberg and Lugtenberg, 2001). Some of the genes reportedly influencing the rhizospheric colonization are (*a*) The M genes specifically expressed in rhizosphere (Rainey and Preston, 2000) (*b*) Flagellar syntheses, motility and chemotaxis genes d and c B (Matthysse and McMahan, 1998) (*c*) The nuo4 gene encoding part ofNADH dehydrogenase (*d*) A site-specific recombinase (sss) (Dekkers *et al.*, 1999) (*e*) The att genes (Matthysse and McMahan, 1998) (*f*) bacterial cellulose synthesis (cel genes (*g*) genes for outer membrane protein OprF that functions as adhesion (*h*) The genes of pot operon, encoding proteins involved in putrescine uptake and the PyrR protein (*i*) Genes required for synthesis of the lipopolysaccharide O-antigens (*j*) the pilus genes, (*k*) The genes htrB and secB (Dekkers *et al.*, 1999) and (*l*) The genes involved in the synthesis of amino acids and vitamin B1, nutrient acquisition, utilization of major exudates of carbon sources. Synthesis of bacterial agglutinins, homologues of a calcium-binding protein hemolysin, role of Levan in the aggregation of root-adhering soil, stress related proteins, a potential multi-drug efflux pump (Bolemberg and Lugtenberg, 2001) are some other factors influencing the rhizospheric colonization (Bolemberg *et al.*, 2002).

Bio-fertilization

The gene products of both the nif and the fix genes involved in nitrogen fixation have been well documented in some PGPR. Soil microorganisms possessing phytase activity contribute to plant phosphorus nutrition (Idriss *et al.*, 2002). Another beneficial effect due to bacterial phytase activity in the rhizosphere is elimination of chelate forming phytate, which binds nutritionally important minerals (Reddy *et al.*, 1989). Siderophores help by chelating iron from the rhizosphere to make it bio-available.

Bio-stimulation

The prominent PGRs and their analogues produced by PGPR are auxins, cytokinins, and gibberellins. Besides having nitrogen fixing ability, *Azospirillum spp.* secretes PGRs. Some PGPR release a blend of volatile components that promote growth (Ryu *et al.*, 2003). Though genes for PGR synthesis have been identified, in several cases, the complete bio synthetic pathways and their regulatory cascades leave many details wanted yet. Few of the genes involved in biosynthesis of PGR are: (*a*) Auxins indole-3-acetic acid (LAA CYP79B2, CYP 7983, nit to nit4) (*b*) Cytokinins trans-zeatin (ipt), cis-zeatin (ZOGI, ZOXI); and (*c*) Gibberellins GA, (CYPSSA Cytochrome P450 Gene, ga3-1, ga3-2).

Bio-control

Important among the group of PGPR are the broad-spectrum antibiotic producing fluoresent Pseudomonas spp. This group of rhizobacteria is an ideal candidate for biocontrol applications (Mark *et al.*, 2003). Pseudomonas putida is a metabolically versatile saprophytic soil bacterium that has been certified as a bio-safe host for the cloning of foreign genes. Ten genes (pitA to pUG, pp and pltR required for the biosynthesis of pyoluteorin were identified within a 24-kb genomic region of *Pseudomonas fluorescens* Pf-5 (Nowak Thompson *et al.*, 1999). Genes responsible for the production of I-ICN have been identified in the hydrogen cyanide synthase gene operon hcnABC. Genes (ph/A, to ph required for 2, 4-diacetylphloroglucinol synthesis are encoded by a 6.5-kb fragment of genomic DNA (Bangera and Thomashow, 1999). Introduction of alien genes such as crylAc7, chiA, and ACC-deaminase gene into rhizobacterial strains enhanced biocontrol and plant growth promotion (Bolemburg and Lugtenberg, 2001).

Rhizobium leguminosarurn, Azospirillum spp. are some of the lesser characterized candidate PGPR for genetic enhancement. One strategy to enhance rhizosphere competence is the development of genetically modified (GM) microbial inoculants (Walsh *et al.*, 2001). The focus is on developing strains in which the timing of the metabolite production is altered. If the relevant biosynthetic genes are uncoupled from their regulatory controls, it could facilitate early production of antifungal metabolites. However, public concerns as to the biosafety of GM bacterial strains in the environment must be addressed satisfactorily before the ultimate utilization of the GM-PGPR.

Bio-remediation

Bacteria engage in important metabolic activities in the environment, including element cycling and the degradation of biogenic and xenobiotic pollutants. They can also act as bio indicators and bio-catalysts. The bph genes together with Pex promoter boosted the contaminant poly-chlorinated biphenyl (PCB) degradation capabilities of *Pseudomonas* (Brazil *et al.*, 1995). Addition of such capabilities can enhance the usefulness of PGPR and broaden their spectrum of utility.

Gene Regulation

Control of gene expression at the transcriptional level is a primary mechanism for modulating the production of secondary metabolites. Key factors in the regulation of the biosynthesis of most anti-fungal microbial are global regulation that simultaneously affects multiple biosynthetic pathways along with pathway-specific regulators that control linked biosynthetic genes, and the quorum sensing (Bolemburg and Lugtenberg, 2001). Global regulation is directed by the gacS/gacA genes. The quorum sensing (QS) involves the production of N-acyl homoserine lactone (AHL) signal molecules byan AHL-synthase. At a threshold concentration of AHL, reached only when a certain density of bacterial cells is present, the AHL will sufficiently bind and activate a transcriptional regulator. The cinR and cinI appear to be at the top of a regulatory cascade or network influencing several ARL regulated QS loci. CinR regulates cinI expression in a cell-density dependent manneL and c expression is positively autoregulated by its own product. The co and coiR genes regulate the colonization in several cases

(Dekkers *et al.*, 1999). The phytopathogen *Pythium* produces diffusible factors that can down-regulate genes important for the survival of the bio agent *Pseudomanas* (Walsh *et al.*, 2001). The fungal toxin fusaric acid (FA) produced by the phytopathogen *Fusarium* can repress expression of the gene phiA

Plant Growth Promoting 'Meta-bug'

Notwithstanding the state of art of the gene technology, it would be a far cry to wish for internalization of all the relevant traits of plant growth promotion into the host plant genome. Alternatively, the PGP 'meta-bug' approach would involve developing a 'core-competent bug envisaging rhizosphere colonization and survival related functions, and stacking it with the genes for bio-stimulation, bio-fertilization, bio-control etc. Hypothetically the core- competent bug would have as many compatible genes relating to rhizospheric competence as possible, loaded into its genophore. To list is a few genes *rhi, col, chv, nuo4, att, pot, htr sec*, etc. This core competent bug then can be decorated with genes for production of phyto stimulatory agents (*aux* (*nit, CYP*) *ipt/ZOG, ZOX; ga3*) bio-control agents (*phz, plt, hen, etc.*), bio-fertilizing activities (*nif, fix, phy*, etc.) and bio-remediation effectors (*bph, pex*, etc.). With appropriate fine tuning and balancing of regulatory systems involved, such as *gacS/ gacA, cinR/cinI, colR/cols*, the meta bug should be ideally a 'single window' delivery system, capable of not only establishing itself in wide range of rhizosphere, but also competent in efficient plant growth promotion. A meta-bug each targeted for different agro-ecology could be developed in the most appropriate strain/s of *Pseudomonas, Rhizobium, Azospirillum*, etc.

PGPR Research in India

In India, the phosphate solubilizing and nitrogen fixing bacteria have been developed as effective bioinoculants and are being used for seed inoculation of important crops like wheat, rice, barley, millet, soybean, sunflower, potato, sorghum, maize and fruits and vegetables. It seems that the potential of rhizobacteria particularly symbiotic nitrogen fixing rhizobia and free living *Azotobacter* to increase crop yield was known in India. The seed inoculants for solubilization of natural phosphate in soils has been recommended in former Soviet Union and India as commercial bioinoculants like 'Phosphobacterin' and 'Microphos', 'Nitragin', 'Azotobacterin' respectively and their application in fields have been shown to increase crop yields. The bioinoculants of *Rhizobium, Frankia* and *Bradyrhizobium* for legumes, *Azotobacter* and *Azospirillum* for cereals as well as BGA for paddy application has been developed by Indian Agricultural Research Institute (IARI), New Delhi. Those biofertilizers have been constantly used by Indian farmers and thus helped for country's sustainable agriculture. Significant improvement observed due to bacterization with PGPR (which includes genera *Acetobacter, Acinetobacter Arthrobacter, Azotobacter, Azospirilium, Bacillus, Berkholderia, Enterobacteria, Klebsiella, Proteus, Pseudomonas, Rhizobium, Serratia* and *Xanthomonas*) during various field trials indicates the possibility evolving eco-friendly input for farmers with limited sources. In the 10th plan (2002-2007) major thrust by Government of India has been given on exploiting biodiversity of nitrogen fixers and PGPR biofertilizers technology improvement.

Indian Institute of Spices Research (IISR), Calicut has standardized fluorescent PGP *Pseudomonas* for large scale production of spices, especially, black pepper and transferred the technology to entrepreneurs for distribution. In order to harness potential benefits of biofertilizers in commercial agriculture, the consistency of their performance improved. This requires research in many diverse areas as these biological systems involve complex interactions among the host, other rhizosphere micro flora and fauna and the environment.

Currently all agricultural universities and IARI are involved in research on PGPR biocontrol by rhizobacteria. In addition to these, many research groups in Indian universities are doing extensive research on occurrence, distribution, mechanism and role of PGPR strains for field applications. Critical research work done at ICRISAT (International Crops and Research Institute for Semi Arid Tropics), Patancheru, and Hydrabad has been a constant source of our knowledge of biological nitrogen fixation and other aspects of PGP and bacteria associations with crop plants. So in the country like India research on diversity of PGPR and their prospect as future fertilizers to increase toil productivity has bright future and important role in the development of country's sustainable agriculture to meet demand of food for growing population.

Conclusion and Future Prospects

More attention is needed to orient our future research on some of the areas *viz.* PGPR survival potential, identification of crop specific strains, mixture formulation, biochemical molecular markers for identification of effective PGPR and genes activated in octadecanoid pathway and Phenyl propanoid pathway due to bacterial treatment in order to enhance the performance of biocontrol bacterial inoculants for managing plant pest and diseases sustainable manner. Most PGPR strain selected from the rhizosphere do not fulfill initial promise. Failures are-often due to poor rhizosphere competence, and to difficulties associated with the instability of bacterial strains in long term culture. The intimate relationship between endophytic bacteria and their hosts could make them good candidates for successful biocontrol agents.

The beneficial effects of PGPR include direct plant growth promotion, biological control and inducing systemic resistance in host plants. Specific PGPR strains bring about ISR against multiple pathogens attacking the same crop. In addition to disease suppression, application of PGPR also reduces the insect and nematode damage. The broad spectrum of control using PGPR strains can provide an effective, economical and practical way of plant protection. The endophytic nature of some PGPR make them suitable for the use in vegetatively propagated crops because of their capability to colonize and persist in the intercellular space of epidermal cells thereby reducing the need for further application if the same vegetative parts are used as propagative material. Furthermore, certain PGPR strain mixtures have showed synergistic action in plant protection and growth promotion, indicating different mechanisms are involved in disease control. So, selecting such combinations of strains would be beneficial in crop production. Stable formulations using different carriers such as peat and talc have been developed for the delivery of the PGPR stains for field level application.

To conclude, the bioinoculants have a tremendous potential in India as a cheap source of plant nutrients, provided the quality of product is assured. The successful development of bioinoculants that meet farmers' expectation must deliver function and value. The subject is so challenging that one should approach it hopefully and with optimism not with skepticism and doubts. As Samuel Johnson observed long ago, "Nothing will ever be attempted if all possible objections must first be overcome". Each successful application, no matter how limited, increases the level of familiarity with bioinoculants among user groups, governmental regulatory agencies, private investors, and the public. Looking ahead, it is relatively easy to visualize thousands of microbes in use worldwide for control of specific diseases on specific crops in specific cropping systems and produced by public or private enterprises as necessary. It requires development of new technologies on a constant basis and assimilates the same in their production set up. To succeed those manufacturing the bioinoculants must deliver the knowledge. Indian farmer is conscious of the fact that chemical fertilizers adversely affect soil health.

Therefore they prefer to use green manure and farm yard manure to preserve the soil health. It may probably be more important for us to convince our farmers about the esuriently and soil rejuvenating nature of bioinoculants along with limitations. For example farmers must understand that bioinoculant application is prone to environment like any other living object, unlike chemical fertilizers. Once these limitations are understood, the benefits of bioinoculants such as efficacious season long control, flexible timing of application, crop, and soil health and environment safety may be established. Limitations and benefits may then be combined to arrive at consistently attainable expectations value of the bioinoculants. Finally, the success of the wider use of bioinoculants will partly depend on the total commitment of companies for maintaining the quality of the bioinoculants and partly on our government regulators to implement regulations to ensure that companies are responsible for maintaining consistently high standards, as the bitterness of poor quality remains long after the sweetness of low price is forgotten. Bioinoculant products that meet and exceed farmer's expectations will drive productivity gains of benefit to all.

References

Arshad, M. and Frankenberger, W.T.J. (1998). Plant growth regulating substances in the rhizosphere microbial production and functions *Adv. Agron.*, 2: 145–151.

Bangera, G.M. and Thomashow, L.S. (1999). Identification and characterization of a gene cluster for synthesis of the polyketide antibiotic 2, 4-Diacetylphloroglucinol from *Pseudomonas fluorescens* Q2–87. *J. Bacteriol.*, 181: 3155–3163.

Bashan, Y. and de-Bashan, L.E. (2004). *Azospirillum*-plant relationship: II, Interaction with plants, agricultural and environmental advances (1997–2002). *Can. J. Microbiol* (In Press).

Benhamou, N., Belanger, R.R. and Paulitz, T.C. (1996a). Induction of differential host responses by *Pseudomonas fluorescens* in Ri T-DNA transformed pea roots after challenge with *Fusarium oxysporum* f. sp. *pisi* and *Pythium ultimum*. *Phytopathology*, 86: 114–178.

Benhamou, N., Kloepper, J.W. and Tuzun, S. (1998). Induction of resistance against *Fusarium* wilt of tomato by combination of chitosan with an endophytic bacterial strain: ultra structure and cytochemistry of the host response. *Planta*, 204: 153–168.

Binns, A.N. (1994). Biochemical, genetic, and molecular approaches. *Annu. Rev. Plant Phys. Plant. Mol., Biol.*, 5: 173–176.

Bolemberg, G.V., M. M., Camacho-Caravajal, T.F.C. Chin-A-Woeing, L.C., Dekkers, A.M., Cees, J.J. vanden Hondel, I. Kuiper, A.L., Lagopodi, G.E.M. Lamers, I. Mulders, A.F.J. Ram, S. deWeert,. A.H.M. Wijfejes, and Lugtenberg, B.J.J. (2002). Rhizosphere colonization by biocontrol *Pseudomonas* spp. In: *Proc. 5th Intl. Plant Growth Promoting Rhizobacterial Workshop*, 29 October–3 November, Cordoba, Argentia.

Bolemburg, G.V. and Lugtenberg, B.J.J. (2001). Molecular basis of plant growth promotion and biocontrol by rhizobacteria. *Curr. Opin. Plant Biol.*, 4: 343–350.

Bong, C.F.J. and Sikorowski, P.P. (1991). Effects of cytoplasmic polyhedrosis virus and bacterial contamination on growth and development of the corn earworm, *Helicoverpa zea*. *J. Invertebr. Pathol.*, 57: 406–412.

Brazil, G.M., L. Kenefick, M. Callanan, A. Haro, V. doLorenzo, D.N, dowling, and FO'Gara (1995). Construction of rhizosphere pseudomonad with potential to degrade polychlorinated biphenyls and detection of *bph* gene expression in the rhizosphere. *Appl. Environ. Microbiol.*, 6: 1946–1952.

Cameron, R.K., Dixon, R. and Lamb, C. (1994). Biologically induced systemic acquired resistance in *Arabidopsis thaliana. Plant J.*, 5: 715–25.

Chatterjee, A., Valasubramanian, V., Vachhani, W.L., Gnanamanickam, S.S. and Chatterjee, A.K. (1996). Isolation of ant mutants of *Pseudomonas fluorescens* strain Pf 7–14 altered in antibiotic production, cloning of ant+ DNA and evaluation of the role of antibiotic production in the control of blast and sheath blight of rice. *Biol. Control*, 7: 185–195.

Chen, C., Bauske, E.M., Musson, G., Rodriguez-Kabana, R. and Kloepper, J.W. (1995). Biological control of *Fusarium* wilt on cotton by use of endophytic bacteria. *Biol. Control*, 5: 83–91.

Chester, K. (1933). The problem of acquired physiological immunity in plants. *Quart. Rev. Biol.*, 8: 129–151.

Corbell, N. and Loper, J.R. (1995). A global regulator of secondary metabolite production *Pseudomonas fluorescens* pf-5. *J. Bacteroil.*, 7: 6230–6236.

Costacurta, A. and Vanderleyden, J. (1995). Synthesis of phytohormones by plant associated bacteria. *Crit. Rev. Microbiol.*, 2: 2001–2013.

Dalisay, R.F. and Kuc, J.A. (1995). Persistence of induced resistance and enhanced peroxidase and chitinase activities in cucumber plants. *Physiol. Mol. Plant Pathol.*, 47: 315–327.

De Meyer, G. and Hofte, M. (1997). Salicylic acid produced by the rhizobacterium *Pseudomonas aeruginosa* 7NSK2 induces resistance to leaf infection by *Botrytis cinerea* on bean. *Phytopathology*, 87: 588–593.

De-Bashan, L.E., Hernandez, J.R. and Bashan, Y. (2003). Microalgae growth promoting bacteria 'helpers" for rnicroalgae a noval approach for removing ammonium and phosphorus from wastewater. In: *Phosphate Solubilizing Bacteria*, (Ed.) F. Velazquez. Kluwer Academic Publishers, Dordrecht, The Netherlands (In Press).

de-Bashan, L.E., Moreno, M., Hernandez, J.P. and Bashan, Y. (2002). Removal of ammonium and phosphorus ions from synthetic waste water by the microalgae *Chlorella vulgaris* coimmobilized in alginate beads with the microalgae growth promoting bacterium *Azospirillum brasilense. Water Res.*, 36: 2941–2948.

Dekkers, L.C., Phoeiich, CC. and Lugtenberg, B.J.J. (1999). Bacterial traits and genes involved in rhizosphere colonization. In: *Microbial Biosystems*, (Eds.) Bell, C.R., Brylinsky, M., Johnson P. and Green. Atlanta Canada Society for Microbial Ecology, Halifax, Canada.

DoMeyer, G., Capieau, C., Audenaert, K., Buchola, A., Mefraux, J.R. and Hofte, M. (1999). Nanograms amounts of salicylic acid produced by the rhizobacterium *Pseudomonas aeruginosa* TNSK: Activate the systemic acquired resistance pathway in bean. *Mol. Plant Microb. Interact.*, 12: 450–458.

Duffy, B.K. and Defago, G. (1997). Zinc improves biocontrol of *Fusarium* crown and root rot of tomato by *Pseudomonas fluorescens* and represses the production of pathogen metabolites inhibitory to bacterial antibiotic biosynthesis. *Phyotpathology*, 87: 1250–1257.

Duffy, B.K. and Weller, D.M. (1995). Use of *Gaeumannomyces graminis* var. *graminis* alone and in combination with fluorescent *Pseudomonas* spp. to suppress take-all of wheat. *Plant Dis.*, 79: 907–911.

Elad, Y. and Baker, R. (1985). The role of competition for iron and carbon in suppression of chlamydospore germination of *Fusarium oxysporum. Phytopathology*, 75: 190–195.

Enebak, S.A. and Carey, W.A. (2000). Evidence for induced systemic protection to *Fusarium* rust in Loblolly pine by plant growth promoting rhizosphere. *Plant Disease*, 84: 306–308.

Frindlender, M., Inbar, J. and Chet, I. (1993). Biological control of soil borne plant pathogens by a β-1,3–glucanase producing *Pseudomonas cepacia*. *Soil Biol. Biochem.*, 25: 1211–1221.

Glick, B.R., Patten, C.L., Holguin, G. and Penrose, N.M. (1999). *Biochemical and Genetic Mechanisms Used by Plant Growth Promoting Bacteria.* London Imperial College Press.

Gaind, S. and Qaur, A.C. (1999). Thermotolerant phosphate solubilizing microorganism the interaction with mungbean. *Plant and Soil*, 133: 141–149.

Gaudin, V., Vrain, D. and Louanin, L., 1994. Bacterial genes modifying hormonal balance in plant. *Plant Physiol. Biochem.*, 3: 11–29.

Glick, B.R. (1995). The enhancement of plant growth by free-living bacteria. *Can. J. Microbiol.*, 1: 109–117.

Glick, B.R., Jacobson, C.B., Schwarze, M.M.K. and Pasternak, U. (1998). 1-Aminocydopropane-1 carboxylic acid deaminase mutants of the plant growth promoting rhizobacterium *Pseudomonas putida* GR 12–2 do not stimulate canola root elongating. *Can. J. Microbial.*, 4: 911–915.

Haas, D., Keel, C. and Reimmann, C. (2002). Signal transduction in plant beneficial rhizobacteria with biocontrol properties. *Antonie Van Leeuwenheek*, 8: 385–395.

Hall, J.A., Peirson, D., Ghosh, S. and Glick, B.R. (1996). Root elongation in various agronomic crops by the plant growth promoting rhizobacterium *Pseudomonas putida* GR 12–2 Isrl. *J. Plant Sci.*, 4: 37–42.

Hoffland, E., Hakulinem, J. and Van Pelt, J.A. (1996). Comparison of systemic resistance induced by avirulent and nonpathogenic *Pseudomonas* species. *Phytopathology*, 86: 757–762.

Hurbert, J.J. and Helton, A.W. (1967). A translocated resistance phenomenon in *Prunus domestica* induced by initial infection with *Cytospora cineta*. *Phytopathology*, 57: 1094–1098.

Idriss, B.E., Makarewicz, O., Farouk, A., Rosner, K., Greiner, R., Bochow, H., Richter, T. and Borriss, R. (2002). Extra cellular phytase activity of *Bacillus amyloliquefaciens* FZB45 contributes to its plant growth promoting effect. *Microbiol.*, 148: 2097–2109.

Jackson, M.B. (1993). Are plant hormones involved in root to shoot communication? *Adv. Bot. Rev.*, 1: 103–186.

Kado, C.I. (1992). Plant pathogenic bacteria. In: *The Prokaryotes*, (Eds.) Balows, A., Truper, H.G., Dworkin, M., Harder, W. and K.H. Schleifer. Springer, New York, pp. 660–662.

Kloepper, J.W., Lifshitz, R. and Novacky, A. (1980). *Pseudomonas* inoculation to benefit plant production. *Anim. Plant Sci.*, p. 60–64.

Kloepper, J.W. and Schroth, M.N. (1981). Development of a powder formulation of rhizobacteria for inoculation of potato seed pieces. *Phytopathology*, 71, 590–592.

Kloepper, J.W., Schroth, M.N. and Miller, T.D. (1980). Effects of rhizosphere colonization by plant growth promoting rhizobacteria on potato plant development and yield. *Phytopathology*, 70: 1078–1082.

Kloepper, J.W., Tuzun, S., Liu, L. and Wei, G. (1993). Plant growth-promoting rhizobacteria as inducers of systemic disease resistance. In: *Pest Management: Biologically Based Technologies*, (Eds.) Lumsden, R.D. and J.L. Waughn. American Chemical Society Books, Washington, DC, pp. 156–165.

Kloepper, J.W., Tuzun, S., Zehnder, G.W. and Wei, G. (1997). Multiple disease protection by rhizobacteria that induce systemic resistance: Historical precedence. *Phytopathology*, 87: 136–137.

Kombrink, E. and Somssich, I.E. (1995). Defense responses of plants to pathogens. *Adv. Bot. Res.*, 2: 1–34.

Kulkarni, S., Surange, S. and Nautiyal, C.S. (2000). Crossing the limits of *Rhizobium* existence is extreme conditions. *Curr. Microbiol.*, 41: 402–409.

Kulkarni, S. and Nautiyal, C.S. (2000). Effects of soft and pH stress on temperature-tolerant *Rhizobium* sp. NBRI 330 nodulating *Propsopis juliflora*. *Curr. Microbiol.*, 40: 221–226.

Kumar, G.N. (2003). Improvement of PGPR by transformation: Prospects and challenges. In: *6th International PGPR Workshop* held at IISR Calicut, India, 5–10 Oct, p. 46.

Kunc, F. and Macura, J. (1998). Mechanisms of adaptation and selection of micro-organisms in the soil. In: *Soil Microbial Associations*, (Eds.) Vancura, V. and R. Kunc. Elsevier, Amsterdam, pp. 281–299.

Leeman, M., Van Pelt, J.A., Den Ouden, F.M., Heinsbroek, M., Bakker, P.A.H.M. and Schippers, B. (1995). Induction of systemic resistance against *Fusarium* wilt of radish by lipopolysaccharides of *Pseudomonas fluorescens*. *Phytopathology*, 85: 1021–1027.

Leeman, M., Van Pelt, J.A., Den Ouden, F.M., Heinsbroek, M., Bakker, P.A.H.M. and Schippers, B. (1996). Iron availability affects induction of systemic resistance to *Fusarium* wilt of radish by *Pseudomonas fluorescens*. *Phytopathology*, 86: 149–155.

Leeman, M., den Ouden, F.M., Van Pelt, J.A.. Drik, F.P.M., Steiil, H, Bakker, P.A.H.M and Schippers, B. (1996). Iron availability affects induction of systemic resistance to *Fusarium* wilt of radish by *Pseudomonas fluorescens*. *Phytopathol.*, 6: 149–155.

Liu, L., Kloepper, J.W. and Tuzun, S. (1993a). Induction of systemic resistance against cucumber bacterial angular leaf spot caused by *Pseudomonas syringae* pv. *lachrymans* with two plant growth promoting rhizobacteria. *Phytopathology*, 82: 1340 (Abstr.).

Liu, L., Kloepper, J.W. and Tuzun, S. (1993b). Induction of systemic resistance to *Fusarium oxysporum* by PGPR strains. *Proceedings of the sixth International Congress of Plant Pathology*, 24 (Abstr.).

Liu, L., Kloepper, J.W. and Tuzun, S. (1995a). Induction of systemic resistance in cucumber against *Fusarium* wilt by plant growth promoting rhizobacteria. *Phytopathology*, 85: 695–698.

Liu, L., Kloepper, J.W. and Tuzun, S. (1995c). Induction of systemic resistance in cucumber by plant growth promoting rhizobacteria: duration of protection and effects of host resistance on protection and root-colonization. *Phytopathology*, 95: 1064–1068.

Lorito, M., Hayes, CL, Zoina, A., Scala, F., Del Sorbo, G., Woo, S.L. and Harman, G.E. (1994). Potential of genes and gene products from *Trichoderma* sp. and *Gliocladiun* sp. for development of biological pesticides. *Mol. Biotechnol.*, 2: 209–217.

Lyer, R. (1983). Rhizosphere microflora in crop mixed coconut soil. *Philippine J. Coconut Studies*, p. 42–45.

Matthyssee, A.G. and McMahan, S. (1998). Root colonization by *Agrobacterium tumifaciens* is reduced in *cel, attB, attD* and *attR* mutants. *App. Environ. Microbiol.*, 64: 2341–2345.

Maurhofer, M., Hase, C., Meuwly, P., Metraux, J.P. and Defago, G. (1994). Induction of systemic resistance of tobacco to tobacco necrosis virus by the root-colonizing *Pseudomonas fluorescens* strain CHAO: influence of the gacA gene and of pyoverdine production. *Phytopathology*, 84: 139–146.

Maurhofer, M., Reimmann, C., Sacherer, S.P., Heeb, S., Haas, D. and Defago, G. (1998). Salicylic acid biosynthetic genes expressed in *Pseudomonas fluorescens* strain P3 improve the induction of systemic resistance in tobacco against tobacco necrosis virus. *Phytopathology*, 88: 678–684.

M'Piga, P., Belanger, R.R., Paulitz, T.C. and Benhamou, N. (1997). Increased resistance to *Fusarium oxysporum* f. sp. *radicis-lycopersici* in tomato plants treated with the endophytic bacterium *Pseudomonas fluorescens* strain 63–28. *Physiol. Mol. Plant Pathol.*, 50: 301–320.

Nair, S.K. and Subba Rao, N.S. (1977). Microbiology of the root region of coconut and cocoa under mixed cropping. *Plant Soil*, 46: 511–519.

Nandakumar, R. (1998). Induction of systemic resistance in rice with fluorescent pseudomonads for the management of sheath blight disease. *M.Sc. (Agric.) Thesis*, TNAU, Coimbatore, India, pp. 105.

Nautiyal, C.S., Bhaduria, S., Kumar, R., Lal, H., Month, R. and Verma, D. (2000). Stress in phosphate solubilization in bacteria isolated from alkaline soils. *FEMS Microbiol. Lett.*, 182: 291–296.

Nayar, K. (1996). Development and evaluation of a biopesticide formulation for control of foliar disease of rice. *Ph.D. Thesis*, TNAU, Coimbatore, India, pp. 223.

Neilands, J.B. and Nakamura, K. (1991). Detection, determination, isolation, characterization and regulation of microbial iron chelates. In: *CRC Handbook of Microbial Iron Chelates*, (Ed.) G. Winkelmann. CRC Press, London.

Norman-Setterblad, C., Vidal, S. and Palva, E.I. (2000). Interacting signal pathways control defense gene expession in *Arabidopsis* in response to cell wall degrading enzymes from *carotovora*. *Mol. Plant Microbe Interact.*, 3: 430–438.

Nowak-Thompson, B., Chaney, N., Wingh, J.S., Gould, S.J. and Loper, J.E. (1999). Characterization of the Pyoluteorin biosynthetic gene cluster of *Pseudomonas fluorescens* Pf–5. *J. Bactriol.*, 181: 2166–2174.

Oostendorp, M. and Sikora, R.A. (1990). *In vitro* interrelationship between rhizosphere bacteria and *Heterodera schachtii*. *Rev. Nematol.*, 13(3): 269–274.

Panwar, J.D.S. and Sirohi, G.S. (1989). *Azospirillum*: An important biofertilizers for improving crop productivity. *Farmers and Parliament*, 24: 23–24.

Panwar, J.D.S., Saikia, S.P. and Naidu, V.S.G.R. (2001). Biofertilizers for enhancing crop productivity and environmental security. *Indian Farming* (special issue) during 88th session of the Indian Science Congress, 50(10): 56–60.

Pennickx, I.A., Eggermont, K., Terra, E.R., Thomma, B.R, De Samblanx, G.W., Buchala, A., Metraux, J.R., Manners, J.M. and Broekaert, W.Y. (1996). Pathogen-induced systemic activation of a defensive gene in *Arabidopsis* follows a salicylic acid-independent pathway. *Plant Cell*, 8: 2309–2323.

Persello-Cartieaux, F., Nussaume, L. and Robaglia, C. (2003). Tales from the underground: Molecular plant rhizobacteria interactions. *Plant Cell Environ.*, 2: 189–199.

Pierson, L.S. and Pierson, B.A. (1996). Phenazine antibiotic production in *Pseudomonas aureofaciens*: Role in rhizosphere ecology and pathogen suppression. *FEMS, Microbiol. Lett.*, 2: 101–108.

Pierson, E.A. and Weller, D.M. (1994). Use of mixtures of fluorescent pseudomonads to suppress take–all and improve the growth of wheat. *Phytopathology*, 84: 940–947.

Pleban, S., Ingel, F. and Chet, I. (1995). Control of *Rhizoctonia solani* and *Sclerotium rolfsii* in the greenhouse using endophytic *Bacillus* spp. *Eur. J. Plant Pathol.*, 101: 665–672.

Potter, S., Uknes, S., Lawton, K., Winter, A.M., Chandler, D., DiMaio, J., Novitzky, R., Ward, and Rvals, J. (1993). Regulation of a heveine like gene in *Arabidopsis*. *Mol. Plant Microbe Interact.*, 6: 680–685.

Press, C.M., Wilson, M., Tuzun, S. and Kloepper, J.W. (1997). Salicylic acid produced by *Serratia marcescens* 90–166 is not the primary determinant of induced systemic resistance in cucumber or tobacco. *Mol. Plant Microbe. Interact.*, 10: 761–768.

Qingwen, Z., Ping, L., Gang, W. and Qingnian, C. (1998). On the biochemical mechanism of induced resistance of cotton to cotton bollworm by cutting of young seedling at plumular axis. *Acta Phytophylacica Sinica*, 25: 209–212.

Raazijmakers, J.M., Leeman, M., Van Oorschot, M. Van der slwis, I., Schippers, B. and Bakker, P.A.H.M. (1995). Dose-response relationships on biological control of *Fusarium* wilt of radish by *Pseudomonas* spp. *Phytopathol.*, 8: 1075–1081.

Rabindran, R. and Vidhyasekaran, P. (1996). Development of a formulation of *Pseudomonas fluorescens* PfALR2 for management of rice sheath blight. *Crop Prot.*, 15: 715–721.

Raguchander, T., Jeyashree, K. and Samiyappan, R. (1997). Management of *Fusarium* wilt of banana using antagonistic microorganisms. *J. Biol. Control*, 11: 101–105.

Rainey, P.B. and Preston, G.M. (2000). *In vivo* expression technology strategies: Valuable tool for biotechnology. *Curr. Opin. Biotechnol.*, 11: 440–444.

Rangarajan, S., Loganathan, P., Saleena, L.M. and Nair, S. (2001). Diversity of pseudomonads isolated from three different plant rhizospheres. *J. Appl. Microbiol.*, 91: 742–749.

Rangarajan, S., Saleena, L.M., Vasudevan, P. and Nair, S. (2003). Biological suppression of rice diseases by *Pseudomonas* spp. under saline soil conditions. *Plant Soil*, 251: 73–82.

Raupach, G.S. and Kloepper, J.W. (1998). Mixtures of plant growth promoting rhizobacteria enhance biological control of multiple cucumber pathogens. *Phytopathology*, 88: 1158–1164.

Raupach, G.S., Liu, L., Murphy, J.F., Tuzun, S. and Kloepper, J.W. (1996). Induced systemic resistance in cucumber and tomato against cucumber mosaic cucumovirus using plant growth promoting rhizobacteria (PGPR). *Plant Dis.*, 80: 891–894.

Reddy, N.R., Pierson, M.D., Sathe, S.K. and Salunke, D.K. (1989). *Phytases in Cereals and Legumes*. CRC Press, Boca Raton.

Rehman, A. and Nautiyal, C.S. (2002). Effect of drought on the growth and survival of the stress-tolerant bacterium *Rhizobium* sp. NBRI 2505 *Sesbania* and its drought-sensitive transposan Tn5 mutant. *Curr. Microbiol.*, 45: 368–377.

Rosales, A.M. and Mew, T.W. (1997). Suppression of *Fusarium moniliforme* in rice by associated antagonistic bacteria. *Plant Dis.*, 81: 49–52.

Ross, A.F. (1961). Systemic acquired resistance by localized virus infection in plants. *Virology,* 14: 340–358.

Ryu, C.M., Farag, M.A., Hu, C.H., Reddy, M.S., Wei, H.X., Pare, P.W. and Kloepper, J.W. (2003). Bacterial volatiles promote growth in *Arabidopsis. Proc. Natl. Acad Sci.,* 100: 4927–4932.

Samac, D.A. and Shah, D.M. (1994). Effect of chitinase antisense RNA expression on disease susceptibility of *Arabidopsis* plants. *Plant Mol. Biol.,* 2: 587–596.

Samiyappan, R. (2003). Molecular mechanisms involved in the PGPR mediated suppression of insect pests and plant pathogens attacking major agricultural and horticultural crops in India. In: *6th International PGPR Workshop*, 5–10 October.

Santhi, A. and Sivakumar, V. (1995). Biocontrol potential of *Pseudomonas fluorescens* (Migula) against root-knot nematode, *Meloidogyne incognita* (Kofoid and White, 1919) Chitwood, 1949 on tomato. *J. Biol. Control,* 9: 113–115.

Sarma, Y.R., Rajan, P.P., Paul, D., Beena, N and Anandaraj, M. (2000). Role of Rhizobacteria on disease suppression in spice crops and future prospects. In: *Seminar on Biological Control with Plant Growth Promoting Rhizobacteria for Sustainable Agriculture.* University of Hyderabad, Hyderabad.

Schisler, D.A., Slininger, P.J. and Bothast, R.J. (1997). Effects of antagonist cell concentration and two–strain mixtures on biological control of *Fusarium* dry rot of potatoes. *Phytopathology,* 87: 177–183.

Schnideir U., Keel, C., Blumer, C., Troxler, J., Defago, G. and Haas, D. (1995). Amplification of the housekeeping sigma factor in *Pseudomonas fluorescens* CHAO enhances antibiotic production and improves biocontrol abilities. *J. Bacteriol.,* 7: 5387–5392.

Schroth, M.N. and Hancock (1982). Disease-suppressive soil and root colonizing bacteria, *Science,* 2: 1376–1381.

Sikora, R.A. (1988). Interrelationship between plant health promoting rhizobacteria, plant parasitic nematodes and soil microorganisms. Med. Fac. Landbouww. Rijksuniv. *Gent.,* 53/2b, 867–878.

Sikora, R.A. (1992). Management of the antagonistic potential in agricultural ecosystems for the biological control of plant parasitic nematodes. *Annu. Rev. Phytopathol.,* 30: 245–270.

Sikora, R.A., Holmann-Hergarten, S. (1992). Importance of plant health-promoting rhizobacteria for the control of soil-borne fungal diseases and plant parasitic nematodes. *Arab. J. Plant Prot.,* 10: 53–58.

Singh, H.P. and Singh, T.A. (1993). The interaction of rock phosphate, *Bradyrhizobium,* vesicular arbuscular mycorrhizae and phosphate-solubilizing microbes on soybean grown in a Himalayan mollisol. *Mycorrhiza,* 4: 37–43.

Smith, J.L., Papendick, R.L., Bezdkek, D.R. and Lynch, J.M. (1993). Soil organic matter dynamics and crop residue management. In: *Soil Microbial Ecology,* (Ed.) F.B. Melting, NewYork, p. 65–94.

Spiegel, Y., Cohn, E., Galper, S., Sharon, E. and Chet, I. (1991). Evaluation of a newly isolated bacterium, *Pseudomonas chitinolytica* sp. nov., for controlling the root-knot nematode *Meloidogyne javanica. Biocontrol Sci. Technol.,* 1: 115–125.

Stock, C.A., Mcloughlin, T.J., Klein, J.A. and Adang, M. (1990). Expression of a *Bacillus thuringiensis* crystal protein gene in *Pseudomonas cepacia* 526. *Can. J. Microbiol.,* 36: 879–884.

Suslow, T.V. (1980). Growth and yield enhancement of sugar beet by pelleting with specific *Pseudomonas* spp. *Phytopathol. News,* 12: 40 (Abstr.).

Swarnakumari, N., Lakshmanan, P.L. and Samiyappan, R. (1999). Screening of different isolates of *Pseudomonas fluorescens* against rice-root nematode, *Hirschmanniella oryzae*. In: *International Seminar on Integrated Pest Management,* held at Hyderabad, India, pp. 102.

Thomashow, L.S., Weller, D.M., Bonsall, R.E. and Pierson, E.S. (1990). Production of the antibiotic phenazine-1-carboxylic acid by fluorescent *Pseudomonas* species in the rhizosphere of wheat. *Appl. Env. Microbiol.,* 56: 908–912.

Thompson, D.C. (1996). Evaluation of bacterial antagonist for reduction of summer patch symptoms in Kentucky blue grass. *Plant. Dis.,* 80: 856–862.

Tripathi, A.K., Mishra, B.M. and Tripathi, P. (1998). Salinity stress responses in the plant growth promoting rhizobacteria *Azospirillum* spp. *J. Bio. Sci.,* 23: 463–47.

Tuzun, S. and Kuc, J. (1991). Plant immunization: an alternative to pesticides for control of plant diseases in green house and field. In: *The Biological Control of Plant Diseases,* (Ed.) J. Bay-Peterson. Food and Fertilizer Technology Centre, Taiwan, pp. 30–40.

Van Loon, L.C., Bakker, P.A.H.M. and Pieterse, C.M.J. (1998). Systemic resistance induced by rhizosphere bacteria. *Annu. Rev. Phytopathol.,* 36: 453–483.

Van Peer, R., Niemann, G.J. and Schippers, B. (1991). Induced resistance and phytoalexin accumulation in biological control of *Fusarium* wilt of carnation by *Pseudomonas* sp strain WCS 417r. *Phytopathology,* 81: 728–734.

Van Peer, R. and Schippers, B. (1992). Lipopolysaccharides of plant growth promoting *Pseudomonas* spp. strain WCS417r induce resistance in carnation to *Fusarium* wilt. Neth. *J. Plant Pathol.,* 98: 129–139.

Van Wees, S.C.M., Pieterse, C.M.J., Trijssenaar, A., Van'T westende, Y., Hartog, F. and Van Loon, L.C. (1997). Differential induction of systemic resistance in *Arabidopsis* by biocontrol bacteria. *Mol. Plant Microbe Interact.,* 10: 716–724.

Velazhahan, R., Samiyappan, R. and Vidhyasekaran, P. (1999). Relationship between antagonistic activities of *Pseudomonas fluorescens* isolates against *Rhizoctonia solani* and their production of lytic enzyme. *J. Plant Dis. Prot.,* 106: 244–250.

Vidhyasekaran, P. and Muthamilan, M. (1999). Evaluation of powder formulation of *Pseudomonas fluorescens* Pf1 for control of rice sheath blight. *Biocontrol Sci. Technol.,* 9: 67–74.

Vidhyasekaran, P., Rabindran, R., Muthamilan, M., Nayar, K., Rajappan, K., Subramanian, N. and Vasumathi, K. (1997a). Development of powder formulation of *Pseudomonas fluorescens* for control of rice blast. *Plant Pathol.,* 46: 291–297.

Vidhyasekaran, P., Sethuraman, K., Rajappan, K. and Vasumathi, K. (1997b). Powder formulations of *Pseudomonas fluorescens* to control pigeon pea wilt. *Biol. Control,* 8: 166–171.

Viswanathan, R. (1999). Induction of systemic resistance against red rot disease in sugarcane by plant growth promoting rhizobacteria. *Ph.D. Thesis,* TNAU, Coimbatore, India, 175 pp.

Viswanathan, R. and Samiyappan, R. (1999a). Induction of systemic resistance by plant growth promoting rhizobacteria against red rot disease caused by *Collectotrichum falcatum* went in sugarcane. In: *Proceedings of Sugar Technology Association of India,* 61: 24–39.

Walsh, U.F., Morrissey, J.R. and FO'Gara. (2001). *Pseudomonas* for biocontrol of phytopathogens: from functional genomics to commercial exploitation. *Curr. Opin. Plant Biol.,* 12: 289–295.

Wei, G., Kloepper, J.W. and Tuzun, S. (1991). Induction of systemic resistance of cucumber to *Colletotrichum orbiculare* by select strains of plant growth promoting rhizobacteria. *Phytopathology,* 81: 1508–1512.

Wei, L., Kloepper, J.W. and Tuzun, S. (1996). Induced systemic resistance to cucumber diseases and increased plant growth by plant growth promoting rhizobacteria under field conditions. *Phytopathology,* 86: 221–224.

Wei, Z.M., Laby, R.J., Zumoff, C., Bauer, D.W., He, S.Y., Colimer, A. and Beer, S.V. (1992). Harpin, elicitor of the hypersensitive response produced by the plant pathogen. *Erwinia amylovora, Science,* 7: 85–88.

Wyss, U. (1989). Video assessment of root cell responses to Dorylaimid and Tylenchid nematodes. In: *Vistas on Nematology,* (Eds.) Veech, J.A. and D.W. Dickson. Society of Nematologists, Inc., Hyattsville, Maryland, pp. 211–220.

Zdor, R.E. and Anderson, A.J. (1992). Influence of root colonizing bacteria on the defence responses in bean. *Plant Soil,* 140, 99–107.

Zehnder, G., Kloepper, J., Yao, C. and Wei, G. (1997). Induction of systemic resistance in cucumber against cucumber beetles (Coleoptera: Chrysomelidae) by plant growth promoting rhizobacteria. *J. Econ. Entomol.,* 90: 391–396.

Plant Diseases Management for Sustainable Agriculture *Pages 328–338*
Editor: **Dr. Shahid Ahamad**
Published by: **DAYA PUBLISHING HOUSE**

Chapter 26

Potential Donors for Resistance Against Biotic Stresses in Major Agri-Horticultural Crops

S.K. Mishra[1], Ashok Kumar[1], J.C. Rana[2], Gunjeet Kumar[1] and S.K. Sharma[1]

[1]*National Bureau of Plant Genetic Resources, New Delhi – 110 012*
[2]*NBPGR Regional Station, Shimla*

Introduction

Evaluation of plant genetic resources deals with assessing the potential agronomic value of a collection, while detailed evaluation deals with assessing their potential against various major constraints such as biotic and abiotic stresses. Detailed evaluation of plant genetic resources is the key to establish the potential value of the accessions and to facilitate their use in crop improvement programmes. In India, systematic efforts were made to evaluate the plant genetic resources of agri-horticultural plants with greater emphasis in some of the major crop groups like cereals and millets, pulses and legumes, oil seeds, vegetable crops etc.

Cereals and Millets

In rice, Directorate of Rice Research (DRR) has evaluated about 75,000 entries against biotic and abiotic stresses and other agronomic features during 1969-90. A network project entitled "Germplasm Evaluation of Rice for Biotic Stresses" was one of the most important activities on rice germplasm evaluation specifically against biotic stresses. Recently, NBPGR has launched a multi-location evaluation of rice germplasm. In this programme, 1056 accessions of rice were evaluated for agro-morphological parameters at seven locations and also screened for biotic stresses at 14 hot spot locations, respectively (Gautam *et al*, 2007). Over the years of evaluation, the promising accessions of rice identified for different biotic stresses are given in Table 26.1.

Table 26.1: Potential Donors for Important Biotic Stresses in Cereals

Trait	Potential Donors Identified and Utilized
	Rice
Blast	ARC 1250, BJ 1, Carreon, Co 4, Dawn, Manoharsali, MNP 36, MTU 5, Raminad Strain 3, Sigadis, Tadukan, Tetep, Zenith
Sheath blight	ARC10531, ARC10532, ARC10635, ARC15362, Athebu, BCP3, Buhjn, KRC355, KRC356, Laka, Nangmons 4, OS4, Phoure 1, Ramadja, Saibham, Suduwee, T141, Ta-Poo-cho-z
Bacterial leaf blight	BJ1, Cemposelak, Chinsura Boro II, DV85, Java 14, Lacrose/Zenith Nira (LZN), MTU 15, *O. longistaminata*, Sayaphal, Sigadis, TKM6, Wase-Aikoku, Zenith
Rice tungro virus	ADT 21, BJ 1, Ambemohar 159, ARC10599, ARC11554, ARC12981, ARC14320, ARC14766, ARC15570,Latisal, PTB 2,PTB 18,PTB 21
Grassy stunt virus	ARC13820, ARC13901
Stem borer	ADT2, ARC10257, ARC10598, ASD10, Ptb15, Ptb18, SLO17, TKM6, Tadukan, Phouren
Gall midge	ARC5984, Bengle, Bhumansam, Eswarakora, PTB 10, PTB 18, Leaung 152, NHTA8, Siam 29, W1263,W 12708,R 68-1
Brown plant hopper	ARC6650, Babawee, Chemban, Co25, HR19, IR 34, IR 36, IR 50 Mudgo, PTB 10, PTB 18, PTB19, PTB 21, PTB 33, Ratu Heenati
Green leaf hopper	ADT 14, ARC 10313, ARC 6006, ARC 7012, ASD7, Co 9, Gam Pai 15, Pankhari 203, Peta, PTB 18, PTB 21, SLO 4, W 1263
White backed plant hopper	ARC5752, ARC10239, ARC11316, Chempar, HR22, JBS34, Kalubmati, N22, Pankhari 203, SLO 12, Senawee, WC1240
	Wheat
Black, brown and yellow rust	45 accessions
Karnal bunt	30 accessions
Leaf blight	BL1883, BW14973, BW362, HPW57, HS317, HS345, HW2023, JWS17, K9441, PATO(R)/TRPPSN-SM64, PBW378, WH913, BOW'S, CHIRYA1, CHIRYA3, CHIRYA7, MAYUR, QUAN FENG 2, SUZHONES, VEE'S, YANGMAI 6
Loose smut	HI8498, NP200, PBW465, Raj 1555, PBW65, MACS9, VL639, VL614, PDW216 (d), PDW (d)
Powdery mildew	CMH76, DW694, DW698, DW774, DW776, DW777, DW812, DW844, DWL5010, DWR187, HD2189, HPT-6, TL2980, Sonora
	Maize
Maydis leaf blight	CM103, CM 105, LM 5, CML 9, CM 211
Turcicum leaf blight	VM 103, CM 120, NAI 142, CM 210, Poza Rica 8223, NAI 155
Banded leaf and sheath blight	CM 104, CML-3
Sorghum downy mildew	CM 104, CM 115, CM 118, CM 120, CM 211, CM 501
Brown stripe downy mildew	CM 102, CM 104, CM 111, CM 115, CM 118, CM 120, CM 201, CM 211, CM 501, XCM 4
Erwinia stalk rot	CM 119, CM 211, CM 131
Post Flowering Stalk Rot	CM 119, CM 125, CM 209, CML 381
Maize stalk borer	EC3108, PRO340, CM 500, FBR, EMR, AEB, (Y), AEB (W), CM–400, IPA 34-f-106-2, Composite-9116, Syn.A III, IPA 9-7, IPA 34-10-13-3, IC 9801, C-633

In wheat, to cater the national need, the materials of diverse origin were obtained from CIMMYT, USDA, ICARDA and other countries and distributed to various centres. From 1980-81, NBPGR is also involved in the evaluation of germplasm and all the introduced wheat materials are grown at the NBPGR field for quarantine and evaluation. Over 2200 accessions of indigenous and over 73,000 of exotic germplasm of wheat, mainly international nurseries/trials of CIMMYT and ICARDA have been characterized and evaluated at NBPGR. Recently, NBPGR has initiated the multilocation germplasm evaluation programme. The promising accessions of wheat identified for different traits are given in Table 26.1.

In maize, the germplasm evaluation programme was initiated during 1950s with the introduction of materials from Latin America and Southeastern United States. The concerted efforts were made to characterize the indigenous and exotic germplasm and large number of accessions of indigenous as well as exotic origin was evaluated under AICMIP. Some promising sources of resistance against major biotic stresses are listed in Table 26.1.

In India, the world collection of sorghum assembled under the Rockefeller Foundation Project was screened for insect and disease resistance. Murty *et al.* (1967) catalogued a comprehensive world collection (10,000 accessions) from different countries identifying lines resistant to diseases/insect-pests. Several trait-specific A/B lines have been developed at ICRISAT. Some of these are given in Table 26.2.

Table 26.2: Trait Specific A/B Lines in Sorghum Developed at ICRISAT

Biotic Stress	*Resistant Lines (A/B lines)*
Downey mildew	ICSA-/B-201 to ICSA-/B259 (59 accessions)
Anthracnose	ICSA-/B-260 to ICSA-/B-295 (36 accessions)
Leaf blight	ICSA-/B-296 to ICSA-/B-328 (33 accessions)
Rust	ICSA-/B-329 to ICSA-/B350 (22 accessions)
Grain mould	ICSA-/B-351 to ICSA-/B-408 (58 accessions)
Shoot fly (rainy season)	ICSA-/B-409 to ICSA-/B-436 (28 accessions)
Shoot fly (post-rainy season)	ICSA-/B-437 to ICSA-/B-463 (27 accessions)
Stem borer (rainy season)	ICSA-/B-464 to ICSA-/B-474 (11 accessions)
Midge	ICSA-/B-488 to ICSA-/B-545 (58 accessions)
Head bug	ICSA-/B-546 to ICSA-/B-565 (20 accessions)
Striga	ICSA-/B-566 to ICSA-B-599 (34 accessions)

The evaluation of germplasm under All India Coordinated Sorghum Improvement Project (AICSIP) resulted in identification of several sources of resistance to biotic stresses (Table 26.3). Genetic stocks such as IS14332, IS14375, IS14384 and IS18758 have been extensively used for incorporation of the grain mould resistance in national programmes.

In pearlmillet, a catalogue on "Evaluation of Pearlmillet Germplasm" was brought out through NBPGR- ICRISAT collaborative evaluation programme (Mathur *et al.,* 1993 a, b). The extensive evaluation of germplasm at ICRISAT has resulted in the identification of lines for sweetness, dwarfness, glossiness, high protein, early maturity, midrib-less, seedling thermo-tolerance, photo-insensitivity, brown midrib, cytoplasmic- nuclear male sterility, resistance to smut, downy mildew, ergot and rust.

Potential donors identified for different traits. A number of male sterile lines with disease resistance were developed using genetic stocks of Nigerian origin (5054 A and MS5141). Using these lines, the downy mildew resistant hybrids Pusa 23, Pusa 322, Pusa 444 and Pusa 605 were developed.

Table 26.3: Disease Resistant Sorghum Germplasm Accessions and Parental Lines

Diseases	*Accessions*	*Parental Lines*
Grain moulds	IS625, IS3612, IS3691, IS6365, IS 8283, IS8763, IS9470, IS9487, IS10892, IS12622, IS14332, IS14375, IS14384, IS18758, E 35-1	SRT18B, SRT26B, SR155-12, SR330-20, SR367, SR384, SR839, TNS30, RS29, CS3541, SB1085, NSV13. GRMP4, GRMP5, GRMP8, GRMP9, GRMP13, GRMP27, GRMP28, GRMP30, GRMP35, GRMP44, C43, C85
Downy mildew	IS84, IS2195, IS2328, IS2217, IS2415, IS3443, IS6265, IS8283, IS12622, IS18348, IS18551, IS18758	SRT26B, SB101B, RS29, M12B, TNS30, MR750, C43
Rust	IS84, IS2217, IS3443, IS3691, IS8283, IS14332	CS3541, SB101, SB401BB, SRT18B, MR750, RS29
Charcoal rot	IS84, IS1054, IS1085, IS3443, IS14332, IS18432, E36-1, BP53	RS29, CS3541, SB101B, SB401B, SB1085
Anthracnose	IS3443, IS1085, IS14332, IS18758	RS29, C43, CS3541, SB101B, SB 401B

Pulses

After rigorous evaluation of pulses germplasm, several promising accessions were identified for different traits including resistant to biotic stresses including lines with multiple disease resistance (Table 26.4). After evaluation of chickpea germplasm, catalogue on "Characterization and Evaluation of World's Collection on Chickpea" was brought out by ICRISAT (Pundir *et al.*, 1988). Subsequently under NBPGR-ICRISAT, a catalogue on 'Joint Evaluation of Chickpea Germplasm' was published (Mathur *et al.*, 1993). Among wild species, *C. judaicum* was found resistant to *Botrytis* grey mould and has high methionine content, while *C. pinnatifidum* was resistant to *Botrytis* grey mould and has high tryptophan content. Promising accessions were identified for different agro-morphological traits including biotic and abiotic stresses including lines with multiple disease resistance. In Pigeonpea, a catalogue on 'World Collection on Pigeonpea' was brought by the ICRISAT (Ramanandan *et al.*, 1988). Several potential donors for resistance/tolerance against various diseases, insect-pests and abiotic stresses have been identified. In mungbean, stable sources of resistance against MYMV, powdery mildew, wilt and cyst nematode have been identified. Incorporation of multiple disease resistance resulted in the development of varieties like Pusa 105, Co 4, TT9E (Dhauli), ML 131, S-8, Pant Moong-1, Pant Moong-3, BM 4, ML 613, Pusa 9531 and Pusa Vishal. In urdbean, a wide range of variability has been recorded in germplasm and the lines with desirable agro-morphological traits, including resistance against major biotic and abiotic stresses have been identified. In Lentil, 4000 lines received from ICARDA were evaluated at IARI (Mishra, 2002). These efforts resulted in identification of a core collection of 287 accessions, which was evaluated for morphological and physiological parameters (Tullu *et al.*, 2001) and resistance against different biotic stresses. Sources for resistance to powdery mildew and rust (JP 179, JPBB 7, JPU 496, P 16, P20, P 43) and wilt (L4 15/92) have been identified. A variety, HFP 4 has been developed through crossing exotic germplasm, with afila (leaves are converted in to tendrils) dwarf and powdery mildew resistance.

Table 26.4: Sources of Resistance Against Major Biotic Stresses Identified in Pulses

Crop/Stresses	*Resistant Sources*
	Chickpea
Wilt	45 accessions
Ascochyta blight	30 accessions
Multiple disease	ICC 12237, ICC 12269 (wilt and dry root rot); ICC 1069 (wilt, ascochyta and gray mould); ICC 10466 (wilt, stunt virus and dry root); ICC 858, ICC 959, ICC 4918, ICC 8933, ICC 9001 (wilt and stem rot), ICC 11317, ICC 12259, ICC 12275 (wilt and black root rot)
Viral diseases	JG 221
Pod borer	Desi 3108, GL 645, ICC 3580, 1CC 12483, ICC 506, LHR 69, P 202, P 927, P 696-1, PDG 84-10
	Pigeonpea
Sterility mosaic	ICP 7867, ICP 9145, ICP 10976, ICP 10977
Wilt	15 accessions
Phytophthora stem blight	KPBR 80-2-2, ICP 9252
Multiple diseases	ICP 8094, ICP 9174, ICPL 227 (wilt and sterility mosaic virus); ICP 11302 ICP 11303, ICP 11304 (wilt, sterility mosaic virus and phytophtora blight); ICP 8861, ICP 8862, ICP 10960 (wilt, SMV, alternaria blight)
Pod borer	GA 4750, ICPL 151, ICPL 7035, ICPL 87075,
	Mungbean
MYMV	L 80, LM 214, LM 294, KB 5, LBG 622, ML 5, ML 337, ML 405, ML 408, Mung 2, Pusa 87, Tarai local, VB 3, VGG 3
Powdery mildew	JRUM 1, ML 323, ML 395, TARM 1, TARM 22, WGG 46, WGG 47, WGG 48, WGG 62
Bacterical leaf spot	AKM 9105, GM 8835, ML 337, RMG 344
Multiple diseases	L 87, LM 755, LM 784, ML 3, ML 9, SML 70, V 4852 (MYMV and Cercospora leaf spot); ML 4, LM 162 (Leaf spot, root rot and stem rot); NP 52 (Powdery mildew, root rot and stem rot); V 3661 (MYMV and blight)
Bruchid	TC 1966
Thrips	Pusa 107, SML 77, SML 99, SML 100, UPM 82-4
	Urdbean
MYMV	JKU 1, JKU 9, K 66-110, Line 100, Line 7382, Mash 1-1, NP 16, NP 21, Pant U 26, PH 50, PH 55, PLU 131, PLU 220, PLU 277, PLU 340, UPU 2
Powdery mildew	LBG 17, P 115, Pant U 19,
Macrophomina blight	PLU 137, PLU 241, Sel 1
Cercospora leaf spot	HPBU 51, IC 1008, P 38, PDU 2, PDU 85-5, Pusa 3, TEU 95-1, UG 407, UPU 91-7
Leaf crinkle	MM 5-60, NDU 94-6, S 210, S 250, Sialkot
Multiple diseases	PLU 289, ShU 9511, ShU9513, ShU 9606, ShU 9716(MYMV& Leaf crinkle)
	Lentil
Wilt	JL 33, LG 8, LP 199, Pusa 6-9, Pant L 209, Pant L 220, Pant L 234
Rust	E 153-258, LG 120, Pant L 406, UPL 175, DPL 52, L 4, L 47
Blight	FLIP 84-43, L 84-851, LL 858, L 3328, Lenka, SS 159
Multiple diseases	LC 202, LC 270, SL 420, SL 896(wilt and rust)
Bruchids	LL 56, Pant L 406, Pant L 639

Oilseed Crops

Over the years of evaluation of rapeseed-mustard germplasm, promising genotypes have been identified in different species of oilseed Brassicas; namely, Tobin, KBS 3 white rust resistant, PT 77, Bhawani, *Alternaria* blight resistant in *toria*; PYS 16 *Alternaria* blight resistant, YSB 9, RAURDYS 9702, YS 7, YS 8, YS 9, YS 11, YS 12 white rust resistant, YS 841 *Sclerotinia* wilt resistant in *yellow sarson;* BIOYSR, BIO 467-95, JYM 10, PAB 2002, JGM 9011, TM 20, DLM 81, PBR 9, RN 505, RN 510, NDR 9902, RK 9902 white rust resistant, PBR 2012, PAB 534, PAB 2002, SR 142-2, PHR 1, KYM 95-5 resistant to *Alternaria* blight, PYSR 5, PYSR 7, PYSR 10, PBR 91 downy mildew resistant, TM 18, RN 505 powdery mildew resistant, RN 505, RN 510, PCR 10, PWR 9541, PW 4810 *Sclerotinia* wilt resistant in mustard; PBN 2001, PBN 2002 resistant to *Alternaria* blight and downy mildew, WBBN 1, WBBN 2 resistant to *Sclerotinia* wilt in *B. napus*; PPCS 1, PBC 9427 downy mildew resistant in *B. carinata* (Kumar, 1987; Rana and Singh, 1992; Singh, 1996). In soybean, over the years of evaluation, several accessions have been identified for resistant to various diseases and insects/pests are; EC 106991, EC 241691, EC 251308, IC 118005, IC 118008 and Bragg for resistance to bacterial pustules; EC 24046, EC 75193, EC 232082, EC 251362 and EC- 251526 for resistance to pod blight; EC 106991, EC 106998, EC 127501, EC-281030 and EC 291393 for multiple disease resistance. Besides, 23 genotypes, *viz.* DS 97-11, DS 97-13, DS 97-14, DS 98-11, DS 98-12, DS 98-14, DS 98-19, DS 96-7, DS 76-1-2-1, DS 76-1-37-2, SL 427, PK 1029, PK 725, JS 9029, PK 1162, G 2121, G 2130, G 2132, G 225, EC 439599, EC 439609 and DS 98-17 exhibited resistance to soybean rust. In sesame, germplasm collections were evaluated at NBPGR Regional Stations at Jodhpur and Akola. The catalogues on sesame germplasm were brought out (Chopra and Thomas, 1982; Chandra *et al.*, 1983). A catalogue consisting of 2068 accessions evaluated for 39 descriptors was also published (Loknathan *et al.*, 1993). Among the exotic introductions, EC 342782-83 (ex-Korea) were dwarf and early maturing, EC 355656 (ex. Thailand) was blight tolerant with medium long capsules, EC 352814 (ex. Israel) was high yielding and tolerant to bacterial blight. Over 6000 accessions of sesame are being maintained at Project Coordinating Unit (Sesame and Niger) at JNKVV, Jabalpur (MP). Recently, NBPGR and AICRP have evaluated over 4000 accessions and have published four catalogues (Duhoon *et al.*, 2000). In Groundnut, using standard descriptors of IBPGR/ICRISAT, over 10,000 accessions were screened for various biotic and abiotic stresses resulting in identification of a number of sources for various diseases (Table 26.5). Systematic work on evaluation of sunflower germplasm started with the establishment of research centers under AICORPO. Based on evaluation data, two catalogues were published (Virupakdhappa and Sindagi, 1987; Jayaramiah *et al.*, 1995). Promising lines have been identified for different economic traits including sources of resistance for different diseases (Table 26.5). In safflower, 7317 accessions were characterized and evaluated (Mehtre and Hegde, 1997) and were screened against *Alternaria* and aphid. The germplasm tolerant to *Alternaria* (117 accessions) and aphid (65 accessions) have been identified. In addition, 1200 germplasm accessions were evaluated for 13 descriptors (Hegde *et al.*, 1999). In castor, 2750 accessions were evaluated by the Directorate of Oilseed Research (DOR) under rainfed conditions and 2100 accessions were also evaluated through multilocation evaluation by AIRCPO (Castor) under irrigated as well as rainfed conditions. The potential donors for biotic stresses were also identified and listed in Table 26.5.

Vegetable Crops

Systematic work on evaluation and utilization of germplasm in vegetable crops was initiated with the creation of the Division of Horticulture at IARI and the establishment of State Agricultural Universities. A large number of germplasm of these crops have been characterized and evaluated at NBPGR, IARI, IIVR, SAUs etc. Potential donors identified for agronomic as well as resistance against biotic and abiotic stresses. A large number of varieties have been developed in these crops utilizing the indigenous as well as exotic genetic resources indicating the real utilization of germplasm in vegetable crops.

Table 26.5: Sources of Resistance Biotic Stresses in Oilseeds

Crop/Disease	*Accessions*
	Groundnut
Aspergillus flavus	48 accessions
Bacterial wilt	ICG1609, 1703, 1705, 5272, 5273, 5276, 5313, 7343, 7893, 7894, 7968, 8666, 11325, 15220, 15221, 15222, 15224, 15226, 15227, 15228, 15230,
Early leaf spot	NRCG1086, 1123, 1146, 1575, 2300, 2358, 5298,ICG1703, 1707, 1710, 2711, 3527, 4995, 5501, 6280, 6284, 6330, 6709, 6902, 7756, 7878, 8298, 8339, 9294, 9988, 9989, 10000, 10040, 10450, 10756, 10883, 10900, 10904, 10920, 10946, 10948, 10954, 110099
Late leaf spot	80 accessions
Rust and pod rot	ICG4477, 5125, 5175, 12083, 12087, 15223, 15234, 15235, 15236
	Sunflower
Downy mildew	35, 82, 133, 212, 244, 325, 333, 358, 367, 391, 471, 664, 705, 715, 777, 916, 1147, 1172, 1180, 1185-1, 1187-6, 1195, 1228, 1256, 1279-1, 1288, 1405, 1506, 1552, 1559, 1583, 1860, 1876
Alternaria	70 accessions
	Safflower
Alternaria leaf spot	120 Accessions
Aphid	54 accessions
	Castor
Fusarium wilt	45 accessions
Whitefly and jassids	35 accessions

A large number of okra germplasm including wild species were characterized and evaluated for various qualitative and quantitative characters at NBPGR (Bisht *et al.*, 1993, 1995, 1997). At IIVR lines resistant/tolerant against major diseases and pests such as NIC 9303A, NIC 6308, NIC 3322, NIC 9408, NIC 3325, EC 329375, K 4409, *A. crinitus, A. angulosus* for YVMV; *A. tetraphyllus, A. angulosus* for powdery mildew; *A. crinitus, A. moschatus, A. angulosus* for *Cercospora* blight; *A. crinitus, A.ficulneus, A. manihot* for enation leaf curl virus; *A. tuberculatus, A. moschatus* for fruit borer; EC 305656, EC 305664, EC 305696 for mites; *A. moschatus, A. crinitus,* EC 305656, EC 305694, EC 305695, EC 305714, EC 06731 for Jassids and IC 90186, U 4307, U 4365, U 4579 for *Microphomina*. A wild species *A. manihot* ssp. *manihot* (now identified as *A. caillei*) was utilized to develop many resistant lines against YVMV *viz.*, G 2 and G 2-4, Punjab Padmini, P 7, Parbhani Kranti, Arka Anamika and Arka Abhay. Apart from this, NBPGR has also developed Sel-2 from multiple cross [(Pusa Sawani x Best I) x (Pusa Sawani x IC 7194)] in 1989. During 1999, HRB-9-2 and HRB-55 were developed from HAU, Hissar. An open-pollinated YMV resistant variety Pusa A-4 was developed which is suitable for commercial cultivation in spring/summer and kharif seasons.

In Tomato, NBPGR has assembled around 2000 accessions from different parts of the world including some wild species. The wild species namely, *L. pimpinellifolium, L. hirsutum, L. peruvianum* have been used to transfer the genes for disease and pest resistance. TMV resistant variety, Marglobe was developed involving wild species *L. pimpinellifoliun* as a donor parent. Other important varieties developed were Pusa Upahar, SL 120, DT 10 etc. Pusa Red Plum is a small-fruited variety developed through utilization of South American wild species. Hisar Lalit (NT 8) was developed through hybridization of the resistant genetic stock from Bangalore x HS 101.

In brinjal, 319 germplasm was evaluated for 43 descriptors at NBPGR. An effort is also being made to develop core set at NBPGR. These efforts resulted in identification of sources for tolerance/ resistance to various diseases and stress. Some of these include, Harris 468 Special, Hibush, Harris Hybrid 77631 for *Verticillium* wilt; *S. torvum*, S. *melongena* var. *insanuum*, *S. xanthocarpum*, SM-81, SM-6, MS-48, SM-56, SM-71, SM-72, SM-74, H-8, Arka Keshav, Arka Nidhi, Arka Neelkant, Swetha and Surya, Pusa Purple Cluster, *S. nigrum*, *S. sisymbriifolium* for bacterial wilt; Florida Market, Bangalore *S. torvum* and *S. sisymbriifolium* for root knot nematode and *Phomopsis* fruit rot; Pusa Bhairav and Pusa Purple for *Phomopsis; S. khasianum, S. sisymbriifolium, S. macrocarpon, S. gilo, S. incanum, S. indicum* Aushey, Black Beauty, H- 165, H-407, H-408 for shoot and fruit borer.

In Hot pepper, a large number of accessions have been evaluated at NBPGR for different agro-morphological descriptors identifying promising lines. Several cultivated species other than *C. annuum* are resistant/tolerant source of biotic stresses. *C. baccatum* is resistant to Anthracnose, *Phytophthora* rot, Cucumber mosaic virus whereas *C. chinense* to Anthracnose, Leaf curl virus, Tomato mosaic virus, *Verticillium* wilt and *C. frutescens* to Leaf curl virus. Other sources of disease resistance include IHR 517, A 32, Dondaicha for powdery mildew; *C. frutescens*, PI 201234 for fruit rot and leaf blight; *C. baccatum* var. *baccatum*, *C. chinense* accessions 1555, 1554, 906, PBC 932 for anthracnose fruit rot; Ujala, White Khandari, Manjari for bacterial wilt; California Wonder, Hungarian Wax for *Cercospora* leaf spot; Jalapeno, PI 163192, 260435 for bacterial spot; *Capsicum Chinense* accessions EC 533896, EC 533897 for root knot nematode and Pant C 1 for leaf curl virus.

A large number of germplasm of cucurbitaceous vegetables have been characterized and evaluated at NBPGR, IARI, IIVR, SAUs. Potential donors have been identified for agronomic as well as resistance against diseases (Table 26.6). A large number of varieties have been developed through utilization of germplasm in cucurbitaceous vegetables. In bottlegourd, 58 collections were evaluated during summer and 124 collections during *kharif* at NBPGR. Fourteen accessions were evaluated at IIVR and promising accessions for different traits including resistant to red pumpkin beetle (U 10–316 and NIC 1225), resistant to leaf miner (U10–316, IC 92362 and IC 92392) and downy mildew resistant (U10–316, I C 92362, Laxmipur Patna l). In bittergourd, a set of 33 accessions at NBPGR and121accessions at IIVR has been evaluated and promising lines has been identified. Accessions showing resistant against CMV are IC 44428B, NIC 72285, IC 85604A, IC 85608B, IC 85611, IC 85636, V 89/0-103C, VRBT 29 and VRBT50 and fruit fly resistance (IC 85641 and V-89/0-103C). In cucumber, source for resistance to Anthracnose (*Colletotricum laginarium*) was found in the accession PI-197087 and resistance to downy mildew (*Pseudopernospora cubensis*) was found in PI-197087. The source for resistance to powdery mildew (*Sphaerotheca fuliginia*) was found in Poinsette, Yomaki, Sparton Salad and in PI-197087. The source for resistance to angular leaf spot (*Pseudomonas lachrymans*) has been found in PI-169400, MSU-9402, GY-14A and to bacterial wilt (*Erwinia tracheiphila*) in PI-200815, PI-200818, 79-48, 79-42-24. Resistance to cucumber green mottle mosaic virus (CGMMV), was found in *Cucumis anguria* and for cucumber mosaic (CMV) in cultivar TMG-l. Another resistant source for CGMMV was identified in cultivar Brasil at BCKV, Kalyani. In Muskmelon, promising accessions identified for different traits include downy mildew resistant (Seminole, Edisto, AC70-154, IIHR-190) and *Fusarium* wilt resistant (*C. melo* var. *reticulates*, inodorus, chito, flexuosus, Delicious-51).In Ivy gourd, two local varieties (Aligarh Local and Allahabad Local) were evaluated for their ability to resist powdery mildew. Cultivar

Allahabad Local was found susceptible and Aligarh Local was found resistant. IGKVV has evaluated and characterized 35 genotypes and AC. No. 59 was found excellent. A total of 39 genotypes were evaluated and few accessions were found resistant to diseases and insects.

Table 26.6: Potential Donors Identified for Various Traits in Cucurbitaceous Vegetables

Crop/Trait	*Potential Donors*
	Cucumber
Anthracnose	PI 197087, Pointsett
Cucumber scab	Wisconsin SMR 9
Angular leaf spot	MSU 9402, PI 169400, MSU 9402
Bacterial wilt	PI 200815, PI 200818
Cucumber green mottle mosaic	*C. africans, c. ficifolius, C. anguria*
	Melons
Downey mildew resistant	*Cucumis callous*, Edisto, Harvest queen, Seminole,
Fusarium wilt resistant	*C. inodorus, C. flexuosus, C. melo* var. *chito, C. reticulates*
Powdery mildew	Campo, Edisto, Georgia 47, PMR 47
Watermelon mosaic virus	*C. metuliferus*, PI 414723
	Watermelon
Anthracnose	Africon citron 8, Charleston Grey, Congo, Crimson Sweet
Fusarium wilt	Citron, Shipper, Smokylee, Summit
Powdery mildew	Arka Manik
	Bottlegourd
Red pumpkin beetle	U 10-316, NIC 1225
Downey mildew	U 10-316, IC 92362, Laxmipur Patna 1
	Pumpkin
Bacterial wilt	*C. ficifolia, C. loundalliana*
Cucumber mosaic virus	*C. pepo* var. Cinderella, Gold Rush
Powdery mildew	*C. laundelliana, C. okeechobeensis*

In leguminous vegetables, the germplasm of vegetable cowpea was evaluated at NBPGR and IIVR, which resulted in identification of promising accessions for biotic stresses. At IIVR, BC 24074, NC 63382, EC 169371, BC 244009, BC 240770, IC 20215, EC 15567, DCB 253 and IC 39890, IV 41209, MS 9804 was found resistant to yellow mosaic virus; VS 389, BC 244002, Arka Garima, BC 240787, RKT 89/743, BC 7096, BC 2409-17-A, BC 240770, EC 243948, EC 240750, PDCP192 for resistant to *Pseudocercospora*; and Local l, PDCP II, PDCP 13, Sel. 2-1, PDCP 16, PDCP 17, EC 4860 for resistant to Pod borer (*Etiella zinckenella*) were identified. In addition, about 1200 accessions have been evaluated for different characters including biotic and abiotic stresses and the donors for desirable traits have been identified (Mishra *et al.*, 2004; Mishra *et al.*, 2005). In vegetable pea, Pusa Pragati, a powdery mildew resistant variety has been developed through selection from the exotic material Maestro (ex-Mexico). Some of the sources for resistance identified against diseases include Arka Ajit, Arka Karthik and CHPMR 1 for rust (*Uromyces fabae*); Arka Ajit, Arka Karthik, CHPMR 1, NDVP 250 for powdery mildew (*Erisiphae pisi*); FC 2 for Fusarium root rot (*Fusarium solani*) and Batri Brown, Kalanagni, JP 4 and JP 83 for Fusarium wilt (*Fusarium oxysporum*).In lab-lab bean, three varieties have been developed namely, Pusa Early Prolific, Pusa Sem 2 (tolerant to viral disease and insect pests like aphids, jassids and pod borer and also tolerant to frost) and Pusa Sem 3 (tolerant to viral disease and insect pests like aphids, jassids and pod borer). Breeding for resistance Screening of *Dolichos lablab* (113 cultivars) of for pod-borer resistance at UAS, Bangalore resulted in the identification of two cultivars, PLS 24 and PLS 16-1 as resistant. They contain a significant degree of antibiosis. The cultivars, PLS- 93 exhibited a high degree of tolerance, while M 39579 and IC-661-1 were moderately tolerant. *Lablab purpureus* var. *lignosus* is also highly resistant to pod-borer.

References

Bisht, I.S., Patel, D.P., Sapra, R.L., Thomas, T.A. and Rana, R.S. (1993). *Catalogue on Okra (Abelmoschus esculentus L. Moench.) Germplasm Part–III.* National Bureau of Plant Genetic Resources, New Delhi, India, p. 77.

Bisht, I.S., Patel, D.P., Mahajan, R.K., Thomas, T.A. and Rana, R.S. (1995). *Catalogue on Wild Abelmoschus Species Germplasm.* National Bureau of Plant Genetic Resources, New Delhi, India, p. 51.

Bisht, I.S., Patel, D.P. and Mahajan, R.K. (1997). Classification of genetic diversity in *Abelmoschus tuberculatus* Germplasm collection using morphometric data. *Ann. Appl. Biol.*, 130: 325–335.

Chandra, Umesh, Sapra, R.L. and Mehra, K.L. (1983). *Catalogue on Sesame (Sesamum indicum L.) Germplasm.* National Bureau of Plant Genetic Resources, Regional Station, Shimla, India, p. 294.

Chopra, D.P. and Thomas, T.A. (1982). *Catalogue on Sesame Germplasm.* National Bureau of Plant Genetic Resources, Regional Station, Jodhpur, India, p. 68.

Chopra, D.P., Thomas, T.A., Mehra, K.L. and Singh, Ranbir (1983). *Catalogue on Mungbean Germplasm.* NBPGR, Regional Station, Jodhpur, India, p. 21.

Duhoon, S.S., Singh, B.R., Jharia, H.K., Goswami, U., Srivastava, N. and Jain, S.C. (2000). Genetic *Resources of Sesame (Sesamum indicum L.).* All India Coordinated Project (Sesame and Niger) Publication, Jabalpur, India, p.60.

Gautam, N.K., Mishra, S.K., Raiger, H.L., Mahajan, R.K. and Sharma, S.K. (2007). *Multi-location Evaluation of Rice Germplasm.* National Bureau of Plant Genetic Resources, New Delhi, p. 410.

Hegde, D.M., Yadav, W.S., Karkeli, M.S., Koli, B.D., Pisal, A.A., Mehetre, S.P., Deshmukh, D.P. and Sawant, P.H. (1991). *Evaluation of Safflower Germplasm.* Project Coordinating Unit, All India Coordinated Safflower Improvement Project, Solapur, Maharashtra, India, p. 56.

Jayaramaiah, H., Virupakshappa, K., Gowda, J.N. and Jagadish, B.N. (1995). *Evaluation and Characterization of Sunflower (Helianthus annuus L.)* Germplasm. Project Coordinating Unit (Sunflower), Bangalore, India.

Kumar, P.R. (1987). Genetic resources activities in Oilseed Brassicae. In: *Plant Genetic Resources: Indian Perspectives,* (Eds.) Paroda, R.S., Arora, R.K. and K.P.S. Chandel. National Bureau of Plant Genetic Resources Publication, New Delhi, India, pp. 201–209.

Loknathan, T.R., Patel, D.P., Verma, V.D., Mahajan, R.K., Singh, B., Koppar, M.N. and Rana, R.S. (1993). *Catalogue on Sesame (Sesaum indicum L.) Germplasm.* National Bureau of Plant Genetic Resources, New Delhi, India.

Mathur, P.N., Pundir, R.P.S., Patel, D.P., Rana, R.S. and Mengesha, M.H. (1993). *Evaluation of Chickpea Germplasm, Part–I.* (NBPGR–ICRISAT Collaborative Programme). National Bureau of Plant Genetic Resources, New Delhi, India, p. 194.

Mathur, P.N., Apparao, S., Agarwal, R.C., Mengesha, M.H. and Rana, R.S. (1993a). *Evaluation of Pearlmillet Germplasm, Part–1.* National Bureau of Plant Genetic Resources, New Delhi, India, p. 200.

Mathur, P.N., Apparao, S., Sapra, R.L., Mengesha, M.H. and Rana, R.S. (1993b). *Evaluation of Pearlmillet Germplasm, Part–2.* National Bureau of Plant Genetic Resources, New Delhi, India, p. 215.

Mehetre, S.P. and Hegde, D.M. (1997). Genetic resources of safflower in India. In: *Proc. Fourth International Safflower Conference*, (Eds.) A.Corleto and H. H. Mundel. BARI, Italy, pp. 205–206.

Mishra, S.K. (2002). *Technical Report on Collection, Maintenance, Characterization and Classification of Germplasm of Major Pulse Crops of India* submitted to ICAR, New Delhi, India.

Murty, B.R., Arunachalam, V. and Saxena, M.B.L. (1967). Classification and cataloguing of world collection of sorghum. *Indian J. Genet.*, 27: 1–312.

Pundir, R.P.S., Reddy, K.N. and Mengesha, M.H. (1988). *ICRISAT Chickpea Germplasm Catalogue: Evaluation and Analysis.* ICRISAT Patancheru, Andhra Pradesh, India, p. 324.

Rana, R.S. and Singh, R. (1992). Present status of rapeseed and mustard germplasm in India. In: *Advances in Oilseeds Research, Vol. 1*, (Eds.) D. Kumar and M. Rai. Scientific Publishers, Jodhpur, India, pp. 189–200.

Ramanandan, P., Sastri, D.V.S.S.R. and Mengesha, M.H. (1988). *ICRISAT Pigeonpea Germplasm Catalogue: Evaluation and Analysis.* ICRISAT, Patancheru, A.P., India.

Tullu, A.K., Ushmenoglu, I., McPhee, K.E. and Muehlbauer, F.J. (2001). Characterization of core collection of lentil germplasm for phenology, morphology, seed and straw yields. *Genet. Resour. Crop Evol.*, 48: 143–162.

Singh, R. (1996). Germplasm resources. In: *Oilseeds and Vegetable Brassicas: Indian Perspective*, (Eds.) Chopra, V.L. and S. Prakash, pp. 279–291.

Virupakashappa, K.and Sindagi, S.S. (1987). Characterization, evaluation and utilization of sunflower germplasm accessions. P.C. Unit (Sunflower) Report, Bangalore, India.

Plant Diseases Management for Sustainable Agriculture *Pages* **339–347**
Editor: **Dr. Shahid Ahamad**
Published by: **DAYA PUBLISHING HOUSE**

Chapter 27

Recent Trends in Management of Fungal Diseases of Grapevine and their IDM

C. Ravindran and Shahid Ahamad

RARS, SKUAST-J, Rajouri – 185 131

Grapes (*Vitis vinifera* L.) were well known in India since the 11th century BC and Commercial varieties of grapes were introduced in India mostly by invaders of Iran and Afghanistan. The commercial viticulture on south India started around mid thirty, where as in north India it dates back to mid fifties.

The total area under grapes is estimated at 50,000 hectares with an estimated annual production of 10-12 lakhs tonnes and average grape yield per unit area of tropical India is highest in the world. Its cultivation is confined to the states of Andhra Pradesh, Haryana, Karnataka, Maharashtra, Punjab and Tamil Nadu, which jointly contribute nearly 90 percent of area and production of grapes in India of which Maharashtra alone the area under grapes is over 20,000 hectares and production exceeding 300,000 tonnes with a productivity averaging 15 tonnes per hectare. India is fast emerging as one of the grape growing countries in the world.

The high demand on quality of the grapes requires intense plant protection measure against insect pest and fungal diseases. Fungal diseases cause considerable reduction in the yield of grapes both in southern and northern India. Some of the disease even leads to complete failure of production. It is estimated that the fruit production could be increased at least by 26 per cent provided the protection from various diseases is achieved (Rawal, 1993)

The continued use of traditional chemical insecticide and fungicide has led to environmental and economical problem. Use some of fungicide like copper salt solution as pernosporacide (against downy mildew) are prohibited because of the potential eco-toxicity. As a consequence, integrated disease management (IDM) is favoured for plant production. Systemic induced resistance is considering being one of the important strategies to achieve IDM through manipulation of natural defense mechanism in plants. Synthetic chemicals for systemic induced resistance have provided practical possibilities for induction of resistance to control of fungal diseases in grapevine.

For many years plant entomologist and pathologist have worked to understand the specific mechanism of insect pest and disease resistance and how their expression is controlled. These research efforts have gained importance in recent years and restriction have been placed on the chemical that have been used to control insect pest and fungal diseases because of consumer concern about the effect of these chemicals on human health.

Organic grape growers are faced with the same complex of disease as are conventional grape growers. Growers of both organic and conventional grapes will likely be faced with the need to manage grape disease powdery mildew and downy mildew, the two major diseases that are widely distributed and common throughout grape growing areas of southern parts of India.The difference in conventional and organic management methods for disease largely rests on what actions a grower is willing to take (or materials he is willing to apply) when confronted with an economically-important diease problem.

This chapter describes the key components common to successful IDM programs- whether conventional or organic, what types of spray materials are available to organic growers, outline of organic management options available for the major grape diseases- and briefly touch on how other disease incidence may be affected by current organic practices.

Components of IDM

Organic and IDM approaches to disease management are approaches that rely on knowledge and informed decision making. The key components of the IDM approach are:

Disease Identification

Proper identification of disease and an understanding of their biology are essential for successful management. Many disease affect vineyards, but only a few cause economic damage. Some cause very conspicuous disease injury but has no effect on vine productivity. Proper identification of the disease is an essential first step to appropriate management.

Monitoring Disease Incidence

Economic effects of disease infestations depend on incidence levels of infection. Especially in vineyards, disease incidences vary greatly from year to year and vineyard to vineyard. Monitoring vineyards to determine incidence levels is the key to avoiding unnecessary spray applications–and for timely application of spray materials when necessary to prevent economic damage.

Economic Thresholds

The economic threshold is defined as the incidence or injury level (determined by monitoring) at which a treatment should be applied to prevent economic losses due to the disease.

Risk Assessment

Risk assessment means using information about the vineyard site, weather conditions, and crop condition (such as cropping level and vigor) to forecast or predict the likelihood that disease incidence will cause economic injury. These four elements are common to both conventional IDM and organic management programs. Using them is the key to taking full advantage of the natural factors (biological and non-biological) that often keep disease infection well below the economic threshold.

Organic Management for Grapes

Over the past several years, it is extensively monitored disease in vineyards that have received no fungicides treatments. Studies have consistently shown that economically significant disease incidence failed to develop in well over 50 per cent of vineyards surveyed. Moreover, high disease incidence tends to occur at the same small proportion of vineyard sites year after year. What this means is that disease pressure is not a major impediment to organic production for many growers. Organic alternatives are available for the major disease of grapes.

Powdery Mildew

Grape powdery mildew is caused by *Uncinula necator*. The pathogen is native to North America, where it coevolved with wild ancestors of *Vitis labrusca* grape cultivars. The disease was of minor importance in viticulture until its introduction to Europe in 1848. It rapidly spread throughout the continent, and nearly destroyed the European industry before it was controlled by the use of sulfur in the 1850s. It is now distributed worldwide. Powdery mildew became more destructive in southern viticulture with the planting of more susceptible *Vitis* interspecific hybrid cultivars in the latter half of this century. The most recent introduction and widespread planting of cultivars of the European wine grape (*Vitis vinifera*) has further increased the acreage planted to mildew-susceptible cultivars.

Variation in resistance to powdery mildew among grape species and cultivars is considerable, and should be the first criterion considered in developing an organic disease control program. In general, cultivars of native North American *Vitis labrusca*, and hybrids with substantial *V. labrusca* parentage and phenotype are most resistant. There are some minor differences in susceptibility to powdery mildew among cultivars of *V. vinifera*, but none approach the substantial resistance of the most resistant labrusca or hybrid cultivars

Downy Mildew

Downy mildew disease in grapes is caused by *Plasmopara viticolas*, a biotrophic oomycete belonging to the family Peronosporaceae. It was introduced into US during the 19th century along with accessions of American wild *Vitis* species and it rapidly spreaded in Europe, causing severe yield losses (Galet, 1977). The first symptom is the appearance of light yellow and translucent spots of variable sizes and forms showing an oily appearance on the upper surface of the leaf. Under warm weather conditions, the under surface of leaves gets covered with tuffs of white downy sprongiophores and sporangia on the underside of such spots. The affected portion of leaves turn brown and can not support the bunch development as their photosynthetic ability is reduced. The pathogen infects all green shoot tissues including leaves, tendrils, shoots, inflorescences and fruit bunches, and significantly depresses the productivity and quality (Lafon and Clerjeau, 1988).

This disease is prevalent throughout peninsular India wherever grape cultivation is done, but more serious is the disease in Karnataka, Andhra Pradesh and Maharashtra. Heavy loss of vine yield was reported by Caruso (1985) in Italy and total crop failure was usually encountered in South India under favourable weather conditions like 20-25° C and RH above 80 per cent (Sohi, 1983). Srinivasan and Jeyarajan (1983) found substantial crop losses when flower or young berries were killed. In recent years, heavy losses were incurred due to downy mildew in most districts of Maharashtra and Karnataka. Even as high as about 100 per cent losses of grapevine productivity due to this disease were recorded in and around Hyderabad during 1992-1995 and 1997-1998 (Ramarethinam *et al.*, 2001)

Figure 27.1: Downy Mildew in Grape Caused by *Plasmopara viticola*

The American and Asian *Vitis* species show varying levels of resistance to *P. viticola*, ranging from moderately resistant (*Vitis rupestris, Vitis amurensis*) to highly resistant (*Vitis cinerea, Vitis riparia, Vitis rubra, Vitis candicans*); however variations were also observed between accessions within the same species (Boubals 1959, Staudt and Kassemeyer 1995 and Brown *et al.*, 1999). In contrast, almost all the commercial cultivars of *Vitis vinifera* were highly susceptible (Boubals 1959, Galet 1977, 1995, and Pearson and Goheen 1996). However, the susceptibility of pathogen has differed between cultivars and even between clones of a single variety (Ravaz 1914 and Perez Marin 1992). Similarly, many studies have indicated that different varieties of *Vitis vinifera* vary in their level of resistance to downy mildew (Boubals 1959, Galet 1977, Staudt and Kassemeyer 1995 and Staudt, 1997).

Control of disease through chemicals started in the beginning of 19th century, using Bordeaux mixture. In recent times many chemical fungicides, including antifungal antibiotics were evaluated against downy mildew. However, one of the major problems in grapevine disease control is the indiscriminate use of chemicals such as fungicides, bactericides and nematicides, which not only raises the production costs, but also leads to the contamination of soil and water resources. In general, 8-10 fungicide applications are necessary to control this disease depending on the meteorological conditions (Viret *et al.*, 2001). It causes not only environmental pollution but also threatens the future of viticulture industry.

Anthracnose

This is by far most widerspread and distructive disease in the vineyards. This is caused by a fungus *Elsinoe ampelina* with continue to survive in affected vine and is disseminated in the rainy season through rains. The fungus infects all the green parts of the vine including fruits. Dark brown spots with darker darker margin are formed on leaves around mid-rib and vines. However the lesion on canes is elongated, sunken, dark brown with dark purple raised margins. On the berries, the spot initially light brown in colour but as they grown in size they turn grey at the centre surrounded by reddish or purple margin.

The most effective control measure include, the affected parts destroyed just after pruning and the vine should be sprayed immediately with 0.2 per cent copper oxy chloride containing fungicides. Spray vineyard on leaf emergence with 0.2 per cent benlate or bevistin. Repeat the same spry after one month. At least 4 sprays of these fungicides should be given during the rainy season (after harvest) at 15 days intervals.

Botrytis Bunch Rot

Botrytis bunch rot is caused by the fungus *Botrytis cinerea.* The pathogen has a very broad host range, and can also subsist on decaying vegetation. Thus, there are several possible sources of overwintering inoculum. Under unfavorable conditions, the pathogen forms a resting stage called a sclerotium. Sclerotia can germinate to produce two types of airborne spores in spring, both of which can infect grapevines. In India, occurrence of this disease not to be reported in the literature

Neither sulfur nor copper fungicides are effective against this disease. However, adequate control can generally be obtained on loose clustered cultivars (such as Concord) by avoiding poor air circulation within the grapevine canopy, both through proper canopy management, and through proper site selection for the vineyard. Site selection is also important with respect to proximity to woodland edges, which may place the vineyard at risk for damage from berry moth. Fruit damaged by berry moth are eventually colonized by *Botrytis cinerea*, which may then spread to the rest of the cluster. In addition to the site selection, more intensive management is required for consistent control on cultivars with compact clusters, such as Riesling and Delaware. Hand removal of leaves surrounding the cluster has been demonstrated to effectively reduce losses due to bunch rot. Nitrogen fertilization, especially with manures, must be performed in such a way as to avoid lush canopy growth. Many of the above cultural practices will also reduce losses due to black rot and downy mildew.

Elements of an Effective Disease Control Program

Cultivar Selection

This is an important factor in determining the risk of disease in a vineyard. Certain cultivars are relatively trouble-free *in most years* with respect to the major fungal diseases. However, it is important to recognize that in some years, significant losses due to one disease or another can potentially occur on any cultivar grown under organic management practices. Copper and sulfur fungicides simply will not control botrytis bunch rot, downy mildew and powdery mildew under the most severe conditions; and no cultivar is immune to all of the major diseases. Market forces will largely dictate which cultivars can be grown profitably. Nonetheless, a grower can minimize losses by selecting cultivars with some measure of resistance to the major diseases. In general, all cultivars of *Vitis vinifera* are highly susceptible to powdery mildew, downy mildew. Most cultivars of *Vitis labrusca* are less susceptible to powdery mildew, but susceptibility to downy mildew and black rot varies within the species. The greatest variation in resistance occurs within the interspecific hybrid cultivars. In general, those hybrid cultivars that most resemble *Vitis vinifera* are more susceptible to the major diseases, while those that resemble *Vitis labrusca* are less susceptible.

Site Selection

Gently sloping land with a southeast or southwest exposure, absence of woodland borders and wild or abandoned vineyards, and excellent air-drainage will all lessen the severity of most fungal diseases, both directly and indirectly. Rapid drying of vines after rain or dew is a major factor affecting growth of the downy mildew pathogens, which requires free water for infection. Woodland borders shade nearby rows, restrict air drainage, and increase the time that foliage remains wet. Wild or escaped cultivated grapes inhabiting woodland edges and abandoned vineyards serve as reservoirs of inoculum for the major diseases. Abandoned vineyards can produce dense clouds of powdery mildew spores by mid-summer, and may overwhelm the disease control program in adjacent plantings. Gradients of powdery mildew resulting from the influx of spores from abandoned vineyards may be clearly visible up to 300 ft from the source, so removal or avoidance of these vines can contribute to disease control. As mentioned above, the risk of berry moth infestations and secondary development of botrytis bunch rot is also most severe near woodland edges.

Cultural Practices

Pruning and training practices have both direct and indirect effects of disease development. Open, well aerated canopies are less conducive to development of downy mildew, botrytis bunch rot. Similarly, weed and ground cover management, both under the trellis and within the alley will effect the environment for disease. Although the greatest reductions of phomopsis fruit rot occur when vines are Umbrella-Kniffen trained and hand-pruned, even hedge-pruned vines will benefit from hand follow-up pruning to remove excess dead wood from the trellis. Some common practices can lead to problems with one or more of the major diseases. For example, the practice of dumping pomice within vineyard rows can create an ideal environment for an epidemic of downy mildew. Seeds from cold-pressed grapes can produce a dense carpet of highly susceptible seedlings on the vineyard floor. Establishment of downy mildew on these seedlings often leads to later severe infection of the vines.

Fungicides

The original copper fungicide was copper sulfate mixed with lime (Bordeaux mixture), and this is still used today, although other formulations are available and are easier to use. The effective component of this mixture is the copper ion in copper sulfate. This is also what injures copper-sensitive vines. Injury due to copper application varies among cultivars, but is also dependent on environment. Even less sensitive cultivars may be injured when cool, wet conditions persist after application, resulting in a high concentration of copper ion for long periods on plant tissue. Copper fungicides are very effective against downy mildew when used properly. They must be used as protectants, and have little or no curative activity. Organic growers have no fungicide options for control of black rot, botrytis bunch rot, and phomopsis fruit rot, and should weigh the risks involved in selecting certain cultivars, sites, and cultural practices accordingly. Fungicides are therefore timed for optimal control of powdery mildew and downy mildew. Powdery mildew infection can occur as soon as shoots are 1 inch long. Initial downy mildew infection does not occur until 2 weeks before bloom. If both copper and sulfur may be used, use sulfur alone in applications until downy mildew is expected. Depending upon the cultivar sensitivity to powdery and downy mildew, a combination of both materials could be used (best for cultivars highly susceptible to powdery and downy mildews), or copper could be used alone (powdery mildew resistant cultivars only). The interval between applications can vary between 7-14 days. No applications should be needed after veraison. Of all sprays applied, those used between pre bloom and fruit set contribute the most to control of fruit infection. Even in minimal-spray programs, these applications should not be neglected.

Growing Grapes without Fungicides in India

It is entirely possible to grow grapes in India without any insecticides and fungicides. However, very few cultivars are suited to this approach, and the risk of substantial crop loss in some years is a near certainty. Powdery mildew resistance is required, since this disease is destructive nearly every year on unsprayed vines of susceptible cultivars. Downy mildew resistance would also be required, because serious fruit losses can occur 1 year in 3, and late-season defoliation nearly every other year. Careful attention to sanitation may delay its introduction for many more years.

Traditional breeding for resistance is an excellent way to create fungus resistant grape cultivars, which allows reduction in the number of fungicide applications and the time consuming processes.

In search of alternative strategies in plant protection, many attempts have been made to find solutions for this disease. In order to achieve an induction of systemic resistance and, therefore, and to enhance the resistance against pathogenic micro organism, several chemicals, elicitors and various compounds have been tested for their efficiency. Compounds like potassium and phosphate salts like KH_2PO_4, K_2HPO_4, K_2SO_2, $NH_2H_2PO_4$, extract from some plants (Daaf *et al.*, 1995) and micro organisms (Strobel *et al.*, 1996) are well documented as plant protecting agents and efficient in systemic resistance inducers in many pathosystems (Reuveni *et al.*, 1998).

This emerging technology of plant disease control is based on the activation of the plant's own defense system with the aid of low molecular weight synthetic molecules. Compounds such as salicylic acid (SA-suggested as an endogenous signal substance in SAR (Systemic Acquired Resistance) activation), 2,6-dichloroiso-nicotinic acid (INA) (Hijwegen and Verhaar, 1993 and Kogel *et al.*, 1994), benzo(1,2,3)-thiadiazole-7- carbothionic acid-S-methyl ester (BTH) (Anfoka, 2000, Colson- Hanks and Deverall, 2000)) and DL-β-aminobutyric acid (BABA (Zimmerli *et al.*, 2000; Conrath *et al.*, 2002; Ton and Mauch-Mani, 2004) are able to induce systemic acquired resistance (SAR) through synthesis of number of pathogenesis–related proteins (PRP_S) in a variety of plants against a wide range of microbial pathogens (Sticher *et al.*, 1997).

'Pathogenesis related' (PR) proteins are induced in plants after infection with fungi, bacteria, and viruses. These proteins were characterized by their acidic nature, resistance to proteases and extra cellular location (Van Loon, 1976). PR proteins in plants have been classified into groups, some having unknown function, and others such as peroxides (PR9) β-1, 3-glucanase (PR2), chitinase, (Class III in *Vitis* spp.) etc., exhibiting specific enzyme activities. It is believed that chitinase and β-1, 3-glucanase play a role in plant defense mechanisms against fungal infection (Schumbaum *et al.*, 1986). Expression of these proteins such as PR2 and PR9 seems to depend on leaf age (Reuveni, 1998).

The modulation of the identification of preformed anti-fungal compound and PR protein level induced disease resistance by biological, physical and chemical means may increase the resistance of existing native *Vitis vinifera* cultivars (mostly American species that shows resistance which are not easily crossed with *Vitis* species) by conventional breeding, or through genetic modification by transformation with introduced genes and thus form a new option for effective control of grape diseases.

References

Anfoka, G.H. (2000). Benzo-(1,2,3)-thiadiazole-7-carbothioic acid S-methyl ester induces systemic resistance in tomato (*Lycopersicon esculentum*. Mill cv. Volledung) to cucumber mosaic virus. *Crop Protection*, 19: 401–405.

Boubals, D. (1959). Contribution a l'etude des causes de la resistance des vitacees au mildiou de la vigne et de leur mode de transmission hereditaire. *Ann Amelior Plantes.*, 9: 1–236.

Colson-Hanks, E.S. and Deverall, B.J. (2000). Effect of 2,6-ichloroisonicotinic acid, its formulation materials and benzothiadiazole on systemic resistance to Alternaria leaf spots in cotton. *Plant Pathology*, 49: 171–178

Conrath, U., Pieterse, C.M.J. and Mauch-Mani, B.(2002). Priming in plant-pathogen interactions. *Trends Plant Sci.*, 7: 210–216.

Daaf, F., Schmitt, A. and Belanger, R.R. (1995). The effects of plant extracts of *Reynoutria sachalinensis* on powdery mildew development and leaf physiology of long English cucumber. *Plant Disease*, 79: 577–580

Galet, P. (1977). les Maladies et les Parasites de la Vigne. Tome I. Les Maladies dues a des Vegetaux. Imprimerie Le Paysan du Midi., Montpelbrwier.

Hijwegen, T. and Verhaar, M.A.(1993). Induced resistance to *Peronospora parasitica* in red cabbage. Netherlands Journal of *Plant Pathology*, 99 (Suppl 3): 103–107

Kogel, K.H., Beckhove, U., Dreschers, J., Munch, S. and Romme, Y. (1994). Acquired resistance in barley-the resistance mechanism induced by 2,6-dichloroisonicotinic acid is a phenocopy of a genetically based mechanism governing race-specific powdery mildew resistance. *Plant Physiol.*, 106: 1269–1277.

Lafon and Clerjeau.(1988). Downy mildew. In: *Compendium of Grape Diseases*, (Eds.) Pearson, H.M. and C. Goheen. APS Press, St. Paul Minn, p. 11–13.

Pearson, R.C. and Goheen, A.C. (1996). Plagas y enfermedades de la vid. Ed. Mundi–Prensa, Madrid.

Perez Marin, J.L.(1992). Los Parasitos de la Vid. Estrategias d eproteccion Razonada. Ministerio de Agricultura, Pesca y Alimentacion. Ed. Mundi–Prensa, Madrid.

Ravaz, L.(1914). LeMildiou. Masson Ed, Paris.

Reuveni, R., Dor, G. and Reuveni, M. (1998). Local and systemic control of powdery mildew (*Leveillula taurica*) on pepper plants by foliar spray of mono-potassium phosphate. *Crop Protection*, 105: 703–709.

Reuveni (1998). Relationship between leaf age, peroxidase and β-1,3-glucanase activity, and resistance to downy mildew in grapevines. *Phytopathology*, 146: 525–530.

Schumbaum, A., Mauch, F., Vogeli, V. and Boller, T. (1986). Plant chitinase are potent inhibitors of fungal growth. *Nature*, 324: 365–367.

Staudt, G. and Kassemeyer, H.H. (1995). Evaluation of downy mildew resistance in various accessions of wild *Vitis* species. *Vitis*, 34(4): 225–228.

Staudt, G. (1997). Evaluation of resistance to grapevine downy mildew (*Plasmopara viticola*) resistance in various accession of wild vitis species. *Vitis*, 36: 151–154.

Sticher, L., Mauch-Mani, B. and Metraux, J.P. (1997). Systemic acquired resistance. *Annu. Rev. Phytopathol.*, 35: 235–270.

Strobel, N.E., Ji, C., Gopalan, S., Kuc, J.A. and He, S.Y. (1996). Induction of systemic acquired resistance in cucumber by *Pseudomonas syringae* pv *syringae* 61 HrpZ(Pss) protein. *Plant J.*, 9: 431–439.

Ton, J. and Mauch-Mani, B. (2004). Beta-amino-butyric acid-induced resistance against necrotrophic pathogens is based on ABA-dependent priming for callose. *Plant J.*, 38: 119–130.

Ton, J. and Mauch-Mani, B. (2004). β-amino-butyric acid-induced resistance against necrotrophic pathogens is based on ABA-dependent priming for callose. *Plant Journal*, 38: 119–130.

Van Loon, L.C. (1976). Specific soluble leaf protein in virus-induced tobacco plant are not normal constituent. *Journal of General Virology*, 309: 375–379.

Zimmerli, L., Jakab, G., Metraux, J.P. and Mauch-Mani, B. (2000). Potentiation of pathogen-specific defense mechanisms in Arabidopsis by β-amino-butyric acid. *Proceedings of the National Academy of Science*, USA, 97: 12920–12925.

Plant Diseases Management for Sustainable Agriculture *Pages* ***348–366***
Editor: **Dr. Shahid Ahamad**
Published by: **DAYA PUBLISHING HOUSE**

Chapter 28

Advances in Management of Carnation Wilt

Sunita Chandel

Department of Mycology and Plant Pathology,
Dr. Y.S. Parmar University of Horticulture and Forestry, Nauni, Solan – 173 230, H.P.

Wilt induced by different species of *Fusarium oxysporum* is one of the most widespread and destructive diseases of many ornamental and horticultural crops. *Fusarium* species are invariably distributed in the soil environment and possess a wide host range for their survival. The soil borne fungus causes vascular wilts by infecting plants through the roots and growing internally through the cortex to the stele (Beckman, 1987 and Tjamos and Beckman, 1989). *Fusarium* wilt of carnation is one of the most prevalent and damaging diseases of carnation in all parts of the world.

Carnation is grown globally but commercially its cultivation is practiced in France, Netherlands, Holland, Italy, Israel, Germany and USA covering 6000 hectare with replanted area of 4000 hectare (Strider, 1985 and Van-Ruiten, 1992). However, the phenomenon growth of floriculture in India during the last couple of decades has led the world floriculture pundits to decide the strategy for its sustainability in expanding market. In India the carnation culture is in budding stage. These are being grown commercially in Solan, Shimla, Mandi, Kullu, Chandigarh, Ludhiana, Delhi, Gurgaon, Bangalore and Kalimpong. The mid-hill zones of Himachal Pradesh, Jammu and Kashmir, Uttar Pradesh, West Bengal, Nilgiri hills, Bangalore and Nasik are considered as potential areas for its cultivation (Gupta and Bhattacharjee, 2002).

Distribution of Disease

Like other crops this flower also succumbed to large number of pathogens, some of which are frequently observed and are responsible to cause enormous losses. Fungi constitute a major share and wilt disease (*Fusarium oxysporum* f. sp. *dianthi*) amongst all other reported diseases is considered to be the most serious problem associated with the carnation as it is continuously causing considerable economic losses in the crop (Lahdenpera, 1987). Diseases caused by Fusarium have been cited worldwide as a major problem in the production of carnation (Nelson *et al.*, 1983). The evidence of its occurrence was very well known to occur as early as 1897 in United States and Great Britain causing huge lose in carnation (Hood and Stewart, 1957). The disease was also encountered from Nepal, Columbia, South Africa, Israel, Veramin areas of Iran and other parts of the world including India (Villiam *et al.*, 1972; Evans, 1987; Ben-Yephet *et al.*, 1993 and Katoch, 1999).

The carnation industries in Victoria suffered worth \$10m per annum losses and \$2 m in production due to disease problem (Say *et al.*, 1992). In Germany, about 46 per cent disease incidence while in Italy 60 per cent reduction in flower production was observed. However, in India disease incidence recorded as high as 40 per cent but in Himachal Pradesh, 79 per cent incidence was found in variety " Purple Choppin".

Morphological Characters of the Fungus

Gilman studied a morphological character of *Fusarium oxysporum* f.sp *dianthi* in 1957 that reported two types of conidia, micro and macro. Conidia occurred in bright orange coloured sporodochia. The microconidia are hyaline, one–celled or rarely septate, globose or oval, produced in abundance in white to marvel and rosy floccose aerial mycelium. Macroconidia are spindle-sickle shaped, curved a bit at the top, weakly pedicellated and often hooked at the tapered end, somewhat thicked in upper third than in central portion and 3-4 septate. Other species like *Fusarium graminearum* Schwabe, *F. avenaceaum* (Fr.), *F. culmorum*(W.G. Smith) Sacc. and F.*poae* (Perk) Wollens have been reported on carnation causing stem, basal and bud rot (Tammen, 1954; Nelson *et al.*, 1983 and Elad,1906). In addition, *F. roseaum* was found associated with carnation plants causes wilt like symptom. Three fungi *viz. F. oxysporum*, *F. graminacearum* and *F avenaceaum*, however, were reported to cause major losses to carnation production in Victoria (Say *et al.*, 1992).

Symptomatology

Bickerton (1942) described the general symptoms of Fusarium wilt of carnation. Initial symptoms appeared as chlorosis and distortion of lower leaves, often one side of the plants. The stem often shriveled and turned grayish at the affected side of plant. Foliar chlorosis, necrosis and plant stunted became more pronounced as the disease progresses. While the remaining leaves remain green, wilting occurs on the affected side of the plant followed by vascular discoloration and stem necrosis. The wilted plants became stunted yellow and dry with hollow moves into the stem (Bowers and Locke, 2000). Similar kind of symptoms were also observed by Baayen and Elgersma(1985) and Baayen and Maat(1987).

Pathogenicity and Physiological Studies

Hood and Stewart (1957) conducted for the first time the pathogenicity test of fungus. According to them root dip and stem wound inoculations with *Fusarium oxysporum* f. sp. *dianthi* induced symptoms of carnation wilt more quickly than did inoculation by infestation of the soil.

Ben- Yephet *et al.* (1992) also proved the pathogenicity of the fungus by immersing rooted stem cutting in spore suspension (10^{6} spores/ml). Disease symptoms produced within 2-3 weeks was observed by Rattink (1974). Chen *et al.* (1994) in similar way proved the pathogenic nature of *Fusarium oxysporum* f. sp. *gladiolus* and *F. solani* on gladiolus inciting yellow under *in vitro* conditions. Harling *et al.* (1988) reported a consideration role of temperature in disease development and symptom expression, when they inoculated six commercial cultivars with *Fusarium oxysporum* f. sp. *dianthi.* No wilt symptoms along with very little colonization of the fungus observed in all the cultivars at 14-15° C. At 26°C all the cultivars except the most resistant Carrier 929 rapidly become diseased. The incidence of Fusarium wilt of carnation influenced by factors like solar radiation, intensity, temperature in addition to growth substrate which was also reported to influenced the incidence of Fusarium wilt of carnation (Ben-Yephet *et al.*, 1996b). Pargola *et al.* (1979) found that in glasshouse experiment, the soil pH of 7.5 or higher decreased the infection of wilt pathogen while the acidic conditions favoured the same.

Physiological studies of the carnation wilt pathogen reported that pathogen could grow fast and reproduce heavily on potato dextrose agar and carnation leaf agar while Richards solution supported maximum dry weight of the fungus amongst the liquid media tested. Optimum temperature of 25° C, pH 5-6.5 produced abundant spores but no spores observed at 4.0 and above 8.0.

Disease Management

Plant diseases can be effectively managed by employing various methods of plant disease control. Methods of control vary considerably from one disease to another, depending on kind of pathogen, the host and interaction of the two and many other variables. Since *Fusarium* is a soil borne fungus, the management practices generally employed include cultural and chemical methods. The cultural methods aim at eradicating or reducing the amount of pathogen present in an area, a plant or plant parts while the chemical methods are most promising and are easily and quickly achievable. They are generally used to protect plant surfaces from infection or to eradicate a pathogen that has already infected a plant. A review of various management studies is given below:

Non-chemical Methods in Disease Management

Cultural Methods

Cultural methods are very helpful in management all types of a soil diseases. They include crop rotation, general sanitation, selection of seed and planting material. Heins and Wilkins (1977) study the effect of plant spacing on disease development of carnation wilt and flower yield. Kitchener (1977) also investigated the effect of plant spacing on the wilt progress where different spacings like 15x12.5 cm, 15x20 cm and 17x22.5 cm and found 15x20cm spacing had a maximum disease. It was noticed that population of *Fusarium* spp. increased with Chinese cabbage, welse onion, radish stalk, crushed wheat straw and sweet potato stalk and decreased with wheat straw, gingseng leaves and cabbage as compared with untreated control Sun and Huang(1985). The amendment of soil with oil cakes like mustard, groundnut and margosa was effective in reducing cotton wilt as per the investigations of Sharma and Bedi(1988). Tu (1988) reported that crops like oat, maize, sorghum and sudan grass promoted the saprophytic bacterial activity which is helpful in suppressing the activity of root rot and wilt fungi. According to Orozoco *et al.* (1993), ammonium nitrate and lower potassium doses gave control of carnation wilt. It was found that pine needles followed by mustard and neem cakes gave maximum control of the disease when applied two months before planting of cuttings in soil over wheat straw, poultry manure and castor cake (Katoch, 1999).

Organic Amendment in Disease Management

Soil is an excellent medium for the growth of micro-organism. Organic and inorganic amendments of soil produce profound effect on the soil microbes as described by Alexander (1971) and Garrett (1960). Several workers have studied the qualitative changes in the soil amended with different organic and inorganic substances (Adam *et al.*, 1968; Bhagyaraj and Rangaswami, 1967; Chandra *et al.*, 1981; Chatopadhyay and Mustafee, 1980; Singh *et al.*, 1991; Huber and Watson, 1970).

Srivastava and Sinha (1971) observed that addition of organic amendments to the soil infested with *Fusarium oxysporum* f.sp. *corianderii*, causing wilt of coriander, reduced the plant mortality to 25.5 per cent as compared to 46.4 per cent in the control. Similar reduction in the per cent mortality of grams infested with *Fusarium* wilt was observed by Chauhan (1960) when the soils were amended with oil cakes. Khanna and Singh (1974) observed that neem cake caused inhibition of sporulation of *Fusarium* spp. in rhizosphere soil. Out of the nine amendments tested against *Fusarium* wilt of tomato in pot and field, Homma *et al.* (1979) observed that rape seed and soybean oil cakes reduced the disease severity in pot experiments. Ramakrishnan and Jeyarajan (1987) reported the control of *Fusarium* wilt of crossendra, tomato and onion through addition of neem cakes. Chakrabarti *et al.* (1991) studied the effect of oil cakes on wilt of muskmelon caused by *Fusarium solani* and reported that neem cake reduced the wilt incidence by 80 per cent, mustard cake by 65 per cent and ground nut cake by 37 per cent. Neem cake and sunflower cake reduced the mortality of tomato and onion due to *Fusarium* wilt by 85 and 75 per cent, respectively (Alice, 1996). Katoch (1999) found the pine needles and neem cake, most effective against *Fusarium* wilt of carnation. Neem cake reduced the wilt of muskmelon caused by *Fusarium solani* by 90 per cent, followed by mustard cake (65 per cent). Navas *et al.* (2000) reported that several treatments of including methyl bromide and souil steaming reduced inoculum densities of *Fusarium oxysporum* f. sp. *dianthi* to undetectable or very low levels and flowers yield in most effective treatment was more than 10 times greater than that of the untreated plots. Pera and Filippi (1987) observed that in greenhouse protection, the popular amended bark compost soils have less intensity of wilt infection.

Soil Solarization

Soil solarization has been used for the management of soil borne pathogen since 1976 (Katan *et al.*, 1976) and still widely used in many area of the world (Chellemi and Mirusso, 2002). Soil solarization involves the use of heat as a lethal agents for controlling pests in soil through the use of traps for capturing solar energy by means of transparent polyethylene mulches effectively increasing soil temperature for several hours a day (Katan, 1981).Various methods of using solar irradiation for controlling pests have been developed much before the development of plasticulture. These include heating of seed tubers or soil (by exposing to sunshine) for controlling plant pathogen. Plastic provided a new and effective test for solar heating of soil. Polyethylene popularly known as polythene, one of the most important material used in agriculture, was first introduced on a commercial scale in 1939 (Byrdson, 1970). However, Jones *et al.* (1966) for the first time used polyethylene traps to increase the soil temperature, which proved lethal to soil borne pathogens of tomato.

Mulching with polyethylene sheets increased soil temperature and resulted in pronounced reduction in the population of *Verticillium dahliae* and *Fusarium oxysporum* f.sp. *lycopersici*. Soil solarization using clear, transparent polyethylene (0.1 mm thick) for 30 days gave satisfactory control of *Fusarium* wilt (F. *oxysporum* f.sp. *dianthi*) of carnation and *Verticillium* wilt (*V. dahilae*) of eggplant (Wong, 1985). Elena and Tjamos (1997a) found that soil solarization for nearly seven weeks significantly reduced or eliminated the inoculum of *F. oxysporum* f.sp. *dianthi* added to soil as a conidial suspension prior to mulching. Elena and Tjamos (1997b) further reported that soil solarization alone or in combination with antagonists drastically reduced the percent disease incidence of *Fusarium* wilt of carnation. Fujinaga *et al.* (2005) reported that soil sterilization with hot water treatment as a highly effective against the pathogen *Fusarium oxysporum* f.sp. *appi* causing fusarium yellows *of celery* and also effective against, *Fusarium oxysporum* f.sp. *callistephi*, *F. o.* f. sp. *dianthi* and *F. solani*. The combination of soil solarization with and burial of infected soil at depth of 15 cm, 30 cm and 45 cm in untreated plots also reduced the viability of *Fusarium oxysporum* f. sp. *dianthi* (Suarez et al., 2005).

Biological Control

Biological control of soil borne pathogens by introduced micro-organisms has been studied for over 65 years (Baker, 1987; Cook and Baker, 1983), but during most of that time, it has not been considered commercially feasible. Since about 1965, however, interest and research in this area have increased steadily (Baker, 1987) as reflected by number of books (Baker and Cook, 1974; Chet, 1987 and Papavizas, 1981). An ecological balanced soil is the best medium for growing healthy plants, such soils can be maintained by following sound cultural practices. Though biological antagonists are subjected to numerous ecological limitations under field conditions, yet this method can be exploited to become an important control measures against soil borne diseases.

Biocontrol Agents from Rhizosphere of Carnation

Microorganisms that can grow in the rhizosphere are ideal for use as biocontrol agents, since the rhizosphere provides the front line defense for roots against attack by pathogens (Weller, 1988).Various attempts have been made by Papavizas (1981) to study occurrence and distribution of antagonists in their natural habitats and to study population dynamics, survival in soil and other habitat. Despite many efforts, scarcity of information do exists on the survival of the antagonists in the soil, which was estimated probably due to the lack of precise techniques and the appropriate culture media for isolation and enumeration. Johnson (1957) gave the dilution plate method for the isolation of micro-organisms from soils. Although, earlier Warcup (1950) developed the soil plate methods for the isolation of micro-organisms from the soil.

Screening of micro-organisms

Dennis and Webster (1971) and Johnson and Curl (1972) developed the dual culture technique for screening the antagonists. Gilman (1957) and Rifai (1969) gave methods of identification of antagonists. Subramanian (1971) developed the identification keys for the identification of isolated fungi while Buchanan and Gibbons (1988) produced the manual for the identification of isolated bacteria.Various mechanisms of antagonism were suggested by D'ercole *et al.* (1984). Mukhopadhyay and Mukherjee (1996) reported *Trichoderma* as destructive parasite, employing different modes of action to fungal hyphae and their resting structures. Kharakrang *et al.* (2002) also observed strong antagonisms of *Trichoderma harzianum, T. viride* and *Gliocladium virens* against *Fusarium* species.

Potential Biocontrol Agents

Microorganisms rank among our best allies in the control of plant diseases. Indigenous fungi and bacteria are recognized as common contributors to disease suppression in cropping systems and mediate effects of many cultural practices in reducing pathogen populations and limiting their severity (Ness and Baliga, 1999).

Fungal Antagonists

Amongst the natural enemies, antagonistic fungi have received the maximum attention (Muthusamy, 1999). These include several species of *Trichoderma, Gliocladium virens, G. roseum, Streptoverticillium* sp., *Latesaria arvalis, Coniothyrium minitants, Aspergillus* sp. and *Penicillium* sp. Of these, potential species *Trichoderma* and *Gliocladium* have been extensively exploited by the pathologists due to their high efficacy, broad spectrum activity and ease in isolation and cultivation (Mukhopadhyay and Mukherjee, 1996).

Trichoderma spp. has been reported to possess antagonistic properties that might have made them promising biological control agents (Chet *et al.*, 1981 and Weindling and Emerson (1936). Cho *et al.* (1989) observed that *T. harzianum* had the highest mycoparasitic and low antibiosis to *Fusarium oxysporum* f.sp. *cucumerinum* causing wilt of cucumber. *Fusarium* wilt of strawberry caused by *Fusarium oxysporum* f.sp. *fragariae* was also effectively controlled by *Trichoderma harzianum* (Basak and Lee, 2002).

The potential use of the antagonists, *T. harzianum, Streptomyces griseovirides* and non-pathogenic isolate of *Fusarium* spp. for the wilt disease of carnation, caused by *F. oxysporum* f.sp. *dianthi* was investigated by Rattink in 1992 and observed that these antagonists reduced the disease more than 50 per cent. Tamietti *et al.* (1991) indicated that *G. virens* is a good colonizer of heat treated soil where its population increases very quickly. Locke *et al.* (1984) suggested the biological control of *Fusarium* wilt of green house chrysanthemum by different isolates of *T. viride* alone or in combination with *Aspergillus orchraceus* which reduced the disease by atleast 50 per cent in vegetatively maintained plants. Harman *et al.* (1981) demonstrated the seed application of antagonists and adopted it as an alternate approach in introducing them in soil. Papavizas (1985) stated that effectiveness of antagonistic fungi as seed treatment is determined by their bio-control quality and performance in the rhizospheres. Species of *Trichoderma* and *Gliocladium* controlled the occurrence of pathogenic fungi in soil (Kowalik, 1996). Hazarika *et al.* (2000) reported seed coating with conidial suspension of biocontrol agents more effective against soil borne plant pathogens.

Bacterial Antagonists

Bacteria having potential as biocontrol agents occur in many genera including *Pseudomonas, Bacillus, Agrobacterium, Actinomycetes, Enterobacter, Erwinia, Arthrobacter* etc. However, a few of these *viz. B. subtilis, P. fluorescens, Agrobacterium* are used for controlling the diseases (Muthusamy, 1999). Interest to use bacterial flora for plant disease control or yield improvement has increased steadily since mid 1960s (Baker, 1987).

Fusarium oxysporum f.sp. *dianthi*, pathogen of carnation, is highly sensitive to inhibition by *Bacillus subtilis* (Filippi *et al.*, 1987). Duijff *et al.* (1993) found that treatments of carnation roots with *Pseudomonas* sp. significantly reduced the wilt caused by *F. oxysporum* f.sp. *dianthi*. Kloepper *et al.* (1980a) were the first to demonstrate the importance of siderophore production as a mechanism of bio-control. Antagonism of *Pseudomonas* strain towards many other pathogens can be attributed to production of siderophores pseudobactin by the *Pseudomonas* strain (Kloepper *et al.*, 1980b). Subsequently siderophores have been shown to be involved in the suppression of formae speciales of *Fusarium oxysporum* (Elad and Baker, 1985). Mycolytic enzymes chintinase are antifungal and significantly contribute to the anatogonistic activity of fluorescent pseudomonas belonging to plant growth promoting rhizobacteria against the *Fusarium oxysporum* f. sp. *dianthi* (Ajit *et al.*, 2006).

Application Methods and Field Performance of Biocontrol Agents

Successful biological control of soil-borne pathogens by inoculating infested soils with Trichoderma species have been described by many scientists. Chet and Elad (1982) suggested different methods of their application. They found that coating of iris bulbs as well as broadcast application of *Trichoderma harzianum* significantly reduced the incidence of diseases caused by *Rhizoctonia solani* and *Sclerotium rolfsii*. Maximum number of plants survived when *Trichoderma harzianum* was applied as a seed treatment or applied to soil at sowing time (Muthamilan and Jeyarajan, 1996).Pennock-vos *et al.* (1990) concluded that *T. hamatum, Gliocladium roseum* and *Streptomyces* species were effective against tulip bulb disease. Ramarao *et al.* (1987) observed effective biological control of soil borne diseases caused by *Fusarium oxysporum, Sclerotium rolfsii, Rhizoctonia solani* by species of *Trichoderma*. An antagonistic microorganism isolated from the soil infected with *Fusarium oxysporum* f. sp. *dianthi* reduced the wilt infection in hydroponic system. Close control and frequent exchange of the circulating nutrient solutions with heating to 18-22°C are recommended (Jacob and Krebs, 1986). Treatment with *Pseudomonas putida* WCS358r, rifampicine resistant derivatives of the strain WCS 358 significantly reduced the Fusarium wilt grown in rockwool if the disease incidence is moderate. Siderophore–mediated competition for the iron was shown to be the mechanism of suppression. Arbelaez *et al.*(1993) used integrated control for the Fusarium wilt of carnation by considering Pink Calypso, Improved White, White CC and Nova Barlo cultivars, fungal and bacterial antagonists (*Trichoderma harzianum*, *Psuedomonas putida* and *Serratia liquefaciens*) alone and in combinations and obtained positive role of them in managing the disease. Different doses of N and K and soil disinfections with vapour+ dazomet (as Basamid) at lower K gave more effective disease control. Postma and Rattink(1991) reported that a spore suspension of the non-pathogen, *Fusarium oxysporum* isolate 618-12 just before or just after planting rooted cutting in soil infested with *Fusarium oxysporum* f. sp. *dianthi* gave best control normally more than 80 per cent. *Fusarium moniliforme* was also found effective in controlling the wilt infection in carnation. Inoculation with *Trichoderma* spp. controlled the pathogen multiplication first 3 months on seedling growth (Rumine, 1989). Postma (1992) showed that wilt can be easily suppressed by *Fusarium* non-pathogenic isolate *Fusarium* 618-12 when applied along with the pathogen in the soil but not when they were inoculated at different locations (stem or soil). Obieglo(1992) recommended that combination of antagonist strain *Bacillus subtilis* and 0.025 per cent potassium permeganate to the perlite solution before planting check the wilt during rooting of the carnation. Manka and Fruzynska-Jozwiak, 1992) found that soil of old pine stands were suppressive to the pathogen isolates of *Trichoderma* and *Penicillium* spp. from these soils were used for the biological control of the pathogen of carnation wilt and showed that there was drastic reduction to 4.1–20.8 per cent compared with 100 per cent in the inoculated control after 5 months. Chandel and Pathania (2004) concluded that *T.virens* suppressed maximum growth *in vitro* in dual culture while *T.harzianum* gave excellent control of wilt in pot experiments and field when applied twice as soil drench.

Use of Botanical Extracts and Neem Formulations in Disease Management

The hazardous effects of synthetic pesticides have resulted in exploring the possible use of herbal pesticides (Jacobson, 1989). More than 200 plant species belonging to 170 families are known to have pesticidal and insecticidal properties (Kumarera, 2002). The use of natural products for the control of fungal diseases in plants is considered interesting alternative to synthetic fungicides due to their lower native impact on the environment (Brunelli, 1995).

The toxicity and various other ill effects in human and animals, synthetic fungicide face their restricted use in plant disease control. Among the alternate safe methods, several plant extracts and their active principles are attracting much attention in this direction (Mishra *et al.*, 1993; Singh *et al.*, 1990; Singh *et al.*, 1995 and Singh *et al.*, 2001).Currently, pre-plant soil fumigation and fungicides application are used to control the wilts and other diseases caused by soil borne pathogens on high value crops. However, the major fumigant used "Methyl bromide" has been defined by the Montreal Protocol of 1991 as a chemical that contributes to the depletion of the ozone layer. The U.S. Clean Air Act (Section 602) requires that production and importation of any substance defined as an ozone depletor by Montreal Protocol should be phased out (Bowers and Locke, 2000).

Many plants and plant products have been reported to possess pest control properties (Grange and Ahmed, 1988; Grayer and Harborne, 1994). Although much of the literature on natural products in the agricultural field concerns insect control, a small but emerging body of paper reported that plant extracts and plant essential oils are effective antimicrobial agents against soil borne fungi (Awuah, 1994; Bianchi *et al.*, 1997; Dubey and Kishore, 1987). Siddiqui and Kaushal (2002) investigated the antifungal properties of some extracts such as *Ocimum sanctum* L, *Vinca rosea* L, *Datura alba* Nees,

Launea asplenifolia Hk., *Euphorbia hirta* L. and observed more than 40 per cent of disease inhibition of *Fusarium oxysporum* f.sp. *zingiberi*.

Kumar *et al.* (2002) studied antimicrobial properties of ten plants *viz. Azadirachta indica* A. Juss, *Ocimum sanctum* L, *Calotropis procera* (Ait.) R.Br., *Ricinus communis* L, *Catharanthus roseus* (L.) G. Don., *Carria tora* Baker, *C. occidentalis* L, *Eucalyptus ovelis* Endl., clove *of Allium sativum* L. and *Lawsonia incerme* against *Fusarium oxysporum* f.sp. *udum* (Butler) Snyder and Hansen and reported a higher per cent of disease control specially with *Allium sativum* and *Eucalyptus orvelis* which gave 94 and 90 percent disease control, respectively.

Amaresh and Nargund (2001) also studied the thirty commonly available plant extracts against rust of sunflower and found that *Allium cepa* and *Azadirachta indica* were most effective plant extracts giving the per cent inhibition of 70 and 95 per cent, respectively. Neem formulation gave excellent results in the management of carnation wilt. Bhat and Srivastava (2003) observed the effectiveness of nimbicidine and neemazol on *Fusarium* wilt of carnation. Neem leaves in the form of green manure help in controlling the fungal diseases. Singh and Pandey (1966) reported the effect of green and matured plant residues and compost on the population of *Pythium aphanidermatum* in soil.

The fungicidal effect of neem (*Azadirachta indica*) on some soil-borne pathogens of chickpea (*Cicer arietinum*) was studied by Singh *et al.* (1980). The growth of test fungi, *Fusarium oxysporum* f.sp. *ciceri, Rhizoctonia solani, Sclerotium rolfsii* and *Sclerotinia sclerotium* was inhibited in liquid medium containing extracts of leaf, trunk bark, fruit pulp and oil of neem. Oil-treated seeds sown in soil infested with pathogens, singly or mixed, produced disease free seedling. The inhibitory effect of neem leaf extract along with other extracts was tested on *Sacrocladium oryzae* and *Fusarium oxysporum* f.sp. *cepae* (Eswaramurthy *et al.*, 1989). Neem extract inhibited the fungi by 34.9 and 24.1 per cent, respectively.

The toxicity of neem leaf extracts along with the leaf extracts of *Calotropis procera* and *Datura stramonium* was tested against five fungal pathogens of rice *viz. Pyricularia oryzae, Rhizoctonia solani, Curvularia lunata, Aspergillus niger* and *Fusarium monilforme* (Mishra and Tiwari, 1990). All extracts possessed toxic principles against one or more of these pathogens. The bio-efficacy results had attracted the attention of the pesticide industry in India and abroad. Nearly five dozen products are either marketed or are awaiting commercialization in India. Most of the products are either oil-based or based on various extractives. Neem products have been reported to act as insecticide, fungicide, nematicide, insect antifeedant, repellent or deterrent, growth regulator etc. Godrej Achook, Neemark, Nimbicidine and Neemactine have been reported to act as fungicide against many soil-borne diseases (Parmar and Ketkar, 1996).

Screening of Germplasm for Resistance

Recently commercial propagators are beginning to breed carnation with resistance to the vascular wilt pathogens. Little progress has been made in this regard. Carnation cultivars like Sam's Prode, Revanda, Novanda were reported to be resistant to Fusarium wilt (Glas, 1984). Revanda and Novanda cultivars showed about 80 per cent resistance against the disease. Baayen and Elgersma (1985) reported typical wilt symptoms of the disease with stem colonization and vascular browning on Erly Sam's cultivar after inoculation with conidial suspension. However, Novanda cultivar did not produce such type of symptoms after inoculation. Rooted cuttings of these cultivars were planted in infested soil and observation on the symptoms was recorded after 5-6 weeks. It was concluded that Novanda was resistant as its stem was not colonized and remained healthy whereas Early Sam's was susceptible because fungus could colonize the stem, which resulted in death of the saplings. Ben–Yephet *et al.* (1993) also gave the explicit classification of carnation cultivars towards resistance. They considered

75 per cent disease incidence as cut off point for separating cultivars into resistant (0-74 per cent) and susceptible (75-100 per cent) types. They further minutely classified the cultivars into highly resistant, resistant, moderately resistant, fairly resistant, susceptible and highly susceptible cultivars. Scarlette, Eveline, Galit, lior and Herman were reported to be completely resistant, highly resistant, moderately resistant, susceptible and highly susceptible respectively.

Mediterranean lines appear promising in possessing the resistance to wilt pathogen and development of these lines relied strongly on the use of wild species suggesting the use of gene pool in selection of resistant lines (Schiva *et al.*, 1985). On screening carnation cultivars, resistance was obtained with cv. Duca (Garibaldi and Rossi, 1987). Out of 249 clones obtained from the crossing of numerous cultivars many of them found tolerant to the wilt while two Polish cultivars, Jurand and Kimicic showed good tolerance. Harling *et al.* (1988) found that at temperature of 26° C all six commercial cultivars except Carrier 929 rapidly become diseased and died within 23 day of inoculation due to vascular wilt and stem and root rooting symptom development. Be- Yephet (1993) reported that the level of resistance varied according the resistance possessed in the cultivars in root–dip inoculation when compared in field and greenhouse tests. The severity was counted on 0-4 scales according the disease intensity in different cultivars.

Chemical Control

Fungicides are most popular among the chemicals for the control of fungal pathogens of seed, soil or air borne nature. The fungicides have been employed as sprays, soil drench or seed dresser. *In vitro* study of benzimidazole fungicides against *Fusarium* spp. it was found that (MBC) cabendazim was most effective followed by benomyl and TBZ (Valaskova, 1983). A significant inhibition in germination of conidia obtained with various fungicides *viz.* Aliette, Aureofungin, Dexon, Dithane Z-78, Rovral, Syllite, Vitavax and Bayleton at concentrations of 50, 100 and 150 ppm, respectively. Bavistin and Aureofungin found most effective while Benlate recorded least effective in inhibiting the conidial formation (Prasad *et al.*, 1988). Thiolute at concentration of 10 ppm or less was active against *Fusarium solani* and if applied to soil resulted in increased population of antagonistic organism mainly *Trichoderma* and *Penicillium* spp. (Deb and Dutta, 1984). Fletcher and Martin (1972) studied the carnation wilt and reported the beneficial effect of benomyl as a drench to potted plants @ 1.5 or 3.0g in 4.5 l water. Application of 200ml of chemical suspension per pot was found to result in suppression of Syptoms. Maximum activity of benomyl observed after 10-14 days of application. Two application of benomyl as soil drench at 3.6 g a.i per 11.1 m^2 bed proved superior against the disease (Evans, 1976).

Satisfactory control of carnation wilt was obtained in commercial crops by drenching of benomyl at 3.6 g a.i per 5.5l m^2 when applied soon after planting and subsequently at three months interval. Bud rot of carnation caused by *F. sporotriciella* var. *poae* can be well managed by application of caryptol (50 per cent) at 0.02-0.04 per cent and fundazole at 0.1-0.2 per cent as sprays during growth or wetting the soil around the plant and orthocide (captan) at 50-100 g/m^2 per planting soil treatment, were found most promising (Georgieva *et al.*, 1979). Application of thiophanate methyl at higher concentration was recommended by Mirkova (1984) for the treatment of glasshouse carnation. The fungus, *F. solani* was best controlled by Beyleton (thiadimeton) followed by Fenaminosulf, Difolatan (captafol), syllite (dodine) and Ridomil (metalxyl) as stated by Mathur *et al.* (1984). In the field trial procloraz used as soil drench at 6.6 g a.i/m^2 gave very good results better than carbendazim at 6.6 g (Ganbosa and Larroque, 1985, 1986). Chemicals such as benomyl, zineb methyl tosin, thiabendazole and vapam have been suggested by various workers for the management of the vascular diseases (Kutova and Petkova, 1975; Obieglo *et al.*, 1990). Similarly, one or two application of benomyl 10g/m^2 were found to decrease incidence of wilt along with increase in crop yield (Garibaldi *et al.*,1988). However, decrease in its effectiveness was observed in subsequent years. Ebben (1997) also studied the efficacy of carbendazim in managing carnation wilt and found it to be highly efficacious.

Besemer (1980) reported that best control measure could be achieved with soil sterlization with chemical fungicides, uses of pathogen free planting material and general sanitation in the greenhouse. For the effective disease control, integrated disease management practices by use of suppressive soil, resistant cultivars and fungicidal treatment had been suggested by Garibaldi and Gullino (1986). Findings of Kovacikova and Altmanova (1989) reported that a single application of Benlate effectively controlled the wilt of carnation whereas combined application of minerals (Ca, K, and N) and benomyl increased the average infection level. Bavistin and kri-benomyl completely inhibited the vegetative growth at 100 and 200ppm. Roko, aureofungin and baycor were next best fungicides amongst systemic whereas kavach from non- systemic type inhibited the fungal growth to greater extent compared to fytolan. In field bavistin and kribenomyl when drench one month after transplantation and at bud formation stage reduced the wilt incidence besides enhancing the flower yield (Chandel and Katoch,2001).

References

Adam, P.B., Papavizas, G.C. and Lewis, J.A. (1968). Survival of root-infecting fungi in soil. The effect of cellulose amendment on chlamydospore germination of *F. solani* f.sp.*phaseoli* in soil. *Phytopathol.*, 58: 373–377.

Ajit, W.S., Rajni, V. and Shanmugham, V. (2006). Extracellular chintinase of fluorescent pseudomonas antifungal to *Fusarium oxysporum* f. sp. *dianthi* causing carnation wilts. *Current Microbiology*, 52(4): 310–316.

Alexander, M. (1971). *Introduction to Soil Microbiology*. John Wiley and Sons Inc., N.Y. and London, 472 p.

Alice, D. (1996). Study on *Fusarium* wilt of crossendra, tomato and onion. *Ph.D. Thesis*, Tamil Nadu Agric. Univ., Madurai, 267 p.

Aloi, C., Conti, A. and Garibaldi, A. (1994). *Fusarium* antagonistic against *Fusarium* wilt of carnation. *Colture-Protette*, 23: 79–82.

Amaresh, Y.S. and Nargund, V.B. (2001). Effect of plant extracts on *Puccinia helianthi* causing rust of sunflower. *Indian Journal Plant Pathology*, 19: 61–63.

Anonymous, (1989). *Annual Report, 1989–90*. Department of Mycology and Plant Pathology, UHF, Nauni, Solan (HP).

Arbelaez, G., Guzman, S., Leon, J., Gonzalez, M, Molina, J.C., Parra, J., Ferncy Angulo, J., Alvarez and Awuah, R.T. (1994). *In vivo* use of extracts from *Ocimum gratissimum* and *Cymbopogon citratus* against *Phytophthora palmivora* causing black pod disease of cocoa. *Annual Applied Biology*, 124: 173–178.

Baayen, R.P. and Elgersma, D.M. (1985). Colonization and histopathology of susceptible and resistant carnation cultivars infected with *Fusanum oxysporum* f.sp. *dianthi*. *Netherland Journal Plant Pathology*, 91: 119–135.

Baayen, R.P. and Maat, W. (1987). The elegans *Fusaria* causing wilt of carnation. *Netherland Journal Plant Pathology*, 94: 273–288.

Baayen, R.P. and Egersma, D.M. (1985). Colonization and histopathology of susceptible and resistant carnation cultivars infected with *Fusarium oxysporum* f. sp. *dianthi*. *Neth. J. Pl. Pathol.*, 91: 119–135.

Baker, K.F. and Cook, R.J. (1974). *Biological Control of Plant Pathogens.* Annual Phytopathological Society, St. Paul, 433 p.

Baker, K.F. (1987). Evolving concepts of biological control of plant pathogens. *Annual Review of Phytopathology*, 25: 67–85.

Bambarkar, S. (1990). Neem: A vast potential for agrochemicals. *Pesticides*, 24: 36–38.

Bansal, R.K. and Gupta, R.K. (2000). Evaluation of plant extracts against *Fusarium oxysporum* wilt pathogen of fenugreek. *Indian Phytopathology*, 53: 1407–1408.

Basak, A.B. and Lee, M.W. (2002). Antagonistic effect of *Trichoderma* sp. on the three soil borne diseases of cucumber. *Indian Journal Plant Pathology*, 20: 15–21.

Beckman, C.H. (1987). *The Nature of Wilt Disease of Plants.* American Phytopathological Society, St. Paul, MN.

Ben-Yephet, Y., Shtienberg, D., Reven, M. and Mor, Y.(1993). Response of carnation cultivars to cultivars to *Fusarium oxysporum* f. sp. *dianthi*. *Plant Pathology*, 45: 98–105.

Ben-Yephet Y., Reuven, M., Lampel, M., Szmulewich, Y. and Nitzani, Y. (1992). Susceptibility reaction of carnation cultivars to *Fusarium oxysporum* f.sp. *dianthi* in an infested field and in spore suspension. *Acta Horticulturae*, 307: 83–88.

Ben-Yephet, Y., Shtienberg, D., Renven, M. and Mor, Y. (1993). Response of carnation cultivars to *Fusarium oxysporum* f.sp. *dianthi* in the field. *Netherland Journal Plant Pathology*, 99: 3–12.

Ben-Yephet, Y., Reuven, M. and Mor, Y. (1993). Selection methods for determining resistance of carnation cultivars to *Fusarium oxysporum* f. sp. *dianthi*. *Plant Pathology*, 42: 517–521.

Besemer, S.T. (1980). In: *Introduction to Floriculture,* (Ed.) R.A. Larson. Acedmic Press.

Bhagyaraj, J. and Rangaswami, G. (1967). Effect of fertilizers on the microflora of soil and rhizosphere of ragi (*Eleusine coracana*). *Indian Journal of Microbiology*, 7: 29–38.

Bhat, N.M. and Srivastava, L.S. (2003). Evaluation of some fungicides and neem formulations against six soil-borne pathogens and three *Trichoderma* spp. *in vitro*. *Plant Disease Research, Ludhiana*, 18: 56–59.

Bianchi, A., Zambonelli, A., Zechini Daulerio, A. and Bellesia, F. (1997). Ultrastructural studies of the effects of *Allium sativum* on phytopathogenic fungi *in vitro*. *Plant Disease*, 81: 1241–1246.

Bickerton, J.M. (1942). *Fusarium* wilt of carnation caused by *Fusarium dianthi* Prill, and Del M.Y. (Cornelia). *Agri. Exp. Stn. Bull.* 788: 31. Bilai var. poae (PK). Gradinarska I Lozarska Nauka 16: 71–78. Biocontrol Science and Technology 4: 279–288.

Booth, C. (1971). *The Genus Fusarium.* McGraw and Hill Company, London, 273 p.

Bowers, J.H. and Locke, J.C. (2000). Effect of botanicals extracts on the population density of *Fusarium oxysporum* in soil and control of *Fusarium* wilt in the green house. *Plant Disease*, 84: 300–305.

Brunelli, A. (1995). I Prodotti natural! nella Lotla alle malattie Fungie. *La Difesa delta Piante*, 18: 57–59.

Buchanan, R.E. and Gibbons, N.E. (1988). *Bergey's Manual of Determinative Bacteriology*. The William and Wilkins Com., Baltimore.

Byrdson, J.A. (1970). *Plastic Materials*. Lliffe Books, London, 597 p.

Caius, S. (1937). *J. Bombay Net. Hort. Soc.*, 39: 563.

Chakrabart, S.K., Sen, B. and Sen, B. (1991). Suppression of Fusarium wilt of muskmelon by organic and cellulose amendment on chlamydospore germination of *F. solani* f.sp.*phaseoli* in carnation cultivars infected with *Fusarium oxysporum* f. sp. *dianthi*. *Neth. J. Pl. Pathol.*, 91: 119–135.

Chandel, S.S. and Katoch, R.(2001). Chemical control of *Fusarium oxysporum* f. sp. *dianthi*, an incitant of carnation wilt. *Indian J. Microbiol.*, 41: 135–137.

Chandel, S.S. and Pathania, N. (2004). Biological control of major soil borne diseases of carnation using fungal antagonists, particularly *Trichoderma* spp. *Eco. Envion. and Cons.*, 10: 475–478.

Chandra, S., Raizada, M. and Khanna, K.K. (1981). Effect of organic amendments on the rhizosphere microflora of tomato. *Proceeding Indian Academy of Science*, 90: 189–197.

Chang, Y.C., Baker, R., Kleifeld, O. and Chet, I. (1986). Increased growth of plants in the presence of the biological control agents. *Trichoderma harzianum. Plant Dis.*, 70:145–148.

Chatopadhyay, S.B. and Mustafee, T.P. (1980). Influence of organic amendments on the growth of soil inhabiting fungal pathogens. *Indian Jouranl of Microbiolology*, 18: 69.

Chauhan, S.K. (1960). Effect of oil cakes on *Fusarium* wilt of gram. In: *Proc. 47th Indian Science Congress*, 329 p.

Chellemi, D. and Mirusso, J. (2002). Non chemical management of soil borne pests in fresh market vegtetable production system. *Phytopathology*, 92: 1367–1372.

Chen, L.Z., Gan, X.B., Seng, J.Y. and Gu, W. (1994). A study on gladiolus root rot. *Journal of Shanghai Agricultural College*, 12: 240–246.

Chet, I., Harman, G. E. and Baker, R. (1981). *Trichoderma hamatum:* its hyphal interactions with *Rhizoctonia solani* and *Pythium* spp. *Microbial Ecology*, 7: 29–38.

Chet, J. and Elad, Y. (1982). Prevention of plant infection by biological means: La selection Des Plants. Institute National de la Recherche Agronomique, Paris, France, p. 195–204.

Chet, I. (1987). *Innovative Approaches to Plant Disease Control*. Wiley, New York, 372 p.

Cho, C.T., Moon, B.J. and Ha, S.Y. (1989). Biological control of *Fusarium oxysporum* f.sp. *cucumerinum* causing cucumber wilt of *Gliocladium virens* and *Trichoderma harzianum*. *Korean Journal Plant Pathology*, 5: 239–249.

Cochran, W.G. and Cox, G.M. (1963). *Experimental Designs, 3rd edn*. Asia Publishing House, Bombay. commercial conditions. Bulletin OILB/SROP 15:127–129.

Cook, R.J. and Baker, K.F. (1983). The nature and practice of biological control of plant pathogens. St. Paul, *Ann. Phytopathol. Soc.* 539 p. cultivars to *Fusarium oxysporum* f. sp. *dianthi*. *Plant Pathology* 42: 517–521.

Deb, P.R. and Dutta, B.K. (1984). Activity of thiolutin against certain soil borne plant pathogen. *Curr. Sci.*, 53: 659–660.

Dennis, C. and Webster, J. (1971). Antagonistic properties of species group of *Trichoderma* III. Hyphal Interaction. *Trans Brit. Mycol. Soc.*, 57: 363–369.

D'ercole, M., Sportelli, M. and Nipoti, P. (1984). Different types of antagonism of *Trichoderma* spp. towards plant pathogenic soil fungi. *Informatose Fitopathologico*, 34: 43–47.

Dohroo, N.P. and Sharma, S.L. (1984). Biological control of rhizome rot of ginger in storage with *Trichoderma viride. Indian Journal Plant Pathology*, 2: 185–186.

Dubey, N.K. and Kishore, N. (1987). Fungitoxicity of some higher plants and synergistic activity of their essential oils. *Tropical Science*, 27: 23–27.

Duijff, B.J., Bakker, P.A..H.M. and Schippers, B. (1994). Suppression of Fusarium wilt of carnation by emphasis on colonization and inoculation methods. Bulletin OILB/SROP, 15:8–10.

Duijff, N.J., Meijer, J.W., Bakker, P. and Schippers, B. (1993). Siderophore-mediated competition for iron and induced resistance in the suppression of *Fusarium* wilt of carnation by fluorescent *Pseudomonas* spp. *Netherland Journal Plant Pathology*, 99: 5–6.

Dwivedi, S K. (1991). Studies on population of dynamics of *Fusarium oxysporum* f.sp. *lycopersici* in solar heated soil. *National Academy of Science Letter*, 14: 235–237.

Ebben, M.H. (1997). Symposium on carnation. *Acta Horticulturea*, 71: 298.

Elad, Y. and Baker, R. (1985). Influence of trace amounts of cations and siderophore-producing pseudomonads on chlamydospore germination of *Fusarium oxysporum. Phytopathology*, 75:1047–52.

Elena, K. and Tjamos, E.C. (1997a). Soil solarization for the control of *Fusarium* wilt of greenhouse carnation. *Phytopathologia-Mediterranea*, 36: 87–93.

Elena, K. and Tjamos, E.C.(1997b). Evaluation of soil solarization method for control of *Fusarium* wilt of carnation in field. *Annales-de-I' Institut-Phytopathologique-Benaki*, 18: 13–24.

En-Yephet,Y, Reuven, M., Mor, Y. (1993). Selection methods for determining resistance of carnation cultivars to *Fusarium oxysporum* f. sp. *dianthi. Plant Pathology*, 42: 517–521.

Eswaramurthy, S., Muthusamy, M., Muthusamy, S. and Mariappan, V. (1989). Inhibitory effect of neem, *Acacia, Prosophis* and *Ipomoea* extracts on *Sacrocladium oryzae* and *Fusarium oxysporum* f.sp. *cepae. Neem Newsletter*, 6: 4–5.

Etabarian, H.R. (1996). Occurrence of *Fusarium* wilt of carnation caused by *Rhizoctonia solani. Plant Disease Research*, 6: 83–84.

Evans, S.G. (1978). Chemical only a partial answer to carnation *Fusarium* wilt. *The Grower* 89: 113–117.

Filipp, C. and Baganoli, G. (1992). Carnation. *Zenztrealbtalt Fiir Miknobiologie* 147: 345–350.

Filippi, C., Baganoli, G., Volterrani, M. and Picii, G. (1987). Antagonistic effects of soil bacteria on *Fusarium oxysporum* Schlecht f.sp. *dianthi* (Prill, and Del.) Snyd. and Hans. *Plant and Soil*, 98: 161–167.

Fletcher, J.T. and Martin, J.A. (1972). Spread and control of Fusarium wilt of carnation. *Plant Pathology*, 21: 182–187.

Fujinaga, M., Kobayashi, H., Komatsu, K., Ogisu, H., Uehara, T., Onn, Y., Tomita, Y. and Ogawara, T. (2005). Control of Fusarium yellows on celery by soil sterilization with hot water injection using a portable broiler. *Annual Report of the Kanto Tosan Plant Protection Society*, 52: 25–29.

Ganbosa, B. Sand and Larroque, O.R. (1985-86). Trail *in vitro* and in the field to control *Fusarium oxysporum* f. sp. *dianthi* (Prill. And Del) Synd and Hans. Revista del al Plats, 61/62: 153–15.

Garibaldi, A. and Gullino, M.L. (1987). *Fusarium* wilt of carnation: Present situation, problems and perspectives. *Acta Horticuturae*, 216: 45–54.

Garibaldi, A. and Gullino, M.L. (1986). Management of Fusarium wilt of carnation: an integration of *Fusarium* isolates. Medelelingen van de Faculterit Landbouwwetenschappen, Rijksuniveriteit Gent.

Garibaldi, A., Aloi, C. and Gullino, M.L. (1988). Further research on the possibility of chemical control of vascular wilt of carnation. *Att. Giornate Fitopatologiche,* 1: 469–480.

Garibaldi, A. and Rossi, G. (1987). Observations on the resistance of carnation to *Fusarium oxysporum* f.sp.*dianthi*. *Ponorama Floritcola,* 12: 5–9.

Garrett, S.D. (1960). *Soil Fungi and Soil Fertility*. Pergamon Press, Oxford, 165 p.

Georgieva, M.D., Kotev, S. and Wanova, M. (1979). Carnation bud rot (*Fusarium sporotrichiella* germination of *Fusarium semitectum* causing fruit rot of papaya (*Carica papaya*).

Gilman, J.C. (1957). *A Manual of Soil Fungi*. Oxford IBH Pub. Co., New Delhi, 450 p.

Glas, R. (1984). A step further with Fusarium resistant carnation cultivars, Revanda and Novanda.

Grange, M. and Ahmed, S. (1988). *Handbook of Plants with Pest Control Population*. John Wiley and Sons, New York.

Grayer, R.J. and Harborne, J.B. (1994). A survey of antifungal compounds from higher plants, 1982–1993. *Phytochemistry,* 37: 19–42.

Gullino, M.L. and Garibaldi, A. (1986). Vascular Fusarium disease of carnation: Epidemiology and control.

Gupta, Y.C. and Bhattacharjee, S.K. (2002). Carnation: Technical bulletin No. 15 of All India Coordinated Research Project on Floriculture, ICAR, New Delhi (India).

Harling, R., Taylor, G.S, Methews, P. and Arthur, A.E.(1988). The effect of temperature on symptom.

Harman, D.E., Chet, I. and Baker, R. (1981). Factors affecting *Trichoderma harzianum* applied to seeds as a biocontrol agents. *Phytopathology,* 71: 569–572.

Hazarika, D.K. and Das, K.K. (1998). Biological management of root rot of fresh bean caused by *Rhizoctonia solani*. *Plant Disease Research,* 13: 101–105.

Hazarika, D.K., Sarmah, R., Paramanick, R., Hazarika, K. and Phookan, A.K. (2000). Biological management of tomato damping–off caused by *Pythium aphanidermatum*. *Indian Journal Plant Pathology,* 18: 36–39.

Heald, F.D. (1906). Bud rot of carnation. *Science,* 23: 620.

Homma, Y., Kubo, O., Ishi, M. and Onata, K. (1979). The effect of organic soil amendments on the suppression of disease severity of *Fusarium* wilt of tomatoes. *Bulletin Shikou Agri. Expert. Station,* 34: 89–101.

Hood, J.R. and Stewart, R.N. (1957). Factor affecting symptom expression in *Fusarium* wilt of *Dianthus*. *Phytopathology,* 47: 173–178.

Huber, D.K. and Watson, R.D. (1970). Effect of organic amendments on soil-borne plant pathogens. *Phytopathology,* 60: 22–26.

J.P. 1993. Integrated control of vascular wilt of carnation caused by *Fusarium oxysporum* f. sp.

Jacob, M. and Krebs, B.N. (1985). Occurrence and control of *Fusarium* wilt in choice carnations. *Nachrichtenblatt-fur-den-pflanzenschutz-in-der-DDR,*39: 16–19.

Jacob, M. and Krebs, B. (1986). Improved control of Fusarium wilt in carnation. *Gartenbau*, 33: 214–215.

Jacobson, M. (1989). Botanical pesticides: Past, present and future. In: *Insecticides of Plant Origin*, (Eds.) Arnason, J.T., Philogene, B.J.R. and P. Morand. ACS Symp. Ser. 387, American Chemical Society, Washington, D.C., pp. 1–10.

Jaypal, S., Singh, R.S. and Singh, H.V. (1996). Selection of substrates for growth and sporulation of different antagonists of *Rhizoctonia solani* in mass-culture. *Plant Diseases Research*, 11: 132–136.

Johnson, L.F. (1957). Effect of antibiotics on the numbers of bacteria and fungi isolated from soil by dilution plate method. *Phytopathology*, 47: 630.

Johnson, L.F. and Curl, E.A. (1972). *Methods for Research on the Ecology of Soil Borne Plant Pathogens.* Burgess Pub Co., Minneapolis, 247 p.

Jone, J.P., Querman, A.J. and Geraldsen, C.M. (1966). Effect of fumigants and plastic film on the control of several soil borne pathogens of tomato. *Phytopathology*, 56: 929–932.

Jyotsna, S. (2001). Studies on retriment of virus free plant through bio-virus inhibitors. *Ph.D. Thesis*, Dr. Y.S. Parmar University of Horticulture and Forestry, Nauni, Solan (HP), India, 112 p.

Katan, J. (1981). Solar heating (solarization) of soil for control of soil borne pests. *Annual Review of Phytopathology*, 19: 211–236.

Katan, J., Greenberger, A., Alon, H. and Grinstein, A. (1976). Solar beating by polyethylene mulching for the control of diseases caused by soil borne pathogens. *Phytopathology*, 66: 683–688.

Katoch, R. (1999). Studies on *Fusarium* wilt of carnation. *M.Sc. Thesis*, Dr. Y.S. Parmar University of Horticulture and Forestry, Nauni, Solan (HP), India, 53 p.

Khanna, R. N. and Singh, R S. (1974). Rhizosphere populations of *Fusarium* species in amended soils. *Indian Phytopathology*, 27: 331–339.

Kharakrang, A., Kabitarani, A., Upadhyay, S. and Upadhyay, D.M. (2002). Disease control and growth promotion in tomato, potato and paddy by *Trichoderma viride* and *T. harzianum*. *Indian Journal of Plant Pathology*, 29(1–2): 25–29.

Kloepper, J.W., Leong, J., Teintze, M. and Schroth, M.N. (1980). Enhanced plant growth by siderophores produced by plant growth promoting rhizobacteria. *Nature*, 286: 885–886.

Kloepper, J.W., Leong, J., Teintze, M. and Schroth, M.N. (1980a). *Pseudomonas siderophores* mechanism explaining disease suppressive soils. *Current Microbioliolgy*, 4: 317–320.

Kovacikova, E. and Altmanova, O. (1989). Experience with application of fertilizers and fungicides to control Fusarium wilt of carnations. Shornik UVT12 Zahradnictis, 16:39–45.

Kowalik, M. (1996). *Trichoderma* spp. and *Gliocladium* spp. as factors controlling the occurrence of pathogenic fungi in stands of mixture of alfalfa and grasses. *Phytopathologia Polonica*, 11: 59–66.

Kumar, D., Sharma, M. and Singh, T. (2002). Bioprospecting of plant extracts against seed–borne *Fusarium oxysporum* f.sp. *udum* in pigeon pea. *Indian Journal of Plant Pathology*, 29(1–2): 55–57.

Kutova, I. and Petkova, M. (1975). Fusariosis of carnation and its control. *Bulgarian Scientific Literature*, 11: 37–40.

Lahdenpera, M.L. (1987). The control of *Fusarium* wilt of carnation with streptomyces-preparation. *Acta Horticulturae*, 216: 85–92.

Locke, J.C., Marois, J.J. and Papavizas, G.C. (1984). Biological control of *Fusarium* wilt of greenhouse, grown chrysanthemum. *Plant Disease,* 69: 167–169.

Lurie, A., Kiplinger, D.C. and Nelson, K.S. (1968). *Commercial Flower Forcing.* McGraw Hill Book Company.

Manka, M., and Fruzynska-Jozwiak, D. (1992). Forest soil fungi for he biological control of *Fusarium* soil fungi for biocontrol of *Fusarium oxysporum dianthi. Bulletin OILB/SROP,* 15: 15–17.

Manku, M. and Fruzysiska-Jozwick, D. (1992). Studies on carnation. *Bulletin OILB/SROP,* 15: 15–17.

Mathur, K.B., Lodha, C. and Dorjees, C. (1988). Management of rhizome rot of zinger by organic soil amendment and *Trichoderma viride. Indian Phytopathology,* 41: 265–302.

Mirkova, E. (1984). The resistance of *Fusarium oxysporum* f. sp. *dianthi* on carnation to some fungicides. *Rastite Lna Zashchta,* 32: 21.

Mishra, M. and Tiwari, S.N. (1990). Ethanolic extract toxicity of three botanicals against five fungal pathogens of rice. *Natational. Acad. Sci. Letters,* 13: 409–412.

Mishra, A.K., Dubey, N.K. and Mishra, L. (1993). A fungitoxic principle from the leaves of *Prunus persica. Pharmaceutical Pharmacological Letters,* 2: 203–206.

Mukhopadhyay, A.N. and Mukherjee, P.K. (1996). Fungi as fungicides. *International Journal of Tropical Plant Diseases,* 14: 1–17.

Muthamilan, M. and Jeyarajan, R. (1996). Integrated management of Sclerotium root rot of groundnut involving *Trichoderma harzianum, Rhizobium* and *carbendazim. Indian Journal of Mycology and Plant Pathology,* 26: 204–209.

Muthusamy, M. (1999). Biofungicides: Effective tool for the management of plant disease. *Pestology,* 23: 185–187.

Navas, B.L.R., Vela, D.M.D, Lopez, R.M., Basallote, U.M.J., Prados, L.A., Zea, B.T., Lopez, M.C.J. and Melero, V.J.M. (2000). Comparative effectiveness of several methods of soil disinfestations to control Fusarium of carnation. *Acta Hoticulturae,* 332: 247–251.

Nelson, P.E., Toursoun, T.A. and Marasas, W.F.O. (1983). *Fusarium Speices: An Illustrated Manual for Identification.* Pa State University Press, University Park.

Ness, A. and Baliga, R.P. (1999). Observation on environment and antagonistic effects of *Trichoderma harzianum* and *Trichoderma viride* and role of biocontrol in integrated pest management. *Pestology,* 23: 59–61.

Obieglo, U., Krebs, B. and Junge, H. (1990). Use of microbial antagonists of Fusarium wilt of carnation (*Fusarium oxysporum* f. sp. *dianthi*) in hydroponic culture system. *Nachrichtenblatt Pflanzenschntz,* 44: 169–170.

Obieglo, U. (1992). Biocontrol experiment with *Bacillus subtilis* during the rooting period of carnation under commercial cultivation. *Bulletin OILB/SROP,* 15: 127–129.

Orozco De Amezquita, M., Garces de Granada, E. and Arbetaez, G. (1993). Effect of different levels of *oxysporum dianthi. Bulletin OILB/SROP,* 15: 15–17.

Panse, V.G. and Sukhatme, P.V., 1985. *Statistical Methods for Agricultural Workers.* ICAR, New Delhi, India.

Papavizas, G.C. (1981). *Biological Control in Crop Production.* Allanheld Osmum, London,461 p.

Papavizas, G.C. (1985). *Trichoderma* and *Gliocladium* biology, ecology and potential for biocontrol. *Annaul of Review of Phytopathology,* 18: 389–413.

Parmar, B.S. and Ketkar, C.M. (1996). Commercialization. In: *Neem,* (Eds.) Randhawa, N.S. and B.S. Parmar. New Age International (P) Ltd., New Delhi, pp. 318–332.

Patel, S.I. and Mishra, A. (1994). Evaluation of cheap and easily available substrates for mass-multiplication of *Trichoderma harzianum* Rifai. *Gujarat Agric. Univ. Res. J.,* 20: 185–186.

Pawliczule, Z. and Orlikowski, L.B. (1987). Polish carnation cultivars to tolerant to *Fusarium oxysporum* f. sp. *dianthi. Acta Hort.,* 216: 345–347.

Pennock-Vos, M.G., Reobroeck, J.A. and Skrzypezak, C.Z. (1990). Preliminary results on biological control of *Botrytis cinerea* in forced tulips. *Acta Hort.,* 266: 425–428.

Pera, A. and Filippi, C. (1987). Controlling of Fusarium wilt in carnation with bark compost. Biological Pflanzenschntz, 44: 169–170. *Phytopathology,* 58: 373–377.

Postma, J. and Rattink, H. (1991). Biological control of Fusarium wilt of carnation with non-pathogenic.

Postma, J. (1992). Biological control of Fusarium wilt of carnation with non-pathogenic *Fusarium* isolates.

Prasad, J., Pathak, V.N. and Pathak, A.K. (1988). Efficacy of certain fungicides to submit conidial *Pseudomonas putida* WCS 358 at different levels of disease incidence and iron availability.

Raghava, S.P.S. and Dadlani, N.S. (2000). Revival of Fortunes. *The Hindu Survey of Indian Agriculture,* Chennai, 151 p.

Raj, H. and Gupta, V.K. (1995). Soil solarization for controlling mango wilt. *Indian J. Agric. Sci.,* 66: 258–262.

Ramakrishnan, P. and Jeyarajan, S. (1987). Neem cake in control of *Fusarium* wilt of vegetable crops. *Andhra Agric. J.,* 47: 120–123.

Ramarao, P., Manoharachary, C. and Veridehi, B.K. (1987). Rhizosphere microflora and root-borne fungal pathogens. *ISCA Symposium,* pp. 239–258. *Rastite Lna Zashchita,* 32:21.

Rattink, J. (1974). Research on vascular disease of carnation. *Onderzoek ove vatziekten pijanjer,* 29: 15–29.

Rattink, J. (1992). Biological control of *Fusarium* wilt of carnation by a non-pathogenic isolate of *Fusarium oxysporum. Acta Hort.,* 307: 37–42.

Rifai, M.A. (1969). A revision of genus *Trichoderma. Mycology Papers,* 116: 56.

Rumine, P. (1989). Studies of antagonistic activity of *Trichoderma* spp. against *Fusarium oxysporum* f. sp. *dianthi* on carnation. *Annali dell' Istituto Sperimentale per la Floricolture,* 20:57–67.

Say, M., Wright, G.K. and Wimalajeewa, S. (1992). Epidemiology and control of *Fusarium* disease of carnation in Victoria. *Acta Hort.,* 307: 89–94.

Schiva, J., Guda, C. Dalla and Mercuri, A. (1985). Genetic analysis of resistance to *Fusarium oxysporum* f. sp. *serimentale per la Floricolture,* 16: 23–38.

Sharma, D.D., Gupta, Y.P. and Chandrashekhar, D.S. (1999). Compatibility of certain biocontrol agents with chemical pesticides and fertilizers. *Indian J. Sericulture,* 38: 79–82.

Sharma, J.R. and Bedi, P.S. (198). Effect of soil amendments with oil cakes on wilt of cotton. *J. Res. Punjab Agric. Univ.*, 23: 414–416.

Sharma, S.N. (2001). Studies on biological control of *Fusarium* yellows of gladiolus. *M.Sc. Thesis*, Dr. Y.S. Parmar University of Horticulture and Forestry, Nauni, Solan (HP), India, 63 p.

Shekhawat, P.S. and Prasad, R.(1971). Antifungal properties of some plant extracts in inhibition of spore germination. *Indian Phytopathol.*, 24: 800–802.

Shin, G.C., Yu, S.H. and Park, J.S.(1987). Biological control of sesame soil borne disease by antifungal micro-organisms. *Korean J. Plant Prot.*, 26: 229–237.

Siddiqui, M.Z. and Kausal, V.K. (2002). Efficacy of water soluble extracts of some weeds against growth and sporulation of *Fusarium oxysporum* f.sp. *zingiberi* in Bundelkhand region. *Indian J. Plant Path.*, 20: 90–92.

Singh, A. and Sindh, D. (1994). Biological control of *Sclerotium rolfsii.* Sacc. causing cotton rot of Brinjal. *J. Biol. Control*, 8: 105–110.

Singh, K.P., Bihari, K. and Singh, V.K. (1991). Effect of culture filtrates of fungi colonizing neem cake on *Heterodera cajani. Current Nematology*, 2: 9–14.

Singh, K.P., Bihari, S.S., Vaish, S.S. and Gupta, R.C. (2002). Effect of neem oil-cake amendment on soil mycoflora. *Indian J. Plant Pathol.*, 20: 99–103.

Singh, R. and Pandey, K.R. (1966). Effect of green and matured plant residues and compost on population of *Pythium aphanidermatum* in soil. *Indian Phytopathol.*, 19: 368–371.

Singh, R.S. (1983). Organic amendments for root disease control through management of soil microbiota and the host. *Indian J. Mycol. Pl. Pathol.*, 13: 1–16.

Singh, S.K., Singh, U.P. and Leepika, T. (2001). Effect of some plant extracts on *Alternaria tenuissima* (Kunze en Pers.) Wiltshire, the causal agent of leaf blight of Pigeon pea (*Cajanus cajan* L). *Indian J. Plant Pathol.*, 19: 29–33.

Singh, S.K., Singh, U.P. and Tuli, L. (2001). Effect of some plant extracts on *Alternaria tenuissima* (Kunez ex Pers.) Wiltshire, the causal agent of leaf blight of pigeon pea (*Cajanus cajan* L). *Indian J. Plant Pathol.*, 19: 29–33.

Singh, U.P., Pandey, V.M, Wagner, K.G. and Singh, K.P. (1990). Antifungal activity of ajoene, a constituent of garlic (*Allium sativum*). *Canadian Journal of Botany*, 68: 1354–1456.

Singh, U.P., Prithiviraj, B., Wagner, K.G. and Plank-Schumacher, Karin.(1995). Effect of ajoene, a constituent of garlic (*Allium sativum*) on powdery mildew (*Erysiphe pisi*) of pea (*Pisum sativum*). *Journal of Plant Diseases and Protection*, 102: 399–406.

Singh, U.P., Singh, H.B. and Singh, R.B. (1980). The fungicidal effect of neem (*Azadirachta indica*) extracts on some soil-borne pathogens of gram (*Cicer arietinum*). *Mycologia*, 72: 1077–1093.

Snyder, W.C., Schroth, M.N. and Christon, T. (1959). Effect of plant residues on root rot of bean. *Phytopathol.*, 49: 775–776.

Srivastava, U.S. and Sinha, S. (1971). Effects of various soil amendments on the wilt of coriander (*Coriandrum sativum* L). *Indian J. Agric. Sci.*, 41: 779–782.

Staby, G.L., Robertson, J.L., Kiplinger, D.C. and Conover, C.A. (1976). *Proceedings in National Floriculture Conference Commod*, Handl. Colombus, Ohio, pp. 1–77.

Strider Davis, L. (1985). *Diseases of Floral Crops, Vol. II*. Praeger Publishers, New York, USA, 629p.

Sturgis, W.C. (1897). Preliminary investigations on a disease of carnation. *Can. Agric. Exp. Stn. Ann. Rep.*, 21: 175–181.

Suarez, B.F., Lopez, H.C.J., Melero, V.J.M. and Prados, L.A.H. (2005). Thermal inactivation of *Fusarium oxysporum* f. sp. *dianthi* and its eradication by soil solarisation at different depths. *Acta Horticulturae*, 698: 161–165.

Subramanian, C.V. (1971). Hyphomycetes: account of Indian species except *Cercospora. Ind. Coun. Agri. Res., New Delhi*, 903 p.

Tamietti, G., Benzi, D. and Ferraris, L. (1991). Studies on the antagonistic activity of *Gliocladium virens* against *Sclerotium cepivorum. Petria Glornale dIPathologia delle Plant*, 1: 155–156.

Tammen, J. (1954). Pathogenicity of *Fusarium roseum* to carnation and to wheat. *Phytopathol.*, 48: 423–426.

Tjamos, E.C. and Beckman, C.H. (1989). *Vascular Wilt Diseases of Plants: Basic Studies and Control.* Springer-Verlag, Berlin.

Utkhede, R.S. and Rahe, J.E. (1983). Interactions of antagonists and pathogens in biological control of onion white rot. *Phytopathol.*, 73: 890.

Valaskova, E. (1983). Reaction of Fusarium species to simultaneous use of benzimidazole preparations and some aspects of resistance and cross resistance. *Shornik UVTIZ Ochrana Rosthn*, 19:37–48.

Van-Ruiten, J.E.M. (1992). Quality assurance of carnation propagation material. *Acta Hort.* 307: 117–123.

Villiam, R., Manum, G. and Rameri, G. (1972). Two year trials against vascular fusariosus of carnation. *Anal. Delta Facalta di Science Agane dell.*, 6: 200–213.

Vincent, J. M. (1947). Distortion of fungal hyphae in the presence of certain inhibitors. *Nature*, 150: 850.

Warcup, J.H. (1950). The soil plate method for isolation of fungi from soil. *Nature*, 166: 117–118.

Weindling, R. and Emerson, O.H. (1936). The isolation of toxic substance from the culture of a *Trichoderma. Phytopathol.*, 26: 1068–1070.

Weller, D.M. (1988). Biological control of soil borne plant pathogens in the rhizosphere with bacteria. *Ann. Rev. Phytopathol.*, 26: 379–407.

Wickens, G.M. (1935). Wilt stem rot die-back of the perpetual flowering carnation. *Ann. App. Biol.*, 22: 630–633.

Wong, P.T.W. (1985). Soil solarization: effects on *Fusarium* wilt of carnation and *Verticillium* wilt of egg plant. In: *Ecology and Management of Soil Borne Plant Pathogens*, (Eds.) Hardy, G.S., Srivasithamparam, S., Parker, C.A., Ravira, A.D., Moore, K.J. and P.T.W. Wong. School of Agriculture, University of Western Australia, Nedlands, pp. 279–281.

Index